International Conference on Chemical Engineering UNPAR (ICCE 2019)

IOP Conference Series: Materials Science and Engineering
Volume 742

Bandung, Indonesia
28 - 29 November 2019

ISBN: 978-1-7138-3497-7
ISSN: 1757-8981

Printed from e-media with permission by:

Curran Associates, Inc.
57 Morehouse Lane
Red Hook, NY 12571

Some format issues inherent in the e-media version may also appear in this print version.

Institute of Physics
Dirac House, Temple Back
Bristol BS1 6BE UK

Phone: 44 1 17 929 7481
Fax: 44 1 17 920 0979

techtracking@iop.org

Additional copies of this publication are available from:

Curran Associates, Inc.
57 Morehouse Lane
Red Hook, NY 12571 USA
Phone: 845-758-0400
Fax: 845-758-2633
Email: curran@proceedings.com
Web: www.proceedings.com

TABLE OF CONTENTS

PROCESS ENGINEERING

PRODUCT ENGINEERING, FOOD TECHNOLOGY AND BIOTECHNOLOGY

MATERIAL ENGINEERING

IOP Conf. Series: Materials Science and Engineering **742** (2020) 011001 doi:10.1088/1757-899X/742/1/011001

Preface

The International Conference on Chemical Engineering (ICCE) 2019 was held in Bandung, Indonesia from November 28-29, 2019. ICCE 2019 was organized by the Department of Chemical Engineering, Parahyangan Catholic University. The objective of this conference was to highlight trends and research focuses in various areas of chemical engineering disciplines, especially when related to the innovation in chemical engineering fundamental concepts and applications.

The proceedings are a compilation of the accepted papers and represent important results of the conference. We would like to acknowledge all people who were involved in ICCE 2019. Each individual and institution's support was very meaningful for the success of this conference. Especially, we would like to thank the organizing committee for their valuable advice in the organization and helpful peer review of the papers.

We sincerely hope that ICCE 2019 would serve as an ideal meeting point for academia, industrial practitioners, and authorities from around the globe, where every participant can share and at the same time update their most recent ideas and discoveries.

Arenst Andreas Arie, PhD. Parahyangan Catholic University

IOP Conf. Series: Materials Science and Engineering **742** (2020) 011002 doi:10.1088/1757-899X/742/1/011002

Peer review statement

All papers published in this volume of *IOP Conference Series: Materials Science and Engineering* have been peer reviewed through processes administered by the proceedings Editors. Reviews were conducted by expert referees to the professional and scientific standards expected of a proceedings journal published by IOP Publishing.

IOP Conf. Series: Materials Science and Engineering **742** (2020) 012001 doi:10.1088/1757-899X/742/1/012001

A Kinetic Study on Supercritical Carbon-dioxide Extraction of Indonesian *Trigona sp.* Propolis

B A Fachri[1], M F Rizkiana[1], M Muharja[1]
[1]Department of Chemical Engineering, Faculty of Engineering, University of Jember, Jember, Indonesia.

E-mail: fachri.teknik@unej.ac.id

Abstract. This work is to investigate the supercritical CO_2 extraction of Indonesian *Trigona sp.* raw propolis. The kinetic study was conducted at CO_2 mass flowrate of 15 g/min, the temperature of 50 °C and pressure (250-350 bar). The process was run for 3 h and the extract was accumulated every 30 min. The proposed mathematical extraction models were Sovova, Logistic and Reverchon. The result shows that Sovova model gives a good fit between calculated and data value. The Sovova model also has lower SSE than Logistic and Reverchon model. The highest extract of 46 wt% was achieved at extraction time of 60 min, the temperature of 50 °C, the mass CO_2 flowrate 20 g/min and the pressure 300 bar. The extract constituent consisting of galangin, p-coumaric acid, ferulic acid, caffeic acid, cinnamic acid, kaempferol, and chrysin was detected by HPLC analysis. The antioxidant activity analyzed by DPPH method was expressed as IC_{50} and the value is comparable to ascorbic acid.

1. Introduction

In recent decades, propolis has attracted increasing attention and use in foods and beverages, cosmetics, pesticides, and medicine (Oldoni, 2015). Propolis contains bioactive flavonoid and phenolic acid compounds providing antioxidants (Aga et al., 1994; Bankova et al., 2000; Sforcin and Bankova, 2011). Supercritical fluid extraction with CO_2 is well-known as an advanced alternative method in the extraction process (Biscaia and Ferreira, 2009; del Valle, 2015). A CO_2 supercritical fluid is mostly occupied as a supercritical solvent to extract natural bioactive compounds for bio-nutrition and pharmaceutical foods (de Melo et al., 2014; Garcia-Mendoza et al., 2015). In supercritical conditions, the fluid accommodates properties of gas and liquid. The supercritical fluid is dense like a liquid and viscous like a gas. Thus, supercritical fluid can diffuse rapidly into solid matrices and dissolve the bioactive compounds. These properties lead to the advantages of supercritical extraction such as regulating the solvent power, giving high selectivity and yielding high free-solvent extract (del Valle, 2015). Some studies on the extraction of propolis with supercritical CO_2 fluids have been published (De Zordi et al., 2014; Machado et al., 2016) and informs promising.

To perform a kinetic study on supercritical CO_2 extraction, some models are mathematically proposed. The model mostly was formulated based on the mass transfer model. In this work, the most common models proposed by Sovová (1994), Reverchon (1997), and Martinez et al. (2003) were applied. These models involve the mass balance partial differential equations. Further, this initial and boundary conditions were needed to solve the equation (Priyanka and Khanam, 2018).

Based on the literature review, the kinetic study on supercritical CO2 extraction of Indonesian Trigona Sp has never been performed. Thus, this work is to investigate the supercritical CO_2 extraction of Indonesian *Trigona sp.* raw propolis as well as to study the effect of process variables such as pressure, mass flow rate and temperature on extraction mass.

2. Methods

2.1. Chemicals

The *Trigona sp.* raw propolis was obtained from the Institute for Non-Timber Forest Product Technology, Mataram, West Nusa Tenggara. It was grounded and sifted into 140 mesh. Ultra-high purity (UHP) Carbon dioxide (CO_2) was employed as an extraction solvent. Ethanol, gallic acid, ferulic acid, caffeic acid, cinnamic acid, CAPE, galangin, quercetin, and p-coumaric acid was also utilized for the HPLC standard (Sigma-Aldrich, Singapore).

2.2. Supercritical CO_2 extraction

The supercritical extraction apparatus system (KIST-Korea) was prepared. Raw propolis of 50 g was milled and poured into the extraction chamber. Consecutively, the CO_2 fluid was fed into the extraction chamber with flowrate 10-20 g/min. The extractor was regulated until it reached a supercritical condition (P = 250 bar, T= 250 °C and CO_2 flow rate = 20 g/min), then the process was run. The kinetic study was conducted for 3 h and the extract was accumulated every 30 min. The process variables were CO_2 pressure (250-350 bar).

2.3. Analysis

The liquid product was separated from residue by centrifuging, filtered and prepared to analysis. The HPLC analysis was employed to identify the extract composition. Qualitative analysis of propolis extract was carried out using the HPLC Water Alliance, column C18, UV detector (280 nm), phase gradient of methanol and 1% acetic acid.
The antioxidant activity was measured using the DPPH method.

2.4. Kinetic model

Three models were applied to investigate the extraction kinetic. These models are developed based on mass transfer phenomenon. In this section, the models are presented in the final equation terms, as seen in equation 1 for the Sovova model, equation 2 for the Logistic model, and equation 3 for the Reverchon model (Martínez et al., 2003).
Sovova Model:

$$m_{ext} = Y^*\left(1 - \exp(-Z)Q_{CO_2}t_{ext}\right) \qquad (1)$$

Logistic Model:

$$m_{ext}(h = H, t) = \frac{m_t}{exp(bt_m)}\left\{\frac{1+exp(bt_m)}{1+exp[b(t_m-1)]} - 1\right\} \qquad (2)$$

Reverchon Model:

$$m_{ext} = m_0\left[1 - \frac{6}{\pi^2}\sum_{n-1}^{\infty}\frac{1}{n^2}exp\left(\frac{-n^2\pi^2 D_t}{r^2}\right)\right] \qquad (3)$$

with m_{ext} is the extracted mass (g), Y* is the solubility of the extract in the fluid phase (M/M), Z is CER time parameter, t_{ext} is extraction time (min), h is adjustable parameter, m_t is the initial solid mass (g), b is adjustable parameter (1/s), t_m = adjustable parameter (s), m_o is the initial component mass in solid (g), D_t is the diffusion coefficient and r is the particle diameter.
The suitability of the model was determined based on the sum squared error (SSE). The model chosen was the model that had the smallest SSE value.

$$SSE = \sum\{[m_{ext}]_{calculated} - [m_{ext}]_{data}\}^2 \qquad (4)$$

3. Result and discussion

In the earlier study, the highest yield of propolis extract was achieved at a pressure of 250 bar and a temperature of 50 °C. Therefore, in this study, extraction was observed and studied at the optimum operating conditions. The profile of the supercritical CO_2 extraction of propolis was depicted in Figure 1. As shown in Figure 1, the extraction rate increases with the extraction time increasing during 60 min extraction time. Afterward, the extraction rate approaches constantly. It can be explained that at the beginning of extraction, the solubility of the extract in supercritical CO2 increases, but after 60 min, equilibrium begins to form and consequently the extraction rate decreases.

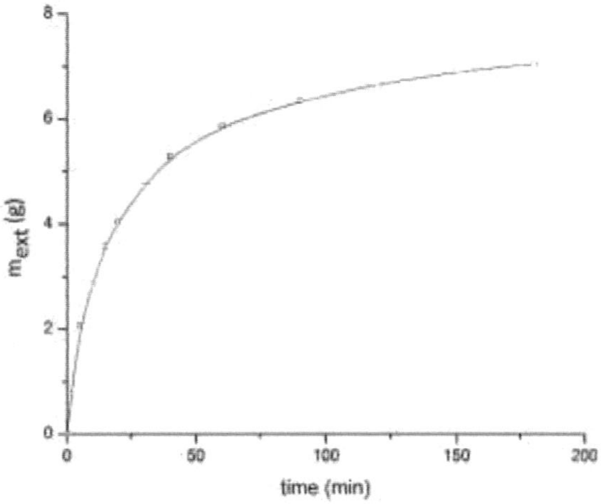

Figure 1 Profile of propolis extraction by using SFE CO_2 method at 250 bar and 50°C

The models proposed by Sovová (1994), Reverchon (1997) and Martinez et al. (2003) were fitted with the experimental data of *Trigona sp.* Propolis and shown in Figure 2. As shown in Figure 4, a model proposed by Sovová (1994) provided a better approximation with the experimental data than the Reverchon and the Logistic model. The SSE of Sovova model is 0.13 % which is less than Reverchon model of 0.21 % and Logistic model of 0,35 %.

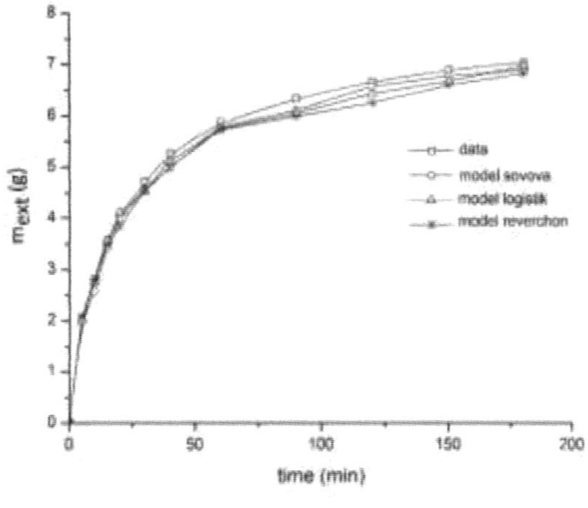

Figure 2. Relationship between extraction mass and extraction time on a different model (P=250 bar, T=250 °C)

Both Sovova and Reverchon models accommodated the mass transfer process in the model which considered the internal mass transfer. Consequently, the models fitted with the experimental data. Meanwhile, the Logistic model neglected the mass transfer resistance in supercritical extraction model. It caused the model slightly fitted with the experimental data.

It is well-known that in supercritical extraction, the pressure is mostly the significant variable that plays an important role. It is in line with our previous study. Thus, this work is also to check the effect of pressure on extraction mass. It was shown in Figure 3.

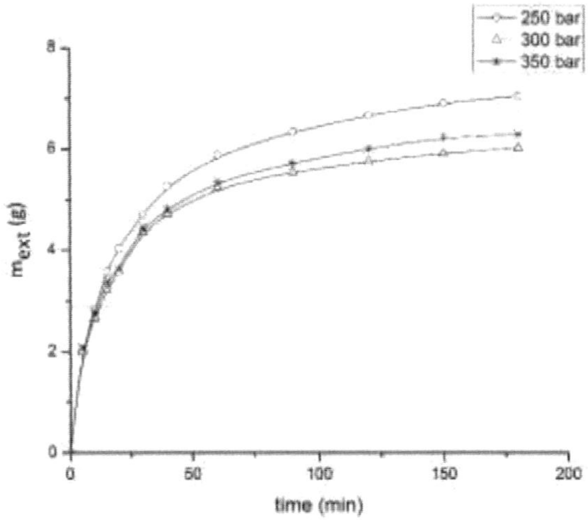

Figure 3. Relationship between extraction mass and extraction time on different pressure

In figure 3, it appears that the extraction carried out at a pressure of 250 bar gives a higher yield than at pressures of 300 and 350 bar. This can be caused because at higher pressures, the system is saturated with supercritical CO2 fluid, so it is unable to dissolve more extracts. Meanwhile, at the same pressure, increasing the extraction time will increase the extract yield.

The result of the qualitative analysis from HPLC analysis was presented in Table 1. Propolis produced by the supercritical method contains chemical compounds that function as antioxidants such as galangin, ferulic acid, caffeic acid phenyl ester (CAPE), p-coumaric acid, chrysin, caffeic acid, and cinnamic acid. Antioxidant test (DPPH test) also strongly indicates that the extract of propolis produced has very strong antioxidant properties, which is indicated by a very small IC_{50} value (0.983 mg/g).

Table 1. Chemical composition of propolis extract

Retention Time (min)	Compounds	Retention Time (min)	Compounds
11.75	Caffeic acid	65.32	Chrysin
17.01	p-coumaric acid	66.46	CAPE
19.22	Ferulic acid	67.88	Galangin
31.80	Cinnamic acid		
56.45	Kaempferol		

4. Conclusion

This work summarizes that supercritical CO_2 extraction can be applied to extract the antioxidant compounds from Indonesian Trigona Sp raw propolis which provides the IC_{50} value of 0.983 g/g. This value indicates that the extract shows promise as a super antioxidant.

This work also indicates that the Sovova model fits with the supercritical CO2 extraction of Indonesian Trigona Sp raw propolis. This model gives the SSE of 0.13 %.

Acknowledgment
The authors would like to thank the Ministry of Research, Technology, and Higher Education for funding through Strategis Nasional Institusi Scheme.

References

[1] Aga H, Shibuya, T, Sugimoto, T, Kurimoto, M, Nakajima, S 1994. Biosci. Biotechnol. Biochem. **58** 945.
[2] Bankova V S, Castro S L, de Marcucci M C 2000. Apidologie **31** 3.
[3] Biscaia D, Ferreira S R S 2009. J. Supercrit. Fluids **51** 17.
[4] de Melo M M R, Silvestre A J D, Silva C M 2014. J. Supercrit. Fluids **92** 115.
[5] De Zordi N, Cortesi A, Kikic I, Moneghini M, Solinas D, Innocenti G, Portolan A, Baratto G., Dall'Acqua S, 2014. J. Supercrit. Fluids **95** 491.
[6] del Valle J M 2015. J. Supercrit. Fluids **96** 180.
[7] Garcia-Mendoza M P, Paula J T, Paviani L C, Cabral F A, Martinez-Correa H A 2015. LWT - Food Sci. Technol. **62** 131.
[8] Machado B A S, Silva R P D, Barreto G A, Costa S S, Silva D F, Brandão H N, Rocha J L C, Dellagostin O A, Henriques J A P, Umsza-Guez M A, Padilha F F 2016. PLoS One **11** e0145954.
[9] Martínez J, Monteiro A R, Rosa P T V, Marques M O M, Meireles M A A 2003. Ind. Eng. Chem. Res. **42** 1057.
[10] Oldoni T 2015. J. Braz. Chem. Soc. **26**.
[11] Priyanka K S 2018. Sep. Sci. Technol. **53** 71.
[12] Reverchon E 1997. J. Supercrit. Fluids **10** 1.
[13] Sforcin J M, Bankova V 2011. J. Ethnopharmacol. **133** 253.
[14] Sovová H 1994. Chem. Eng. Sci. **49** 409.

IOP Conf. Series: Materials Science and Engineering **742** (2020) 012002 doi:10.1088/1757-899X/742/1/012002

Synthesis of Tert-butylhydroquinone Derivative as a Soluble Biodiesel Antioxidant

Felita Dwinanda Putri[1], Hery Sutanto[1], Akhmad Darmawan[2] and M Nasikin[3]

[1] Department of Chemical Engineering, Faculty of Life Sciences and Technology, Swiss German University, Kota Tangerang, Banten, Indonesia

[2] Research Center for Chemistry, Indonesian Institute of Sciences, Kawasan PUSPIPTEK Serpong, Tangerang Selatan, Banten, Indonesia

[3] Department of Chemical Engineering, Faculty of Engineering, Universitas Indonesia, Kota Depok, Jawa Barat, Indonesia

E-mail: felitadwinandap@gmail.com

Abstract. Biodiesel has been widely used as a substitution for petroleum diesel for years. It offers some various advantages such as higher cetane number, higher efficiency, and low sulfur content. Despite the advantages, biodiesel has one major drawback which is prone to oxidation. The oxidation degrades its quality and causes disturbances to the machine during operations. Therefore, antioxidant was added into biodiesel. However, Tert-butylhydroquinone (TBHQ), as the most effective antioxidant for biodiesel, has low oil solubility. Hence, in this research, TBHQ was modified with methyl linoleate via coupling reaction using benzoyl peroxide. TLC, HPLC, and LC-MS results confirmed that the reaction had successfully produced a new product and its isomer, methyl (10*E*,12*Z*)-9-(4-(*tert*-butyl)-2,5-dihydroxyphenyl)octadeca-10,12-dienoate and methyl (9*Z*,11*E*)-13-(4-(*tert*-butyl)-2,5-dihydroxyphenyl)octadeca-9,11-dienoate. The yield was found to be 10.4%. The product mixture was subjected to a solubility test and an antioxidant activity evaluation. The result showed that the reaction product had higher oil solubility than TBHQ indicated by a steady increase in product absorbance up to 7000 ppm. TBHQ had reached its solubility limit at 5500 ppm. DPPH dot-blot assay, acid value, and rancimat test results revealed that the product mixture had antioxidant activity.

1. Introduction

In these recent years, the world has been facing an energy crisis due to the increase in energy demand and the scarcity of fossil fuels. According to International Energy Agency or IEA, the global energy demand increased by 2.1% in 2017 [1]. On the other hand, fossil fuel reserves are diminishing and predicted to run out after 65 years [2]. Besides the energy crisis, global warming is also one of the biggest problems in the world. As stated by IEA recently, global energy-related carbon dioxide emissions increased by 1.4% in 2017 [1]. Due to energy crisis and global warming issue, the use of renewable energy as an alternative energy continues to grow.

In 2017, the share of renewables reached a new level of 10.2% in total OECD primary energy supply and the largest proportion of the renewable supply comes from biofuels and waste, which accounts for 53.5% [3]. Biofuels, such as biodiesel, have already been implemented as transportation fuels in many countries including Indonesia. Biodiesel can be used directly as fuel or blended with diesel fuel. Biodiesel is long chain fatty acids which is produced from vegetable oil or animal oil through a transesterification reaction.

IOP Conf. Series: Materials Science and Engineering **742** (2020) 012002 doi:10.1088/1757-899X/742/1/012002

Biodiesel as transportation fuel could offer various advantages. Some main advantages are more environmentally friendly than diesel fuel, renewable and also biodegradable. In addition, biodiesel has higher cetane number, higher efficiency, and no sulfur content. Regardless of the advantages, biodiesel has also weakness which is susceptible to oxidation resulting in quality degradation. The unsaturated fatty acids in biodiesel react with oxygen leading to formation of insoluble gums and deposits that later can cause engine blocking and corrosion [4]. Therefore, biodiesel antioxidant is needed to minimalize the oxidation.

Phenolic compounds are commonly used as antioxidants, for instance propyl gallate, pyrogallol, gallic acid, TBHQ, BHA, and others [5]. The problem arises from using phenolic compound as antioxidant is its low solubility in biodiesel. Many studies have been carried out to improve the solubility of antioxidants in biodiesel. Some of them are the modification of TBHQ with palmitic acid [6], pyrogallol modification with palmitic acid [7], and synthesis of pyrogallol derivatives as antioxidant [8]. Unfortunately, the results showed that the solubility increased but inversely the antioxidant activity decreased.

In this research, TBHQ was used as the antioxidant molecule. TBHQ, as the most effective antioxidant for biodiesel, was modified with methyl linoleate using benzoyl peroxide as the polymerization agent. Methyl linoleate was used as it is one of the components contained in biodiesel. The aim of this research was to obtain a new product from the modification process (TBHQ derivative) with high oil solubility and has antioxidant activity.

2. Methodology

Synthesis of TBHQ derivative was done chemically by combining a TBHQ molecule, which have a relatively high polarity, with one molecule of fatty acid methyl ester, which is methyl linoleate as a representation of the nonpolar group. As these two molecules combined, the new product was expected to have a better solubility in oil compared to the original TBHQ molecule.

2.1. Materials and equipment

The chemicals used for the reaction were TBHQ (obtained from PT Sumber Berlian Kimia), methyl linoleate (obtained from Xian Plant Bio-Engineering Co.), benzoyl peroxide (obtained commercially) as polymerization agent, and ethyl acetate (EMSURE® Merck, Germany) as the solvent. For analysis using TLC, n-hexane (EMSURE® Merck, Germany) and ethyl acetate were used as the solvents. While for the TLC plates, TLC silica gel-aluminium plates (Merck, Germany) was used. The biodiesel used for solubility and antioxidant activity tests was obtained from Research and Development Centre for Oil and Gas Technology (LEMIGAS). For solubility test, n-hexane was used for diluting the samples. DPPH (Sigma Aldrich, US) and methanol (EMSURE® Merck, Germany) were used for determining the antioxidant activity qualitatively. In determining acid value, 2-propanol (EMSURE® Merck, Germany) was used as solvent, sodium hydroxide (Merck, Germany) was used as titrant, and phenolphthalein (Merck, Germany) was used as an indicator.

In this research, UV/VIS spectrophotometer (PG Instruments T-60, UK, SGU Lab) was used for solubility test. High-Performance Liquid Chromatography (HPLC) was used for identifying components in mixture after reaction. Afterwards, Liquid Chromatography–Mass Spectrometry (LC-MS/MS) (Waters Xevo G2-S Qtof) was used for further identification of the reaction products.

2.2. Synthesis reaction

The reaction between TBHQ and methyl linoleate was performed in a 250 ml boiling flask equipped with three necks. Firstly, 1 mmol of benzoyl peroxide was dissolved in 10 ml of ethyl acetate in the boiling flask. Afterwards, 1 mmol of methyl linoleate was added. In order for benzoyl peroxide to decompose into benzoyloxy radicals, the mixture was heated. The mixture was heated in water bath at 90°C under continuous stirring for two hours. After two hours, the mixture was let cool for 5 minutes at room temperature. During cooling time, 1 mmol of TBHQ was weighed in a beaker and 1 ml of ethyl acetate was added. TBHQ solution was added to the reaction mixture and the mixture was stirred again for another 15 minutes at room temperature.

The reaction product mixture was first analyzed using Thin Layer Chromatography (TLC) to identify the presence of new compound by comparing the Rf values of each spots with the standards. The presence of new product would be indicated by a new spot in TLC. Further identification of the reaction product mixture was performed using HPLC and LC-MS/MS to determine more specifically the components in the mixture and also to obtain information of the proposed product.

2.3. Application to biodiesel

2.3.1. Solubility test. This test was performed using UV/Vis spectrophotometer to obtain the solubilities of TBHQ and TBHQ derivative in biodiesel and afterwards the solubilities were compared. The method used for this test was adapted from the previous research by Rosa Belinda [7]. TBHQ samples as well as product samples with concentration range between 0 - 7000 ppm were prepared. Sample with 0 concentration of TBHQ was used as a blank sample. All samples were diluted using n-hexane with a dilution factor of 1:100. Prior to the measurement, the wavelength of maximum absorbance was determined by scanning 2 diluted samples, the lowest (2000 ppm) and the middle (4000 ppm) concentrations in the UV-Vis spectrophotometer. Thereafter, the absorbance of each sample was measured at the optimum wavelength. From the measurements, the solubilities of TBHQ and its derivative were obtained.

2.3.2. DPPH dot-blot assay. The dot-blot assay was carried out to determine the antioxidant activity of product qualitatively. DPPH solution in methanol was prepared by adding 0.1 gram of DPPH and 20 ml of methanol to a beaker. The mixture was added to a 50 ml volumetric flask. It was filled with methanol up to 50 ml line to make a 0.2% methanolic DPPH solution. Afterwards, 10 μL of the product solution and 10 μL of TBHQ solution (concentration of 2000 ppm) were spotted on a TLC paper and allowed to dry for a few minutes. Lastly, DPPH-methanol solution was sprayed onto the paper until it was completely covered.

2.3.3. Rancimat test. The rancimat test was done according to the standard test method EN 14112 using 873 Biodiesel Rancimat Oxidation Stability Analyzer [9]. In this method, biodiesel sample was heated up to 110°C in a vessel with a continuous flow of air at rate of 10 l/h passing through the sample. Under this condition, biodiesel underwent oxidation resulting in formation of oxidation products such as short chain carboxylic acids which are water-soluble and volatile. By the time, those acids caused a significant increase in conductivity. The result of this test was induction period (IP) which was determined automatically by the equipment. The induction period is the time needed (h) right before the high rate oxidation starts to occur.

2.3.4. Acid value. The acid value method used in this experiment was adapted from Journal Of Food and Drug Analysis [10] based on AOAC International standard. Three beakers were prepared for biodiesel, TBHQ, and product. In the first beaker, 20 ml of biodiesel was added without any additional substance. In the second beaker, 20 ml of biodiesel was added with 0.03 g of TBHQ. While in the third beaker, 2 ml of product mixture and 18 ml of biodiesel were added. The concentration of both TBHQ and product samples was 1500 ppm. The acid value of all samples were measured weekly within 4 weeks storage time. For the measurement, 1 ml of sample was dissolved in 10 ml isopropyl alcohol. Fifteen drops of phenolphthalein were added as an indicator. The sample was titrated with 0.1M NaOH solution until the color of solution changed to slightly pink. The acid value was calculated using the equation (1) below. Table 1 shows the acid value standards according to ASTM D6751, EN 14214, and SNI 7182.

$$Acid\ Value = \frac{56\ x\ molarity\ of\ NaOH\ x\ (vol\ NaOH\ sample - vol\ NaOH\ blank)}{mass\ of\ sample} \quad (1)$$

IOP Conf. Series: Materials Science and Engineering **742** (2020) 012002 doi:10.1088/1757-899X/742/1/012002

Table 1. Biodiesel acid value standards [11].

Parameter	ASTM D6751	EN 14214	SNI 7182
Acid value (mg KOH / g oil)	0.5 max	0.5 max	0.5 max

3. Result and discussion

3.1. Synthesis reaction
When benzoyl peroxide and methyl linoleate were dissolved in ethyl acetate, the solution was clear and still liquid. Methyl linoleate was mixed first with benzoyl peroxide since it takes longer for methyl linoleate to react with benzoyloxy radicals compared to TBHQ, which is highly reactive to radicals. Ethyl acetate was chosen to be the solvent because it is a semi polar solvent at which all reagents used in this experiment namely benzoyl peroxide, TBHQ, and methyl linoleate are soluble in it. Furthermore, ethyl acetate is a good solvent for polymerization using benzoyl peroxide [12].

During the heating process, the color and viscosity of mixture remained the same as previous, it was clear and non-viscous. However, during the mixing process upon the addition of TBHQ, the color slowly turned to yellow. Figure 1 shows the color of product mixture at the end of reaction. The color change of the reaction product mixture to light yellow was due to the reaction of TBHQ with free radicals such as benzoyloxy and fatty acid radicals. TBHQ underwent oxidation and formed 2-tert-Butyl-1,4-benzoquinone (TBBQ) as an oxidation product [13]. TBBQ has a yellow color which made the solution looked transparent yellow.

Figure 1. Reaction product mixture.

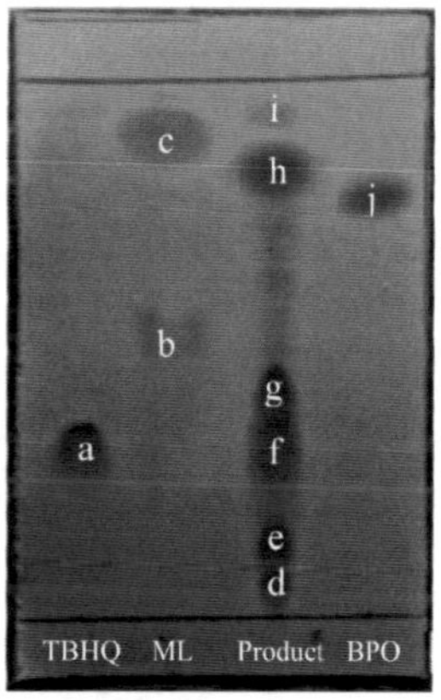

Figure 2. Thin layer chromatography (TLC) result under UV light.

3.2. Product identification
3.2.1. Thin Layer Chromatography (TLC). The reaction product mixture was evaluated using TLC. For the mobile phase, ethyl acetate was blended with n-hexane with ratio of 1:4. This ratio was found to be the most appropriate ratio as the spot of the product could be seen well. Silica gel was used as a polar stationary phase.

In general, TLC separates substances in mixture based on polarity of each substance. Based on its log p value, TBHQ polarity is higher than methyl linoleate and benzoyl peroxide. P value is the ratio between concentration of a compound in organic solvent and concentration of the compound in aqueous solvent. So it quantifies the amount of the compound dissolves in organic solvent versus in aqueous solvent. Low log p value means that the compound is more hydrophilic while high value means the compound is more lipophilic [14]. P values of TBHQ, benzoyl peroxide, and methyl linoleate successively are 2.96, 3.55, and 6.23. The Rf value of TBHQ was expected to be below 0.5 [6], while methyl linoleate spot was expected to be higher than TBHQ spot. According to ChemDraw, the log p value of the desired product is 8.68 and since the value is higher than methyl linoleate log p value, the spot of desired product was expected to be higher than methyl linoleate spot.

The size of TLC paper was 3.5 cm wide and 5.5 cm long with travel distance of 4.5 cm. Figure 2 was taken 5 minutes after the TLC paper had been taken out from the TLC chamber. The UV light was used so that the spot could be clearly seen. In TLC, spot (a) was TBHQ, (b) and (c) were methyl linoleate, (d, e, f, g, h, and i) were reaction products, and (j) was benzoyl peroxide.

The Rf values of TBHQ (a), methyl linoleate (c), and benzoyl peroxide (j) were 0.31, 0.9, and 0.78 respectively. In the reaction product, apparently there were 6 spots appeared. Those 6 spots (d, e, f, g, h, i) from the bottom to top in sequence had Rf values of 0.07, 0.13, 0.31, 0.42, 0.83, and 0.95. The Rf value of the third spot (f) was similar to TBHQ Rf value which indicates it was the unreacted TBHQ spot. The Rf values of 5 other spots (d, e, g, h, i) had no similarity to either benzoyl peroxide or methyl linoleate. This means the reaction had successfully produced some new products. From those new products, there was one product with Rf value higher than methyl linoleate, which was spot (i). This spot was predicted as the spot of the desired product.

3.2.2. High Performance Liquid Chromatography (HPLC). The second identification was done using HPLC to confirm the presence of the new product in reaction mixture. The stationary phase in the column was non-polar, while the mobile phase was polar. Mixture of methanol and water with ratio of 8:2 was used as polar solvent. The measurement was run at constant flowrate of 0.5 ml/min and using wavelength at 280 nm.

The retention time of the desired product was expected to be longer than methyl linoleate retention time. From several peaks appeared in the HPLC chromatogram (figure 3), the peaks with retention time of 3.178 and 3.322 minutes were identified to be the excess of TBHQ. The retention time of benzoyl peroxide and methyl linoleate were obtained to be 7.263 min and 8.399 min respectively. There was no peak found which belonged to either benzoyl peroxide or methyl linoleate in figure 3. This means all benzoyl peroxide has completely decomposed into benzoyloxy radicals and all methyl linoleate has completely reacted.

Small peak with retention time of 4.524 minutes was considered as contaminant. The rest of peaks have retention time of 3.588, 5.121, 5.634, 6.604, 9.490 and 10.052 minutes which did not belong to TBHQ, methyl linoleate, nor benzoyl peroxide but new products. This confirmed the TLC result that there were more than one product formed. As mentioned earlier, the desired product is more non-polar than methyl linoleate meaning the peak would be more on the right of methyl linoleate. There are 2 peaks located at the right side of methyl linoleate. They are peaks with retention time of 9.490 and 10.052 minutes. The peak with retention time of 9.49 minutes was predicted as the peak of other reaction product with log p value of 6.28. Thus, it was concluded that the peak with retention time of 10.052 minutes was probably the peak of the desired product. Based on the HPLC result, the yield of the desired product was 10.4%.

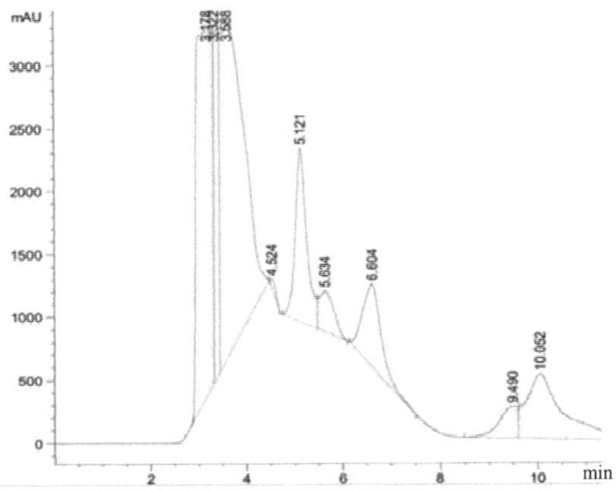

Figure 3. HPLC chromatogram of product mixture.

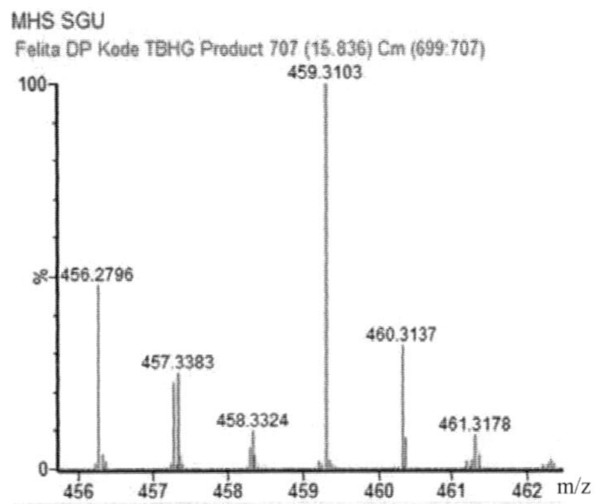

Figure 4. Mass spectrum result of the product mixture at 456 – 466 m/z.

3.2.3. Liquid Chromatography – Mass Spectrometry (LC-MS). Further identification was necessary to ensure that the desired product really existed. Last identification was carried out using LC-MS/MS (Xevo G2-S QTof).

Molecules in sample are separated based on polarity in column chromatography and go to mass spectrometer. Mass spectrometer can only detect ionized molecules, hence molecules were ionized prior to mass analyzer [15]. The instrument uses hydrogen ion to ionize molecule. Due to the attachment of hydrogen ion, then the mass of molecule increase by one. The molecular mass shown in chromatogram is the total mass of molecule itself plus the attached hydrogen. Thus, the mass in chromatogram must be deducted one to obtain the real molecular mass.

Molecular mass of TBHQ and methyl linoleate are 166 g/mol and 294 g/mol respectively. Since the desired product was a combination of TBHQ and methyl linoleate, thus the desired product had a predicted molecular mass of 458 g/mol. Due to the additional hydrogen ion, the predicted molecular mass of product in spectrum became 459 g/mol. Figure 4 is the mass spectrum result of product mixture. As can be seen, the molecular mass of 459 g/mol appeared in mass spectrum result at retention time of 15.836 minutes. The result implies that the desired product was formed.

3.3. Proposed reaction mechanism

The coupling reaction mechanism was started with the decomposition of benzoyl peroxide into 2 benzoyloxy radicals in the presence of sufficient heat [16]. Methyl linoleate, as a polyunsaturated fatty acid, has two double bonds located at C9 and C12. In between, it has a methylene hydrogen, which is very prone to radical attack [5]. Some benzoyloxy radicals attacked this hydrogen resulting in formation of methyl linoleate radical. The electron was delocalized creating resonances with stable conjugated double bonds [17].

Upon addition of TBHQ, some benzoyloxy radicals reacted with TBHQ by capturing one hydrogen from the hydroxyl group. This reaction resulted in the formation of TBHQ radical. TBHQ radical was stable due to the delocalization of the electron. Radical TBHQ in the mixture were present in the form of keto and enol due to tautomerism. However, the enol form of TBHQ radical was more favorable than the keto form [18]. During the termination process (figure 5), TBHQ radical combined with methyl linoleate radical. Since methyl linoleate radical had resonances, thus there were 2 products formed: *methyl (10E,12Z)-9-(4-(tert-butyl)-2,5-dihydroxyphenyl)octadeca-10,12-dienoate* and *methyl (9Z,11E)-13-(4-(tert-butyl)-2,5-dihydroxyphenyl)octadeca-9,11-dienoate.*

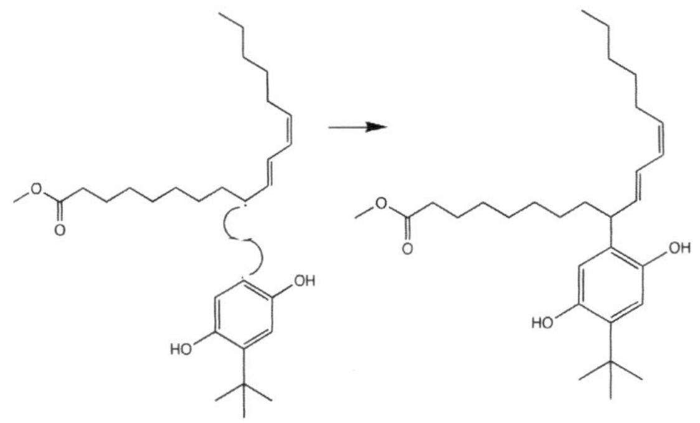

Methyl (10*E*,12*Z*)-9-(4-(*tert*-butyl)-2,5-dihydroxyphenyl)octadeca-10,12-dienoate

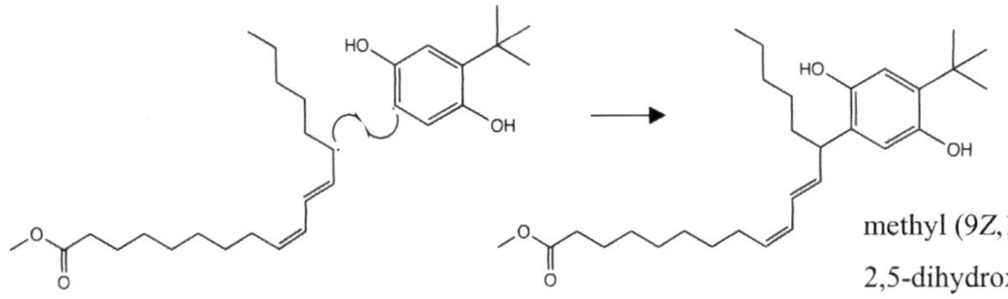

methyl (9*Z*,11*E*)-13-(4-(*tert*-butyl)-2,5-dihydroxyphenyl)octadeca-9,11-dienoate

Figure 5. Termination step: combination of methyl linoleate radical and TBHQ radical.

3.4. Application into biodiesel

3.4.1. Solubility Test. In this test, each sample was diluted with n-hexane with ratio of 1:100 so the absorbance would be in the range between 0 – 2. Prior to the measurement, the optimum wavelengths of TBHQ and product were determined by scanning the samples of 2000 ppm and 4000 ppm at wavelengths 190 – 400 nm. The optimum wavelength of TBHQ was found to be 290 nm, while the reaction product optimum wavelength was 283 nm.

Each sample was measured twice then the average value was calculated. Figure 6 shows the average absorbance values of TBHQ and product of each concentration. For TBHQ, the absorbance increased proportionally with the increasing concentration until it reached concentration of 5500 ppm. The absorbance at 5500 was 1.101 and it slightly decreased to 1.071 at concentration of 6000 ppm. This decrease indicates that TBHQ has reached its solubility limit at 5500 ppm. Adding more TBHQ would not increase the total dissolved TBHQ in biodiesel.

On the other hand, the absorbance of product increases steadily up to 7000 ppm suggesting that the reaction product had higher solubility compare to TBHQ. As expected, the product has higher oil solubility than TBHQ. This is simply due to the attachment of the long chain methyl linoleate to TBHQ. From this result, it could be concluded that solubility of product is higher than solubility of TBHQ in biodiesel. A further study was necessary to obtain the solubility limit of the new product.

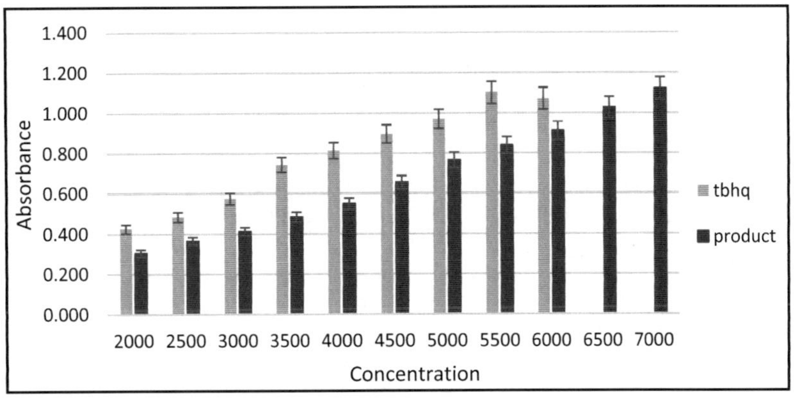

Figure 6. Absorbance of TBHQ and product in various concentrations.

3.4.2. DPPH dot-blot assay. DPPH dot-blot assay could only tell qualitatively whether the sample has antioxidant activity or not based on the DPPH color change. In principle, DPPH is a stable free radical at room temperature and has a purple color in alcoholic solvent. When DPPH radical meets an antioxidant, it attacks one hydrogen from the antioxidant. Because it gains a hydrogen, DPPH is reduced. The reduced form of DPPH is non radical and has yellow color [19].

When the TBHQ and product mixture spots in TLC were sprayed with 0.2% DPPH solution, the color changed immediately to yellow as can be seen in figure 7. TBHQ has been experimentally proven to be the most effective antioxidant for biodiesel [20]. Hence, it was already expected that the color of TBHQ spot would turn to yellow. Similar to TBHQ spots, product mixture spots also turned to yellow evidently. From this test, it was concluded that TBHQ and the product mixture had antioxidant activity. The antioxidant activity of the desired product was nevertheless still unknown.

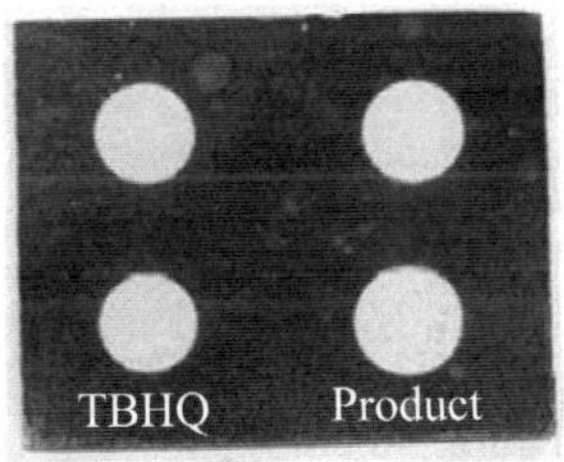

Figure 7. Results of DPPH dot-blot assay of TBHQ and product mixture.

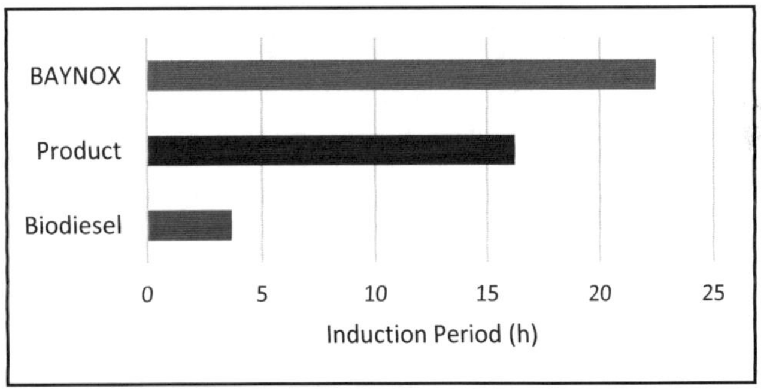

Figure 8. Rancimat test results of biodiesel, product, and Baynox.

3.4.3. Rancimat test. The rancimat test was performed using 873 Biodiesel Rancimat equipment. In this test, oxidation stability of biodiesel, product mixture, and commercial antioxidant BAYNOX were measured. As a requirement, all samples were dissolved in biodiesel to make concentration of 2000 ppm. The results of the rancimat test were depicted in figure 8. The result shows that the induction period of pure biodiesel was 3.68 hours. The induction period of product mixture sample turned out to be surprisingly long. It was 16.22 hours, while the induction period of BAYNOX was 22.48 hours. The long induction period of reaction product might be affected by the presence of other products in mixture. The longer the induction period, the higher the stability of the compound is. From the results, it can be concluded that adding product to biodiesel could increase biodiesel oxidation stability. Among all samples, the induction period of commercial antioxidant BAYNOX was the longest. The result was expected as BAYNOX had already been proven experimentally that it can prevent oxidation of biodiesel.

To meet EN14214 specifications, induction period of biodiesel must be at least 6 hours when tested at 110° with air flowrate of 10l/h [9]. The induction period of pure biodiesel did not meet the specification due to the short induction period of only 3.68 hours. On the other hand, induction period of product mixture sample as well as BAYNOX had successfully met the standard.

3.4.4. Acid value. Acid values of pure biodiesel, TBHQ-biodiesel, and product-biodiesel were measured once a week within 4 weeks storage time. The concentration of TBHQ and product in biodiesel was 1500 ppm with total volume of 20 ml.

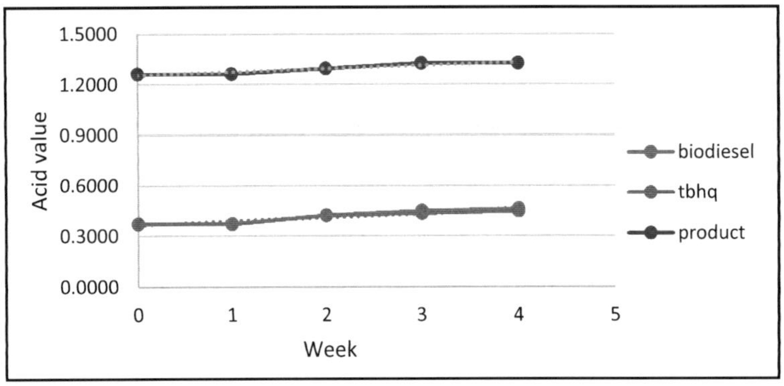

Figure 9. Acid value measurement results of biodiesel, TBHQ, and product.

At the beginning, the acid values of TBHQ and pure biodiesel samples were very similar while acid value of product sample was really high. This condition was probably caused by other reaction products in mixture, which were not observed during this experiment. The acid values of all samples increased over the time. At the end of storage time, acid value of pure biodiesel was slightly higher than TBHQ sample. Within 4 weeks, acid value of biodiesel increased by 0.09 from 0.37 to 0.46. Acid value of TBHQ sample increased by 0.08 from 0.37 to 0.45, while acid value of product sample increased by only 0.06 from 1.26 to 1.32. This increase was caused by the oxidation of biodiesel which resulted in acids formation [9]. According to ASTM D6751, EN14214 and SNI 7182, acid value of biodiesel shall not exceed 0.5 mg KOH/g oil. The acid value of product sample however has far exceeded the international standard starting at week zero. Meanwhile, the acid values of biodiesel and TBHQ samples were still below 0.5 after 4 weeks which means both samples still met the standard requirement. From this result, it could be concluded that the product mixture could not be used directly as an antioxidant for biodiesel due to the high acid content.

4. Conclusions

The desired product and its isomer, methyl (10*E*,12*Z*)-9-(4-(*tert*-butyl)-2,5-dihydroxyphenyl)octadeca-10,12-dienoate and methyl (9Z,11*E*)-13-(4-(*tert*-butyl)-2,5-dihydroxyphenyl)octadeca-9,11-dienoate, were also successfully synthesized. The yield was found to be 10.4%. Results from product identification analysis using TLC, HPLC, and LC-MS/MS showed that the desired products were present in the mixture and had lower polarity than methyl linoleate as expected. The solubility test unveiled that the solubility of reaction product in biodiesel was higher than solubility of TBHQ in biodiesel. The antioxidant activity tests, consisted of DPPH dot-blot assay, acid value, and rancimat test, revealed that the product mixture had antioxidant activity.

Acknowledgments
This research was funded by The Ministry of Research and Technology of the Republic of Indonesia.

References

[1] International Energy Agency (IEA) 2018 *Global Energy & CO2 Status Report 2017*
[2] Huang D, Zhou H and Lin L 2011 Biodiesel: An alternative to conventional fuel
[3] International Energy Agency (IEA) 2018 *Renewables Information: Overview*
[4] Agarwal A K and Khurana D 2013 Long-term storage oxidation stability of Karanja biodiesel with the use of antioxidants *Fuel Processing Technol.* **106** 447-52
[5] Saluja R K, Kumar V and Sham R 2016 Stability of biodiesel – A review *Renewable and Sustainable Energy Reviews* **62** 866-81
[6] Pramastiani A J 2018 Modification of tert-butylhydroquinone with palmitic acid as a soluble antioxidant for biodiesel additive
[7] Belinda R 2018 Modification of pyrogallol with palmitic acid as a soluble biodiesel additive
[8] Susanto B H, Lukman, Sutanto H and Nasikin M 2018 The effect of pyrogallol antioxidant addition and storage temperature to the change of biodiesel quality during storage period *IOP Conf. Series: Earth and Environmental Science* vol 105 (Bali)
[9] Pullen J and Saeed K 2012 An overview of biodiesel oxidation stability *Renewable and Sustainable Energy Reviews* **16** 5924-50
[10] Pan M H *et al.* 2014 Physicochemical properties of Terminalia catappa seed oil as a novel dietary lipid source *J. Food Drug Anal.* **23** 201–9
[11] Etim A O, Betiku E, Ajala S O, Olaniyi P J and Ojumu T V 2018 Potential of ripe plantain fruit peels as an ecofriendly catalyst for biodiesel synthesis: Optimization by artificial neural network integrated with genetic algorithm *Sustainability* **10** 1-15
[12] Owolabi R U, Usman M A, Isanbor C and Kehinde A J 2016 Rationalization of solvent effects in the solution polymerization of styrene *Anadolu Univ. J. Science and Technol. Applied Sciences and Engineering* **17** 13–23
[13] Kurechi T, Aizawa M and Kunugi A 1983 Studies on the antioxidants XVIII: Oxidation product of tertiary butyl hydroquinone (TBHQ) (I) *J. American Oil Chem. Society* **60** 1878–82
[14] Bhal S K 2007 Log P — Making sense of the value
[15] Kang J S 2012 *Tandem Mass Spectrometry* (Rijeka: IntechOpen)
[16] Odian G 2004 *Principles of Polymerization* (New York: John Wiley)
[17] Sutanto H 2016 Synthesis of pyrogallol derivative as antioxidant for biodiesel
[18] Klein D R 2011 *Organic Chemistry as a Second Language* (New York: John Wiley)
[19] Garcia E J, De Alencar S M, Reis A, Loguercio A D, Helena R and Grande M 2012 Antioxidant activity by DPPH assay of potential solutions to be applied on bleached teeth *Brazilian Dental J.* **23** 22–27
[20] Rashed M M *et al.* 2015 Stability of biodiesel, its improvement and the effect of antioxidant treated blends on engine performance and emission *RSC Advances* **5** 36240–61

Modification of Tert-butylhydroquinone with Palmitic Acid as a Soluble Antioxidant for Biodiesel Additive

Hery Sutanto[1], Amanda Jehan Pramastiani[1], Silvya Yusri[1], Akhmad Darmawan[2], Evita H. Legowo[1], and Mohammad Nasikin[3]

[1] Department of Chemical Engineering, Faculty of Life Sciences and Technology, Swiss German University, Tangerang, Banten, Indonesia.
[2] Research Center for Chemistry, Indonesian Institute of Sciences, Kawasan PUSPIPTEK Serpong, Tangerang Selatan, Banten, Indonesia.
[3] Department of Chemical Engineering, Faculty of Engineering, Universitas Indonesia, Depok, Jawa Barat, Indonesia.

E-mail: hery.sutanto@sgu.ac.id

Abstract. Biodiesel is a product resulted from the process of transesterification of vegetable oil or animal fat. Biodiesel is a renewable fuel and uses as alternative to fossil fuel diesel. However, one of the biodiesel drawbacks which is oxidation can affect the storage stability of the biodiesel. One way to solve this issue is by using antioxidant additives to improve the storage stability of biodiesel. Tert-Butylhydroquinone is a common synthetic phenolic antioxidant and has a proven superior antioxidant activity due to its polar characteristic. The difference in polarity between non-polar biodiesel and phenolic antioxidant makes their solubility becomes very low. In this study, modification of TBHQ with palmitic acid was done by using Friedel-Crafts acylation to increase its solubility in biodiesel. The resulting proposed product 1-(2-(tert-butyl)-3,6-dihydroxyphenyl)hexadecane-1-one, was confirmed using TLC, LC-MS/MS and ^{1}H-NMR. The product showed solubility limit at concentration 2500 ppm and antioxidant activity with the IC_{50} value of 675.38 ppm, which both results are lower compared to TBHQ.

1. Introduction

Worldwide energy demand, according to a report by International Energy Agency or IEA in the International Energy Outlook 2016, increases 48% from 2012 to 2050 [1]. Global fossil fuel oil reserves were 1,707 billion barrels in 2016, which would only be sufficient for 50 years or up to year 2067 [2]. As most energy consumption comes from fossil fuels, despite its reserves being limited other alternatives to fossil fuels are needed to anticipate this energy crisis. Biodiesel can be a renewable alternative to reduce the use of fossil diesel fuel. Biodiesel is a product resulted from the process of transesterification of vegetable oil or animal fat. Biodiesel has some advantages as a renewable fuel, such as: reduce greenhouse gas emissions, biodegradable, low smoke emissions, higher lubricity [3]. However, biodiesel has some drawbacks such as thermal and radical oxidations that leads to storage instability of the biodiesel. The stability of biodiesel can affect the changes of color and smell of the biodiesel, as well as deposit formation will also occurs [3]. One way to overcome these issues is by using antioxidant additives [4].

One of the common synthetic phenolic antioxidants used is tert-Butylhydroquinone (TBHQ). TBHQ is a phenolic compound that contain two hydroxyl groups bonded to a benzene ring. Phenolic compounds are known to have the property to reduce the oxidation rates of organic matter due to its reactivity as a hydrogen or electron donating agent from the hydroxyl group to the chain-carrying free radicals [5], which in order to stop the chain propagation step during the process of oxidation. The antioxidant property of phenolic compounds is also related to their ability to delocalize or stabilize the resulting phenoxyl radical within their structures [6]. The reaction mechanism of TBHQ as antioxidant DPPH shown on Figure 1. DPPH was used as the free radical source while TBHQ as phenolic compound donates a hydrogen atom from its phenolic –OH group to the DPPH free radical [7], thus reducing number of DPPH free radical molecules [8].

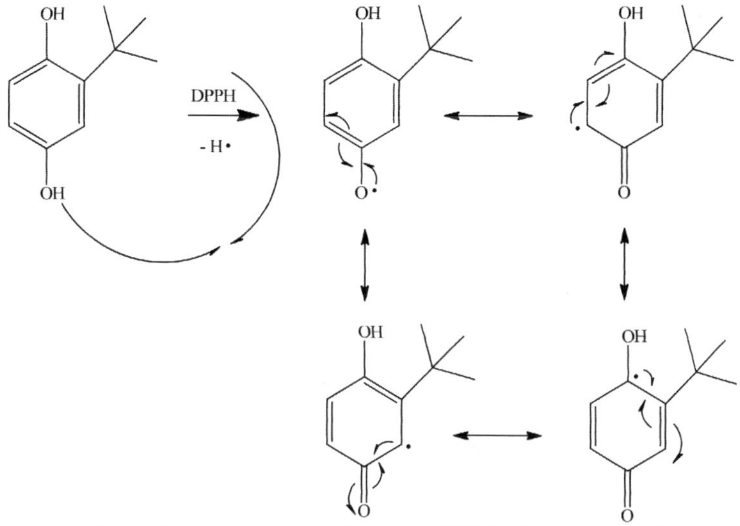

Figure 1. Resonance stabilization of TBHQ by free radical

Study by Weng and Huang [9], reported that steric hindrance affects the antioxidant activity of phenolic compound. The lower the ability of a compound to push electrons to the aromatic ring, the easier DPPH free radical to combine. TBHQ known to possess two hydroxyl groups in the molecule. Steric hindrance makes the hydroxyl group located at position 1 in the aromatic ring being blocked by the tert-butyl due to more crowding in the molecule. This makes the reaction of TBHQ and DPPH free radical is more likely to involve the OH group located at position 4 in the aromatic ring, but by no means impossible for the other hydroxyl group to provide a hydrogen or donate an electron. As for resonance stabilization of product, the mechanism is similar to TBHQ. The difference is only on the acyl group which attached to the aromatic ring. However, this additional of substituent on TBHQ makes the molecule become more crowded. Hence, the steric hindrance in the molecule increases and makes the donation of electrons or hydrogen atoms from the hydroxyl groups more difficult to DPPH free radical. This made the ability of product to reduce an oxidant become lower and thus affecting the antioxidant activity of product.

However, this phenolic antioxidant has weakness such as its solubility in biodiesel. Biodiesel is relatively non-polar due to its long hydrocarbon chain while, phenolic compounds are known to be polar. The electronegativity gap between the hydrogen and oxygen atoms in the hydroxyl group creates polarity [7]. Because of the difference in polarity between biodiesel and phenolic antioxidant, their solubility becomes very low. Therefore, alkyl and aryl group substitution for hydrogen atoms in benzene was developed by Wilson [10]. Wilson developed this method and suggested that the solubility of phenolic compound is increased with this method without lowering its antioxidant activity. In this study, TBHQ antioxidant was modified with palmitoyl chloride to increase its solubility in biodiesel. Palmitoyl chloride was used because this compound is the acid halide form of palmitic acid. Palimitic acid is one of the component that can be found in biodiesel [11].

2. Research Method

As much as 0.8325 gram of TBHQ and 8 mL of nitromethane were mixed in a 50 mL beaker . The solution was stirred for 10 minutes. In a different 50 mL beaker, a spatula tip of around 0.25 gram of aluminium chloride and 1.5 mL of palmitoyl chloride were added then stirred for 10 minutes. The solution of TBHQ dissolved in nitromethane was added to the solution of aluminium chloride and palmitoyl chloride. After the addition was completed, the mix solution was stirred for 20 minutes. The resulted product was then separated and identified by TLC. Followed by structure characterization of the product using LC-MS/MS and ^{1}H-NMR. After confirmation of the product formed, the product was tested for its solubility by using UV/Vis spectrophotometer and was evaluated for its antioxidant activity by using DPPH radical scavenging method to determine the antioxidant activities of TBHQ and the Friedel-Crafts acylation product.

3. Result and Discussion

3.1. Synthesis

From the resulted reaction, the color change of solution was observed from transparent yellow after the addition of nitromethane to TBHQ into orange indicated that the reaction occurred and TBHQ has been transformed into heterogeneous solution. TLC analysis was done by making spots of TBHQ as standard, the solid product, and the liquid product on silica gel TLC plate as the stationary phase. The mobile phase used were 5:1 n-hexane and ethyl acetate due to the properties of TBHQ and the expected product. On the TLC plate, multiple spots were found on TBHQ due to the low purity of TBHQ used as the main material. The proposed product was found at Rf value 0.52 which was higher than TBHQ and the spot wasn't found on TBHQ standard which indicated the presence of a new compound. The TLC chromatogram of the new product is shown in Figure 2.

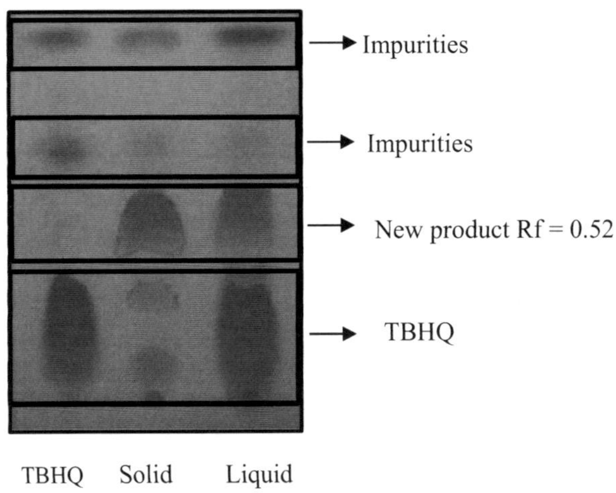

Figure 2. TLC analysis of synthesized product from Friedel-Craft reaction

Further separation of the produce was done by using preparative TLC method. The mobile phase used for this separation process was same as for the identification. The separated product contained on TLC plate was soaked in chloroform to brain the separated product. The dissolved product on chloroform was being separated again by filtering the remaining silica. After the separated product was obtained, the product was placed at 48 °C inside the vacuum oven to vaporize the solvent. This process as done for several times until the amount of product desired was achieved for further analysis.

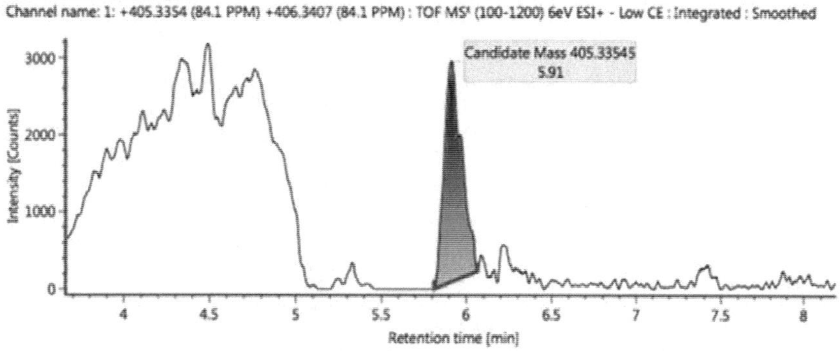

Figure 3. Chromatogram of LC-MS/MS

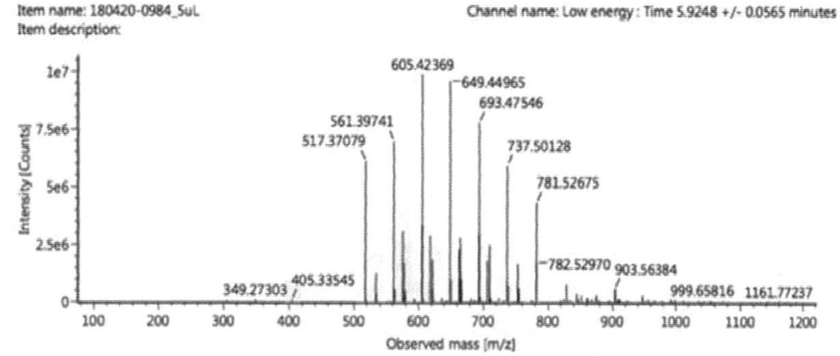

Figure 4. Mass spectrum of LC-MS/MS

LC-MS/MS analysis was performed for molecular weight determination of product using mass spectral fragmentation [12]. The LC-MS/MS results of product are shown on Figure 3 & 4 above. From the resulted chromatogram of LC-MS/MS, the retention time of 5.91 min showed the candidate mass of product $C_{26}H_{44}O_3$ 405.34 g/mol which could be identified as the proposed product due to the adduct of ion $[M+1]^+$. Molecular weight of the product was predicted by using ChemDraw, which is 404.63 g/mol.

^{1}H-NMR analysis was performed to obtain the molecular structure of the product. The peak at δ 0.880, is a triplet signal which corresponds to the protons of the $-CH_3$ group. The peak at δ 1.245 shows a broading signal correspond to the methylene groups $(CH_2)n$ of long alkyl chains. The peak at δ 1.367 is a singlet signal correspond to the $-CH_3$ group in tert-butyl. The peak at δ 1.729 correspond to the protons attached with carbon 3 of $-CO-CH_2-CH_2-$ [13]. The protons of the carbon next to the $-CO-CH_2-$ appeared at a triplet signal with δ 2.516. The hydroxyl group appeared as a singlet signal at δ 5.012. The peaks at δ 6.903 and 6.579 are doublet signals correspond to the protons attached to aromatic ring of the product. From the ^{1}H-NMR results, it was confirmed that the proposed product was formed (Figure 3), with the support of LC-MS/MS result that indicated the presence of molecular weight of the product. On Table 1, the peaks from ^{1}H-NMR results were assigned to each proton in the molecule of product (Figure 5).

Table 1. Chemical shift (δ) assignment of the 1H-NMR results of product.

Proton no.	δ
2	2.516 (2H, t, –CO–CH_2–)
3	1.729 (2H, m, –CO–CH_2–CH_2)
4-13	1.245 (20H, bs, –$(CH_2)_n$–)
16	0.880 (3H, t, –CH_3)
3'	6.903 (1H, d, ArH)
4'	6.579 (1H, d, ArH)
8' - 10'	1.367 (9H, s, –CH_3)
11'	5.012 (1H, s, ArOH)

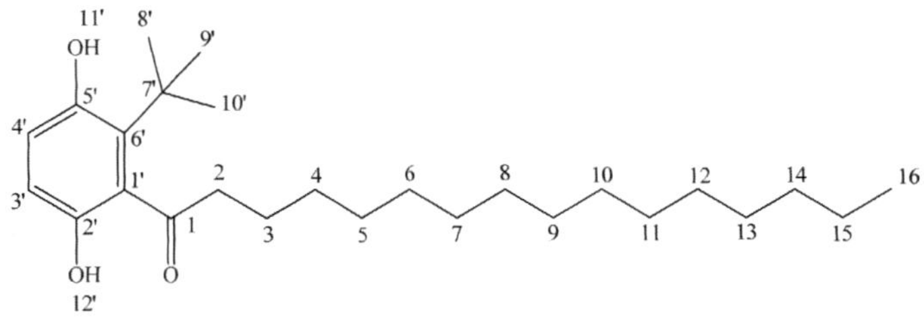

Figure 5. Molecular structure of the product

The formation of product such as positions of where the acyl group would attach in the benzene ring of TBHQ was predicted by the effects of substituents on further electrophilic aromatic substitution. Alkyl and hydroxyl groups in which the atom bonded to the ring are called ortho-para directing [7]. Hydroxyl groups in TBHQ are categorized as ortho-para directing, also the alkyl or tert-butyl which attaches to the ring. Because of this, the acyl group would be directed either to ortho or para position. As predicted the product that was analyzed by LC-MS/MS and ^{1}H-NMR, the analysis results confirmed a product with acyl group attached to the ring at ortho position. The tert-butyl group in the ring as ortho-para directing, also confirmed the directing of the acyl group to the ortho position in the aromatic ring. This supports the analysis of product at 3,4 ortho position. According to IUPAC naming, the product named as 1-(2-(tert-butyl)-3,6-dihydroxyphenyl)hexadecane-1-one (Figure 6). Other possibilities of the formed products from the Friedel-Crafts acylation can be seen on Figure 7 & 8.

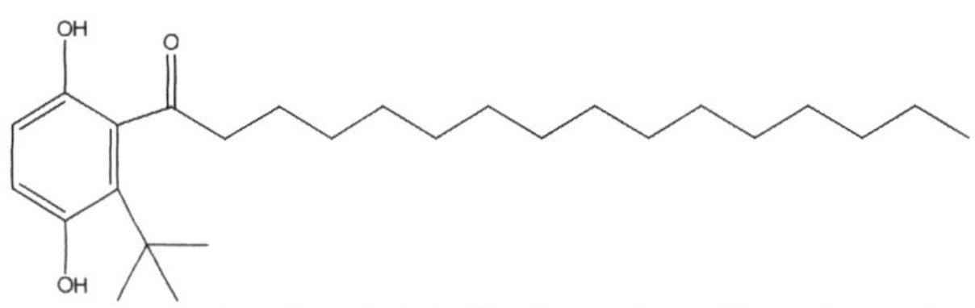

Figure 6. 1-(2-(tert-butyl)-3,6-dihydroxyphenyl)hexadecan-1-one

IOP Conf. Series: Materials Science and Engineering **742** (2020) 012003 doi:10.1088/1757-899X/742/1/012003

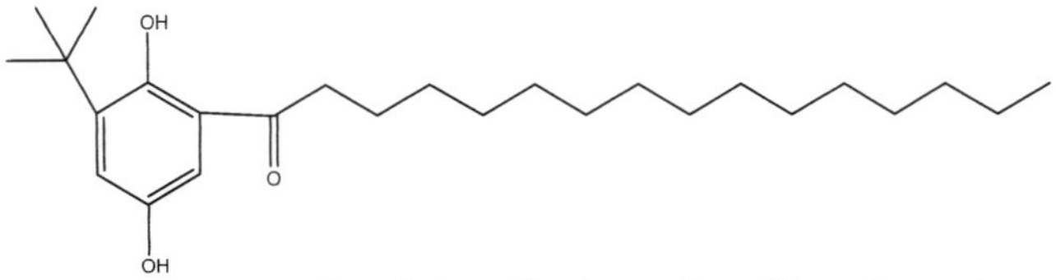

Figure 7. 1-(3-(tert-butyl)-2,5-dihydroxyphenyl)hexadecan-1-one

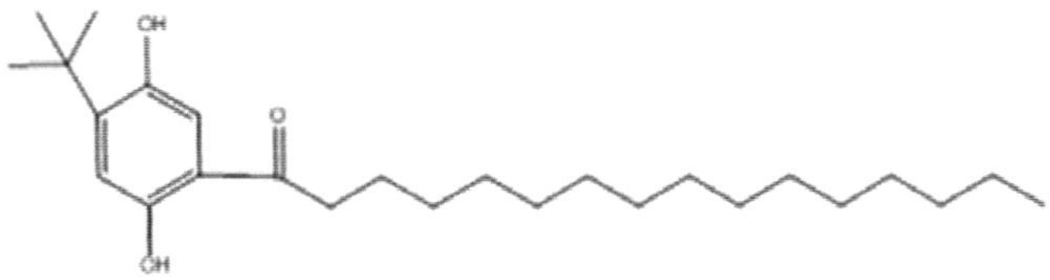

Figure 8. 1-(4-(tert-butyl)-2,5-dihydroxyphenyl)hexadecan-1-one

3.2. Solubility Test

The solubility tests were done on TBHQ and the product in biodiesel at different concentrations (2000, 2500, 3000, 3500 ppm). The sample solutions were diluted in 1:300 methanol. After measuring the blank correction, the sample solutions were scanned from 200-500 nm to determine the maximum wavelength absorbance. The maximum wavelength absorbance of TBHQ was found at 280 nm. The absorbance results of TBHQ in biodiesel was increased as the concentration was increased, which indicated the solubility limit of TBHQ was ≥ 3500 ppm. As for the product, the absorbance was increased at the beginning and decreased at the concentration of 3000 ppm with maximum wavelength absorbance was found at 270 nm, therefore the solubility limit of the product in biodiesel was found at 2500 ppm which was lower than TBHQ. The solubility test data of TBHQ and new product is shown in Figure 9. This decrease in solubility was probably caused by some factors such as the higher molecular weight and the purity of the product.

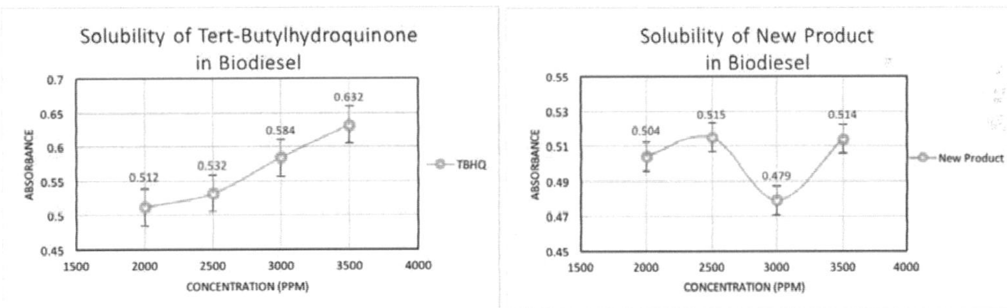

Figure 9. Solubility test of TBHQ and new product

3.3. The Antioxidant Activities

The antioxidant activities of TBHQ and the product were evaluated by using DPPH (diphenylpicrylhydrazyl) radical scavenging method. Evaluation of antioxidant activity to reduce an oxidant, can be indicated by color changes from purple to yellow which shows an indication of decrease in DPPH free radical absorbance [14]. TBHQ and product were analyzed qualitatively. Each compound was dissolved in chloroform then spotted on TLC plate and sprayed by 0.2% DPPH in methanol. From the qualitative results, both compounds showed the loss of color from purple to yellow which indicated the presence of antioxidant activity. The quantitative analysis was done by measuring the absorbance of TBHQ and product after addition of DPPH to the diluted test antioxidants in methanol. The measured absorbance was then used for the calculation of the percentage inhibition which required to plot graphs versus concentration. From the plotted graph, linear equation was obtained for IC_{50} value calculation. IC_{50} value is concentration which is required to inhibit DPPH by 50%.

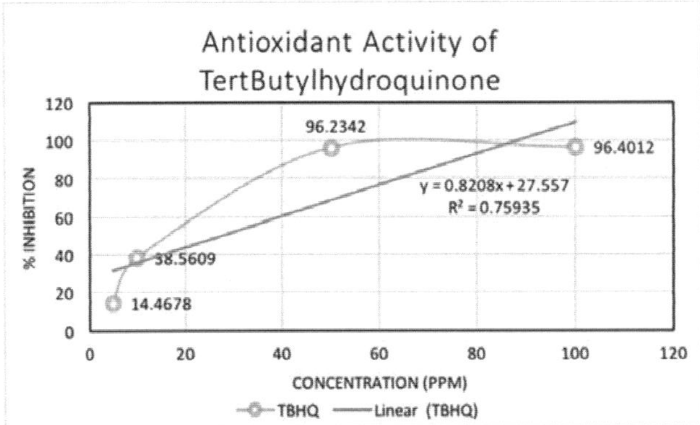

Figure 10. Antioxidant activity of TBHQ

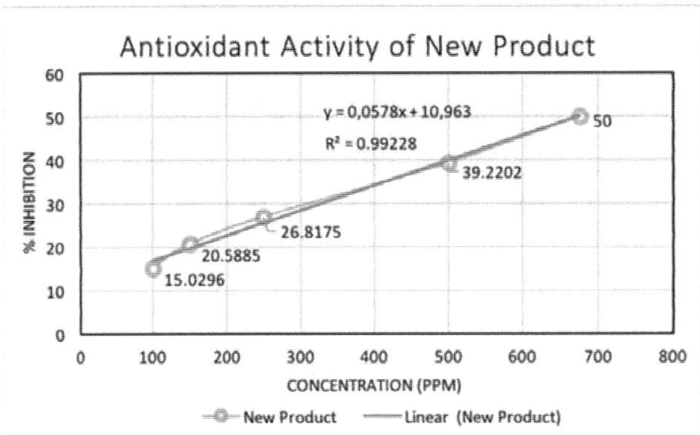

Figure 11. Antioxidant activity of product

From Figure 10, IC_{50} value of TBHQ was calculated from the linear regression equation. The IC_{50} value of TBHQ was 27.34 ppm, which was the concentration of TBHQ to inhibit DPPH free radical by 50%. The IC_{50} value of TBHQ was 27.34 ppm, while the product was 675.38 ppm. For the product, IC_{50} value of product was calculated from the linear regression equation of the graph on Figure 11. The IC_{50} value of product was 675.38 ppm. The lower the IC_{50} value, the higher the antioxidant activity of samples [15]. In comparison between the two compounds, the product showed lower antioxidant activity than TBHQ.

4. Conclusions

This study focused on modification of TBHQ antioxidant with palmitic acid using Friedel-Crafts acylation. The proposed product, 1-(2-(tert-butyl)-3,6-dihydroxyphenyl)hexadecane-1-one was formed and characterized by the LC-MS/MS and 1H-NMR. From the solubility test, the product showed lower solubility compared to TBHQ, with maximum solubility at 2500 ppm. The product showed lower antioxidant activity with the IC_{50} value of 675.38 ppm.

5. Acknowledgment

The authors would like to thank The Ministry of Reseach, Technology and Higher Education Republic of Indonesia for financial support through Penelitian Terapan Unggulan Perguruan Tinggi (PTUPT) research grant (contract no. 2639/L4/PP/2019).

References

[1] International Energy Agency (IEA) 2017 World Energy Outlook: 2017

[2] British Petrolium (BP) 2017 Statistical Review of World Energy June 2017

[3] Saluja, et al. 2016 Stability of biodiesel – A review *Renewable and Sustainable Energy Reviews* **62**, pp. 866-881

[4] Christensen, et al. 2014 Long-term storage stability of biodiesel and biodiesel blends *Fuel Processing Technology* **128**, pp. 339-348

[5] Foti, M. C. 2009 Antioxidant properties of phenols *Journal of Pharmacy and Pharmacology* **59**, pp. 1673-1685

[6] Teixeira, J., et al. 2013 Hydroxycinnamic Acid Antioxidants: An Electrochemical Overview *BioMed Research International* **2013**, pp. 1-11

[7] Brown, W. and Poon, T. 2012 Introduction to Organic Chemistry, 5th edn. Fourth. United States: John Wiley & Sons

[8] Bondet, V., Brand-Williams, W., Berset, C. 1997 Kinetics and Mechanisms of Antioxidant Activity using the DPPH Free Radical Method LWT – *Food Science and Technology* **30** (6), pp. 609-615

[9] Weng, C., and Huang, Y 2014 Relationship structure-antioxidant activity of hindered phenolic compounds *Grasas y Aceites* **65** (4), pp. 5-8

[10] Wilson, J. C. P. (1936). 'Manufacture of Oil-Soluble Polyhydric Phenols', US 2063212 A

[11] Hayyan, A., Mjalli, F. S., Hashim, M. A., Hayyan, M. and Alnashef, I. M. 2013 Conversion of free fatty acids in low grade crude palm oil to methyl esters for biodiesel production using chromosulfuric acid *Bulgarian Chemical Communications*, **45**(3), pp. 394–399

[12] Parasuraman, S., et al. 2014 An Overview of Liquid Chromatography-Mass Spectroscopy Instrumentation *Pharmaceutical Methods* **5** (2), pp. 48-55

[13] Odetoye, T.E., Ogunniyi, D.S., and Olatunji, G.A. 2010 Preparation and evaluation of Jatropha curcas Linneaus seed oil alkyd resins *Industrial Crops* **32** (3), pp. 225-230

[14] Jadid, N., et al. 2017 Antioxidant Activities of Different Solvent Extracts of Piper retrofractum Vahl. using DPPH Assay American Institute of Physics Conference Proceedings 1854, pp. 020019

[15] Lin, L., et al. 2011 Opportunities and challenges for biodiesel fuel *Applied Energy* **88**, pp. 1020-1031

IOP Conf. Series: Materials Science and Engineering **742** (2020) 012004 doi:10.1088/1757-899X/742/1/012004

Modification of Pyrogallol with Palmitic Acid as a Soluble Biodiesel Additive

Hery Sutanto[1], Rosa Belinda[1], Silvya Yusri[1], Akhmad Darmawan[2], Evita H. Legowo[1], and Mohammad Nasikin[3]

[1] Department of Chemical Engineering, Faculty of Life Sciences and Technology, Swiss German University, Tangerang, Banten, Indonesia.
(E-mail:hery.sutanto@sgu.ac.id, teresia.rosabelinda@gmail.com, silvya.yusri@sgu.ac.id, evita.legowo@sgu.ac.id)
[2] Research Center for Chemistry, Indonesian Institute of Sciences, Kawasan PUSPIPTEK Serpong, Tangerang Selatan, Banten, Indonesia.
(E-mail: akhmad.darmawan@lipi.go.id)
[3] Department of Chemical Engineering, Faculty of Engineering, Universitas Indonesia, Depok, Jawa Barat, Indonesia.

E-mail: hery.sutanto@sgu.ac.id

Abstract. Biodiesel, as a promising substitute for fossil diesel oil, has nevertheless a main disadvantage, which is oxidized easily. The oxidation degrades its quality and causes disturbances to the machine during operations. Therefore, antioxidant is commonly used in biodiesel. However, the common antioxidant used, namely pyrogallol, has a poor solubility in biodiesel due to the polarity difference. Therefore, this research focused on the modification of pyrogallol by attaching palmitic acid molecule to the benzene ring of pyrogallol molecule through the Friedel-Crafts acylation. The product was characterized by TLC and ^{1}H NMR spectroscopy. It was found that the reaction produced a new product, 1.1'-(4,5,6-trihydroxy-1,3-phenylene)bis(hexadecane-1-one). The new product was subjected to a solubility test and an antioxidant activity evaluation. The result showed that the solubility limit of the product was at 2500 ppm, which was significantly higher than pyrogallol analyzed using one sample t-test. However, using DPPH method, it was found that the antioxidant activity of the new product was slightly lower than pyrogallol as the IC50 value of the new product was 18.44 ppm, and pyrogallol's was 3.07 ppm.

1. Introduction

Within the next 20 years, the world's population growth and emerging economies would trigger the rise in the energy demand, which would be dominated nevertheless by fossil fuels [1]. On the other hand, with the scarcity issue of fossil fuel and the increasing anthropogenic green house gas (GHG) emission which is caused mostly by the combustion of fossil fuels [2], the use of renewables is growing rapidly [1]. One of the type of renewables that took the largest part in the renewable energy mix in 2016 was bioenergy, that includes biofuels as a substitute of oil, for instance bioethanol and biodiesel [3]. Biodiesel, a fuel made of vegetable oil or animal fats, is a sustainable substitute for fossil diesel fuels. It is renewable, widely available, has a higher cetane number and combustion efficiency, with a lower sulfur content [4]. However, the main disadvantage of biodiesel is its susceptibility to oxidation due to the unsaturated fatty acid content [5]. The oxidation leads to acids and polymer sediments formation, which cause engine blocking and corrosion. Therefore, antioxidant additive is required to delay the degradation [6].

Commonly used antioxidants are phenolic compounds, such as propyl gallate, pyrogallol (PY), gallic acid, TBHQ, and others. PY is found to be among the best antioxidant for biodiesel [7], [8]. However, as they are polar molecules whereas biodiesel is nonpolar, the phenolic antioxidants have poor solubility in biodiesel [7], [9]. Over 80 years ago it was already developed that the solubility of phenolic compounds in oil could be improved by attaching an alkyl or aryl group to the benzene ring without decreasing their antioxidant activity [10]. In this research, palmitic acid was used to be attached to the PY ring by following Friedel-Crafts acylation. Friedel-Crafts acylation is a very common method of creating a carbon-carbon bond between a benzene and an alkyl or acyl group.

2. Research Method

2.1. Friedel-Crafts acylation

5 mmol of PY was dissolved in 7 mL nitromethane and stirred for 10 minutes. Another mixture of 5 mmol palmitoyl chloride and 0.25 g $AlCl_3$ was stirred for 10 minutes. A watch glass was used to cover the beaker. The two were combined and stirred for 15 minutes, also covered by a watch glass. [11-12].

2.2. Product identification

The reaction product was analyzed using thin layer chromatography (TLC) to identify the presence of a new product in the reaction product. Then the component of interest was separated by preparative TLC, and further characterized using proton nuclear magnetic resonance (^{1}H NMR) to identify the molecular structure of the new product.

2.3. Solubility test

Solubility test [13]: The separated product was subjected to solubility test to evaluate the solubility limit of the new product compared to PY in biodiesel. The solubility test was fulfilled by measuring the absorbance value of the antioxidant in biodiesel using UV/Vis spectrophotometer. Four concentrations of antioxidant (1500 ppm, 2000 ppm, 2500 ppm, 3000 ppm) in biodiesel were measured. The absorbance value is expected to increase with the increasing concentration. When the absorbance value is stagnant or decreasing after a certain concentration, then the solubility limit of the antioxidant in biodiesel is reached.

2.4. Antioxidant activity evaluation

Antioxidant activity evaluation [14]–[16]: The antioxidant activity of the new product must also be evaluated. TLC-DPPH method was used for the qualitative analysis. Each PY and the new product with a concentration of 1500 ppm were spotted onto a TLC plate, then the spots were sprayed by 0.2% methanolic DPPH. The color change of the methanolic DPPH would indicate the antioxidant activity. Afterwards, the quantitative analysis was done by using DPPH radical scavenging assay. Five concentrations of PY (1 ppm, 2 ppm, 3 ppm, 4 ppm, 5 ppm) and five concentrations of the new product (10 ppm, 12 ppm, 15 ppm, 17 ppm, 20 ppm) were prepared. After 30 minutes of incubation, the absorbance value was measured using UV/Vis spectrophotometer at 515 nm. Then the % inhibition and IC_{50} value could be calculated and compared.

2.5. Formula used in the analysis

2.5.1. Retention factor$_f$

It is a value obtained by dividing the distance traveled with the total distance in TLC analysis.

$$R_f = \frac{distance\ traveled}{total\ distance}$$

2.5.2. Yield (%)

The amount of the new product contained in the reaction product was obtained by dissolving 25 mg of the reaction product in chloroform, then it was completely spread on a TLC paper. Preparative TLC was done to separate and acquire the new product. So, the yield calculation was:

$$Yield = \frac{weight\ of\ acquired\ product}{initial\ weight\ of\ the\ reaction\ product} \times 100\%$$

2.5.3. One sample t-test

One sample t-test is a method for hypothesis testing for a small-sized sample where the sample standard deviation s might not be close to the population standard deviation σ. The solubility limit of the new product was set as the null hypothesis:

$H_0: \mu = 2500\ ppm.$

Nine solubility limit data of pyrogallol was tested to prove whether the true value, which is 2500 ppm, was really different or not. *P-value* represents the occurrence probability of a given event. The higher the *P-value* means the higher the probability of pyrogallol's solubility limit to be 2500 ppm, which supported H_0 to be plausible, that pyrogallol's solubility limit was really no different than 2500 ppm.

As a general rule of thumb, *P-value* = 0.05 was used as the cutoff for significance, so if the *P-value* was smaller than 5%, it was then convincing that the true value, which was 2500 ppm, was really different.

Prior to finding the *P-value*, the t value must be calculated using the equation below:

$$t = \frac{\bar{X} - \mu}{s/\sqrt{n}}$$

where $\bar{X}$ is the sample mean (2167 ppm), μ is the null hypothesis (2500 ppm), s is the sample standard deviation (250), and n is the sample size. The t value is then transformed into a *P-value* by using a t table.

3. Result and discussion

3.1. Friedel-Crafts acylation

Palmitoyl chloride is acid halide derivative of palmitic acid which was used in this experiment to shorten and simplify the reaction to represent palmitic acid molecule. It acts as the R-X reactant in the *Friedel-Crafts* reaction in the presence of $AlCl_3$.. The acylation reaction produced a product with two phases, liquid and waxy phase. The product mixture had a yellow color because of the use of excess $AlCl_3$ in nitromethane [17]. After a few hours, the mixture color changed to grey with some visible black spots, indicating that the excess $AlCl_3$ might have fumed since colorless malodorous gas kept forming. $AlCl_3$ has a characteristic to fume to form hydrogen chloride when exposed to moist air [18]. The grey color might come from remaining palmitoyl chloride and the product's color, while the black spots might be the remaining PY that had been oxidized by $AlCl_3$-nitromethane, whose combination possessed a strong oxidizing power [19].

3.2. Product identification

The reaction product was analyzed using TLC. The TLC result (Figure 1) of the reaction product showed that PY was still contained in both waxy and liquid product. Moreover, a new spot with an Rf value of 0.77, higher compared to PY's which was 0.34, was identified also in both waxy and liquid product, meaning that the reaction product was more nonpolar compared to PY, because the new spot travelled further with the nonpolar mobile phase, while PY was more attracted to the polar stationary phase, the silica gel TLC plate. The mobile phase was a combination of n-hexane and ethyl acetate with a ratio of 3:2. The widening PY spot in the liquid product (d) indicated a very high PY content.

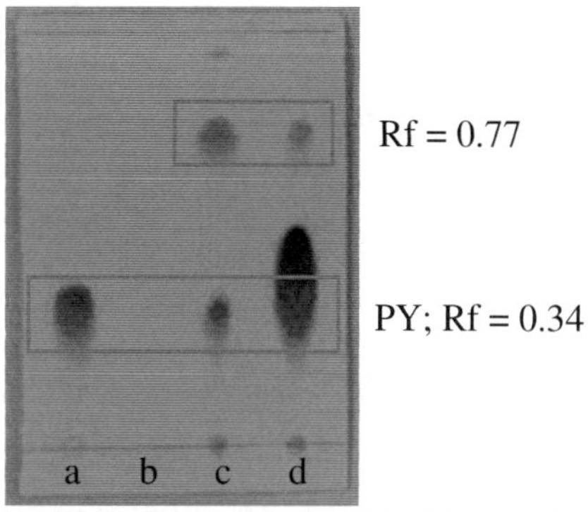

Figure 1. TLC analysis result; (a) PY standard; (b) palmitoyl chloride standard; (c) waxy product; (d) liquid product

The new product (Rf = 0.77) was separated and collected by preparative TLC and further characterized using 1H NMR to identify the molecular structure. The complete chemical shift in the spectrum is shown in Table 1. The spectrum indicated that unreacted PY and palmitoyl chloride were present in the separated mixture. A singlet signal at a chemical shift δ 6.795 suggested that a new product, 1.1'-(4,5,6-trihydroxy-1,3-phenylene)bis(hexadecane-1-one) with a molecular weight of 602.93 g/mol and a chemical formula of C38H66O5 was formed:

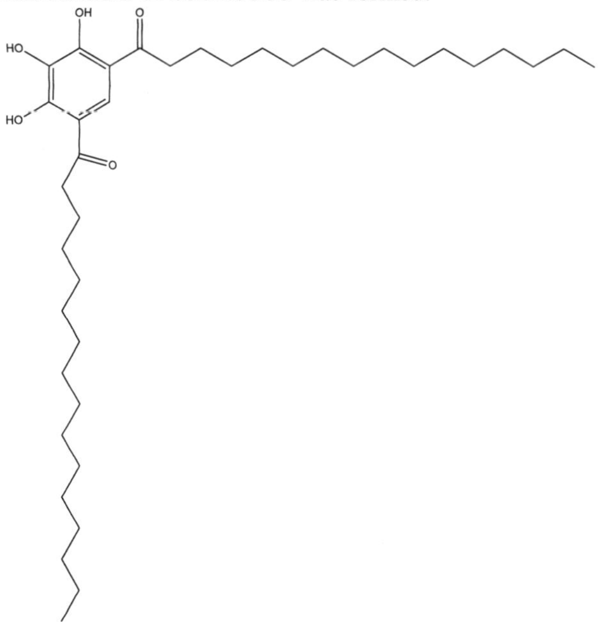

Figure 2. Molecular structure of 1.1'-(4,5,6-trihydroxy-1,3-phenylene)bis(hexadecane-1-one)

Two palmitic groups were substituted to the PY ring instead of only one as initially predicted, due to the strongly activating nature of the existing substituents, the three hydroxyl groups [20].

Table 1. ^{1}H NMR chemical shift data

δ	integration
0.85-0.88	9
1.2-1.5	72
1.6	4
1.75	2
2.3	2
2.6	4
6.6	2
6.78-6.8	1
6.795	1

The reaction yielded 17.6% of the new product, 1.1'-(4,5,6-trihydroxy-1,3-phenylene)bis(hexadecane-1-one). The yield calculation was done by dissolving 25 mg of the reaction product in chloroform, then the mixture was separated using preparative TLC method. After dried and weighed, 4.4 mg of the separated product was obtained.

3.3. Solubility test

In this research, PY was modified with palmitic acid to increase its solubility in biodiesel. Therefore, the new product of the reaction was subjected to a solubility test to compare the solubility of PY and the new product. According to the hypothesis, the new product has a better solubility in biodiesel compared to PY. The solubility test was carried out by measuring the absorbance value of the antioxidant in biodiesel. This approach was used because absorbance is proportional to concentration. The biodiesel-antioxidant mixture was diluted to 1:300 with methanol. The absorbance was measured at 260 nm, referring to the scanning result of PY, palmitoyl chloride, and new product which showed a maximum absorbance at the respective wavelength. The measurement was repeated twice. The result showed that the solubility limit of PY was at 2000 ppm (Figure 3) since the absorbance value reached its highest at the respective concentration, while the new product's solubility limit was at 2500 ppm (Figure 4). This result confirmed that the solubility of the new product in biodiesel was better than PY's.

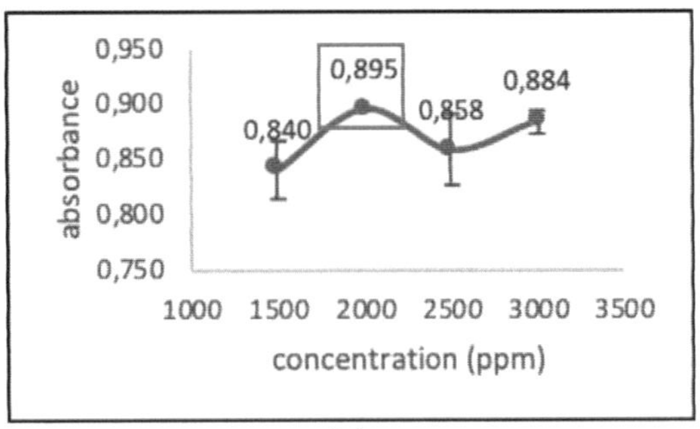

(a)

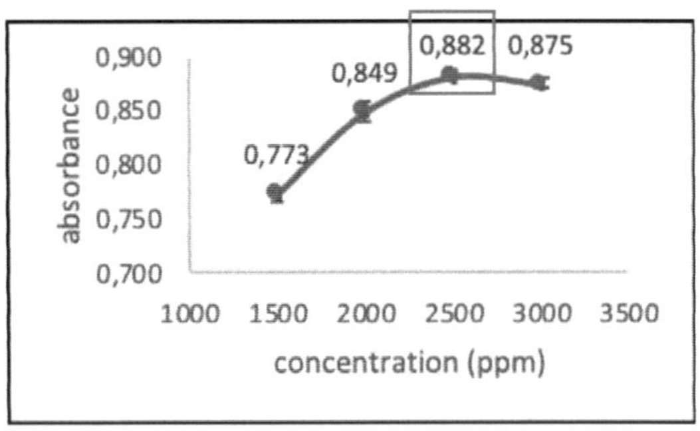

(b)

Figure 3. Solubility limit of PY (a) and new product (b)

However, the solubility limit data set must be further analyzed using one sample t-test hypothesis testing. The purpose was to conclude whether the new product's solubility limit was significantly different from PY's solubility limit. Statistical approach is important in a scientific research to draw a reliable conclusion from repetitive measurements, which often come out differently each time.

Nine PY solubility limit data sample with a mean of 2250 ppm was tested against the solubility limit of the new product, 2500 ppm, as the reference value. Theoretically, when PY is modified by attaching a long chain hydrocarbon to its molecule, the modified PY molecule becomes more nonpolar, thus becoming more soluble in nonpolar solvents, for instance biodiesel. Therefore, a one-tailed test was chosen, because the data set could not have a higher value than the reference value. The test concluded that the solubility limit data of the new product was statistically significant with a P-value of 0.00275.

3.4. Antioxidant activity evaluation

Although the main focus of this research was on improving the solubility of PY in biodiesel, the antioxidant activity of the modified PY must be evaluated as well, to ensure that the new product did not lose its function as an antioxidant.

The principle of the qualitative evaluation was the color change of DPPH. At room temperature, DPPH is a stable free radical which has a violet color in a methanolic solution. In the presence of antioxidants, the free radicals react with a hydrogen so DPPH is losing its free radical property, changing its color to yellow [21-22].

When the 0.2% methanolic DPPH solution was dropped onto the PY and the new product on the TLC paper, both spots immediately turned bright yellow, as seen in Figure 5 below:

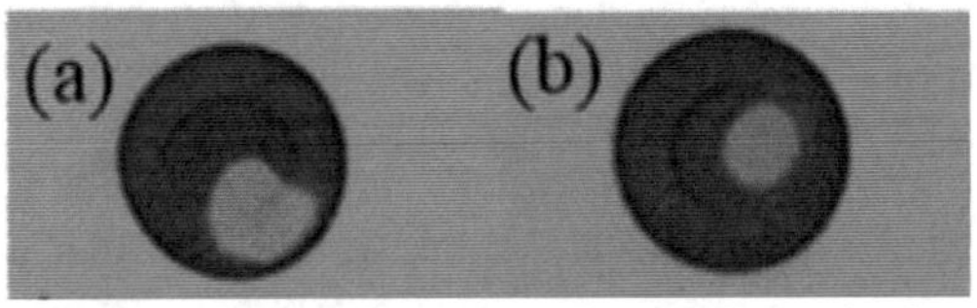

Figure 4. Qualitative TLC-DPPH method;
(a) PY; (b) new product

PY is among the best antioxidant [7], [8], so the result above was expected. As for the new product, from the result above it could be qualitatively concluded that the new product had an antioxidant activity, as it changed the color of the methanolic DPPH from violet to yellow.

After qualitatively proved that the new product also had an antioxidant activity, the quantitative evaluation was performed to compare the antioxidant activity of PY and the new product. The working principle of DPPH radical scavenging assay was the same as explained above, which was based on the color change of DPPH radical due to its reaction with antioxidant, thus changing its color from violet to yellow. However, in the quantitative evaluation, the ability of the antioxidant to inhibit the DPPH free radical, which was evident by the color change, was measured using a UV/Vis spectrophotometer. The absorbance of each sample was measured at 515 nm. The absorbance of DPPH is commonly measured at 515 – 520 nm [15].

Upon DPPH addition, the samples were incubated in a dark container for 30 minutes to allow the DPPH to completely react. They must be kept in the dark because light exposure decreases the absorbance of methanolic DPPH [22]. From the measured absorbance value, the inhibiting percentage was calculated and made into a linear regression plot (Figure 6 and 7). From each graph, the linear regression was used to find the IC_{50} value, which was the amount of antioxidant required to reduce the DPPH radical concentration up to 50% [16].

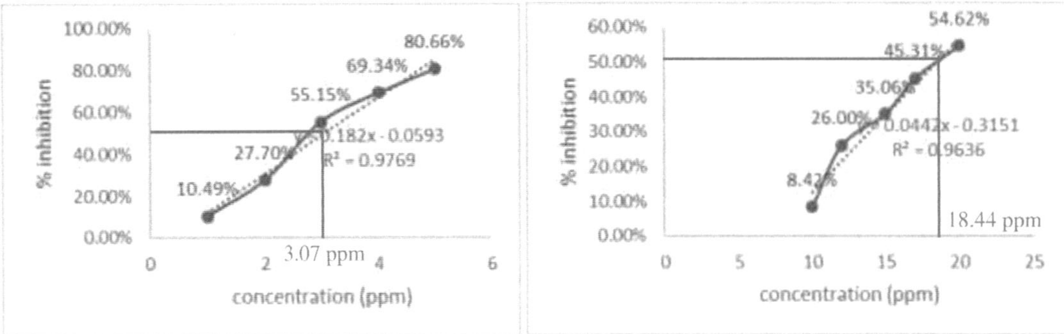

(a) (b)

Figure 5. Inhibiting percentage of PY (a) and new product (b)

The IC_{50} value of PY was 3.07 ppm and for the new product was 18.44 ppm. The IC_{50} value of both PY and the new product was calculated from each linear regression to be 3.07 ppm and 18.44 ppm, suggesting that the antioxidant activity of the new product is significantly lower than PY. However, since the molecular weight of the new product, 602 g/mol, is notably heavier than PY, 126 g/mol, if the IC_{50} value is converted to molar concentration, it will show that the antioxidant activity of PY is only slightly higher than the new product, which are 24μM and 30μM, respectively.

The decreasing antioxidant activity may be caused by some factors. First, since the purification of the new product was only done by preparative TLC, the purity of the product was probably low. Furthermore, the preparative TLC also included a drying process, either by leaving the sample openly to dry or by heating the sample in a vacuum oven. These factors might reduce the antioxidant activity of the sample, which was already 2.5 months old by the time of the antioxidant activity evaluation.

4. Conclusions

Friedel-Crafts acylation reaction successfully modified PY with palmitic acid, producing a new product 1.1'-(4,5,6-trihydroxy-1,3-phenylene)bis(hexadecane-1-one), which was identified using TLC and 1H NMR. The new product has a significantly higher solubility limit at 2500 ppm compared to biodiesel analyzed using one sample t-test, and a slightly lower antioxidant activity than PY with an IC_{50} value of 18.44 ppm.

It is recommended to improve the yield of the reaction product, for instance by finding another alternative solvent for the Friedel-Crafts acylation, considering that palmitoyl chloride is not completely soluble in nitromethane. The product's purity improvement for further characterization by preparative HPLC could give a more accurate characterization result. Moreover, to have a better confirmation of the new product, a liquid chromatography tandem mass spectroscopy (LC-MS/MS) analysis could also be performed.

5. Acknowledgment

The authors would like to thank The Ministry of Research, Technology and Higher Education Republic of Indonesia for financial support through Penelitian Terapan Unggulan Perguruan Tinggi (PTUPT) research grant (contract no. 2639/L4/PP/2019).

6. References

[1] British Petroleum 2017 *BP Energy Outlook 2017 edition*

[2] Intergovernmental Panel on Climate Change (IPCC) 2014 *Climate Change 2014 Synthesis Report*

[3] World Energy Council 2016 *World Energy Resources | 2016*

[4] M.M. Khan and R. U. Khan 2013 Impacts of Biodiesel on the Environment *Int. J. Environ. Eng. Manag.*, vol. 4, no. 4, pp. 345–350

[5] R. Quigley 2007 Biodiesel : The good , the bad … and additives,no. february, pp. 2005–2007

[6] J. Pullen and K. Saeed 2014 Experimental study of the factors affecting the oxidation stability of biodiesel FAME fuels *Fuel Process. Technol.*, vol. 125, pp. 223–235, 2014.

[7] E. Subroto, R. Manurung, H. J. Heeres, and A.A. Broekhuis 2013 Screening of antioxidants as stabilisers for Jatropha curcas L. oil *Eur. J. Lipid Sci. Technol.*, vol. 115, pp. 909–920

[8] S. Park, S. Shin, J. Shin, K. An, and K. Jun 2013 Synthesis of Antioxidant and Evaluation of Its Oxidation Stability for Biodiesel *J. KSTLE*, vol. 29, no. 6, pp. 392–396

[9] K. Varatharajan and D.S. Pushparani 2018 Screening of antioxidant additives for biodiesel fuels *Renew. Sustain. Energy Rev.*, vol. 82, no. July 2017, pp. 2017–2028, 2018.

[10] C. P. Wilson Jr. 1936 *Manufacture of Oil-Soluble Polyhydric Phenols* US2063212

[11] University of Missouri 2007 *Friedel-Crafts Alkylation:1,4-Dimethoxybenzene and Biphenyl* Kansas City

[12] Dixie State University *Friedel-Crafts Acylaton of Ferrocene* pp. 1–6

[13] S. J. I. Wahyudi 2018 *Study of Solubility and Antioxidant Activities of Biodiesel Additives* Swiss German University

[14] SKVM's NMIMS (Deemed-to-be University) *Chapter 7 Screening For Antioxidant* Mumbai.

[15] K. Marxen, K. H. Vanselow, S. Lippemeier, R. Hintze, A. Ruser, and U. P.Hansen 2007 Determination of DPPH radical Oxidation Caused by Methanolic Extracts of Some Microalgal Species by Linear Regression Analysis of Spectrophotometric Measurements," *Sensors*, vol. 7, no. 10, pp. 2080–2095

[16] K. Malinda 2017 *Characterization and Antioxidant Activity of Gallic Acid Derivative* Swiss German University

[17] W. F. Forbes and P. D. Sullivan 1996 The Use of Aluminium Chloride-Nitromethane for the Production of Cation Radicals *J. Am. Chem. Soc.*, vol. 0, no. 9, pp. 2862–2863

[18] A. J. Downs, Ed., 1993 *Chemistry of Aluminium,Gallium,Indium and Thallium* Blackie Academic & Professional

[19] S. H. Pines and A. W. Douglas 1978 Friedel-Crafts Chemistry. A Mechanistic Study of the Reaction of 3-Chloro-4'-fluoro-2-methylpropiophenone with $AlCl_3$ and $AlCl_3$-CH_3NO_2 *J. Org. Chem.*, vol. 43, no. 16, pp. 3126–3131

[20] W. Brown and T. Poon, Eds. 2011 *Benzene and Its Derivatives in Introduction to Organic Chemistry*, Fourth., United States: John Wiley & Sons, Inc., 2011, pp. 316-322; 334-341

[21] A. V. Badarinath, K. Mallikarjuna Rao, C. Madhu Sudhana Chetty, S. Ramkanth, T. V. S. Rajan, and K. Gnanaprakash 2010 *A review on In-vitro antioxidant methods: Comparisions, correlations and considerations Int. J. PharmTech Res.*, vol. 2, no. 2, pp. 1276–1285

[22] S. B. Kedare and R. P. Singh 2011 Genesis and development of DPPH method of antioxidant assay *J. Food Sci. Technol.*, vol. 48, no. 4, pp. 412–422, 2011

IOP Conf. Series: Materials Science and Engineering **742** (2020) 012005 doi:10.1088/1757-899X/742/1/012005

Effect of Manganese Coating on the Sorption Performance of Pyrolusite for Lead in Aqueous Conditions

Yunus Fransiscus[1,2], Carolyn V. Effendy[1] and Herman S. Wijaya[1]

[1]Department of Chemical Engineering, University of Surabaya, Raya Kalirungkut, Surabaya, 60293, Indonesia
[2]Centre for Environmental Studies, University of Surabaya, Raya Kalirungkut, Surabaya, 60293, Indonesia

E-mail : yunus@staff.ubaya.ac.id

Abstract. A process to modify the property of pyrolusite, a natural form of manganese dioxide, was done in order to get a better adsorption performance. Modification was executed by coating the mineral with manganese solution (KMnO4, 0.05 M). Characterization analysis with XRF described significant change on Mn and MnO components, which was higher on the coated material compared to the ore. Meanwhile, although has not perfectly homogenous, it was identified that the distribution of pore diameter reduced from 2.23 – 4.08 μm to 0.65 – 1.07 μm. Another important result from this study was that pH_{pzc} of pyrolusite shifted from 4.2 to 1.9, giving a wider range on pH working solution for the adsorption performance. Manganese coated pyrolusite performed very well in 4 different pH conditions, providing overall removal efficiency above 90% for Lead. Furthermore, from a series of adsorption batch experiments it has been known that the adsorption mechanism of modified pyrolusite for Lead was better explained with Langmuir equation and the kinetics reaction was best modelled by pseudo-second order.

1. Introduction

Pyrolusite, ore form of manganese dioxide, can be enormously found in Indonesia especially in East Nusa Tenggara area. This mineral has been widely known as an important agent for several purposes, such as glass decolorizing, drying paints, coloring ceramics and battery making process. Beside those functions, with its oxides based content pyrolusite also has effective capability to adsorb trace metals in soil and water system. Several researchers have previously investigated the use of natural pyrolusite to eliminate several contaminants in water system. For heavy metals, natural manganese oxide has been reported successfully reduce Copper, Cadmium and Lead as much as 54.35, 6.80 and 98.00 mg/g respectively [1, 2]. Another study provided information that pyrolusite was able to adsorb phosphate in aqueous solution. The maximum adsorption capacity in eliminating phosphate following Langmuir model was as high as 11.40 mg/g [3]. The adsorption capacity of pyrolusite was considered higher if compared to industrial by-products (coal dust, sludge, coal fly ash and lignin) proving that mineral oxides still more effective than any other materials (industrial by-products and/or agricultural/fishing waste) [4]. However, as a common problem for mineral oxides, the poor crystalline morphology of pyrolusite is indicated as barrier for its best adsorption performance [5, 6]. To deal with this problem, several studies proposed modification onto the surface of material and one of the popular methods is by coating process. Coating process is able to create more homogenous surface that may result in smaller pore diameter distribution. In addition, coating compounds (mostly iron or manganese) will enhance the active site of material so that the adsorption capacity can be lifted. Based on those considerations, in this study, modification by applying manganese-coating material onto pyrolusite was carried out. The intention of modification was to upgrade the morphology structure; while manganese substance was chosen to increase the active sites of mineral. The performance of modified pyrolusite was then tested in a series of batch experiments for Lead (Pb) in aqueous solutions. In order to evaluate the adsorption mechanism of coated pyrolusite several variables were applied, such as pH and temperature of solution and initial concentration of Lead.

2. Experimental Set Up

2.1. Coating process

Materials used in this study, including $Pb(NO_3)_2$, HCl, NaOH, $MnCl_2$, $KMnO_4$, H_2SO_4 and KCl were in analytical grade and pyrolusite was supplied by a private company in Surabaya (ECOMEC). Coating process was done by preparing 50 g of pyrolusite (40 mesh) into conical flask with 250 mL of $MnCl_2$ 0.05M in it. The mixture was then shaken carefully in water bath shaker with temperature maintained at 60-70°C. In constant temperature, 200 mL of $KMnO_4$ 0.05M was added slowly to the mixture. Shaking of the mixture was continued for 24 hours and after that the solution was decanted and the material was stored in furnace with temperature of 160°C for 5 hours. Coated material was rinsed with demineralized water prior to characterization analysis and adsorption batch experiment.

2.2. Adsorption experiments

Adsorption experiments were done in batch condition involving several variables such as pH, initial concentration of Lead and temperature. In order to investigate the effect of pH on the adsorption performace of coated pyrolusite, 200 mL of Lead with initial concentration of 10 mg/L was prepared into conical flask prior to the addition of 2.5 g/L adsorbent. The desired pH value (3, 4, 5, 6) was adjust by adding either HCl or NaOH (0.1M). The flask was then mixed constantly at 160 rpm and until the process reached equilibrium condition interval sampling was taken. Measurement of metal concentration was conducted with UV – Spectrometer (HP series 8453). Removal efficiency (%) and adsorption capacity (q_e) were defined by equation (1) and (2).

Removal efficiency (%) :
$$R = \frac{C_0 - C_e}{C_0} * 100\% \quad (1)$$

Adsorption capacity (q_e, mg/g) :
$$q_e = \frac{(C_0 - C_{e)}}{m} * V \quad (2)$$

Where C_0 is the initial concentration of Lead (mg/L), C_e is the concentration of Lead at equilibrium condition (mg/L), q_e (mg/g) is Lead adsorbed at given mass of adsorbent in the working volume (L). The collected data of equilibrium concentrations for each initial concentration of Lead were analysed using two adsorption models, Langmuir and Freundlich.

Langmuir :
$$q_e = \frac{bq_{max}C_e}{1+b.C_e} \quad (3)$$

Freundlich :
$$q_e = k_f C_e^n \quad (4)$$

where C_e is the heavy metal concentration in the solution at equilibrium (mg/L), b is Langmuir constant related to the affinity of binding sites (L/mg), q_{max} is the maximum adsorption capacity (mg/g), k_f is a Freundlich constant related to roughly adsorption capacity (mg/L), n is the constant related to adsorption density. Meanwhile, kinetics reaction was defined by applying pseudo first order (5) and pseudo second order equations (6).

$$\frac{dq_t}{dt} = k_1 (q_e - q_t) \quad (5)$$

$$\frac{dq_t}{dt} = k_2 (q_e - q_t)^2 \quad (6)$$

Another series of batch experiments in four different temperatures (298K, 308K, 318K and 328K) were carried out to evaluate the effect of temperature. Thermodynamic data expressed as values of standard enthalpy change (ΔH°), standard entropy change (ΔS°) and Gibb's free energy (ΔG^{0}). Those values can be determined by applying these equations:

$$K_d = \frac{C_s}{C_e} \quad (7); \qquad \ln K_d = \frac{\Delta S^{\circ}}{R} - \frac{\Delta H^{\circ}}{RT} \quad (8); \qquad \Delta G^{\circ} = \Delta H^{\circ} - T\Delta S^{\circ} \quad (9)$$

where C_s and C_e are the removed and remaining concentrations of Lead respectively. R (8.314 $Jmol^{-1}K^{-1}$) is the ideal gas constant, T is temperature in Kelvin. The values of ΔH^{0} and ΔS^{0} were determined from the slope and y-intercept on the plot between $\ln K_d$ versus ${}^{1}/_{T}$.

3. Results and Discussion

3.1. Characterization of Coated Pyrolusite

Morphology transition of pyrolusite was identified by SEM analysis with 2000 magnifications. Figure 1(a) shows the surface of uncoated pyrolusite that are rough with uneven pores, meanwhile the image of coated pyrolusite (Fig 1(b)) displays a smoother surface with coverage of manganese coat. However the SEM images provide indication that manganese layer has not homogenously cover the material, suggesting either more concentrate of coated compound or more intense mixing during the coating process. Nevertheless, coating process improved the elemental properties of material as proven from XRF analysis. Table 1 describes that Mn and MnO elements in the coated mineral was higher compared to the raw one, this is important for the uplifting of adsorption performance.

(a) (b)

Figure 1. (a) SEM image of raw pyrolusite; (b) SEM image of coated pyrolusite

Table 1. Elements of Pyrolusite from XRF Helium Analysis

Elements	Percentage (%)	
	Raw Pyrolusite	Coated Pyrolusite
Mn	93.1	96.8
Si	1.56	1.18
K	0.199	0.274
Ti	0.18	0.1
MnO	88.4	93.6
SiO_2	2.85	2.16

3.2. Adsorption Experiments

3.2.1. Effect of pH

pH is a crucial factor in determining the nature of attachment between adsorbate and adsorbent in aqueous condition [7]. There were four different pH (3, 4, 5, 6) applied in adsorption experiments, in alkali condition (pH above 6) Lead started to form precipitate therefore pH variation above this value was neglected. Interestingly, among those four pH values total removal for Lead was not significantly different, putting them in the range of 96.20% to 98.55% as can be seen in Fig. 2. This result was unlikely when applying raw pyrolusite, in which the notable gap existed between total removal of Lead in low and high pH value. Comparison between total removal of Lead by using raw and coated pyrolusite in different pH is presented in Fig.3. A plausible explanation for this phenomenon could be related to the pH_{pzc} (point of zero charge) of pyrolusite. Previous study reported that pH_{pzc} of pyrolusite was 4.2 [8], thus if the pH solution is above 4.2 the surface of material will be lack of hydrogen ions resulted in negatively charge [9]. This mean the adsorption performance of pyrolusite will be effective if the working pH solution is above 4.2, expecting a favourable interaction between negative surface charge of the material and positive charge of metal ions. In this study, pH_{pzc} of coated pyrolusite was found at 1.9, shifting from the value of uncoated one, which were 4.2. By that, when the pH solution is gradually increase (above 1.9), the surface charge of coated pyrolusite will tend to be more negative resulting a less electrostatic repulsion to the metal ion. A wider range of pH working solution is very important especially for the implementation of modified pyrolusite in field scale.

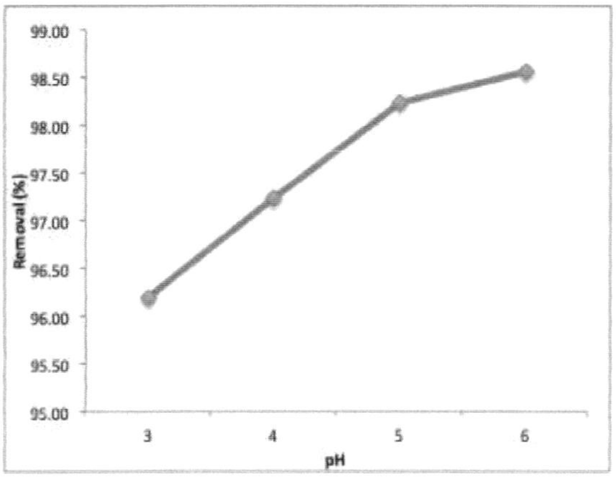

Figure 2. Removal of Lead with Coated Pyrolusite

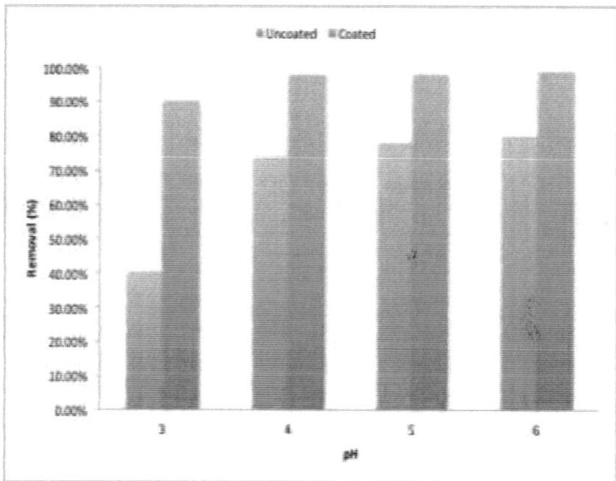

Figure 3. Comparison of Lead removal in various pH values

3.2.2. Adsorption Isotherms

Collected data of equilibrium concentration (C_e) for every initial concentration of Lead were plotted following the Langmuir (3) and Freundlich (4) equations. The results, as displayed in Fig. 4 (a) and (b), propose that the interaction between modified pyrolusite and Lead fit better to Langmuir model. By following Langmuir model, it is indicated that the adsorption is driven by chemical reaction (chemisorption) rather than physical process. Lead is adsorbed creating a smooth monolayer all around the homogenous surface of collector. In addition to that, data analysis with Langmuir equation came up with the information on maximum adsorption capacity of coated pyrolusite which was 221 mg/g, higher than the capacity of the raw one which was 123.45 mg/g.

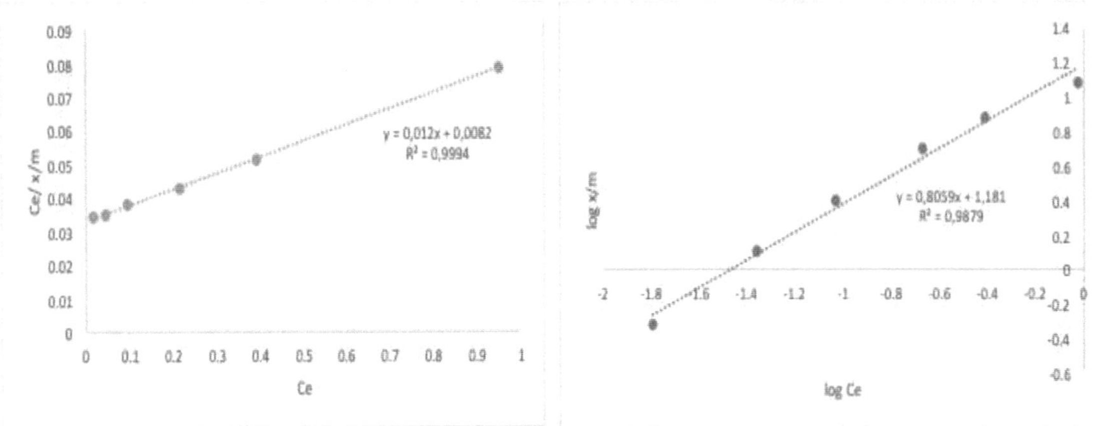

Figure 4. (a) Linear Plot – Langmuir ; (b) Linear Plot - Freundlich

3.2.3. Effect of Temperature

The effect of temperature on process sorption was evaluated by conducting batch experiments in four different temperatures (298K, 308K, 318K and 328K). Figure 5 depicts that the total removal of Lead increases along with the hotter condition of the solution, giving the biggest removal of 98.72% at the highest temperature (328K). Thermodynamic data (Table 2) reveals several important information regarding the adsorption mechanism. The positive value of ΔH° suggests that the adsorption process is endothermic and because the value is bigger than 4200 J/mol it is also indicated that the process determined by chemical reaction [10]. Meanwhile the positive value of ΔS° explains the progress of structural change in the surface of collector when the interaction with Lead occurred. Finally, the negative value of ΔG° proposes a spontaneous reaction between coated pyrolusite and Lead. The lower value of ΔG° in higher temperature describes that the adsorption process is more spontant when the temperature of solution increases [11].

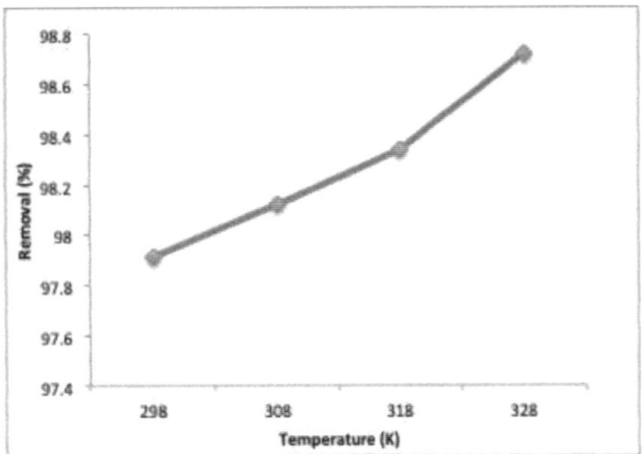

Figure 5. Removal of Pb in different temperature

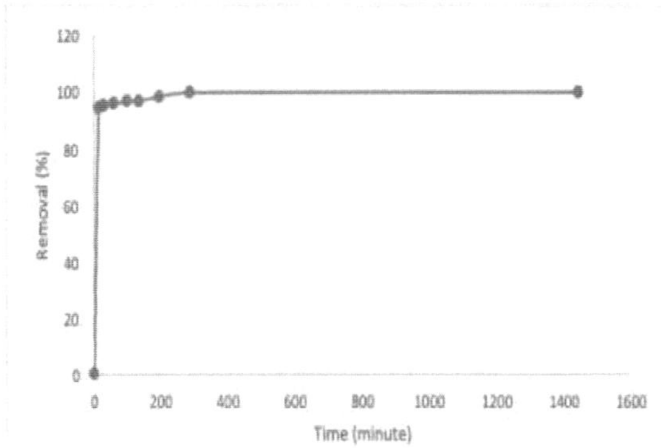

Figure 6. Adsorption rate of Pb

Table 2. Termodynamic data of Pb onto Coated Pyrolusite

Temp (K)	$\Delta H°$ (J/mol)	$\Delta S°$ (J/mol.K)	$\Delta G°$ (J/mol)
298	13,003	69.547	-7,821
308	13,003	69.547	-8,378
318	13,003	69.547	-8,963

328	13,003	69.547	-9,958

3.2.4. Adsorption Kinetic

The spontaneity of adsorption process onto the modified pyrolusite was confirmed by the kinetic study analysis. Figure 6 portrays that the interaction between Lead and collector surface took place in the very beginning of reaction, just after 15 minutes about 95% of Lead was absorbed. Kinetics reaction of the process was evaluated by applying pseudo-first order and pseudo-second order models, the plotting result for both model is presented in Fig. 7 (a) and (b). Based on R^2 value, it is obvious that the pseudo-second order model fit very well to the adsorption process. This model elaborates that both active sites on the collector surface and metal concentration actively determine the adsorption rate [7].

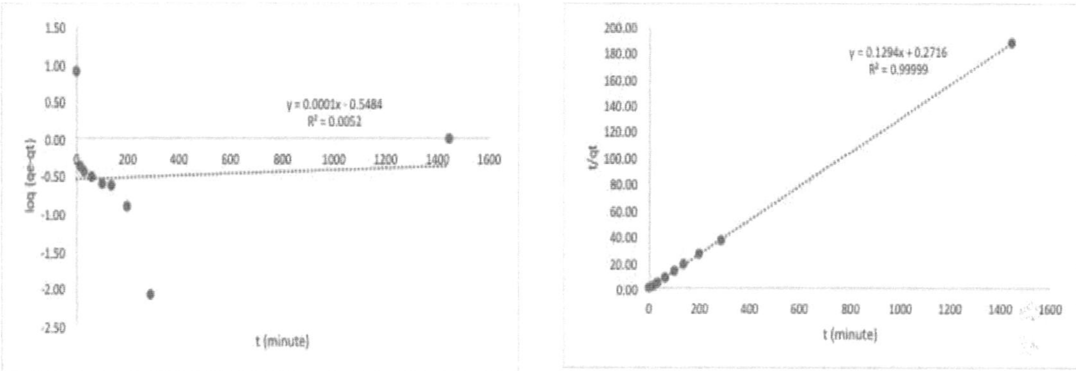

Figure 7. (a) Pseudo-first order and (b) Pseudo-second order model for Pb

4. Conclusions

Modification on pyrolusite by manganese coating gave significant impact on the adsorption performance of the material. The shifting of pH_{pzc} not only expanded the pH working solutions but also increased the adsorption capacity. The adsorption isotherm of Lead onto coated pyrolusite was best modelled by Langmuir equation, indicating a complete monolayer formation. Adsorption process was endothermic, spontaneous and followed the pseudo-second order.

References

[1] Demirkiran N, 2015, *Trans. Nonferrous Met. Soc. China* **25** 647
[2] Sonmezay A, Oncel M S, Bektas N, 2012, *Trans. Nonferrous Met. Soc. China* **22** 3131
[3] Fransiscus Y, Widi R K, Aprilasti G O, Yuharma M D, 2018, *Inter. Jour. Adv. Sci. Eng. & Infor. Tech.* **8(3)** 818
[4] Ahmed M J K, Ahmaruzzaman M, 2016, *Jour. Water Proc. Eng.* **10** 39
[5] Zhu Q, Zhang B, Wang T, Feng X, 2019, *Microporous and Mesoporous Materials*, **273** 90
[6] Tompsett D A, Parker S C, Islam M S, 2014, *Jour. American. Chem. Soc.*, **136(4)** 1418
[7] Chaudhry S A, Khan T A, Ali I, 2016, *Egyptian Journal of Basic and Applied Sciences* **3** 287
[8] Zhao D, Yang X. Yang, Chen C, Wang X, 2010, *Chem. Eng. Jour.* **164** 49
[9] Yang P, Post J E, Wang Q, Xu W, Geiss R, McCurdy P R, Zhu M, 2019, *Environ. Sci. Technol.*, **53(13)** 7453
[10] Zou W, Han R, Chen Z, Jinghua Z, Shi J, 2006, *Colloids and Surfaces A: Physicochem. Eng. Aspects*, **279** 238
[11] Hu C, Zhu P, Cai M, Hu H, Fu Q, 2017, *Applied Clay Science*, **143** 320

Acknowledgments

Authors would like to thank ECOMEC for supplying the raw pyrolusite and to technical staff in the Laboratory of Bioprocess and Environmental Process Technology, Department of Chemical Engineering, University of Surabaya for valuable suggestion in equipment and instrumentation setting.

Cenosphere Separation from Fly Ash Using Modified Gravity Separator: Feed Ratio Assessment and Stream Velocity Optimization

F Y A Saputra[1], Sutijan[1] and H T B M Petrus[1,2]

[1]Departement of Chemical Engineering (Sustainable Mineral Processing Research Group), Faculty of Engineering, Universitas Gadjah Mada, Jl. Grafika 2, Yogyakarta, 55281, Indonesia.
[2]Unconventional Georesources Research Center, Faculty of Engineering, Universitas Gadjah Mada, Jl. Grafika No. 2, Yogyakarta, 55281, Indonesia

E-mail: bayupetrus@ugm.ac.id

Abstract. The increase of the population causing increased energy needs, one of which is electricity. Coal constitutes one of the largest carbon sources to generate electricity where the combustion leaves a by-product in the form of fly ash. Inside the fly ash, there are round-shaped hollow particles whose its density is lower than water of about 0.9 g/cm^3. These particles called cenosphere containing alumino-silicate. Due to its lighter and tougher nature, it makes the cenosphere appealing to use as a lightweight and robust material. Cenosphere can be separated using both dry separation and wet separation. Separation using the wet process produces another impact on the side of the liquid media processing that is used because it potentially carries toxic compounds inside the fly ash. While the separation using the dry separation method still has problems regarding its efficiency. Modified Gravity Separator is a new design for cenosphere separation using air as a fluid medium. This research was conducted to evaluate the performance of the proposed geometry by looking at the effect of the feed velocity to the geometry expansion space using a CFD (Computational Fluid Dynamic) computing software. The feed particles are assumed to be spherical shape and are observed with the Euler-Lagrange equation where the particles interact with continuous airflow. The simulations were carried out at speeds of 0.1 m/s, 0.25 m/s, 0.5 m/s, and 1 m/s. This simulation successfully illustrates the flow pattern and predicts the recovery rate of each collector so that it can be evaluated and optimized for future investigation.

1. Introduction

Coal is one of the biggest sources of carbon used to produce energy. Indonesia has a lot of sources of these resources. In 2018, 548 million tons of coal have been explored and 91.14 million tons (14,78% wt of coal production) had been used to generating electricity [1]. The coal production is increased since 2015 which the production reaches 405 million tons [2]. After burning, this kind of fossil fuel left the by-product such as bottom ash, slag and fly ash. The particle has been collected by Electro-Static Precipitation (ESP), bag filter, stack, and other recovery units to avoid the particle spread to the environment. The utilization of this waste is limited due to the components that are inside the fly ash such as high density, dangerous element and so on.

Inside the fly ash, there are hollow particles that formed from silica and had a density lower than water. This particle called as cenosphere. The fraction is small compared to the other particle, it was 0.01 – 4.8% [3,4]. The diameter has been observed around 20 – 500 μm [5] and usually between 20 – 300 μm [5]. Cenosphere formation and character are different depend on burning temperature and molten drop chemical composition [5,6].

Because has much excellence such as good hardness ability and light [7], the cenosphere should be separated from fly ash for further application. The wet-separation process and Dry-Separation process can be used. The cheapest media used in Wet-Separation is water. But there is an environmental issue due to toxic element which is inside the fly ash. The rare earth element and other dangerous material will be leached to the water. When the toxic material accumulated, this phenomenon will impact the environment [8].

Some research has been developed to collect the cenosphere from the fly ash. Li et al. (2014) were developed a separation tool using a wet-separation process called Inverted Reflux Classifier (IRC). Their work compared IRC with standard RC on a pilot plant scale. The tool used the *Boycott effect* which means that the slope will affect the separation process. The work claiming that the efficiency of separation reaches 80% [9].

Another project has been worked for the dry process, micron separator and closed-type pneumatic separator was evaluated. Closed-type pneumatic separator did not work better than micron separator. The evaluation found that micron separator reaches the Newton Efficiency value of 0.44 and cenosphere recovery reaches 80% [5,10]. This work claim that the separation process can be used to compete with the wet-separation process due to no environmental issue and less space needed.

One side with Petrus et al. (2010) project, Sutijan et al. (2019) was established another type of dry-separator unit called modified gravity separator. This unit was evaluated using CFD simulation with injection particles of cenosphere and fly ash respectively 1:1 in mass. The newton efficiency grade reaches 0.41 with a feed velocity of 0.25 m/s [11]. Their reported work is not using the real injection fraction of the cenosphere so that the major aim of this work is to evaluate the particle track and newton efficiency using a small fraction of the cenosphere inside the fly ash. The particles are assuming sphere in shape and following the Rosin-Rammler distribution.

2. Simulation Setup

2.1. Geometry and Grid Arrangement

The geometry consists of four parts. There are inlet particle chamber, main body, collectors chamber, and an outlet chamber. In the simulation condition, the outlet chamber and the collector chambers are joined together to simplify the result calculation. The detail dimension describes in Fig. 1. Grid arrangement was developed using automatic meshing from ANSYS. There were two examinations to see the better shape of the grid used. The first one is the triangular model and the second shape is the hexagonal model.

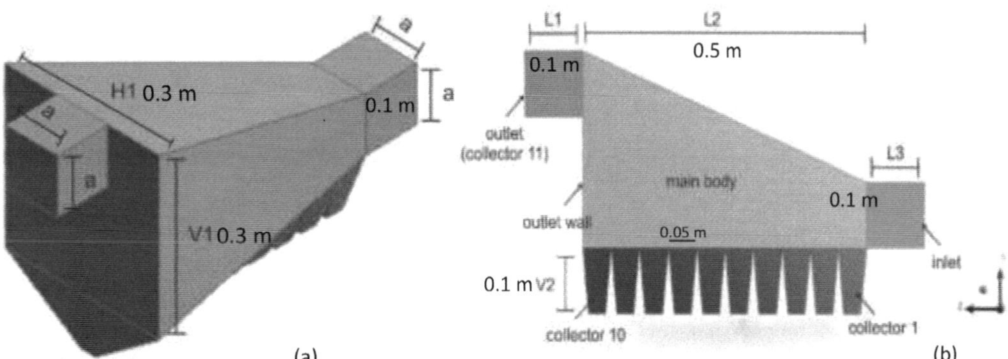

Figure 1. Modified Gravity Separator detail (from [11])

2.2. Simulation Environment

Fly ash from the combustion is variable by the size. For the conventional fly ash, meaning that fly ash coming from the boiler process, the density is about 2.2 – 2.5 g/cm^3 [12]. The diameter range is around 9.63 – 47.6 μm [13] or in the range of 0.01 – 350 μm [12]. Inside the fly ash, the fraction of the cenosphere generally about 0.3 – 1.5 % wt [4] and the density of the cenosphere is in the range of 0.9 – 1.8 g/cm^3 [10]. Like previous work, this sample of simulation is using the same size of the particle. The diameter distribution is using the Rossin-Rammler equation [11]. For the easiest computation, the total mass amount is 8.58e-9 kg/s (cenosphere 0.15% wt), 596 cenosphere individual particles, and 100,909 fly ash particles.

This work simulation is using ANSYS Fluent. The turbulent model is used in this scheme. There is a lot of turbulence models such as Direct Numerical Simulation, Large-Eddy Simulation, Reynolds Stress Model, Two-equation Models, One-equation Models, and zero-equation Models. Direct Numerical Simulation equation has a higher cost of simulation and higher resolution than the others model but has a higher level of approximation. The lower one is the zero-equation model, this model consumes the lowest cost of computational cost but, of course, uses lower approximation and the medium-cost computational is the two-equation models [14]. K-ε and k-ω are kinds of two-equation models. Hard to find the same case of turbulence using this scheme from another work that had been done to choose the better model used. From Andersson et al. (2012), k-ε is the most used model due to its calculation for the turbulence kinetic energy, k, provide the value of the energy-dissipation rate (ε) from the calculation. For the first simulation observation and medium-cost of the simulation, this project uses the standard k-ε model. The following transport equation (1 and 2) has gained the value of the kinetic energy and the rate of dissipation and the turbulence viscosity, μ_t, can be calculated using k and ε which is showed by eq. (3) [15].

$$\frac{\partial}{\partial}(\rho k) + \frac{\partial}{\partial x_i}(\rho k u_i) = \frac{\partial}{\partial x_j}\left[\left(\mu + \frac{\mu_t}{\sigma_k}\right)\frac{\partial k}{\partial x_j}\right] + G_k + G_b - \rho\varepsilon - Y_M + S_k \tag{1}$$

$$\frac{\partial}{\partial}(\rho\varepsilon) + \frac{\partial}{\partial x_i}(\rho\varepsilon u_i) = \frac{\partial}{\partial x_j}\left[\left(\mu + \frac{\mu_t}{\sigma_\varepsilon}\right)\frac{\partial \varepsilon}{\partial x_j}\right] + C_{1\varepsilon}\frac{\varepsilon}{k}(G_k + C_{3\varepsilon}G_b) - C_{2\varepsilon}\rho\frac{\varepsilon^2}{k} + S_\varepsilon \tag{2}$$

$$\mu_t = \rho C_\mu \frac{k^2}{\varepsilon} \tag{3}$$

Where:

G_k	: The generation of turbulence kinetic energy caused by the mean velocity gradients,
G_b	: The generation of turbulence kinetic energy caused by buoyancy,
Y_M	: Value of the fluctuating dilatation in compressible turbulence to the overall dissipation rate,
C	: Constant
σ	: Turbulent Prandtl number
S	: User-defined source term

For tracking the particle, the Discrete Phase Model is used in this simulation condition with the maximum iteration of 10,011. The model used the Euler-Lagrange approach and neglected the particle-particle interaction which means that the dispersed phase occupied a low volume fraction. But high mass loading still acceptable [15].

Table 1. The boundary condition of the simulation

Boundary Zone	Type	Momentum	DPM
Input chamber	Velocity inlet	Velocity: 0.1, 0.25, 0.5, 1 m/s Turbulence: default	Escape
Collector chamber	Pressure outlet	Default	Escape
Outlet chamber	Pressure outlet	Default	Escape
Main body	Wall	Stationary wall; others: default	Reflect
Body interior	Interior	-	-

For the boundary condition, the input chamber was set to be velocity inlet. This velocity is different for every simulation depends on running variables. For Discrete Particle Material condition when walk at this zone is escaping. The boundary condition for the collector and outlet chambers is a pressure outlet. A convergence iteration was not achieved by using the wall type. The SIMPLE model used for the calculation with default Spacial Discretization condition. For the initialization, the computation will begin from the input zone.

2.3. Computing Result

Data of the simulation are collected to the external file. Format of the report based on user-defined function. Then, the data will be treated using MatLab software to know the effectiveness of the Modified Gravity Separator. The diameter, position, and density data are the major information that must be extracted for the calculation. From the particle tracking data, the collectors and overall efficiency can be calculated using the Newton Efficiency formula in equation (4) and (5) [10].

Collector efficiency: $$NE\ (i) = \frac{m_{c(i)}}{m_{ct}} - \left(1 - \frac{m_{ft} - m_{f(i)}}{m_{ft}}\right) \tag{4}$$

Overall efficiency: $$NET = \frac{m_{cs}}{m_{ct}} - \left(1 - \frac{m_{ft} - m_{fs}}{m_{ft}}\right) \tag{5}$$

Where:

NE : Newton Efficiency
NET : Overall Newton Efficiency
$m_{c/f}$: Mass of cenosphere/fly ash
$m_{ct/ft}$: Total mass of cenosphere/fly ash
$m_{cs/fs}$: Concentrated mass of cenosphere/fly ash
(i) : Collector index

ANSYS can extract the mass per particle to the report file. The reported data will be collected to know the total mass of the specific particle in a specific collector. So, its information can be used to know the collector Newton Efficiency. A positive value of the NE is cenosphere concentrated and if the NE is negative that is fly ash concentrated.

3. Results and Discussion

Firstly before trying to run the computation, a configuration of the grid condition is needed. For this geometry, using the tetrahedron shape was better than using the hex-dominant shape. It shows from the skewness and element quality evaluation of the grid. The skewness average reaches 0.4957 by the hex-dominant shape and 0.2663 is achieved by the tetrahedron shape. This evaluation is calculated using automatic medium-mesh generation. If the value is closer to zero, the shape is better to be applied.

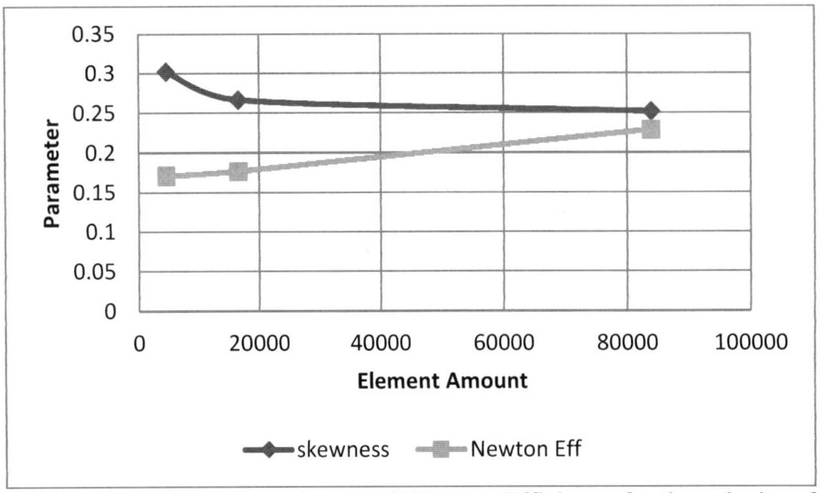

Figure 2. Effect of grid numbers to the overall Newton Efficiency for the velocity of 0.25 m/s

From Fig. 2 above, the result shows that the number of grid elements is affecting the NE. The size of the element is smaller following by increasing the number of the element. This condition makes the iteration more detail for tracking the particle and the rotation of the turbulence pathline is better known. The next discussing simulation was using a fine grid with an amount of 83,942 tetrahedron elements.

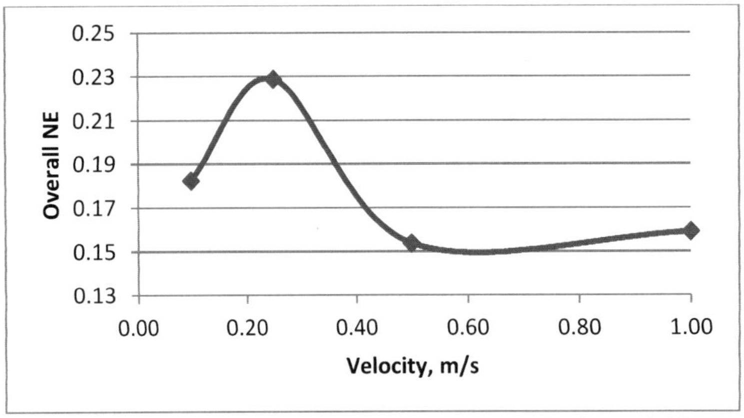

Figure 3. NE (overall) evaluation from each velocity variables

Fig. 3 is describing the effect of the overall NE of the geometry with the stream velocity. It can be seen that the best NE reaches for the speed of 0.25 m/s. Increasing NE happened at 1 m/s after decreasing at 0.5 m/s. This condition occurs because of the turbulence effect. Logically, the speed makes the flow more turbulence and the particles are flying more disordered. Because of the turbulent swirling, some particle collapses to the collectors more random. This hypothesis is supported by Fig 4. At 1 m/s, more particle doesn't reach the terminal velocity and going through with the airflow to the outlet chamber. But different from other velocities, the collector 11 (outlet chamber) reaches a positive NE (cenosphere concentrated).

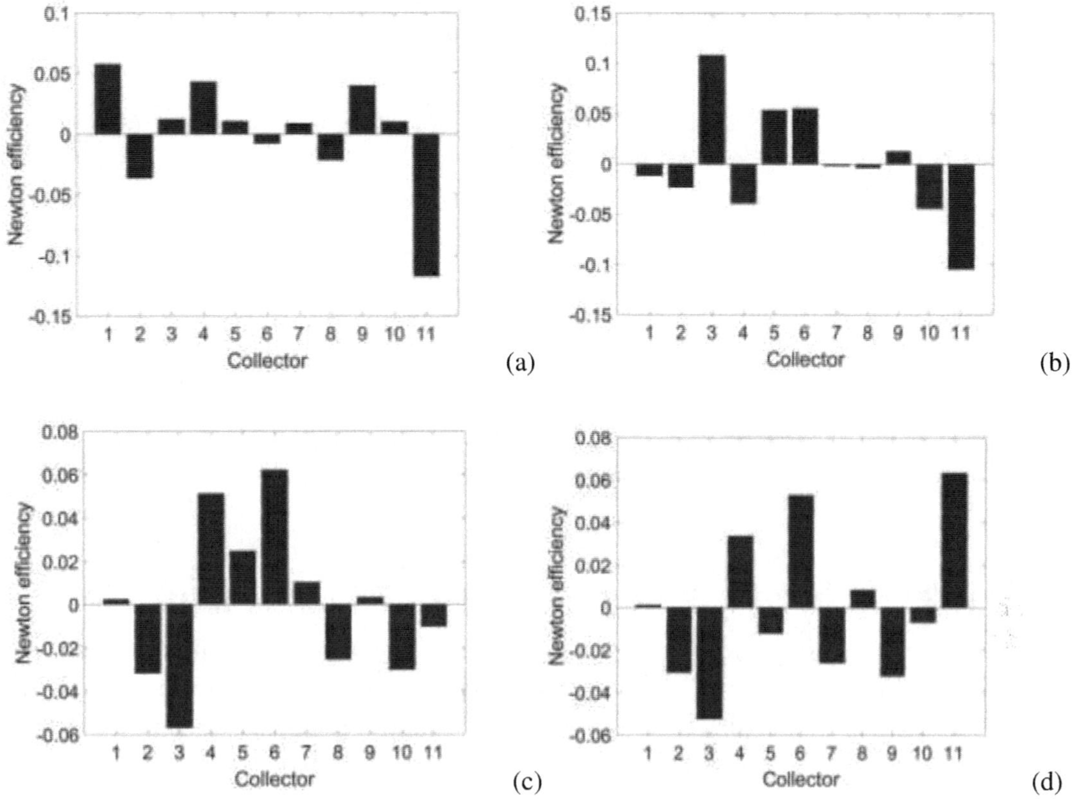

Figure 4. NE for each collector (a) 0.1 m/s; (b) 0.25 m/s; (c) 0.5 m/s; (d) 1 m/s

The size distribution at 0.25 m/s for fly ash evenly distributed in all collectors zone (Fig. 6). In contrast with the cenosphere, the size distribution in every collector is random (Fig. 5). This phenomenon occurs because of the amount of the particle injected. Cenosphere particle size distribution was injected 0.15% inside the total amount of feed. Fly ash particles very numerous compare with cenosphere particles.

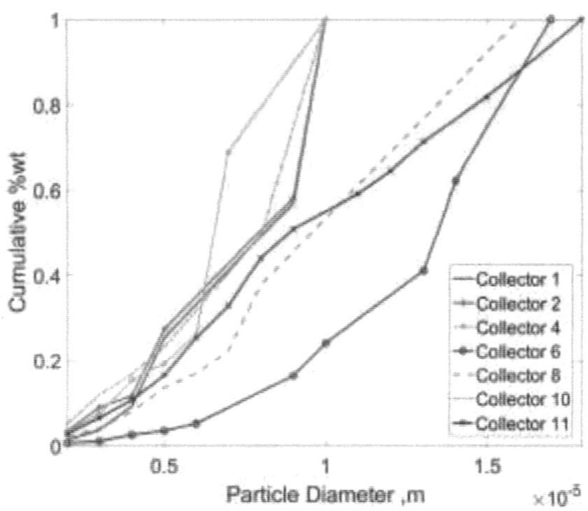

Figure 5. Cenosphere cumulative particle distribution for 0.25 m/s

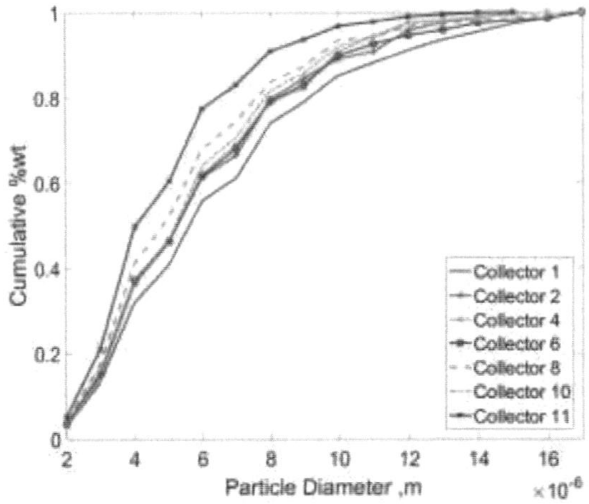

Figure 6. Fly ash cumulative particle distribution for 0.25 m/s

The standard k-ε turbulence model and the DPM (Discrete Particle Model) have been successfully described the tracking particle (Fig. 7a,b). The high concentrated flow which is crash into the wall is conducted above the 10th collector and makes the pressure higher at this location (Fig. 7c). It caused the particle track at this point more random than the tracking pathline at the beginning. There is a particle that has a reverse flow inside the geometry interior after colliding with the wall and there is a particle that is not in turbulence condition mostly at the start point of analysis. This visualization occurs due to the effect of particle diameter and particle mass on the flow of the air.

IOP Conf. Series: Materials Science and Engineering **742** (2020) 012006 doi:10.1088/1757-899X/742/1/012006

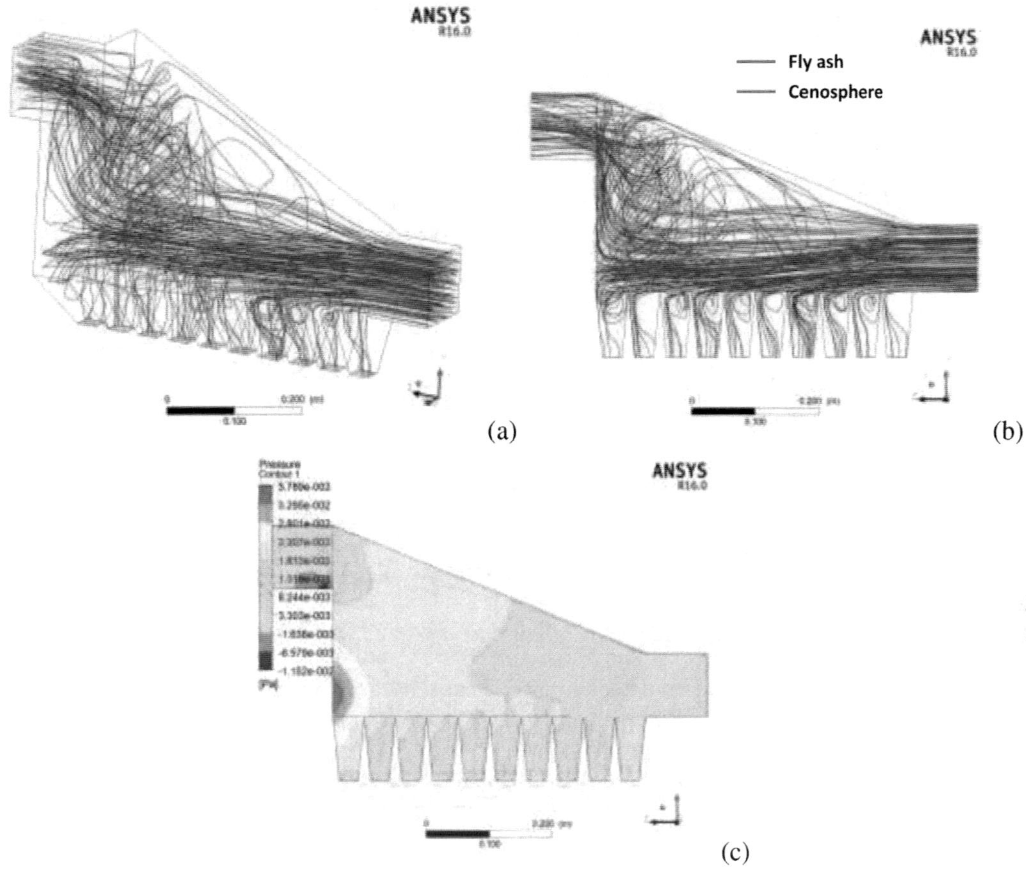

Figure 7. Particle tracking pathline for 0.25 m/s (a) ISO view; (b) positive x-direction view; (c) pressure prediction at positive x-direction

4. Conclusion

Different from previous work, the evaluation based on the real condition of the fly ash sample has been evaluated. The simulation used standard k-ε and resulting a good explanation of particle tracking data. The best NE reaches for speed of 0.25 m/s with NE of 0.229, but this value is not better than Sutijan et al. (2019) works. This means that the fraction of the cenosphere influences the separation quality. A high-velocity speed affected the rheology of the turbulence track where the particle going through on the bottom of the Modified Gravity Separator and crash the wall at the endpoint. This phenomenon induces the pressure increase around the corner at the end of the wall. The pressure makes the particle flowing reverse to the inlet chamber point. In the real condition, this problem may cause a collision between the particles to occur. For the future optimization investigation, the authors suggest comparing another model of turbulence and dense particle injection to see the best model that can describe the better simulation. A lot of variables added, more costly computation occurs.

Acknowledgment

The Minister of Higher Education of Indonesia has been given the funding to conduct this project under PTUPT scheme with contract number 2702/UN1.DITLIT/DIT-LIT/LT/2019, so, the Authors are very grateful for the funding provided.

References

[1] Ministry of Energy & Mineral Resources of the Republic of Indonesia 2019 *Laporan Kinerja Tahun 2018* (Jakarta: Republic of Indonesia)

[2] BPS 2017 Produksi Barang Tambang Mineral *Badan Pus. Stat.*

[3] Sokol E V., Maksimova N V., Volkova N I, Nigmatulina E N and Frenkel A E 2000 Hollow silicate microspheres from fly ashes of the Chelyabinsk brown coals (South Urals, Russia) *Fuel Process. Technol.* **67** 35–52

[4] Fomenko E V., Anshits N N, Solovyov L A, Mikhaylova O A and Anshits A G 2013 Composition and morphology of fly ash cenospheres produced from the combustion of Kuznetsk coal *Energy and Fuels* **27** 5440–8

[5] Petrus H T B M, Hirajima T, Oosako Y, Nonaka M, Sasaki K and Ando T 2011 Performance of dry-separation processes in the recovery of cenospheres from fly ash and their implementation in a recovery unit *Int. J. Miner. Process.* **98** 15–23

[6] Li Y and Wu H 2012 Ash cenosphere from solid fuels combustion. Part 1: An investigation into its formation mechanism using pyrite as a model fuel *Energy and Fuels* **26** 130–7

[7] Ranjbar N and Kuenzel C 2017 Cenospheres: A review *Fuel* **207** 1–12

[8] Borm P J A 1997 Toxicity and occupational health hazards of coal fly ash (CFA). A review of data and comparison to coal mine dust *Ann. Occup. Hyg.* **41** 659–76

[9] Li J, Agarwal A, Iveson S M, Kiani A, Dickinson J, Zhou J and Galvin K P 2014 Recovery and concentration of buoyant cenospheres using an Inverted Reflux Classifier *Fuel Process. Technol.* **123** 127–39

[10] Hirajima T, Petrus H T B M, Oosako Y, Nonaka M, Sasaki K and Ando T 2010 Recovery of cenospheres from coal fly ash using a dry separation process: Separation estimation and potential application *Int. J. Miner. Process.* **95** 18–24

[11] Sutijan, Sandiwan S W, Saputra K I, Mahdi M H, Albarkan M S, Prasetya A and Petrus H T B M 2019 Effect of inlet speed on gravitational air separator for cenospheres accumulation from fly ash: Modeling using computational fluid dynamics (CFD) *IOP Conf. Ser. Mater. Sci. Eng.* **478**

[12] Stoch A 2015 *Fly ash from coal combustion-characterization* (Tecnico Lisboa)

[13] Kumar S, Singh S K and Mishra S C 2018 Processing and Characterization of Fly-ash Compacts *Mater. Today Proc.* **5** 3396–402

[14] Andersson B, Andersson R, Håkansson L, Mortensen M, Sudiyo R and Van Wachem B 2011 *Computational fluid dynamics for engineers* vol 9781107018

[15] ANSYS Inc. 2014 *Ansys Fluent Theory Guide 16* (USA: ANSYS Inc.)

IOP Conf. Series: Materials Science and Engineering **742** (2020) 012007 doi:10.1088/1757-899X/742/1/012007

Kinetics and Thermodynamics Studies of Ketalization of Glycerol and Acetone in the Presence of Basolite F300 as Catalyst

Hary Sulistyo[1*], Indra Perdana[1], Fatimah Tresna Pratiwi[1,2] and Indah Hartati[1,3]

[1]Department of Chemical Engineering, Faculty of Engineering, Universitas Gadjah Mada, Yogyakarta 55281 Indonesia

[2]Agency for the Assessment and Application of Technology, Jakarta 15314 Indonesia

[3]Department of Chemical Engineering, Faculty of Engineering, Universitas Wahid Hasyim, Semarang 50232 Indonesia

E-mail : hary@ugm.ac.id

Abstract. Acetalyzation of glycerol with acetone has been carried out using Basolite F300 as catalyst. Heterogeneous catalysis reaction model was employed and Langmuir Hinshelwood mechanism was simulated and validated to the experimental data. The experiment was conducted at varying temperature (30 – 55°C) with an initial molar ratio of glycerol to acetone of 1:4, agitation speed of 700 rpm and catalyst loading of 1% (w/w glycerol). The agitation speed of 700 rpm was sufficient to neglect the external mass transfer as the rate controlling step.The results showed that glycerol conversion as high as 83.33 % was obtained from one hour reaction time. The glycerol conversion was simulated and the results were compared with experimental data in a temperature range of 30°C to 55°C. From mathematical model, it was found that the pre-exponential factor of 0.0149 min^{-1} and activation energy of 15.7085 kJ·$mole^{-1}$. Comparison of experimental data and calculation results showed that the proposed mathematical model was adequately able to approach the experimental data within the temperature range under investigation. Thermodynamics analysis on the ketalization resulted in ΔS^0, $\Delta H^0$and ΔG^0 which values were 0.0834 kJ·$mole^{-1}$ K^{-1}, 29.7176 kJ·$mole^{-1}$ and -4.8675 kJ·$mole^{-1}$, respectively.

1. Introduction

In recent years, glycerol valorisation via various strategies such as steam reforming, acetalization, hydrogenolysis, selective oxidation, dehydration and pyrolysis have been explored by many researchers to respond for the over production and surplus of glycerol. Among glycerol valorisation products, solketal (4-hydroxymethyl-2,2-dimethyl-1,3-dioxolane) produced from ketalization of glycerol and acetone possesses outstanding properties. These properties are such as high affinity, high miscibility with water and organic substances, and degradable. All those properties lead to a wide range of applications as in polymer, fuel, petrochemical and pharmaceutical industry. Solketal acts as solvent and plasticizer in the polymer and petrochemical industry while in pharmaceutical industry solketal is utilized as carrier, enhancer, solubilizing and suspension agent as an active ingredients. Moreover as fuel additive, solketal is capable of reducing gum formation and particulate emission, decreasing fuel viscosity, enhancing octane number, improving cold flow properties and maintaining fuels oxidation stability [1,2]. Hydrochloric acid, sulfuric acid, hydrofluoric acid, and paratoluene sulfonic acid are some of strong homogeneous acid catalysts typically used for conventional synthesis of solketal by ketalization of glycerol with acetone[1,3].

Some solid catalysts such as Amberlyst-15 resins, Ambelyst 35 resins, Amberlyst-36 resins, modified carbon materials, montmorrilonite K-10, and molybdenum phosphate have been applied in solketal synthesized by ketalization of glycerol with acetone [1,2,4]. Some researchers have investigated ketalization of glycerol using solid catalysts. Reddy et al. [5] used zirconia and promoted zirconia catalysts for glycerol ketalization with acetone. Their results showed that the promoted zirconia catalysts have high catalytic activity. The sulphate is the best for promoting catalysts to achieve the highest glycerol conversion. The experiments can also be investigated in an ambient temperature. Ketalization of glycerol by using ethanol as solvent was undertaken by Nanda et al [1] with Amberlyst-35 as catalysts. They found glycerol conversion of more than 74% at temperature 318K. It was concluded that the reaction temperature can be lower when ethanol was used as solvent. An acid functionalized SBA-15 catalyst was used for solketal formation from crude glycerol and acetone. The optimum reaction temperature was at 343K with 5% catalyst concentration. It was found that 90% solketal yield was obtained. In addition, it was identified that the H^+ of sulfonic acid catalyst could react with the Na^+ in the crude glycerol which led to deactivation of the catalyst [6]. The ZrP-200 solid as catalyst was used for ketalization of glycerol with acetone. Li et al [7] obtained that ZrP-200 is highly active, better performance and excellent stability even under mild condition among various catalyst investigated. They could reach a conversion of glycerol of 85.7% with a selectivity of solketal of 98.6%.

The purpose of the present investigation was to undertake solketal formation form acetone and glycerol using Basolite F300 as catalyst. In addition a kinetics model was proposed which was then verified using glycerol conversion data taken at varying temperature [8].

2. Material and methods

2.1. Materials

Glycerol with 99.5% purity , acetone with a purity of 99.7% from PT Smartlab Indonesia were used as raw materials. Basolite F300 and standard solketal were purchased from Sigma Aldrcih. Synthesis of solketal were performed in a three-neck flask equipped with heating mantle, thermometer, stirrer and condenser as depicted in Figure 1.

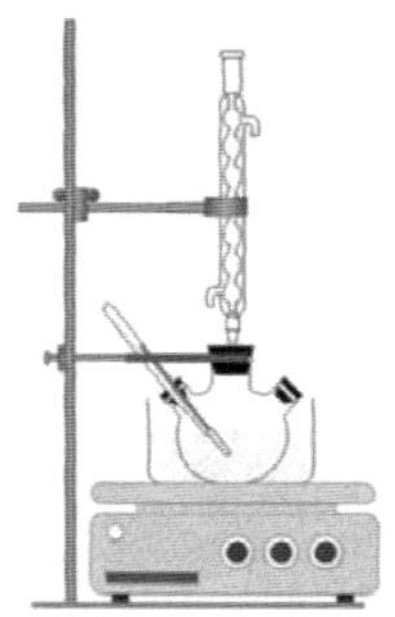

Figure 1. The experimental setup for ketalization of glycerol and acetone

2.2. Catalyst Activation

Basolite F300 catalyst was activated by heat treatment. It was heated in an oven at 150°C for 3 hours and was then cooled to room temperature by keeping it in a desiccator.

2.3 *Solketal synthesis*

0.1 mole of glycerol and 0.4 mole of acetone were place in the three-neck flask. Ketalization of glycerol and acetone over the presence of Basolite F300 were performed at temperature of 30-55°C while the agitation speed was set at 700 rpm. After the desired temperature was reached, sample was taken for initial glycerol concentration (G_0) analysis. 1% (w/w of glycerol) of Basolite F300 was then added to the reaction mixture and the reaction was then carried out for 60 minutes. Samples were taken every 10 minutes and the glycerol concentration(G_s), solketal concentration and acetone concentration were determined using GC-FID Shimadzu QP2010.

2.4. *Determination of Glycerol*

Glycerol concentration was analyzed using GC-FID Shimadzu QP2010 equipped with Rxi-5ms column. The injector and detector temperature were set at 300°C. The column gradient temperature was set at 50°C, maintained for 1 minute and was then raised to 120°C with acceleration speed of 10°C. The helium flowrate was set at 1.5 ml/min with a split ratio of 10:1. Glycerol conversion was calculated by using Eq.1.

$$x_G = \frac{G_0 - G_S}{G_0} . 100\% \qquad (1)$$

3. Results and discussion

Basolite F300, a hybrid of organic and inorganic substance which is categorised as metal organic frameworks (MOF), showed a good performance in the catalysis of ketalization of glycerol and acetone. The synthesis of solketal was performed in a temperature range of 30-55°C with the use of glycerol to acetone molar ratio of 4, agitation speed of 700 rpm and Basolite F300 concentration of 1% (w/w of glycerol). It was found that the highest glycerol conversion of 84.33% was achieved in the glycerol acetalization performed at 50°C. The concentration profile of glycerol, acetone and solketal from ketalization of glycerol and acetone were shown in Figure 2. Figure 2a and 2b show that along with ketalization time, concentration of reactants decreases, while solketal concentration as the ketalization product increases. Figure 2 also shows that the initial concentration of solketal increases with increasing temperature. After a certain time the solketal concentration slightly decreases with increasing temperature. This result is expected since ketalization was an exothermic reaction [1,9,10]

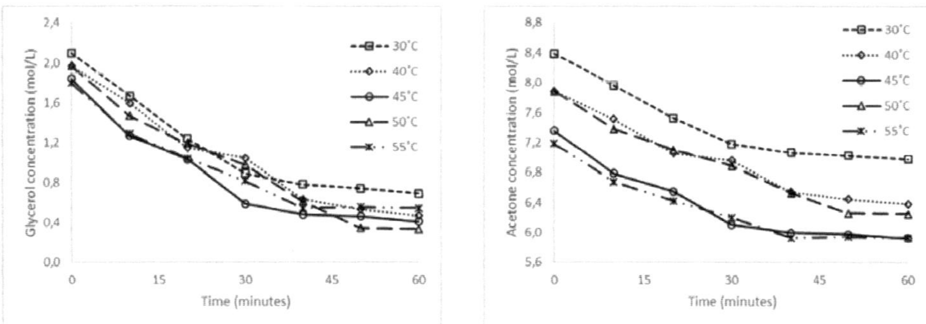

(a) (b)

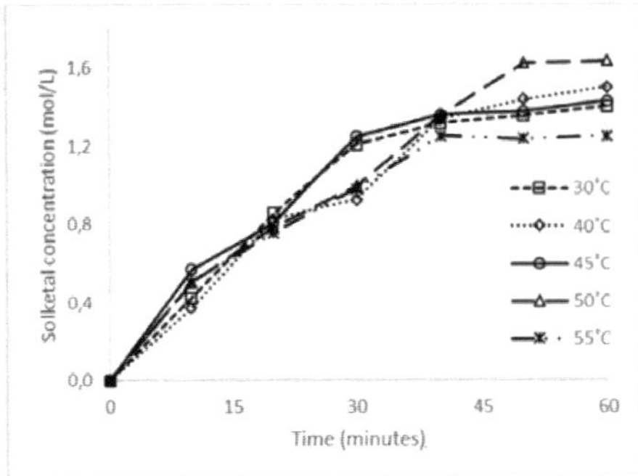

(c)

Figure 2. The concentration profile of (a) glycerol, (b) acetone and (c) solketal

3.1. Kinetics Study

Kinetics studies of ketalization of glycerol and acetone were reported by several researchers. Nanda et al. [1] investigated the kinetics and thermodynamic of solketal synthesis in which utilized Amberlist-35 as the solid catalyst and applied Langmuir-Hinshelwood rate law model for describing the surface reaction mechanism. Moreira et al. [9] also utilized Amberlist-35 as the solid catalyst in a solketal synthesis and applied reaction rate law of Pseudo-Homogeneous (PH), Langmuir–Hinshelwood–Hougen–Watson (LHHW), and Eley–Rideal (ER) in their kinetics study. Meanwhile Rossa et al. [11] used zeolite H-BEA (SAR 19) as a the solid catalyst and applied a reversible kinetics model. In this work, solketal synthesis from ketalization reaction of glycerol and acetone was illustrated as shown in Figure 3.

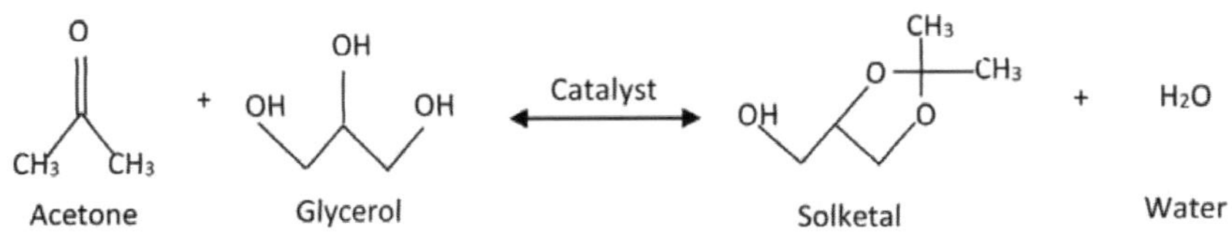

Figure 3. Ketalization reaction of glycerol and acetone to produce solketal

The reaction can be simply written as follow

$$A + B \leftrightarrow C + D, \qquad (2)$$

where A, B, C and D represent for acetone, glycerol, solketal and water, respectively. This reaction was conducted in the presence of Basolite F300.

The reaction mechanism of acetalization has been derived and Langmuir-Hinshelwood-Hougen-Watson mechanism was proposed [1,9]. The reaction steps of this kinetics model can be expressed as follows.

1. Initially the carbonyl groups in acetone are activated by acid sites of the catalyst.
   $$A + s \rightarrow As \qquad (3)$$
2. The OH groups of glycerol also adsorbs at the acid sites of the catalyst.
   $$B + s \rightarrow Bs \qquad (4)$$
3. The absorbed acetone and adsorbed glycerol reacts to form adsorbed solketal and adsorbed water.
   $$As + Bs \rightarrow Ss + Ws \qquad (5)$$
4. Desorption of adsorbed solketal from the surface of catalyst
   $$Ss \rightarrow S + s \qquad (6)$$

5. Desorption of adsorbed water from the surface of catalyst

$$Ws \rightarrow W + s \tag{7}$$

Reaction rate of Langmuir-Hinshelwood model can be derived and the result is expressed as follows

$$r_s = k\frac{\left(C_A.C_B.-\frac{C_C.C_D}{K_{eq}}\right)}{(1+K_A.C_A+K_B.C_B+K_C.C_C+K_D.C_D)^2} \tag{8}$$

where $k = k_s K_A K_B C_t$ and $K_{eq} = \frac{K_A K_B K_S}{K_C K_D}$

The denominator in Equation (8) was then simplified by considering that only water that is strongly adsorbed in the catalyst surface [1] resulting to expression in Eq (9).

$$r_s = k\left[\frac{C_A.C_B-\frac{C_C.C_D}{K_{eq}}}{(1+K_d.C_D)^2}\right] \tag{9}$$

Furthermore, since the reaction were performed in a batch reactor, the mass balance of glycerol in the liquid phase is as follows:

$$N_{B0}\frac{dX_B}{dt} = w_{cat} r_s S_a \tag{10}$$

where N_{B0} is the initial glycerol concentration, w_{cat} is the catalyst mass, r_s is the reaction rate per unit of catalyst surface area and S_a is the catalyst surface area per mass of catalyst.

The value of $\frac{dx_b}{dt}$ were determined from the slope of conversion versus time curve. Multi regression analysis method was applied in order to determine the initial value of the predicted kinetics parameters' values and followed by linier regression to determine the kinetics constants of k, K_{eq} and K_d.

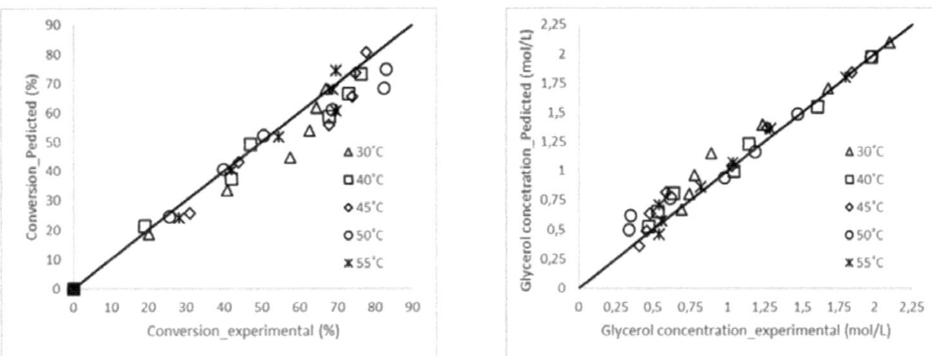

(a) (b)

Figure 4. The relation between experimental data and predicted (a) conversion and (b) glycerol concentration using the proposed kinetics model

The validation of the kinetics model of ketalization of glycerol and acetone with basolite F300 catalyst shows that Langmuir-Hinshelwood mechanism model can fit well with the experimental data as shown in Figure 4a and 4b. The kinetics constants of ketalization of glycerol and acetone performed at temperature varied in a range of 30-55°C, molar ratio of glycerol and acetone of 1:4; agitation rate of 700 rpm and Basolite F300 concentration of 1% (w/w of glycerol) are tabulated in Table 1.

Table 1. Kinetics constants of ketalization of glycerol and acetone

Kinetics constants	Temperature (°C)				
	30	40	45	50	55
k	2.885×10^{-5}	3.606×10^{-5}	4.031×10^{-5}	4.115×10^{-5}	4.741×10^{-5}
K_{eq}	6.345	4.017	2.561	3.290	2.483
K_d	1.029	1.166	1.402	1.550	1.806
SSE	0.1353	0.06347	0.08789	0.12624	0.04197

Since the kinetics rate constant k obtained from the kinetics modelling is a function of temperature, the values can be utilized to calculate the activation energy E_a and the pre-exponential factor A of Arrhenius equation (Eq 11).

$$k = A \exp\left(-\frac{E_a}{RT}\right) \quad (11)$$

where

k = kinetics rate constant (mole L^{-1} min^{-1})
A = pre-exponential factor (min^{-1})
E_a = activation energy (kJ·$mole^{-1}$)
R = universal gas constant, 8.314 J $mole^{-1}$ K^{-1}
T = Temperature (K)

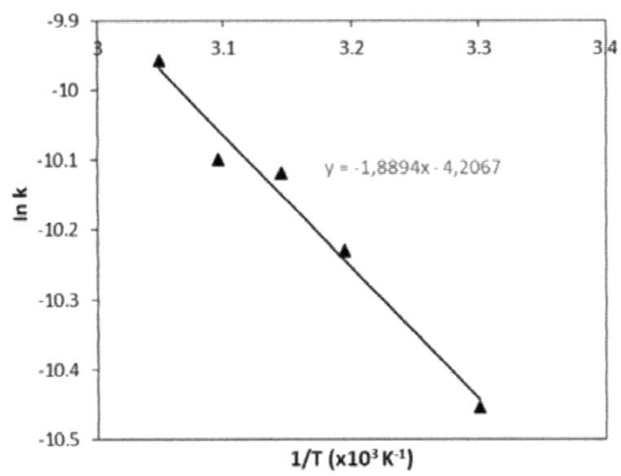

Figure 5. Kinetics rate constant as a function of temperature

Equation 11 was linearized and the result was shown in Figure 5. Pre-exponential factor of 0.014895 min^{-1} and activation energy of 15.7084 kJ·$mole^{-1}$ were obtained. The low of the pre-exponential factor obtained indicates that the product formation occurs mainly due to interaction of reactants and the active site of the catalyst instead of through the collisions between reactants. The activation energy found in this research (15.7084 kJ·mol^{-1}) was slightly lower compared to the ones reported in several literatures such as 28.2 kJ·mol^{-1}, 55.6 kJ·mol^{-1} and 69.6 kJ·mol^{-1} for the activation energy of the ketalization of glycerol and acetone over TTPT [10], Amberlyst-35 [1], Amberlyst-35 [9], respectively. However, the activation energy calculated in this work is still in the same order of magnitude as those reported which indicate the effectiveness of the Basolite F300 in the catalysis of ketalization reaction.

3.2. *Thermodynamic Studies*

The chemical equilibrium constant K_{eq} as temperature function previously calculated from the kinetics study can be utilized to determine the standard Gibbs free energy of the ketalization reaction by applying van't Hoff correlation as written in Eq.12.

$$K_{eq} = e^{\Delta G^0/RT} \tag{12}$$

Meanwhile, the relation of the standard Gibbs free energy with temperature can be expressed as in Eq. 13

$$\Delta G^0 = \Delta H^0 - T\Delta S^0 \tag{13}$$

Substitution of Eq. 13 to Eq. 12 followed by linearization results in the Eq. 14.

$$\ln K_{eq} = \frac{\Delta S^0}{R} - \frac{\Delta H^0}{R}\frac{1}{T} \tag{14}$$

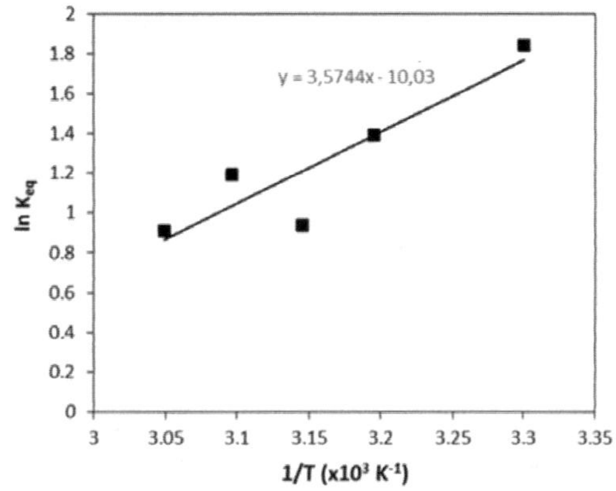

Figure 6. Relation of equilibrium constant K_{eq} with temperature

From the relation of equilibrium constant K_{eq} with temperature as shown in Figure 6, the thermodynamics parameters of ΔS^0 and ΔH^0 can be determined which values are 0.08339 kJ·mole^{-1}K^{-1} and -29.7176 kJ·mole^{-1}, respectively. Meanwhile, the standard Gibbs free energy was found to be -4.86751 kJ·mole^{-1}. The negative value of ΔH^0 indicates that the ketalization of glycerol and aceton over basolite F300 catalyst is an exothermic reaction. Furthermore, the resulting equation which represents the correlation between chemical equilibrium constant K_{eq} and temperature of the ketalization of glycerol and acetone over Basolite F300 are compared with results from other works and listed in Table 2.

Table 2. Equilibrium constants as function of temperature

Equilibrium constant equation	Source
$\ln K_{eq} = \frac{3615.4}{T} - 11.31$	[1]
$\ln K_{eq} = \frac{3003}{T} - 10.793$	[12]
$\ln K_{eq} = \frac{3574.4}{T} - 10.03$	This work

4. Conclusion

The acetalization of glycerol has been carried out using Basolite F300 as catalyst. The highest glycerol conversion was reached as high as 84.33% at 50°C. The kinetics study showed that reaction mechanism of Langmuir-Hinshelwood model fitted well the experimental data. The kinetics parameters including pre-exponential factor of 0.014895 min^{-1} and activation energy of 15.70847 kJ·$mole^{-1}$ were obtained. Thermodynamics parameters of ΔH^0, $\Delta S^0$and ΔG^0 were calculated and the values were -29.7176 kJ $mole^{-1}$, 0.8339 kJ $mole^{-1}$ K^{-1} and -4.86751 kJ·$mole^{-1}$, respectively. The present work confirmed that basolite F300 catalyst was found active in the exothermic reaction of ketalization of glycerol and acetone.

Acknowledgement

The authors would like to thank the Ministry of Research, Technology and Higher Education of Republic of Indonesia for financial support through *Penelitian Dasar Unggulan Perguruan Tinggi (PDUPT)* Research Grant 2019, contract number of 124/UN1/DITLIT/DIT-LIT/LT/2019 in Universitas Gadjah Mada.

References

[1] Nanda M R, Yuan Z, Qin W, Ghaziaskar H S, Poirier M and Charles C 2014 Fuel. 117 470–477
[2] Li Z, Miao Z, Wang X, Zhao J, Zhou J, Si W and Zhuo S 2018 Fuel. 233 377–387
[3] Rossa V and Pergher S B C 2016 Ind. Eng. Chem. Res. 56 479−488
[4] Sulistyo H, Hapsari I, Sediawan W B, Rahayu S S and Mufti M A 2019 MATEC Web of Science 268 07004 1–5
[5] Reddy P S, Sudarsanam P, Mallesham B, Raju G and Reddy B M 2011 J. Ind. Eng. Chem. 17 (3) 377–381.
[6] Vicente G, Melero J A, Morales G and Mart E 2010 Green Chem. 12 899–907.
[7] Li X, Jiang Y, Zhou R and Hou, Z 2019 App. Clay.Sci. 174 120–126.
[8] Pratiwi F T, Sulistyo H and Perdana I 2019 Studi kinetika reaksi ketralisasi gliserol dan aseton menjadi solketal menggunakan katalis Basolite F300, Master Thesis, Chemical Engineering Department, Universitas Gadjah Mada, Yogyakarta, Indonesia
[9] Moreira M N, Faria R P V, Ribeiro A M and Rodrigues E 2019 Ind. Eng. Chem. Res.
[10] Ji Y, Zhang T, Gui X, Shi H, Yun Z, Zhang T and Gui X 2019 Chi. J.Chem. Eng. https://doi.org/10.1016/j.cjche.2019.07.019
[11] Rossa V, Treichel H, Alexandre D and Aranda G 2017 Perspect. Erechim, 41 (155) 101–112.
[12] Meelom A 2014 A Ketalization Of Glycerol With Acetone Catalyzed By Zeolite Beta: Thermodynamics And Kinetics Study, A Thesis Submitt. Partial Fulfillment Requir. Degree Master Eng. Progr. Chem. Eng. Dep. Chem. Eng. Grad. Sch. Silpakorn Univ. Bangkok, Thailand.

IOP Conf. Series: Materials Science and Engineering **742** (2020) 012008 doi:10.1088/1757-899X/742/1/012008

Simulation and Parametric Study of the Innovative Process to Produce High Purity Isopropyl Acetate with Ethylene Oxide Hydration as an Auxiliary Reaction

I G P Wiratama[1], V Novianty[1]

[1]Department of Chemical Engineering, Faculty of Industrial Technology, Parahyangan Catholic University, Bandung, 40141, Indonesia

E-mail: pandega.wiratama@unpar.ac.id

Abstract. Production of high purity isopropyl acetate by esterification is a difficult process since the equilibrium conversion is relatively low and there are azeotropes between isopropyl acetate with water and between isopropyl acetate with the reactants. In this research, an innovative process to produce high purity isopropyl acetate with auxiliary reaction of ethylene oxide hydration was proposed. The process consisted of a reactor and a simple distillation column. The auxiliary reaction abolished the water/isopropyl acetate azeotrope and shifted the esterification equilibrium to the right. Impacts of the reactor type, the feed temperature, the reflux ratio, and the bottoms rate were investigated by simulating the process in Aspen Plus software. The results revealed that the innovative IpAc production with auxiliary reaction of EO hydration gave very high esterification conversion (up to 99.8%) and very high purity of IpAc (up to 99.7%). These results were achieved when the reactor type was isothermal at feed temperature, the feed temperature was 77°C, the reflux ratio was 1, and the bottoms rate was 10 kmol/h.

1. Introduction

Isopropyl acetate (IpAc) is a solvent with a wide variety of manufacturing uses that is miscible with most other organic solvents, and moderately soluble in water. It is a component of some printing inks and perfumes [1]. IpAc commonly produced by esterification of isopropyl alcohol (IpA) with acetic acid (AcH) as

$$CH_3CHOHCH_3 + CH_3COOH \leftrightarrow CH_3COO(CH_3)_2 + H_2O \quad (1)$$

The equilibrium conversion of this reaction is around 79% at room temperature and become smaller as the temperature rises, since it is an exothermic reaction. This makes at the end of reaction process, the purity of IpAc is relatively low. There are also azeotropes in the mixture (Table 1) that makes the separation process much harder to get high purity IpAc.

Table 1. Azeotropes in the mixture

Temp	Mole Fraction			
(°C)	IpA	AcH	IpAc	Water
78.44	0.6053	0	0.3947	0
74.22	0.1848	0	0.4247	0.3905
79.99	0.678	0	0	0.322
99.7	0	0.0999	0	0.9001
74.98	0	0	0.5248	0.4752

Conventionally, IpAc can be readily removed from mixtures containing it, IpA and water, by using extractive distillation in which the extractive agent is a mixture of a polyol and one or higher boiling oxygenated, nitrogenous and/or sulfur containing organic compounds [2]. However, the cost of extractive distillation is relatively high, because there are additional cost to provide the extractive agent and the cost for additional solvent recovery column. Therefore, an innovative IpAc production method that may have lower cost is proposed in this research.

Based on the Table 1, high purity IpAc can be achieved by shifting the reaction equilibrium to the right, so the fraction of AcH and IpA in the mixture can be reduced to near zero, and removal of water. These two steps can be conducted simultaneously, because if water is removed from the mixture, the equilibrium will shift to the right naturally. The idea of innovative IpAc production is that the removal of water can be done by adding ethylene oxide (EO) which can react with water and produce ethylene glycol (EG) as

$$C_2H_4O + H_2O \rightarrow (CH_2OH)_2 \quad (2)$$

Because the equilibrium shifts to the right, a near 100% conversion of IpA and AcH can be achieved. Therefore the mixture after reaction process mainly contain EG and IpAc. Since there is no azeotrope between EG and IpAc, the separation process afterwards can be done in a simple distillation column with IpAc as distillate product and EG as bottom product. The present of auxiliary reaction of EO hydration is also worthwhile, because the product, EG, is valuable chemical which is used as an antifreeze and coolant, in hydraulic fluids, and in the manufacture of low-freezing dynamites and resins [3].

This study is a part of a research about innovative IpAc production. In this study, the innovative IpAc production was simulated using Aspen Plus to investigate its feasibility and to study the effects of the process parameters in steady state condition. Based on the result of this study, the next parts of the research, which are about process design optimization, dynamic behavior, and process control, can be conducted.

2. Base simulation

2.1. Kinetics model

The kinetic model for IpAc production through esterification that takes place in liquid phase, in the presence of Amberlyst 15 as heterogeneous catalyst is adapted as follows [4]

$$r = m_{cat}\frac{k_f\left(a_{AcH}a_{IpA} - a_{AcH}a_{IpA}K_{EQ}^{-1}\right)}{\left(1 + K_{AcH}a_{AcH} + K_{IpA}a_{IpA} + K_{IpAc}a_{IpAc} + K_{water}a_{water}\right)^2} \quad (3)$$

where:

$$k_f(kmol.\,kg_{cat}^{-1}.s^{-1}) = 7.667\times10^{-5}\exp(23.81 - \frac{68620.43}{RT})$$

$K_{EQ} = 8.7$
$K_{AcH} = 0.1976$
$K_{IpA} = 0.2396$
$K_{IpAc} = 0.147$
$K_{water} = 0.5079$

The auxiliary reaction of EO hydration may also produce unwanted by-product of diethylene glycol and triethylene glycol. However, the rate of the side reaction is much lower than that of the main reaction and consequently, the effect of this side reaction can be neglected [5]. The kinetic model of EG production from EO hydration in liquid phase is adapted as follows [5]

$$r(mol.\,cm^{-3}.s^{-1}) = 3.15\times10^9\exp(-\frac{9547}{T})x_{EO}x_{water} \quad (4)$$

2.2. Process simulation

Innovative IpAc production was simulated in ASPEN Plus software, using packed bed reactor and separation process in distillation column as shown in Figure 1. The properties method that was used in this study was NRTL (Non-Random Two Liquid), which has been confirmed to be accurate in simulating a system containing IpA, AcH, water, IpAc, EO, and EG [4, 5].

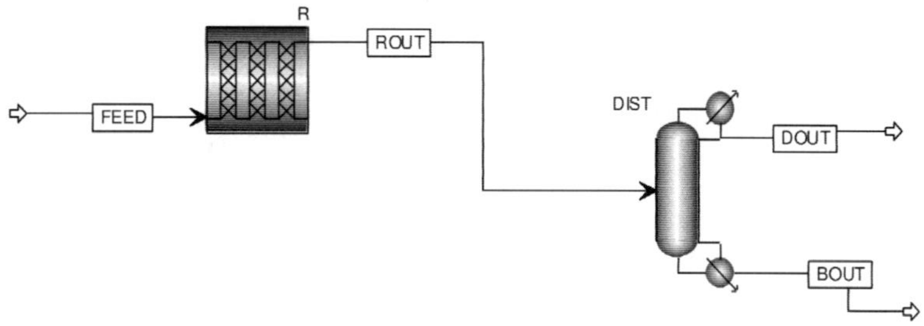

Figure 1. Process configuration of intensified IpAc production

In the reactor, it is assumed that the catalyst bed has bed void fraction of 50% and density of 770 kg/m^3 [4]. In the conventional extractive distillation process, with the extractive agent is a mixture of glycerine, DMFA, and ethanolamine, it is found that the extractive distillation column needs around 18 stages to get IpAc with the purity of 98% [2]. Therefore, in this simulation, number of stages in the distillation column was set to 18, so the result may be compared to the conventional process.

The process variables that were examined first were feed temperature and reactor type (adiabatic or isothermal at feed temperature). The method of examination was by manipulating one variable at the time, by keeping any other variables constant at the initial condition of simulation. The initial conditions of simulation are shown in Table 2. The next variables that were examined were reflux ratio and bottoms rate of distillation column. The method of examination was also by manipulating one variable at the time, at the best feed temperature and reactor type which gave maximum conversion of AcH, and by keeping any other variables constant at the initial condition.

Table 2. Initial conditions of simulation

Parameters	Value	Units
Feed		
AcH mole-flow	10	kmol/h
IpA mole-flow	10	kmol/h
EO mole-flow	10	kmol/h
Reactor		
Diameter	5	m
Length	20	m
Distillation Column		
Number of stages	18	-
Reflux ratio	3	-
Bottoms Rate	10	kmol/h

Note that the initial conditions are not the optimum conditions. The optimum conditions will be covered in the next study.

3. Results and discussion

3.1. The effect of the feed temperature in the adiabatic reactor

In the adiabatic reactor, the reaction system is heated by the heat of reactions. The reactions are quite exothermic so that the heat of reactions is plenty enough to change the phase in the reactor from liquid to vapor. When the feed temperature is increased, the temperature along the reactor is also increased and therefore increases the vapor fraction in the reactor as can be seen in Figure 2.

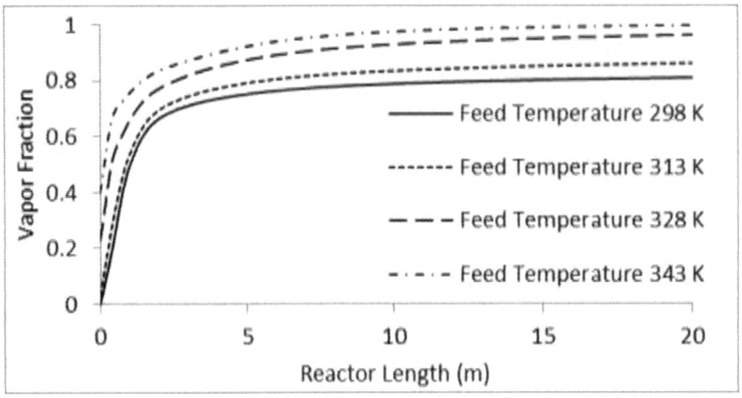

Figure 2. The effect of the feed temperature on vapor fraction profile in the adiabatic reactor

The increasing of vapor fraction decreases the conversion of AcH and EO, because the reactions occurred in liquid phase. Thus, as can be seen in Figure 3, when the feed temperature is increased, the conversion of AcH and EO are decreased.

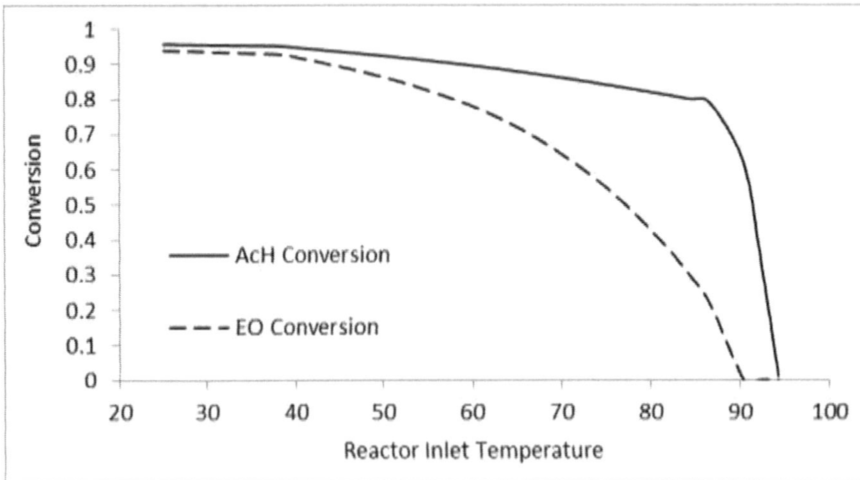

Figure 3. The effect of the feed temperature AcH and EO conversion in the adiabatic reactor

The conversion of AcH in adiabatic reactor is 95.7% at room temperature and not significantly increased at lower temperature, so the effort to provide refrigerant to decrease the feed temperature until below room temperature is not worth it. Therefore, it can be concluded that the room temperature is the best feed temperature that give optimum AcH conversion in adiabatic reactor.

3.2. The effect of the feed temperature in the isothermal reactor

In the isothermal reactor, the heat of reactions is transferred to the cooling jacket, so that the reactor temperature is constant at the feed temperature. It can be seen in Figure 4, when the feed temperature is increased from room temperature to around 77°C, the conversions of AcH and EO are increased, because the rates of reactions are higher at higher temperature. Above 77°C, the phase in the reactor begins to change from liquid to vapor, and because the reactions occurred in liquid phase, the conversion of AcH and EO are greatly decreased.

The maximum conversion of AcH in isothermal reactor is 99.9% at 77°C. This is greater than the conversion of AcH in adiabatic reactor, because in isothermal reactor, the temperature is kept at optimum reaction temperature and the phase in the reactor is maintained as liquid. Therefore, it can be concluded that the best type of reactor is isothermal at the feed temperature of 77°C.

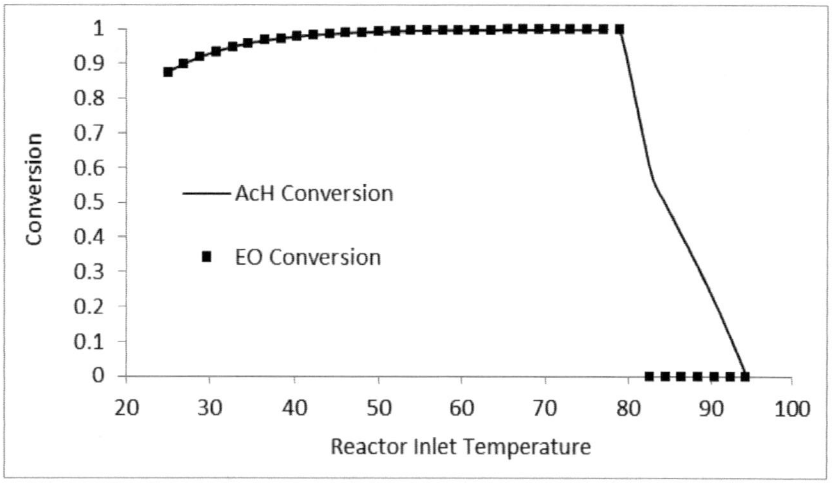

Figure 4. The effect of the feed temperature AcH and EO conversion in the isothermal reactor

3.3. The effect of the reflux ratio

The influence of the reflux ratio was investigated by fixing the reactor condition at optimum condition (isothermal at the feed temperature of 77°C) and any other parameters in distillation column at initial condition. It can be seen in Figure 5, when the reflux ratio is increased from 1 to 20, the reboiler duty is significantly increased, because there is more liquid that is recycled into the column and therefore more energy is needed to boil up the liquid and maintain the bottom flow rate at initial condition (10 kmol/h).

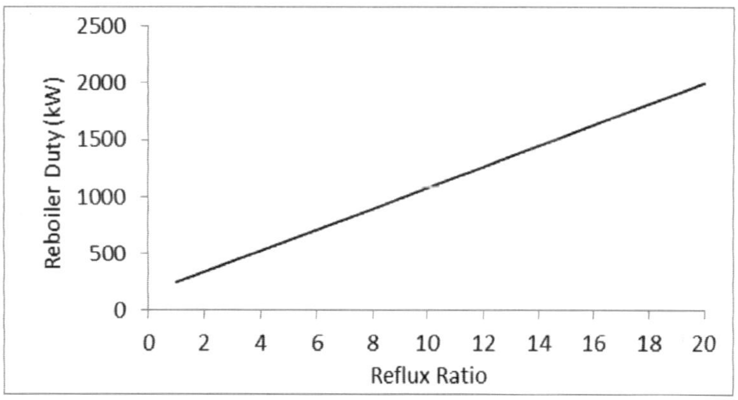

Figure 5. The effect of the reflux ratio on reboiler duty

However, as can be seen in Figure 6, the increasing of reflux ratio does not significantly increase the molar composition and molar flow rate of IpAc at distillate and EG at bottom, because even at a low reflux ratio, the molar composition of IpAc at distillate and EG at bottom are already very close to its maximum value of 100%. Thus, it is more efficient if the distillation is operated at a low reflux ratio.

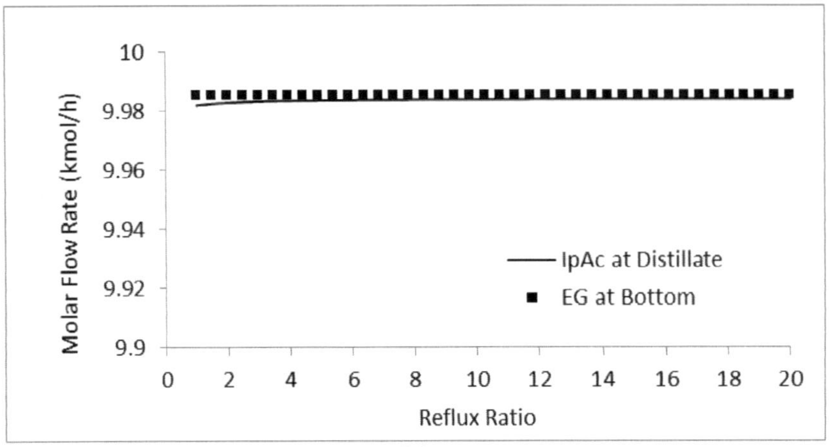

(a)

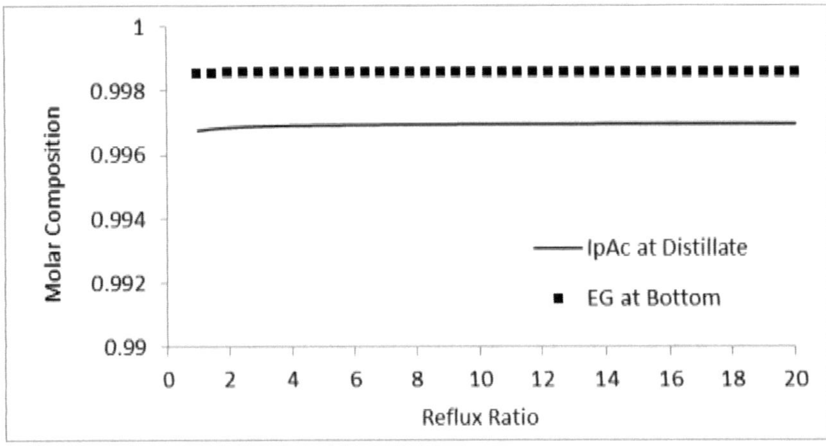

(b)

Figure 6. The effect of the reflux ratio on (a) molar flow rate and (b) molar composition

3.4. The effect of the bottoms rate

Figure 7 shows the influence of bottoms rate of distillation column on reboiler duty. The reboiler duty is decreased when the bottoms rate is increased, because the greater bottoms rate means the lesser liquid in the reboiler should be boiled up.

In Figure 8, it can be seen that when the bottoms rate is increased from 8 to 10 kmol/h, the more EG as heavy component flows to the bottom, leaving IpAc as main component at distillate. Therefore, the molar flow rate of EG at bottom and the molar composition of IpAc at distillate are increased.

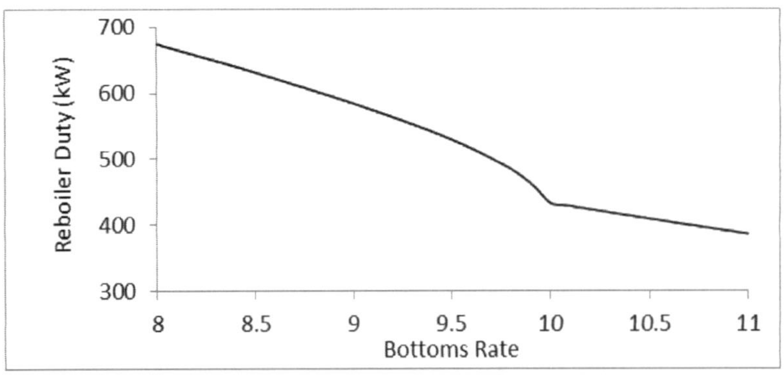

Figure 7. The effect of the bottoms rate on reboiler duty

When the bottoms rate is greater than 10 kmol/h, the molar composition of IpAc at distillate is not significantly increased, since it is already very close to 100%. However, when the bottoms rate is greater than 10 kmol/h, IpAc begins to flow down to the bottom, so that the molar composition of EG at bottom and molar flow rate of IpAc at distillate are decreased.

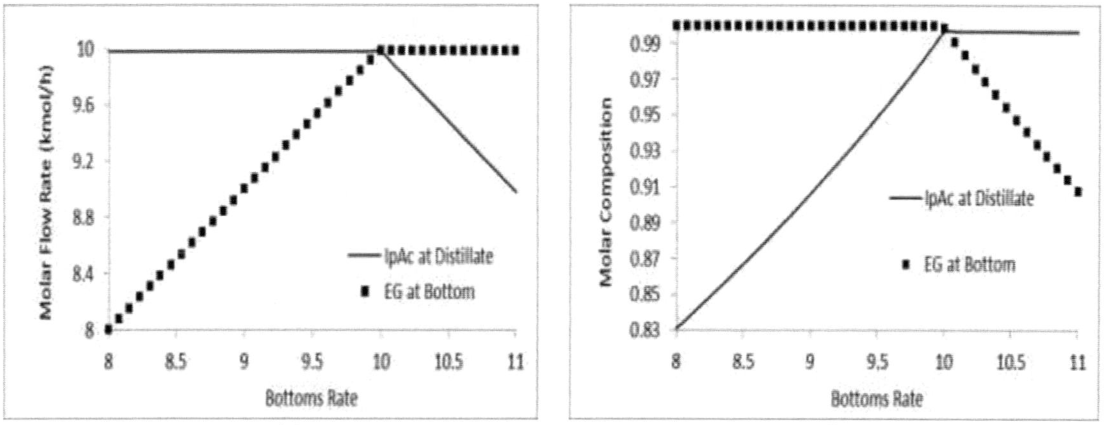

(a) (b)

Figure 8. The effect of the bottoms rate on (a) molar flow rate and (b) molar composition

From Figure 7 and 8, it can be concluded that the best bottoms rate is 10 kmol/h. It gives highest purity and molar flow rate of IpAc at distillate, highest purity and molar flowrate of EG at bottom, and relatively low reboiler duty.

3.5. Comparison

The innovative IpAc production is compared to the conventional configuration which is the combination of reactor and extractive distillation, without the auxiliary reaction of EO hydration. With the operating conditions that are shown in Table 3, by the innovative IpAc production, the AcH conversion of 99.8% and IpAc molar composition of 99.7% can be achieved. These results are better than the results by the conventional process which gives AcH conversion of 79% and IpAc molar composition of 98% [2].

Table 3. Operating conditions that give maximum AcH conversion and maximum IpAc purity

Parameters	Value	Units
Feed		
Temperature	77	°C
Reactor		
Type	isothermal at feed temperature	-
Distillation Column		
Reflux ratio	1	-
Bottoms Rate	10	kmol/h

4. Conclusions

The simulation results revealed that the innovative IpAc production with auxiliary reaction of EO hydration is highly prospective. It gives very high esterification conversion (up to 99.8%) and very high purity of IpAc (up to 99.7%). These results are achieved when the reactor type is isothermal at feed temperature, the feed temperature is 77°C, the reflux ratio is 1, and the bottoms rate is 10 kmol/h.

Nomenclature

a_i	=	activity of component i
k_f	=	rate reaction constant
K_{EQ}	=	reaction equilibrium constant
K_i	=	adsorption constant of component i
R	=	ideal gas constant
T	=	temperature

References

[1] National Center for Biotechnology Information. PubChem Database. Isopropyl acetate, CID=7915, https://pubchem.ncbi.nlm.nih.gov/compound/7915 (accessed on Mar. 26, 2019)
[2] Berg L and Yeh A I 1989 *U.S. Patent* **4826576**
[3] National Center for Biotechnology Information. PubChem Database. 1,2-Ethanediol, CID=174, https://pubchem.ncbi.nlm.nih.gov/compound/174 (accessed on Mar. 26, 2019)
[4] Luyben W L and Yu C C 2008 *Reactive Distillation Design And Control* (Hoboken: John Wiley & Sons, Inc.) p 146-77
[5] Tavan Y, Behbahani R M and Hosseini S H 2013 *Chem. Eng. Process. Process Intensif.* **73** 81-6
[6] Luyben W L 2007 *Chemical Reactor Design and Control* (Hoboken: John Wiley & Sons, Inc.) p 251-65
[7] Altiokka M R and Akyalcin S 2009 *Ind. Eng. Chem. Res.* **48** 10840-4
[8] Ali S H and Merchant S Q 2006 *Int. J. Chem. Kinet.* **38** 593-612
[9] Kumar A and Daoutidis P 2004 *AIChE. J.* **45** 51-68
[10] Othmer D F and Thakar M S 1958 *Ind. Eng. Chem.* **50** 1235-44
[11] Lu X, Xu H, Yan J, Zhou W J, Liebens A and Wu P 2018 *J. Catal.* **358** 89-99
[12] Wisutwattana A, Frauzem R, Suriyapraphadilok U and Gani R 2017 *Comput. Aided Chem. Eng.* **40** 1135-40
[13] Hal J W V, Ledford J S and Zhang X 2007 *Catal. Today* **123** 310-5
[14] Zhao T, Li J, Zhou H, Ma Z and Sun L 2018 *Chem. Eng. Process. Process Intensif.* **124** 97-108
[15] Sanz M T and Gmehling J 2006 *Chem. Eng. J.* **123** 1-8
[16] Sanz M T and Gmehling J 2006 *Chem. Eng. J.* **123** 9-14

IOP Conf. Series: Materials Science and Engineering **742** (2020) 012009 doi:10.1088/1757-899X/742/1/012009

Optimization of Platforming Catalyst Deactivation due to Sintering

Nimrod Alvino[1] and Heri Hermansyah[2]

[1]Post Graduate Student, Chemical Engineering Department Universitas Indonesia, Depok
[2]Professor, Chemical Engineering Department Universitas Indonesia, Depok

E-mail : alvinosinaga21@gmail.com

Abstract. The purpose of this study are to analyze and identify the causes of catalyst deactivation and provide recommendations and solutions to these problems. This research was conducted by identifying phenomena & analyzing the causes of deactivation of catalysts, testing the factors that are the causes of deactivation of catalysts, providing solutions and recommendations and testing these solutions. The results of this study indicate that the decrease in RON values on reformate products is caused by catalyst deactivation due to sintering. The sintering of the catalyst was identified through Regenerator CCR performance which was marked by a decrease in regeneration gas & chlorination gas flow as well as a decrease in the oxygen content of the chlorination gas due to a plugging at the center screen of the CCR Regenerator Tower.

1. Introduction

Reforming unit produces valuable products in the form of reformate or High Octane Mogas Component (HOMC). Reformate is used as a mixing component to produce fuel oil consumed by people in motorized vehicles. The quality of the Reforming unit products must be maintained to meet the fuel specifications. The reduced quality of reformate products will result in the loss of opportunity margin for the refinery unit. Also, it can cause other impacts such as increased naphtha lifting, additional HOMC import cargo, demurrage shipment, and delays in product lifting due to dock occupancy. The quality of the products produced by the Platforming unit is very dependent on the reaction with the help of catalysts that occur in the platforming reactor. Activation of the catalyst to support the reaction in the platforming reactor runs well. Therefore, if there is a decrease in the quality of Platforming products under normal operating conditions, it is necessary to analyze the activation of the Platforming catalyst so that it can restore the performance of the reactor platforming and improve the quality of the products produced.

In general, there are 3 types of causes of catalyst deactivation, namely poisoning, fouling, and sintering [1]. The poisoning of a catalyst is a deactivation of the catalyst caused by a small amount of certain material for a particular catalyst and is related to the adsorption of toxins at the active site of the catalyst, so that it will block the adsorption process of the reactants by the catalyst. Fouling occurs when there are substances in the reactor (can be reactants, products, or intermediates) deposited on the surface of the catalyst and physically close the catalyst active sites. Meanwhile, deactivation of the catalyst due to sintering is caused by the growth or agglomeration of crystals which will change the chemical structure of the catalyst or its support.

2. Methods

The research methodology undertaken in this study are identifying phenomena & analyzing the causes of deactivation of catalysts, testing the factors that are the causes of deactivation of catalysts, providing solutions and recommendations and testing these solutions. The expected result of this study is to identified the causes of deactivation catalyst due to sintering by analyze each parameter condition that is leaven to the catalyst deactivation. Furthermore, the effect of cleaning center screen in the CCR Regenerator will be reviewed as one of the causes of sintered catalyst.

3. Result and Discussion

3.1. Catalyst Deactivation Phenomenon

The deactivation phenomenon of Platforming catalysts can result in catalytic reforming reactions in the platforming reactor not optimally so that the quality of the reformate products produced does not meet the target. Thus, catalyst deactivation can be identified from the decline in the quality of platforming products which is the RON (Research Octane Number) value of reformate. In addition to catalyst deactivation, there are several other factors that also influence the failure to achieve RON target values, including the composition of treated heavy naphtha feeds that do not have enough aromatics component, besides the reactor platforming operating conditions that can be seen from the Weight Average Inlet Temperature (WAIT).

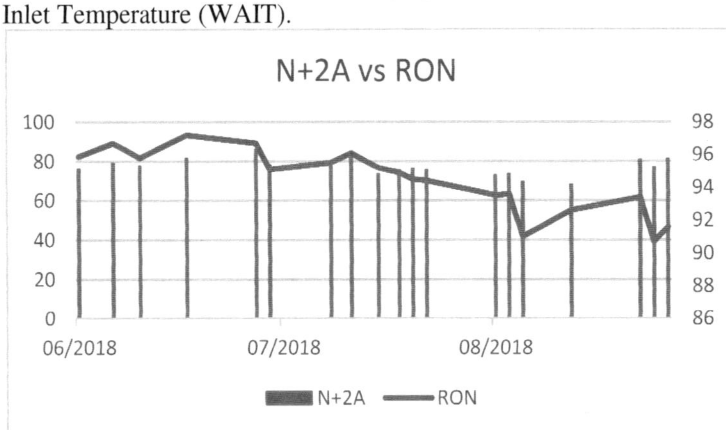

Figure 1. Downward Trend of RON with constant N+2A

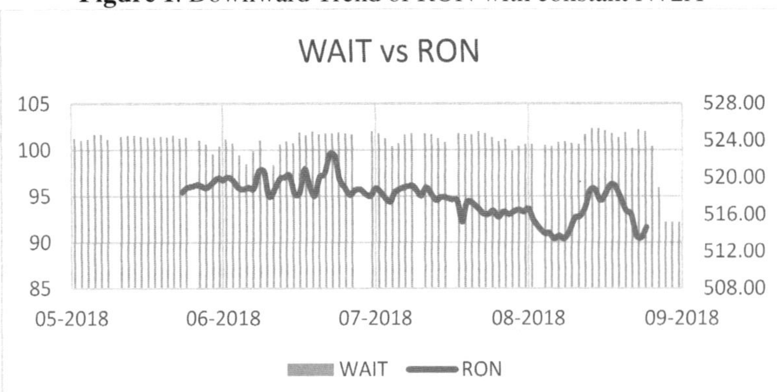

Figure 2. Downward Trend of RON with constant WAIT

Figure 1. showed the decrease of RON value with stable N+2A content if reformer feed. This indicate the decrease of RON value is not caused by low N+2A content. The N + 2A content itself is a variable that cannot be controlled because it is a value already contained in crude oil processed in the Crude Distillation Unit. Meanwhile Figure 2. Showed the decrease of RON value with stable WAIT. The reactor temperature is defined as the weighted average inlet temperature obtained from the catalyst fraction in each reactor multiplied by the temperature inlet [2]. With stable N+2A and WAIT, the RON value should be not decrease.

Other parameter that can be examined related to the decreased RON is hydrogen/hydrocarbon ratio (H_2/HC). The hydrogen to hydrocarbon ratio has a significant effect on catalyst deactivation. The cycle life varies directly with hydrogen to hydrocarbon ratio to approximately the 1.6 power [3]. Figure 3 showed the H2/HC trend in the same period which also stable at approximately 2.0. So we can eliminate these three parameter from the causes of catalyst deactivation.

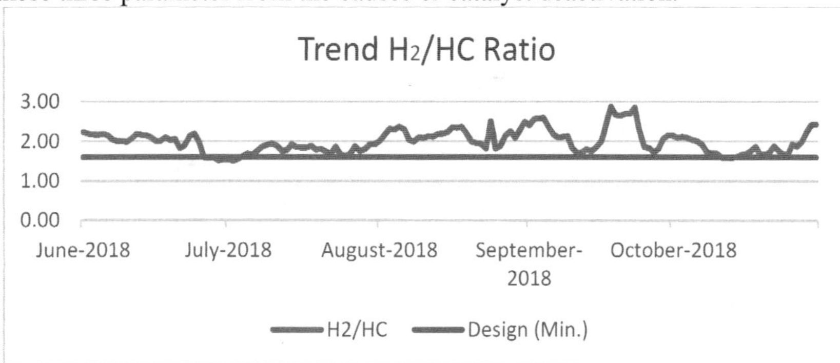

Figure 3. H_2/HC Ratio Trend

3.2. Catalyst Deactivation due to Sintering

Sintering is a thermal deactivation process, which is a physical process related [4]:

- loss of area of carrier material or oxide base
- loss of spread of metal crystals on carrier metal catalysts
- decrease in metal components in the catalyst.

The catalyst used in the reformer reactor is Pt/Al_2O_3. The active side of the catalyst is platinum and acid metal. The acid side in this catalyst is obtained from the addition of chloride to then react with Al2O3 to produce Al_2O_3-Cl. The function of platinum is as a dehydrogenation, dehydrocyclization and dealkylation catalyst, while Al_2O_3-Cl acid functions as an isomerization and hydrocracking catalyst. In this case, the platforming catalyst is sintered due to the loss of the spread of metal crystals on the carrier metal catalyst which ultimately causes a decrease in catalyst activity.

At the regeneration stage of the catalyst in regeneration tower, the oxy-chlorination reaction aims to oxidize platinum and regulate chloride balance. At this stage, the metal catalyst is oxidized at high temperatures (± 510 °C) and high oxygen concentrations (15-18% -mole) [5]. Meanwhile to regulate the balance of chloride, injected organic chloride (Propylene Dichloride-PDC) into the gas circulation. During the burning period, a certain amount of chloride will be lost from the catalyst. With the presence of organic chloride and air, chloride will be re-depositioned to the catalyst to restore its activity. Gas chloride is also formed, and this is vital in the platinum catalyst re-dispersion.

To identify the catalyst deactivation due to sintering, the parameters that can be examined are regeneration & chlorination of gas flow in the regenerator and the oxygen content in chlorination gas.

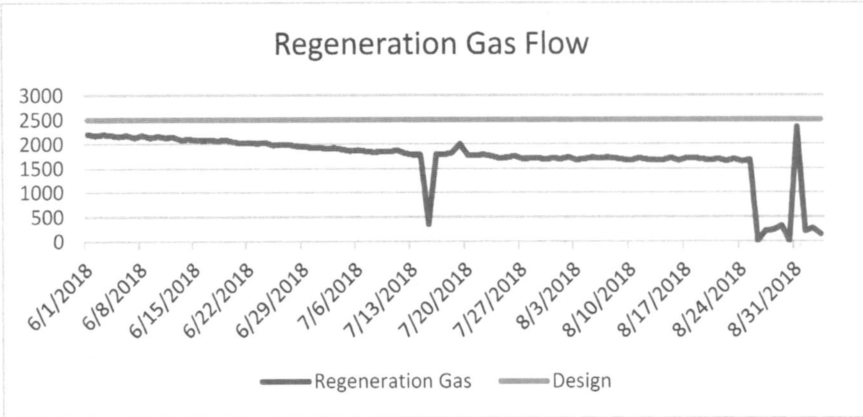

Figure 4. Regeneration Gas Flow Trend

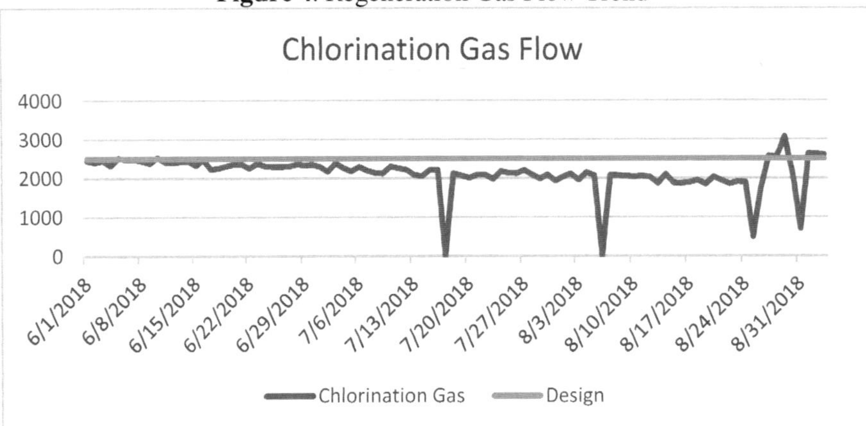

Figure 5. Chlorination Gas Flow Trend

Figure 4 showed that the regeneration gas flow has a downward trend with an average of 1851 Nm^3/hr vs. design 2500 Nm^3/hr. This can have an impact on the regeneration process of the catalyst at the stage of combustion coke does not work well so that it can increase the coke deposit on the catalyst. In the other hand, Figure 5 showed that the chlorination gas flow also has a downward flow with an average 2113 Nm^3/hr vs design 2500 Nm^3/hr. The reduced flow of chlorinated gas means reduced supply of combustion air at the oxychlorination stage in the catalyst regeneration process. Therefore, the redispersion process (distribution) of platinum becomes uneven causing a decrease in catalyst activation.

Decreasing flow chlorination gas will also have an impact on the decrease in oxygen content in chlorination gas. As a result, combustion air that must be supplied for the needs of the oxy-chlorination reaction is reduced. The oxygen content in chlorination gas below 15% -vol will result in platinum re-dispersion in the catalyst is uneven, thereby reducing catalyst activation. In other words, the catalyst is sintered due to imperfect oxy-chlorination reactions in the catalyst regeneration process.

3.3. Cleaning Center Screen

Based on the CCR operating manual, if flow regeneration and chlorination gas had dropped to 20% of the design or around 500 - 600 Nm^3/hr, cleaning center screen in the regeneration tower was needed so that the oxygen content in the chlorination zone could be maintained at a value of 15-18% - mole [5]. Plugging at the center screen is caused by a decrease in gas elutriation flow. Elutriation gas functions to separate spent catalyst with fines in the form of dust, chips and coke pieces before heading towards regeneration tower. Spent catalyst goes into the elutriation tube section (in part in the Disengaging Hopper) and meet nitrogen as a counter-current gas elutriation which will bring fines to the Dust

Collector. Elutriation of gas is flowed using the Lift Gas Blower which also functions to drain the gas elevator to Lift Engager No. 1 to lift the catalyst towards the Disengaging Hopper. Elutriation gas flow has decreased significantly, which is an average of 274 Nm^3/hr vs. design 350 Nm^3/hr, this greatly affects the amount of dust and chips catalyst that passes to the regenerator. The trend of the elutriation gas flow is shown in Figure 6.

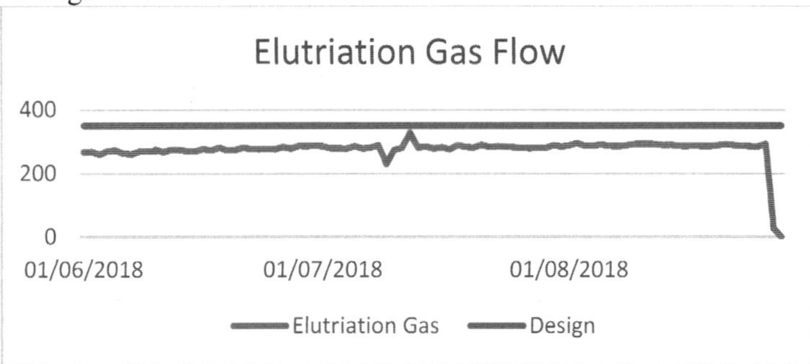

Figure 6. Elutriation Gas Flow Trend

Cleaning center screen is intended to clean the center screen of catalyst dust and catalyst chips that clog so that the regeneration process in the Regenerator CCR does not take place optimally which ultimately decreases the performance of Reforming units. This work is done by shutting down the CCR Regenerator in a cold shutdown mode, which means that the condition of the regenerator is made safe to a certain temperature. This is done because the work carried out requires demolition of the regenerator to remove the center screen from inside the regeneration tower.

3.4. Evaluating the Solution

After the cleaning center screen, regeneration and chlorination gas flow has increased from the previous one, each by 2598 Nm^3/hr vs. design 2500 Nm^3/hr and 2562 Nm^3/hr vs. design 2500 Nm^3/day. This proves that the plugging in the center screen is the cause of the decrease in flow regeneration gas and chlorination gas which ultimately causes uneven re-dispersion of platinum which causes sintering / agglomeration on reforming catalyst. Figure 7 and 8 showed the trend of the regeneration and chlorination gas flow after cleaning center screen.

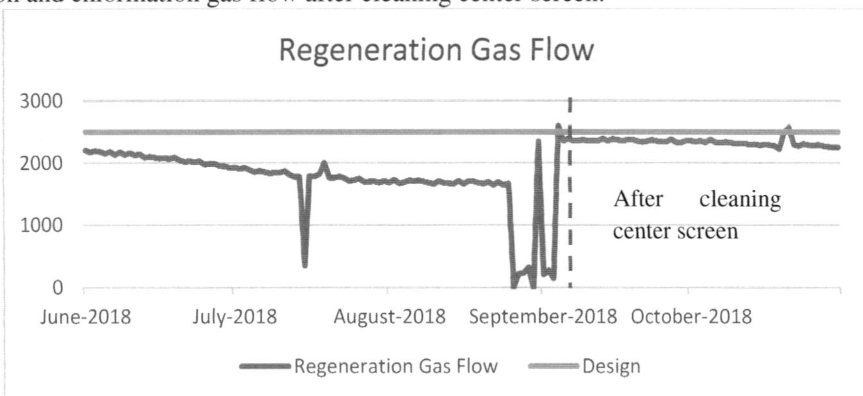

Figure 7. Trend Regeneration Gas Flow after Cleaning Center Screen

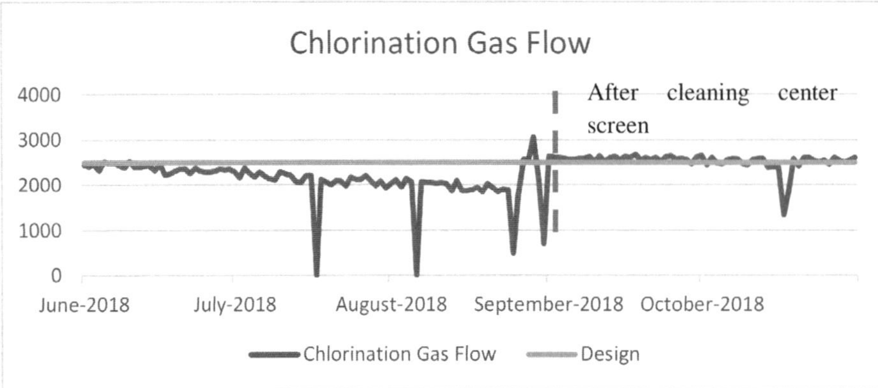

Figure 8. Trend Chlorination Gas Flow after Cleaning Center Screen

After circulation one cycle catalyst, there is an improvement in the condition of the Reforming Unit, namely with RON 94.3 at a capacity of 2448 T/D and N+2A 77.39% vs. before RON 91.0 at a capacity of 2441 T/D and N + 2A 73.57% -wt respectively each at WAIT ± 524 ° C.

TABLE 1. Comparison Table of Performance Reforming Unit After Cleaning Center Screen

Parameter	Desain	Unit	Before cleaning	After cleaning
Capacity	2400	T/D	2441	2448
RIT 014R-1		°C	519.7	523.6
RIT 014R-2		°C	526.0	525.5
RIT 014R-3		°C	523.5	525.3
WAIT 014R-1/2/3	482 - 543	°C	523.5	525.0
N + 2A	Min. 73	%-wt	73.57	73.39
Regen Gas Flow	2480	Nm^3/hr	1866.17	2370.11
Chlor Gas Flow	2480	Nm^3/hr	2194.82	2571.37
Oxygen content	15 - 18	%-vol	3.5	14
Elutriation Gas Flow	350	Nm^3/hr	280	312
RON	Min. 96		91.0	94.3

4. Conclusion

The cause of catalyst deactivation due to sintering can be stated because of the decreased performance of the CCR Regenerator which can be identified by decreasing regeneration and chlorination of gas flow and the oxygen content in chlorination gas flow. The recommendations and solutions provided are conducting cleaning center screen on regeneration towers to restore the performance of the CCR Regenerator so that each step in the catalyst regeneration process can run optimally.

References

[1] Hughes, R. (1984). Deactivation of Catalyst. London: Academic Press Inc.

[2] LLC, U. (2014). CCR Platforming General Operating Manual. Ilionis: Universal Product (UOP).

[3] Ahmed K. Aboul-Gheit, S. A. W. (2008). Catalysis in the Petroleum Naphtha Catalytic

IOP Conf. Series: Materials Science and Engineering **742** (2020) 012009 doi:10.1088/1757-899X/742/1/012009

Reforming Process. Recent Patents on Chemical Engineering, 122-123.
[4] M. Pranjoto Utomo, E.W. (2007). The Study of Catalyst Deactivation on the Heterogeneous Catalysis Reaction. Prosiding Seminar Nasional Penelitian, Pendidikan dan Penerapan MIPA, (hal 113-115). Yogyakarta.
[5] LLC, U. (2002). CCR Platforming Atmospheric Regenerator General Operating Manual. Ilionis: Universal Oil Product (UOP).

Investigation of Ferri-alginate (Fe-Alg) as Environmentally Friendly Catalyst on the Formation of Solketal from Glycerol and Acetone

Mahreni[1] and M.M. Azimatun Nur[1,2]

[1]Chemical Engineering, Faculty of Engineering, Universitas Pembangunan Nasional Veteran Yogyakarta
[2]Energy and Sustainability Research Institute, Faculty of Science and Engineering, University of Groningen

E-mail : mahreni@upnyk.ac.id

Abstract. Alginate is a naturally occurring anionic carbohydrate polymer which could be coupled with metallic cationic molecules to form heterogeneous catalyst. However, the potency of the heterogeneous catalyst on the production of solketal was not explored. This research was proposed to investigate the potency of Ferri-Alginate (Fe-Alg) as a cheap and environmentally friendly catalyst in the solketal reaction. Fe-Alg was synthesized by reacting $FeCl_3$ to sodium alginate with different concentrations (0.1 -0.5 M). Fe-Alg catalyst was characterized both on the physical and chemical activity. By using BET analysis, it was indicated that the addition of FeCl3 concentration increased the surface area of the catalyst. By using TGA/DSC analysis, it was found that Fe-Alg catalyst was stable up to 153°C. From GC/MS analysis, it was found that solketal was formed after the reaction of glycerol and acetone by using Fe-Alg as the catalyst.

1. Introduction

Alginate is a naturally occurring anionic carbohydrate polymer which is generally obtained from the cell walls of brown seaweed. Alginate is widely used in food, cosmetic, and pharmaceutical applications [1]. Alginate has a unique characteristic of an anionic molecule and can be combined with metallic ion molecules such as calcium alginate and sodium alginate to form a new molecule [2].

Currently, the development of low cost and environmentally friendly catalyst has increased [3]. Alginate from brown seaweed can be obtained easily, relatively cheap, and the material source is renewable. Furthermore, alginate is considered as non-toxic material for the environment. Alginate as the organic anionic molecule could be coupled with cationic metallic ion to form solid catalyst which is easily separated from the reaction mixtures. Furthermore, the catalyst can be reused, thus, lowering the cost and minimize the waste. For example, Zhang et al. [4] reported that copper (ii) alginate complexes could enhance the conversion value in the esterification of oleic acid. Cheryl-Low et al. [5] showed that aluminum alginate could be used in the esterification of palm fatty acid. However, to the best of our knowledge, the application of alginate as the anionic catalyst on the ketalization of solketal has not been explored so far.

Previous research mentioned that solketal can be synthesized by using glycerol and acetone on the ketalization process [6]. However, most of the previous researchers were done by using zeolite heterogeneous solid catalyst which is more expensive and could not be renewed [7]. The objective of this research was to investigate a novel ferri-alginate (Fe-Alg) as the cheap and environmentally friendly catalyst on the production of solketal from glycerol and acetone. The catalyst was characterized both in the physical and chemical activity.

2. Methods

2.1. Catalyst preparation

Brown algae were collected fromNgadean beach, Yogyakarta. Fe-Alg was prepared by washing the algae to remove salt and other impurities by using tap water. The cleaned algae (with known weight) was then soaked in pure water with ratio 1:6 (w/w) for 2 h then pulped homogenously by using a blending machine. Sodium carbonate was used to extract the pulped algae to obtain the alginate.

Extraction was performed in an extraction tube with volume 50 L equipped with a heater, temperature controller, timer, and mixer. The pulp brown algae were mixed with sodium carbonate (2% w/v) with a ratio of 1:20 (v/v) for 2 h at 80°C. The mixture of algae was then removed by filtering it to obtain alginate filtrate. The filtrate was then cooled to room temperature.

FeCl3 was varied from 0.1 to 0.5 M to precipitate the alginate extract. The precipitation was done by using titration until the brown color was shown in the precipitated phase. Fe-Alg in the precipitated phase was obtained by using filtration, followed by drying at 80°C until constant weight.

2.2. Surface area determination

Brunauer–Emmett–Teller (BET) is widely used to determine catalyst activity by measuring the specific surface area of the catalyst. In this experiment, the surface area of Fe-Alg (904.5 mg) with different concentration was measured by using BET for 201.8 min, by using nitrogen as the gas carrier at output 300°C.

2.3. Thermogravimetric / Differential thermal analysis

Thermogravimetric analysis (TGA) and differential thermal analysis (DTA) were performed by using TGA-2050 coupled with TGA-2100. Sample (20 mg) was used to determine the values by using aluminum as the holder. The heating rate was set to 10°C min-1 under atmospheric pressure by employing nitrogen gas over the temperature range of 30 to 600°C.

2.4. Ketalization of glycerol and acetone using Fe-Alg catalyst

Fe-Alg was used in the solketal formation by reacting it in the mixture of glycerol, acetone, and ethanol at 3:1:1 mole ratio. The catalyst (2.5 g) was used with a concentration of 0.2 M. The reaction was done at 62°C for 4 h. The result was filtered to obtain filtrate and separate the catalyst. The staged distillation was then employed to separate impurities in the filtrate then analyzedby using GC/MS. The catalyst was reused by washing it using 96% v/v ethanol then dried at 80°C for 2 h. The used catalyst was employed again in the solketal reaction and repeated for three times.

3. Results and discussion

3.1. Catalyst activity

Fe-Alg activity was measured by using BET. Table 1 demonstrated that the addition of $FeCl_3$ in the alginate increased the surface area of the catalyst. The highest surface area (4.38 $m^2 g^{-1}$) was found at F5. This result was in agreement with the previous researcher who mentioned that the structure of alginate changed when interacted with cation molecule and formed Egg Box [8]. The size of the molecule was expanded and hollow spaces was increased. It clearly indicated that the addition of Fe^{3+} increased the hollow spaces, thus, the surface area of Fe-Alg was increased. However, the surface area is relatively low compared to other catalyst such as sulfonated hydrothermal carbons in the solketal reaction [9]. This surface area could be increased by employing calciation at above 600°C.

Table 1. Effect of $FeCl_3$ concentration on the surface area of Fe-Alg

Run	$FeCl_3$ (Molar)	Surface area Fe-Alg ($m^2 g^{-1}$)
F1	0.1	1.45
F2	0.2	2.01
F3	0.3	2.48
F4	0.4	3.69
F5	0.5	4.38

3.2. GTA/DTA profile

Physical analysis of Fe-Alg was shown in Fig. 1. It demonstrated that the degradation rate in TGA was decreasing with the increasing of temperature. Degradation temperature of Fe-alg was recorded at 129 to 221 °C with the peak at 153.25°C. The degradation rate was noted with the increasing of temperature which indicated that the process is exothermic. The exothermic reaction was occurred from 260°C and ended at 400°C. Then, the weight of Fe-Alg significantly degraded from 450 to 600°C and released H_2O, and CO_2.It was recorded that the weight loss was decreasing up to 45.95% above 153.25°C. Generally, the chemical reaction on the thermal degradation employed dihydroxylation, decarboxylation, decarbonylation, and break the macromolecule chain to smaller fragments. The resulting residue could be analyzed using FTIR and MS which will be done in the next project.

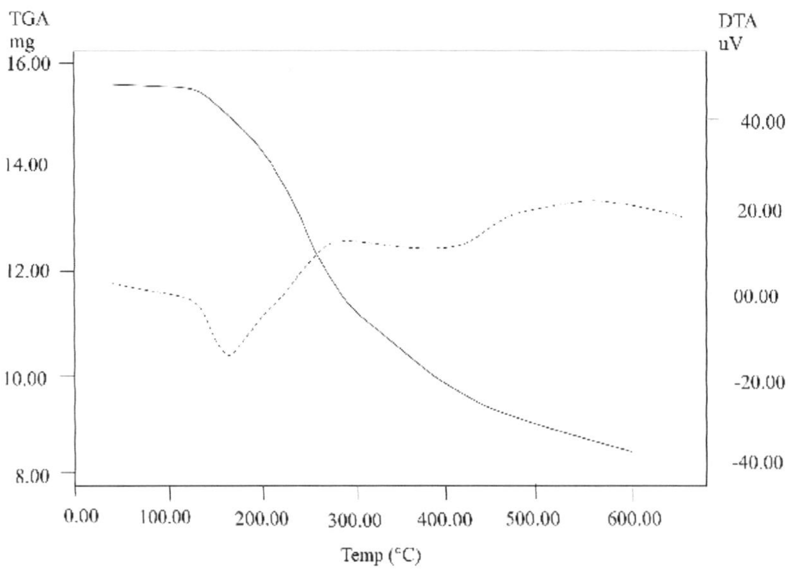

Figure 1. TGA and DTA profile from Fe-Alg under various range of temperatures. Dashed line is DTA. Full line is TGA

3.3. Chemical analysis

IOP Conf. Series: Materials Science and Engineering **742** (2020) 012010 doi:10.1088/1757-899X/742/1/012010

Chemical reaction was done by analyzing the distillate from the glycerol and acetone reaction by using Fe-Alg as the catalyst. From GC/MS analysis, it clearly indicated that the reaction resulted solketal with selectivity 0.568 (Fig 2 Supplementary 1).

Table 2. GC/MS result from the reaction of glycerol and acetone by using Fe-alg 0.5 g.

Number of peak	Retention time (minute)	Component	Selectivity
1	2.2	Vynil Methyl Esther	0.408
2	2.32	Methyl Ethyl Ketone	0.002
3	3.16	Diacetone Alcohol	0.0008
4	**3.94**	**Solketal**	**0.568**
5	4.2	Butanoic acid, 3 hydrxy, 3 Methyl	0.147
6	4.42	Propane 1,1 di propoxy	0.0017
7	4.48	1,2,3 Propanediol (glycerol)	0.0045

However, other byproduct was also appeared such as vynil methyl esther which has 0.408 selectivity (Table 2, Supplementary 2). The selectivity of solketal could be increased by employing calcination of the catalyst at above 600°C to increase the internal pore size. Thus, the active surface area could be increased and higher selectivity could be obtained.

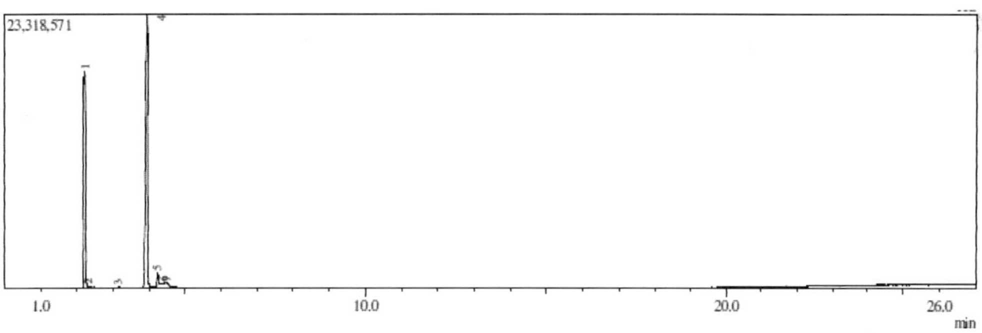

Peak Report TIC

Peak#	R.Time	I.Time	F.Time	Area	Area%	Height
1	2.229	2.160	2.305	65675159	40.81	18434504
2	2.327	2.305	2.390	320888	0.20	132606
3	3.163	3.130	3.210	121435	0.08	61137
4	3.946	3.825	4.130	91449480	56.83	23246523
5	4.240	4.185	4.295	2362539	1.47	1049913
6	4.420	4.400	4.465	276223	0.17	50320
7	4.489	4.465	4.580	718890	0.45	180889
				160924614	100.00	43155892

Figure 2. GC/MS profile of the reaction from glycerol and acetone with Fe-Alg 0.5 M

4. Conclusion

The environmental friendly catalyst, Fe-Alg, was succesfully synthesized by using Sodium alginate and $FeCl_3$. From TGA/DTA result, it showed that organic molecule alginate was interacted with Fe^{3+} and formed Fe-Alg. From BET analysis, the highest surface area was obtained from 0.5 M $FeCl_3$ with surface area 4.38 $m^2\,g^{-1}$. From GC/MS result, it showed that Fe-Alg could be used as the catalyst in the formation of solketal from glycerol and acetone reaction with selectivity of 0.568. Further experiment is needed to increase the selectivity.

References

IOP Conf. Series: Materials Science and Engineering **742** (2020) 012010 doi:10.1088/1757-899X/742/1/012010

[1] Pawar SN, Edgar KJ 2012 Biomaterials33 3279
[2] Lee K Y, Mooney D J 2012 Prog.Polym. Sci. 37106
[3] Laca A, Laca A, Diaz M 2017 J. Enviro. Manage.197 351
[4] Zhang Q, Wei F, Zhang Y, Wei F, Ma P, Zheng W, Zhao Y, Chen H 2017 J. Oleo. Sci.66 491
[5] Low C, Theam KL, Lee HV 2015 Energy. Convers. Manag.106 932
[6] Moreira M N, Faria R P V, Roberiro A M, Rodrigues A E 2019 Ind. Eng. Chem.Res.58 17746
[7] Fatimah I, Sahroni I, Fadillah G, Musawwa M M, Mahlia T M I, Muraza O 2019 Energies122872
[8] Simo G, Fernandez, E F, Vila Crespo J, Ruiperez V, Rodrigues Nogales J M 2017 Carbohydrate polym. 170 1
[9] Fernandez P, Fraile JM, Garcia-Bordeje E, Pires E 2019 Catalysts9804

Supplementary

IOP Conf. Series: Materials Science and Engineering **742** (2020) 012010 doi:10.1088/1757-899X/742/1/012010

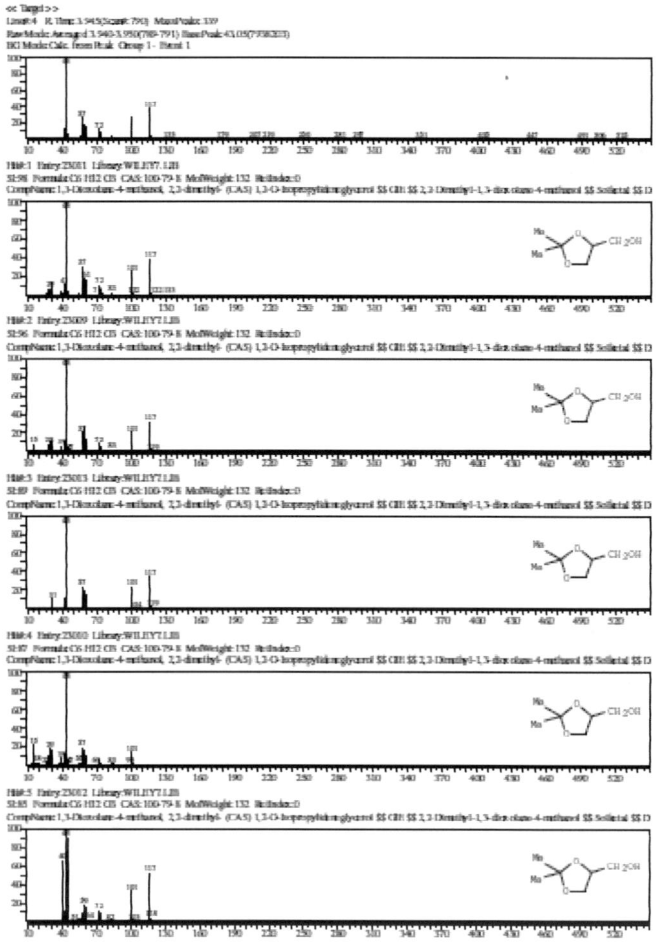

Supplementary 1 . Solketal result from GC/MS at retention time 3.946

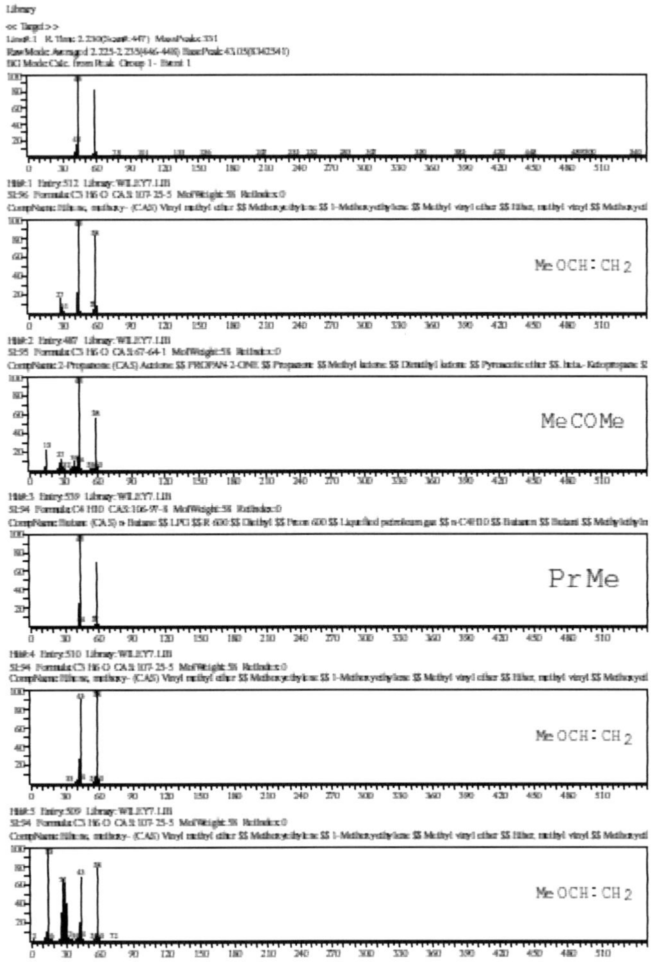

Supplementary 2 . Vynil Methyl Esther result from GC/MS at retention time 3.946

Aluminum-Alginate (Al-Alg) as Green Catalyst in Solketal Synthesis from Acetone and Glycerol

Mahreni[1]

[1]Chemical Engineering, Faculty of Engineering, Universitas Pembangunan Nasional Veteran Yogyakarta

E-mail : mahreni@upnyk.ac.id

Abstract. A catalyst is a chemical that can accelerate a reaction. So far, the catalyst uses precious metals (Pt, Au, Ag) which are heavy and expensive metals. In this research, a biodegradable Aluminum-Alginate (Al-Alg) green catalyst was synthesized using the precipitation method at room temperature. Aluminum is derived from Al $(NO_3)_3$ solution, and the Alginate used is brown algae extract in the form of Sodium-Alginate (Na-Alg). Previous research has reported that Aluminum-Alginate (Al-Alg) has been shown to accelerate the esterification reaction. In this research, Al-Alg was tried as a catalyst in the reaction of Solketal formation from Acetone and Glycerol. The purpose of this research is to prove the Al-Alg catalyst that has been synthesized using the precipitation method which can be used as a catalyst in the reaction of Solketal formation from Acetone and Glycerol. The research was carried out in two stages, namely the extraction of Alginate from brown algae using 2% by weight natrium-alginate solvent, then the second stage was the deposition of Na-Alginate using Al $(NO_3)_3$ solution. The precipitation was carried out at room temperature. The precipitate was dried, then its physical characteristicwas tested using FTIR and TGA/DTA; the chemical characteristic of the catalyst was tested through catalyzed reaction. The FTIR test results showed that the alginate molecule had been bound to Al molecule which was marked by a peak at the wave number of 1383.4 cm^{-1} which indicated the carboxylic group of COOH owned by Alginate, namely the *symmetric stretching vibration of the carboxylate group*. In addition, the peak at the wave number of 827 cm^{-1} showed that the C1 – H deformation vibration of b-mannuronic acid residues from Alginate. The peak at the wave number of 550 cm^{-1} proved that Aluminum (Al) had been bound to Alginate, and forming Al-Alg. The catalyst activity test was carried out by applying the Al-Alg catalyst in the catalyzing reaction of the formation of Solketal from Acetone and Glycerol. The reaction results were analyzed using GC-MS and it showed that the solketal was formed, and the selectivity was quite high at 96 %, proving that Al-Alg had the potential as a catalyst in catalyzing reaction.

1. Introduction

Solketal is an additive material to increase the gasoline octane number or increase the diesel cetane number [1]. The advantages of solketal are: biodegradable; do not contain metals; can be made from renewable raw materials; easy to make, so it can be classified as green chemical. So far Solketal has been synthesized using various types of catalysts, either homogeneous catalysts or heterogeneous catalysts. Homogeneous catalysts use hydrochloric acid, sulfuric acid, Para toluene sulfonic acid (PTSA), $FeCl_3$ and PWA (phosphotungstic acid) [2]. Solid (heterogeneous) catalysts that have been used are Amberlist and Zeolite [3]. The application of Al-Alg heterogeneous biodegradable catalysts has not been much studied.

The Al-Alg catalyst is synthesized using ionic reactions, between anions (carboxylic groups or COO^-) contained in alginate molecules, and Al^{+3}cations contained in Al $(NO_3)_3$ solution, then form molecular structures that resemble an egg box as illustrated in Figure 1. The negative charge of alginate molecules represented by carboxylic groups (COO^-) and OH^- on each Alginate monomer, can bind to the metal cation M^{n+} to form the egg box structure [4]. The valence metals two and three can bind to the Alginate molecule to form an Egg-box structure. Since alginate is a polymer composed of manuronic acid (M) and guluronic acid (G) monomers with the degree of polymerization reaching above 200, it is very effective as catalyst support with the main function of distributing the active component of the catalyst, in this case, metal cations. The catalyst activity increases if the active component of the catalyst can be well dispersed/distributed.

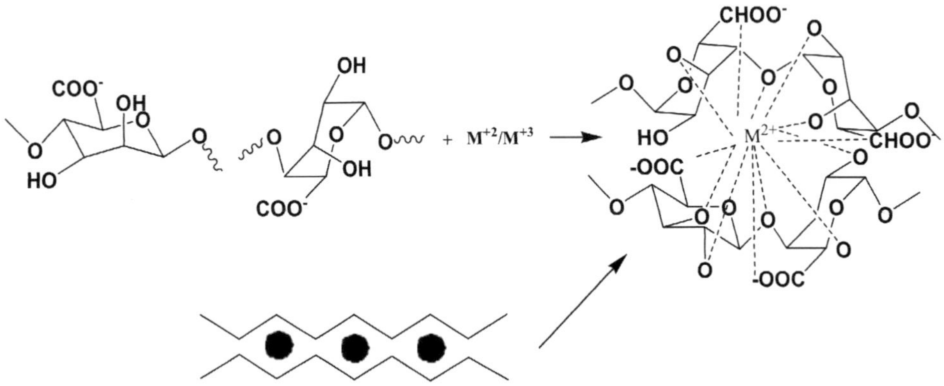

Figure 1. The interaction of Alginate molecules with the M^{2+} cation, forming the egg-box structure.

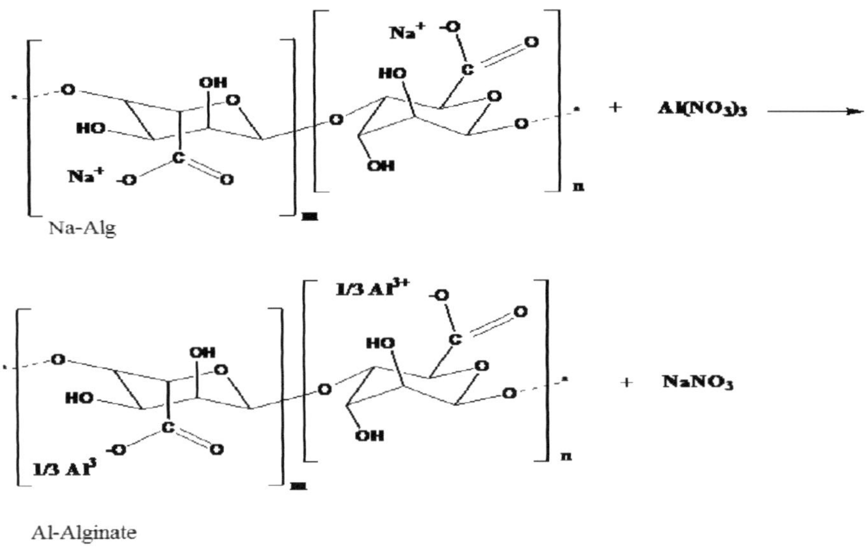

Figure 2. The ionic reaction of Na-Alginate and Al $(NO_3)_3$ to Al-Alg formation.

The purpose of this study is to test the physical properties of the Al-Alg catalyst and investigate the potential of Al-Alginate as a catalyst in the reaction of Solketal formation ftom Acetone and Glycerol

2. Material and Method

2.1. Materials and equipment

The main materials of Al-Alginate are: (1) Brown algae (*Sargassum sp.*), obtained from the southern coast of Yogyakarta (Ngedenan Beach); (2) Al $(NO_3)_6$ H_2O (pa), technical Na_2CO_3, which is directly used without pretreatment; (3) Aquadest; (4) 96 wt. % Ethanol; (5) 96 wt. % Acetone and (6) 96 wt % Glycerol. The equipment used are: (1) 50-liter volume extractor, equipped with temperature control, stirring speed and time; (2) Batch reactors, equipped with temperature control, stirring speed, condenser and time; (3) filtration equipment; (4) Drying ovens; and (5) a set of glassware.

Figure 3. The equipment of Al-Alg catalyst formation (a) Extractor and (b) Reactor

2.2. Flow diagram of the experiment

2.2.1. Production of Al-Alg Catalyst

The ingredients which directly used were $Al(NO_3)_3$ and $NaCO_3$, without any pretreatment. The making of the Al-Alg catalyst, firstly was weighing the brown Algae, then washed it with tap water to remove salt and dirt. The cleaned algae were drained for several hours until there was no more water dripping from the algae. The cleaned algae were soaked in aquadest in a ratio of 1:6 until all the algae were soaked for two hours, then blended it to form a homogeneous mixture (brown algae pulp). The next step was to make a 2%-weight Natrium carbonate solution. The natrium carbonate solution is used to extract Alginate from the brown Algae to form Alginate in the form of Natrium-Alginate salt. Natrium-Alginate is a form of soluble Alginate. The brown Algae extraction was carried out in an extractor with the volume of 50 liters and was equipped with a temperature regulator, stirrer speed regulator, and a timer. The ratio between the volume of Alga pulp and the volume of Natrium carbonate solution was 1:20 (volume/volume). The extraction was for two hours and at 80°C. The extraction product was filtered, and the filtrate was cooled for the precipitation process using a solution of Al $(NO_3)_3$. The filtrate containing Natrium-Alginate would be converted to Al-Alginate in the precipitation process [5]. As the precipitate was the solution of Al $(NO_3)_3$ with concentration that varied from 0.1 to 0.5 M. The precipitation was carried out by titrating the solution of Natrium-Alginate extract with the solution of Al $(NO_3)_3$, drop by drop, and a brown precipitate would be formed. The titration was stopped after no more precipitate had formed. The precipitate was separated by filtration. Furthermore, the precipitate was dried in an oven at 80°C to a constant weight, called the Al-Alg catalyst. The process of making catalyst and the Al-Alg catalys test can be seen in the flowchart in Figure 4.

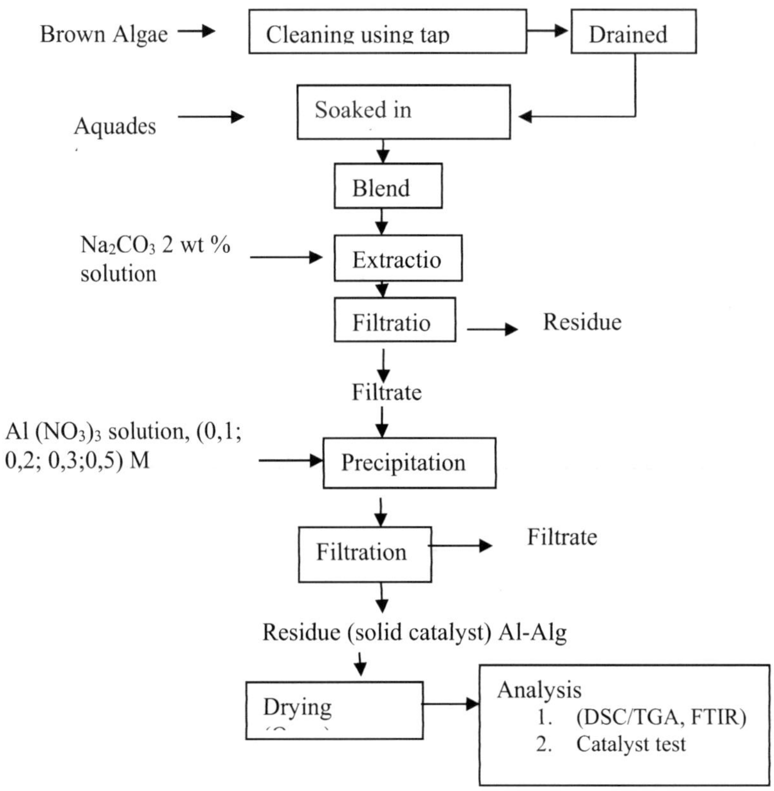

Figure 4. Flowchart of making of Al-Algcatalyst from brown Alga (*Sargassum sp)* and $Al(NO_3)_3$

2.2.2. Activity characterization of Al-Alg catalyst

To test the catalyst activity, the catalyst is used directly in the Solketal formation reaction. The catalyst test was carried out by mixing acetone and glycerol, ethanol and catalyst Al-Alg. After that, heated the mixture in the reactor equipped with a temperature controller, stirrer and condenser. After the temperature reached 63°C, the temperature was kept constant, and the heating was carried out for four hours. After that, the heater was turned off, and the reaction mixture was cooled to reach room temperature. Then the catalyst was separated, and the filtrate distilled to product purification. Purification of the Solketal (the main product) was performed where the distillation was carried out at boiling temperatures of acetone, ethanol, water, solketal, and glycerol. Soketal were analyzed using GCMS to determine the components presented in the distillate phases. The catalyst test process can be seen in 5.

IOP Conf. Series: Materials Science and Engineering **742** (2020) 012011 doi:10.1088/1757-899X/742/1/012011

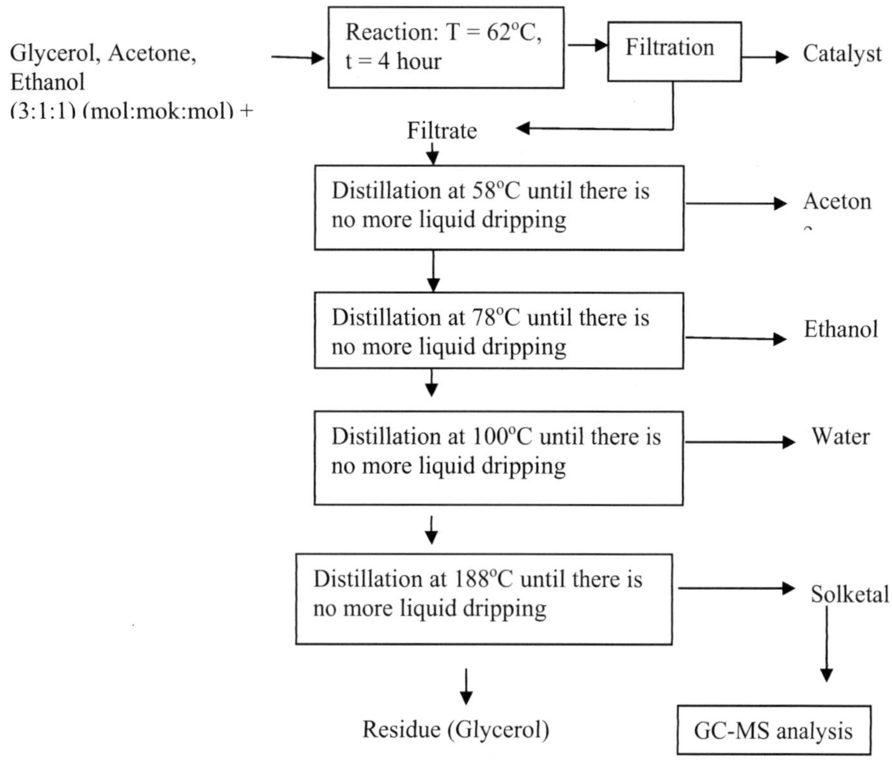

Figure 5. Flowchart of the Al-Alg catalyst test process as a catalyst in the reaction of Solketal formation from Acetone and Glycerol

2.3. Test of Al-Alginate using TGA/DTA
The thermogravimetric (TG) measurement is carried out to determine the thermal properties of the Al-Alginate catalyst. The analysis was performed using TGA-2050 thermogravimetry which was combined into a TGA-2100 thermal analyzer (both from TA Instruments) using a 20-mg sample mass and an alumina sample holder. The heating speed was 10°C min^{-1} below Nitrogen atmospheric. The measurement (DSC) carried out in DSC-910 modulus was combined into the TGA-2100 thermal analyzer (both from TA Instruments) using the sample mass of 20.0 mg in a closed aluminum sample container with a central pin hole with the heating rate of 10°C min^{-1} below Nitrogen atmospheric.

3. Results and Discussion

3.1. FTIR analysis
The results of FTIR analysis of Al-Alginate catalyst can be seen in Table 1 and Supplementary 1 (see supplementary). Based on the FTIR pattern, wavelength around 3453 cm^{-1} contributed to O-H bending from Alginate (Manuronic and Guluronic Acid), and the peak visible in the fingerprint area on 1633 cm^{-1} as scissoring band of O-H. The other peak on 577 cm^{-1} may be contributed to metal oxide (Al-O) vibration.

Table 1. The results of FTIR analysis of Al-Alginate catalyst produced by researcher (Mohamed, *et al.*, 2017) and the results of the author's research

Information	Wave number (cm^{-1})	Research results of [5] [6]	Wave number (cm^{-1})	Author's research results of Al-Alginate
Weak	1316.79	C–C–H		
Weak	1125.53	O–C–H		
Weak	1094.66	C–O stretching		
Strong	1035.6	C-O stretching vibration	1036	C–O stretching vibrations
Weak	948.2	C–O stretching vibration of uronic acid residues		
Weak	902.83	C1–H deformation vibration of b-mannuronic acid residues		
Strong			827	C1–H deformation vibration of b-mannuronic acid residues
Weak	818.76	C1–H deformation vibration of b-mannuronic acid residues		
Strong			577	Metal oxide (Al-O) vibration
Strong	1416	Symmetric stretching vibration of the carboxylate group	1383.4	Symmetric stretching vibration of the carboxylate group
Strong	1611.57		1633	O-H bending from $Al(OH)_3$
Strong			2329	
Strong	2942.16	C-H stretching vibration		
Strong	3421.05	Hydrogen bonded O-H stretching vibration	3453	O-H bending from Alginate

3.2. TGA/DTA analysis

The Al-Alg catalyst test results using TGA/DTA are shown in Supplementary 2. Supplementary 2 shows the Al-Alg TGA/DTG pattern. The thermogravimetric (TG) measurement is carried out using TGA-2050 thermogravimetry, combined into a TGA-2100 thermal analyzer (both from TA Instruments) using a sample mass of 18.961 mg and the alumina sample holder. The heating speed is 10°C min^{-1} below the atmosphere of Nitrogen at a speed of 30 ml/min. The detector DTG-60 serial number is C30565100652TK. The results of the TGA analysis are shown by the thermogravimetric analysis curve (TGA) and the differential thermogravimetric Analysis (DTG) of Al-Alg at a heating rate of 10°C/min below the Nitrogen atmosphere, as presented in Supplementary 2. The Al-alginate weight reduction stage is located at the wide temperature. In the zone (from 30 to 600°C), the decomposition temperature of the onset crystal water content (Tonset, defined as the temperature at which a 5% weight loss occurs) from Al-Alg is at 67 to 179°C, and the peak is at 121.4°C. At the time of the release of crystal water requires a certain amount of heat, and this is seen in the DTG curve, a decrease in temperature indicates the release of water is an endothermic process, and the temperature drops to 25oC from the initial temperature.

The faster degradation rate is characterized by the rising temperature or exothermic process. The exothermic process started at 220°C and ended at 380°C. Then the weight of Al-Alginate decreased more rapidly, and the Alginate degradation produced H_2O, CO_2, and occurred at temperatures of 410 to 600°C where the reaction to release water and carbon dioxide occurred exothermically so that the temperature increased. The exothermic peak occurred again at temperatures of 550 to 600°C. Weight loss in the first stage of thermal degradation was in linewith the intensive evolution of small molecules and the degradation of alginate by the glycosidic bonding fracture to form compoundsbetween releasing small molecules, such as H_2O and CO_2 [7]. In general, the chemical reactions occurred during the process of thermal degradation in the alginates, involving dehydroxylation, decarboxylation, decarbonylation, and breaking down the macromolecular chains into smaller fragments. The reaction produced residues and gas. The estimated residues were Al and carbon. The final sample weight was 2.5 mg, from the initial sample which was 18.961 mg, proved that Aluminum and Alginate had been formed. Alginate is an organic material that will be degraded to produce CO_2 and small molecules, and the residues are Aluminum oxide and Charcoal. Charcoal is an inorganic component, such as SiO_2, and the metal oxides contained in brown algae that dissolved in the solvent when extracted.

3.3. Chemical characteristic test of Al-Alg catalyst using GC-MS

The purpose of testing the catalyst's chemical properties is to determine the potential of the Al-Alg catalyst as the catalyst in the reaction of Solketal formation from Acetone and Glycerol. The results of the analysis showed that the reaction using Al-Alg as catalyst produced a Solketal with a selectivity of 96.83%. Prove that Al-Alg has great potential for the formation of solketal from Acetone and glycerol

4. Conclusion

Research in the synthesis of Al-Alg catalysts has been carried out. Al-Alg is an environmentally friendly catalyst that can be made from renewable sources (*Sargassum sp.*). From the results of the analysis found that Al-Alg proved to be able to act as a catalyst in the reaction of the formation of Solketal from Acetone and Glycerol. The main product selectivity is high at 96% Solketal. Therefore Al-Alg is very important to be a catalyst in the future

References

[1] Maraísa Gonçalves, Raphael Rodrigues, Thalita S. Galhardo, Wagner A. Carvalho. Highly selective acetalization of glycerol with acetone to solketal over acidic carbon-based catalysts from biodiesel waste Fuel 181 (2016) 46–54. http://dx.doi.org/10.1016/j.fuel.2016.04.083.

[2] Roberto Espositoa, Maria Elena Cucciolito, Angela D'Amoraa, Rossella Di Guida, Fabio Montagnaro, Francesco Ruffo. Highly efficient iron (III) molecular catalysts for solketal production. Fuel Processing Technology 167 (2017) 670–673. http://dx.doi.org/10.1016/j.fuproc.2017.08.018

[3] Nuryoto1, Hary Sulistyo, Wahyudi Budi Sediawan, Indra Perdana. Sintesis Bioaditif Gasoline Melalui Ketalisasi Gliserol Menggunakan Katalisator Padat. Jurnal Bahan Alam terbarukan 5(2) (2016) 74-83. DOI 10.15294/jbat.v4i2.7431.

[4]. Guillermo Simóa, Encarnación Fernández-Fernándeza, Josefina Vila-Crespo, Violeta Ruipérez, José Manuel Rodríguez-Nogales. Research progress in coating techniques of alginate gel polymer forcell encapsulation.. Carbohydrate Polymers 170 (2017) 1–14. http://dx.doi.org/10.1016/j.carbpol.2017.04.013

[5] Etik Mardliyati, Damai Ria Setyawati, Sabar Pambudi, Suryandaru, Retno Kusumaningrum, M. Ikhlasul Amal Preparation of Aluminum Hydroxide by Precipitation Method for Vaccine Adjuvant Application, Int. Journal of Engineering Research and Application www.ijera.com ISSN : 2248-9622, Vol. 7, Issue 11, (Part - 5) November 2017, pp.21-25

[6] Mohamed Fertah, Ahmed Belfkira, El montassir Dahmane, Moha Taourirte, Franc¸ois Brouillette. Extraction and characterization of sodium alginate from Moroccan Laminaria digitata brown seaweed. Arabian Journal of Chemistry (2017) 10, S3707–S3714.

[7] Liu Y, Li Z, Wang J, Zhu P, Zhao J, Zhang C, Guo Y, Jin X, Thermal degradation and pyrolysis behavior of aluminum alginate invesitagted by TG-FTIR-MS and Py-GC-MS,*Polymer Degradation and Stability* (2015), doi: 10.1016/j.polymdegradstab.2015.04.010.

Supplementary

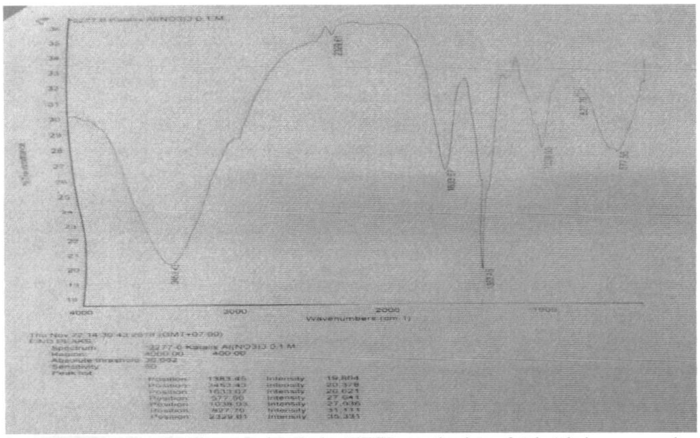

Supplementary 1. The result of the FTIR analysist of Al-Alginate catalyst

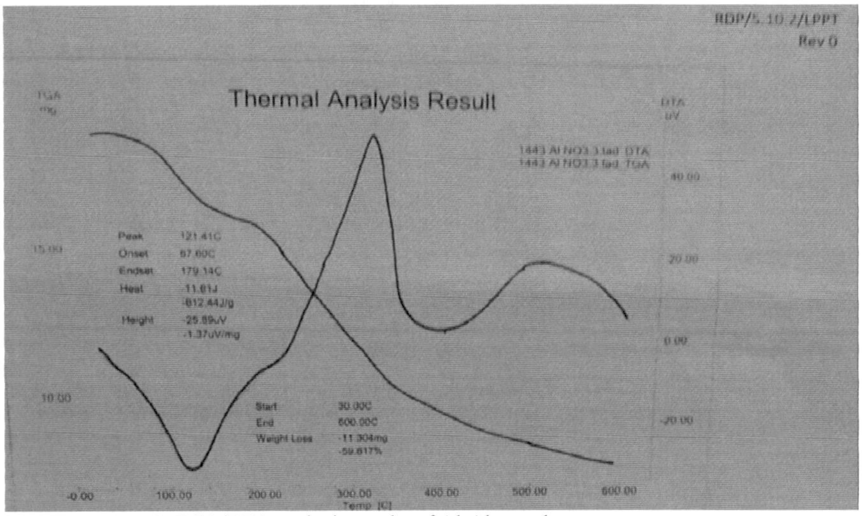

Supplementary 2. TGA/DTA analysis results of Al-Alg catalyst

IOP Conf. Series: Materials Science and Engineering **742** (2020) 012011 doi:10.1088/1757-899X/742/1/012011

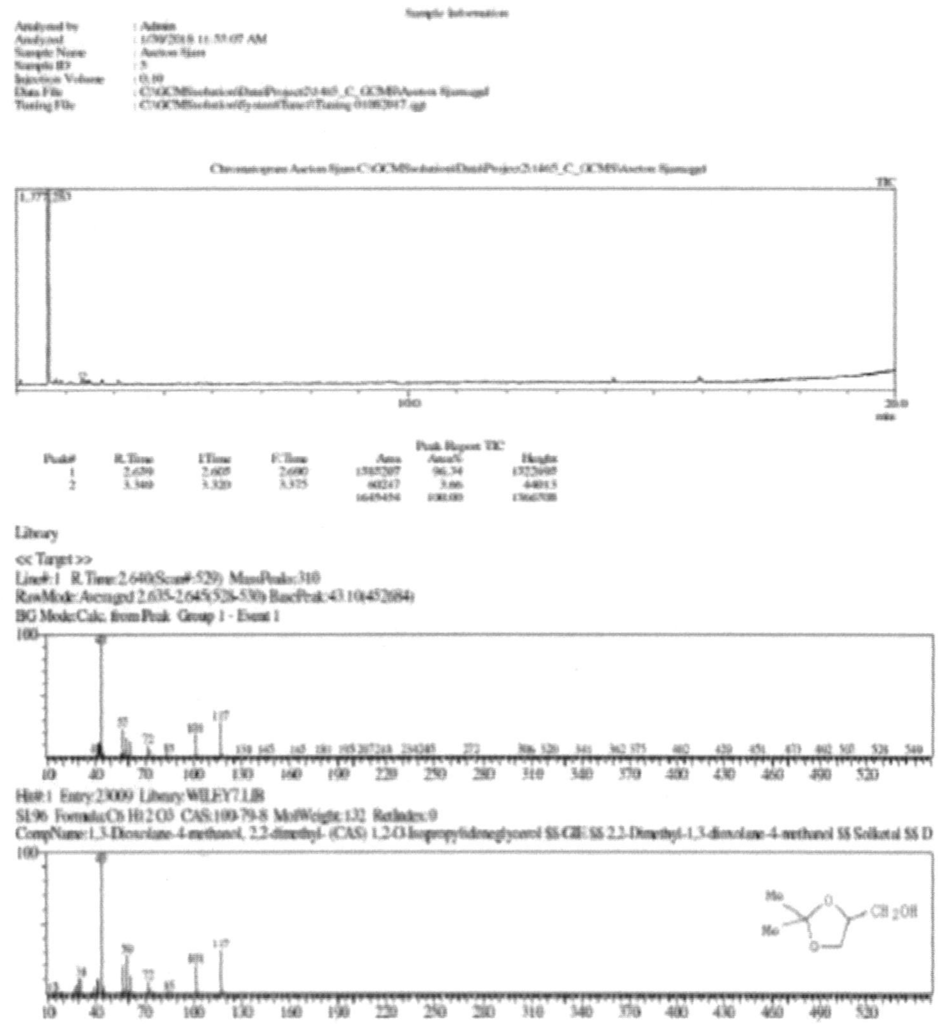

Supplementary 3. The results of GC-MS analysis in the reaction of Acetone and Glycerol using Al-Alg as catalyst at the temperature of 63°C during for four hours with the ration between the acetone mole and the Glycerol = 2:1 (mol/mol).

IOP Conf. Series: Materials Science and Engineering **742** (2020) 012012 doi:10.1088/1757-899X/742/1/012012

Evaluation of CCS Networks Based on Pinch Design Method in the Central Part of Indonesia

Renanto Handogo*[1], Juwari[1], Ali Altway, Annasit[1]

[1]Chemical Engineering Department, Institut Teknologi Sepuluh Nopember, Indonesia.

E-mail : renanto@chem-eng.its.ac.id

Abstract. At least for two decades, various techniques have been developed to solve the greenhouse gas emissions problem. The carbon capture and storage (CCS) technology is the main solution to reduce carbon dioxide emissions from the industrial sector and power plant. CCS involves the capture of CO_2 emissions from an exhaust gas, transporting and storing in geological storage. All of these CCS activities give several planning problems related to geographical conditions that may be an archipelago country, a considerable distance between the CO_2 sources and sinks, and the difference in time of CO_2 capturing and injecting to storage. The process integration technique by pinch design method can be used for CCS networks. This paper presents an application of pinch design method for CCS networks in central parts of Indonesia. The integration of carbon incentive and carbon tax is proposed to evaluate the value of CCS networks. The effect of time difference between sources and sinks on the amount of CO_2 uncaptured was considered to assess the sensitivity of CCS project's value. The results showed that the combination of pinch design method with the carbon trading and carbon tax provided a potential method to assess the effectiveness of CCS networks.

1. Introduction

The growing of energy demand, especially from fossil fuel, has increased over the coming decades. An increase in energy consumption has led to an increase in CO_2 emission. Indonesia as an emerging country ranks seventh in total carbon emissions from fossil fuels. Indonesia's ability to reduce CO_2 emissions will have a significant impact on changes in global emissions [1]. The CCS is one of technology which can reduce CO_2 emissions from fossil fuel combustion and other industrial processes. The CCS consists of several process stages, CO_2 is first captured from the flue gas or hydrocarbon streams, separated CO_2 from the sorbent, and then transported to the geological to stored permanently or reutilized industrially [2]. The planning of CCS systems may occur over several decades due to various reasons, such as discovering new geological sinks, the availability of carbon transportation infrastructure, planned introduction of new carbon sources, and geological sinks being only available at a later date. This leads to the matching among sources and sinks with time consideration [3].

The application of pinch analysis for CCS planning has been used in several studies. The use of pinch analysis technique for carbon-constrained energy sector planning first developed by Tan and Foo [4]. This approach is termed carbon emission pinch analysis (CEPA), uses the energy sector planning scenario with carbon emission constraints, which is a characteristic of fuel or technology. Tan et al. further developed pinch analysis techniques to determine the amount of carbon emission targets for the existing power plant [5]. Tjan et al., presented the use of graphical pinch analysis approach for determining strategies to get carbon footprint reduction in a chemical process [6]. This approach allows for identifying cleaner production options and allows clear visualization of key variables in emission planning. The later improvement of CCS planning was proposed by Diamante et al. This work presents an improvement of pinch analysis based methodology by simultaneously considering injectivity constraint of every sink as well as time availability of various sources and sinks. The latter work by Putra et al., was proposed in the detailed approaches via algebraic technique for multi-period and multi-region CCS planning problem in a realistic situation [7].

The CCS still has a major barrier to be developed in some countries, especially in Indonesia. The main obstacle in developing CCS is high cost. The cost of CCS is a key issue in the public and policymaker [8]. Carbon taxes and emission trading schemes are considered to be the major climate-change policy instrument to reduce carbón emissions [9]. The two policy instruments can be combined to enhance the advantage of the CCS cost.

Based on the previous studies, there were no systematic methods in evaluation of the value of CCS networks in multi-region and multi-period using pinch analysis. Therefore, in this study, a new strategy was developed to design CCS network system in a multi-region using pinch analysis and assess the value of CCS networks. The heuristics of pairing between carbon source and sink in grid diagram analysis was developed based on the work by Putra et al. [7]. The concept of CCS value will be introduced as a parameter to assess the effectiveness of CCS networks by combining CCS cost, carbon incentive, and carbon tax.

2. Research methodology

2.1. Data compilation

In this study, CCS networks are developed in multi-region and multi-period system. Three regions are selected for this study, namely East Kalimantan, South Sulawesi, and North East Java. The sources that will be used in this study are represented by fertilizer plant, cement plant, and LNG Plant. The carbon storage hereinafter referred to as the sink, comes from two places, East Kalimantan, and North East Java[10]. The source and sink data used were listed in table 1 and table 2.

The time availability, operation lifetime, and flow rate of each source and sink points must be known as the parameters of the network design. Time availability for a source is the time while the industry starts to run the CO_2 capturing process for the first time until the end of production, while time availability for a sink is the time while CO_2 is injected to sink for the first time until full capacity. Time availability for a source is assumed based on the consideration that every source cannot start its capture process together, because the planning of capturing CO_2 is different in every source. Time availability for a sink is based on the assumption that planning of CO_2 injection can be done at the same time because some CO_2 storage is ready for use. Operation lifetime is the time duration of carbón exchange operation between the source and sink. Operation lifetime for a source is estimated based on production lifetime of each existing plant, while operation lifetime for a sink is estimated based on fixed normal injection rate of CO_2 which is the injection rate of CO_2 in several pilot storage site in Indonesia can be subsequently developed to absorb 1 – 11 Mt/yr of CO_2 [11]. Therefore, the operation lifetime of the sink would be different for every sink.

This study attempts to show the influence of time difference (t_{diff}) variable on CCS planning. The time difference is schedule delay of CCS implementation because some reasons, such as the sink is not ready to be injected with CO_2, or the availability of transportation facilities especially those using a ship to deliver CO_2 from source to sink. The variables that will be used are 0, 5, and 10 years of a time difference. A sensitivity of CCS cost will be examined to obtain the effectiveness in CCS planning. The variation of the results for alternative storage and unutilized storage has been examined.

IOP Conf. Series: Materials Science and Engineering **742** (2020) 012012 doi:10.1088/1757-899X/742/1/012012

Table 1. Sources characteristic example in the central part of Indonesia

Code	Industrial source	Start time (year)	End time (year)	CO_2 emission rate (Mt/year)	CO_2 load (Mt)
SR1	PupukKaltim	5	30	2.403	60.075
SR2	Badak LNG	5	20	5.514	82.71
SR3	Semen Bosowa	7	27	1.575	31.5
SR4	Semen Tonasa	8	28	3.791	75.82
				Total CO_2 produced	250.105

Table 2. Sinks characteristic in the central part of Indonesia

Code	Geological sink	Earliest time (year)	End time (year)	CO_2 injection rate (Mt/year)	Storage capacity (Mt)
SK1	Kutai-Tarakan Basin	5	35	4.65	139.50
SK2	North East Java	5	30	3.76	94.0
				Total storage capacity	233.5

2.2. Cascade table analysis

The cascade table approach is used to calculate the amount of minimum alternative storage needed, unutilized storage, and pinch point that will be used for generating a grid diagram. The cascade table approach is an algebraic targeting technique to analyze CCS system. This approach has been conducted by Diamante et al., known as carbon capture and storage cascade analysis (CCSCA) [3]. The method was similar to the problem table algorithm in heat exchanger network design. Originally this method was used for calculating energy targets in heat exchanger network directly without the necessity of graphical construction. Putra et al., modified generic cascade table for CCS system by inserting a number of columns to detail the calculation steps [7]. The last improvement of CCSCA was developed by Thengane et al., that complement with graphical technique of grand composite curve (GCC) to determine the actual targets of CCS system [13].

The generic cascade table used in this study is similar to the cascade table was developed by Thengane et al., [13] with slight improvement. The table consists of eight columns with a title as follows: time (t), source (S), sink (D), the time interval (Δt), flow rate CO_2, load CO_2, cumulative CO_2 load, and feasible CO_2 cascade respectively. The procedure is given by steps 1 – 8, with results shown in table 2.

1) Time availability of all sources and sinks are tabulated in the first column with the order from t_1 in the first row until t_T in the last row, where $t_1<t_2<\ldots<t_T$. The source or sink with the same time value, it does not need to be repeated
2) The CO_2 sources located in the second column based on their time availability. Every CO_2 sources have certain flow rate and denoted as $S_{i,t}$, where $S_{i,t}$ is CO_2 emission rate from source *i* at time *t*.
3) The CO_2 sinks located in the third column based on their time availability. Every CO_2 sinks have certain injection rate and denoted as $D_{j,t}$, where $D_{j,t}$ is CO_2 injection rate in sink *j* at time *t*.
4) The fourth column is the time interval of all sources and sinks which is arranged in chronological order and denoted as Δt_T
5) The net CO_2 flowrates are the difference between the CO_2 sinks flowrates and CO_2 sources flowrates at certain time intervals. It is tabulated in the fifth column.
6) The CO_2 load for each time interval is tabulated in the sixth column. The CO_2 load is determined using equation 1.

$$\Delta m_T = \left(\sum_j D_{j,T} - \sum_i S_{i,T} \right) \times \Delta t \quad (1)$$

7) The summation of CO_2 loads from previous rows are tabulated as cumulative CO_2 load in the seventh column. The first entry of cumulative CO_2 load is assigned as zero. Hence, the cumulative CO_2 load in time, T, is expressed as follows:

$$I_T = \sum_T \Delta m_T \quad (2)$$

8) To obtain a feasible solution, all entry values in the seventh column should not be negative values. The largest negative value in the seventh column is used as the first entry in the eight-column as an absolute value. As shown in the eight-column of table 3, the larges negative value denoted as I_N. The first entry in this column indicates uncaptured CO_2 emissions or alternative storage, while the last entry in this column indicates unutilized CO_2 storage.

Table 3. Generic carbon cascade

Time, t (y)	Source, $S_{i,t}$ (Mt/y)		Sink, $D_{j,t}$ (Mt/y)		Δt (y)	Net CO_2 flowrate (Mt/y)	CO_2 load (Mt)	Cumulative CO_2 load (Mt)	Feasible CO_2 cascade (Mt)
1	2		3		4	5	6	7	8
1	SR1							$I_1 = 0$	$F_1 = -I_N$
	↓	SR2	SK1		Δt_1	$\sum D_{j,1} - \sum S_{i,1}$	Δm_1		
2	↓	↓	↓	SK2				I_2 = cum Δm_1	$F_2 = F_1 + I_2$
	↓	↓	↓	↓	Δt_2	$\sum D_{j,2} - \sum S_{i,2}$	Δm_2		
3	↓	↓	↓	↓				I_3 = cum Δm_2	$F_3 = F_2 + I_3$
:		↓	↓	↓		:		:	:
:		↓	↓	↓		:		:	:
:		↓	↓	↓		:		:	:
T		↓	↓	↓				I_T = cum Δm_{T-1}	$F_T = F_{T-1} + I_T$
			↓		Δt_T	$\sum D_{j,T} - \sum S_{i,T}$	Δm_T		
T+1			↓					I_{T+1} = cum Δm_T	$F_{T+1} = F_T + I_{T+1}$

9) The implementation of time scenarios using time difference variable (t_{diff}) of 0, 5, and 10 years was done by shifting down the sink operating time according to time difference variable. The sink operating time in the third column and time availability in the first column should be adjusted according to the time scenario.

2.3. Generating a grid diagram

The design of CCS networks can be developed using a grid diagram. The grid diagram is presented to show the carbón exchange from an individual source to an individual sink. Grid diagram can be generated by following steps:

1) The number of horizontal lines was drawn which represents each source and sink stream. The line was drawn from left (start time) to right (end time) with appropriate time position.
2) The vertical line was drawn which represents the pinch point line.

3) Grid diagram is divided in two zones, below the pinch or left side of the pinch line, and above the pinch or right side of the pinch line.
4) Source-sink stream connection is started and finished in operation lifetime horizons.
5) There are different rules for connecting the source and sink streams in each zone.
    - Above the pinch : focus on the source streams. There is no alternative storage above the pinch and the sink flowrate must larger than source flowrate.
    - Below the pinch : focus on the sink streams. There is no unutilized storage below the pinch and the source flowrate must larger than sink flowrate.
6) In pairing process between source and sink streams, some rules were developed:
    - The most similar start-end time between source and sink were paired at first.
    - The most operating lifetime between source and sink were paired at second order
    - The most similar carbón flowrate between source and sink streams were preferred

2.4. CCS cost calculation

There are three major elements involved in CCS networks, each of which has investment and operating costs: (1) Capture, (2) Transportation, and (3) Storage. In this study CCS cost calculations only consider capture and storage cost. Transportation cost is not taken into account because it still requires further study, especially for ship transportation in archipelago country. The calculation of carbon capture costs uses the carbon capture equation model developed by Hasan et al., [14]. The equations are shown by equation 3 and 4.

$$IACC = \alpha + (\beta x_{CO_2}^{n} + \gamma)F^{m} \tag{3}$$

$$OACC = \alpha' + (\beta' x_{CO_2}^{n'} + \gamma')F^{m} \tag{4}$$

Where IACC and OACC are annual investing costs and annual operating costs carbon capture respectively. The symbols of α, β, γ, *n*, *m* are cost parameters for carbon capture as shown in table 4. The cost unit in equation 3 and 4 are expressed as annual cost.

The cost for carbon injection of CCS includes the cost of CO_2 injection and construction of new wells. The wells number required for injection from source *i* to sink *j* is calculated as follows:

$$n_{well\ i,j} = \left\lceil \frac{\sum_i F_{i,j}}{M_{well}} \right\rceil \tag{5}$$

Where $F_{i,j}$ is the flowrate of CO_2 transported from source *i* to sink *j*, and M_{well} is the maximum capacity of a CO_2 injecting well. When the number of wells, the maximum injection capacity and the well-depth, d_{well}, are known, the injection cost can be calculated using equation model developed by Hasan et al., [14].

$$IIC_j = (m_1 d_{well,j} + m_2) \times n_{well,j} \tag{6}$$

$$IAIC_j = \frac{i(1+i)^n}{(1+i)^n - 1} \times IIC_j \tag{7}$$

$$OAIC_j = O\&M_{well} \times IIC_j \tag{8}$$

Where IIC is investment cost of CO_2 injection, and IAIC is the annual investment cost of CO_2 injection. Annual operating and maintenance cost for carbon injection (IAOC) are expressed in terms of a percentage of the investment cost, $O\&M_{well}$, which is 4%. The parameters m_1 and m_2 are given in table 5. When the well data are not available, the cost estimation for carbon storage published by IPCC is used for estimation [15]. The cost for carbon storage are typically between 0.5 – 8.0 US$/ton CO_2, includes investment and operating cost. The upper limit of cost estimation is selected (8.0 US$/ton CO_2), because carbon storage infrastructure and operation in archipelago country are more challenging.

The total annual cost of carbon capture is the sum of annual investment cost and annual operating cost of carbon capture, while total annual cost of carbon injection is the sum of annual investment cost and annual operating cost of CO_2 injection. Therefore, the total annual cost (TAC) of CCS network is total annual cost of carbon capture plus total annual cost of carbon injection:

$$TAC = IACC + OACC + IAIC_j + OAIC_j \tag{9}$$

Referring to the previous study, we assume that the cost of CCS investment and operation will decrease with technology improvement or change with time, hence:

$$C_I^t = C_I^0 \times e^{-At} \tag{10}$$

$$C_{OM}^t = C_{OM}^0 \times e^{-Bt} \tag{11}$$

Where C_I^0 is the capital cost of CCS retrofitting in the base year, C_{OM}^0 is operating cost at initial time after completing retrofitting. C_I^t and C_{OM}^t denote capital cost and operating cost at time t, respectively. α and β are parameters reflecting the effect of technology improvement on capital and operating cost, respectively. Based on the literature, the value of A is 0.0202 and B is 0.057 [16].

In this study, we considered carbon incentive and carbon tax scenario in operations. The magnitude of a carbon tax was set as US$ 20/ton CO_2 based on the government policy to achieve the target of 26% carbon emission reduction in 2020. The amount of carbon incentive will be varied to measure the sensitivity of CCS project value.

Table 4. Common parameter and description for carbon capture cost equation of CCS

Parameters	Description	Value	Notes
α	Investment cost parameter of carbon capture	7719	From Hasan et al. [14]
β	Investment cost parameter of carbon capture	67871	From Hasan et al. [14]
γ	Investment cost parameter of carbon capture	901	From Hasan et al. [14]
n	Investment cost parameter of carbon capture	0.660	From Hasan et al. [14]
M	Investment cost parameter of carbon capture	0.800	From Hasan et al. [14]
α'	Operating cost parameter of carbon capture	0	From Hasan et al. [14]
β'	Operating cost parameter of carbon capture	24088	From Hasan et al. [14]
γ'	Operating cost parameter of carbon capture	0	From Hasan et al. [14]
n'	Operating cost parameter of carbon capture	1	From Hasan et al. [14]
m'	Operating cost parameter of carbon capture	1	From Hasan et al. [14]

Table 5. Common parameter and description for carbon injection cost equation of CCS

Parameters	Description	Value	Notes
m_1	Cost parameter for well const. and injection	1,560,000	From Hasan et al. [14]
m_2	Cost parameter for well const. and injection	1,250,000	From Hasan et al. [14]
A	Learning effect of investment cost	0.0202	From Zhang et al. [16]
B	Learning effect of operating cost	0.057	From Zhang et al. [16]

3. Results and discussion

The next step pinch cascade table was performed for the case study, and the preliminary targets can be obtained by cascade table, as shown in table 6, 7 and 8. The cascade table is used to determine the amount of alternative storage and unutilized storage, and the location of pinch point. The correct cascade table gives an accurate preliminary target for creating the grid diagram.

The cascade table for zero time difference scenarios is shown in table 6. In the first time interval, at year of 5 to 8, there are positive CO_2 loads, because in this interval the magnitude of CO_2 emissions lower than the sink capacity. The excess of sinks capacity can be used for CO_2 storage in a later time interval. The pinch point of the system is obtained in year 20. The amount of alternative storage with zero time difference is 58.572 Mt, while unutilized storage is 28.017 Mt.

Table 6. The cascade table for zero time difference

t (y)	Source, $S_{i,t}$ (Mt/y)				Sink, $D_{j,t}$ (Mt/y)		Δt	CO_2 rate (Mt/y)	CO_2 load (Mt)	Cum. load (Mt)	Feasib. cascade (Mt)
5	SR1	SR2			SK1	SK2				0.000	58.572
	2.403	5.514			4.65	3.76	2	0.493	0.986		
7	↓	↓	SR3		↓	↓				0.986	59.558
	↓	↓	1.575		↓	↓	1	-1.082	-1.082		
8	↓	↓	↓	SR4	↓	↓				-0.096	58.476
	↓	↓	↓	0.501	↓	↓	12	-4.873	-58.476		
20	↓	↓	↓	↓	↓	↓				-58.572	0
	↓		↓	↓	↓	↓	7	0.641	4.487		**pinch**
27	↓		↓	↓	↓	↓				-54.085	4.487
	↓			↓	↓	↓	1	2.216	2.216		
28	↓			↓	↓	↓				-51.869	6.703
	↓				↓	↓	2	6.007	12.014		
30	↓				↓	↓				-39.855	18.717
					↓		5	4.650	23.250		
35					↓					-16.605	41.967

Figure 1 shows the grid diagram for zero time difference. The pinch point divides the grid diagram into two zones. The first is below the pinch, the zone in the left side of the pinch line, and there is the alternative storage. The second is above the pinch, the zone in the right side of the pinch line, and there is the unutilized storage. It is shown in figure 1, below the pinch the carbon exchange occurs from SR1 to SK2, SR2 to SK1 and SK2, and SR3 to SK2. The excess of CO_2 emission can be obtained from SR3 and SR4 as alternative storage. Above the pinch the carbon is transferred from SR1 to SK1, SR3 to SK1, and SR4 to SK1 and SK2. The excess of CO_2 storage can be obtained from SK1 and SK2 as unutilized storage. There is no carbon transfer from SR2, because the operation has finished in year 20. The total amount of alternative storage and unutilized storage are equal to the result from the cascade table.

The cascade table with 5 years of time difference is shown in table 7. In this scenario, the operating time of sink is shifted for 5 years. The sinks operating time change from year 5-35 to year 10-40 for SK1 and from year 5-30 to year 10-35 for SK2. The amount of alternative storage required for the system increased to 100.622 Mt, while unutilized storage increased to 93.317 Mt. The pinch point for the system remains the same with zero time difference, which is in year 20. However, the sink pinch point is in year 15. The sink pinch point is obtained by subtracting pinch point with time difference. There is no positive CO_2 load in this scenario, it means the CO_2 emission larger than feasible CO_2 storage in all-time interval.

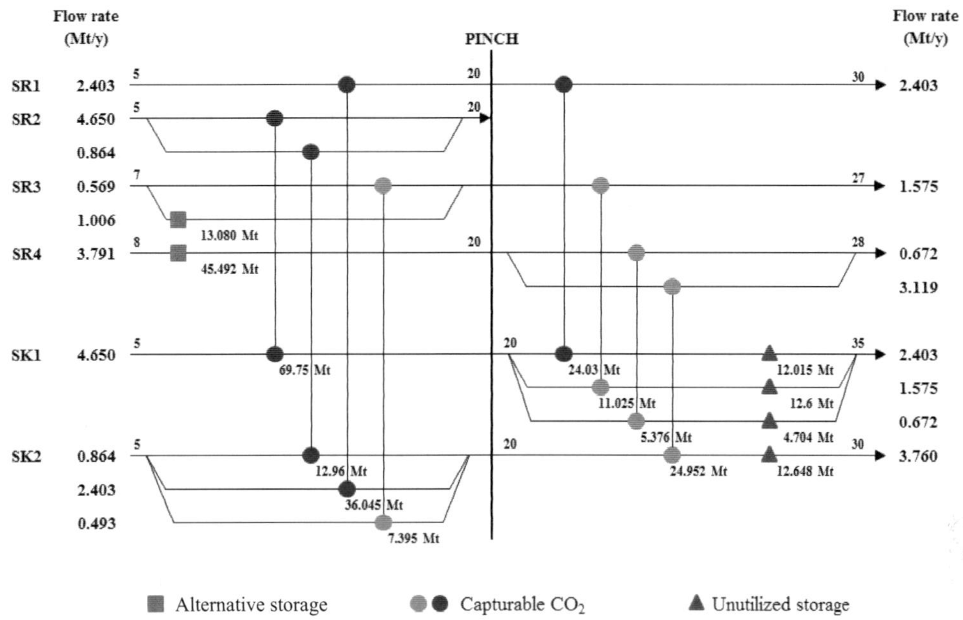

Figure 1. Grid diagram for CCS system with zero time difference

Table 7. The cascade table for five years of time difference

t (y)	Source, $S_{i,t}$ (Mt/y)				Sink, $D_{j,t}$ (Mt/y)		Δt	CO_2 rate (Mt/y)	CO_2 load (Mt)	Cum. load (Mt)	Feasib. cascade (Mt)
5	SR1	SR2								0.000	100.62
	2.403	5.514					2	-7.917	-15.834		
7	↓	↓	SR3							-15.83	84.788
	↓	↓	1.575				1	-9.492	-9.492		
8	↓	↓	↓	SR4						-25.326	75.296
	↓	↓	↓	0.501			2	-13.28	-26.566		
10	↓	↓	↓	↓	SK1	SK2				-51.892	48.73
	↓	↓	↓	↓	4.65	3.76	10	-4.873	-48.730		
20	↓	↓	↓	↓	↓	↓				-100.62	0
	↓		↓	↓	↓	↓	7	0.641	4.487		**pinch**
27	↓		↓	↓	↓	↓				-96.135	4.487
	↓			↓	↓	↓	1	2.216	2.216		
28	↓			↓	↓	↓				-93.919	6.703
	↓				↓	↓	2	6.007	12.014		
30	↓				↓	↓				-81.905	18.717
					↓	↓	5	8.410	42.050		
35					↓	↓				-39.855	60.767
					↓		5	4.650	23.250		
40					↓					-16.605	84.017

Figure 2 shows the grid diagram for five years of time difference. There are some differences between the grid diagram for zero and five years of time difference. Below the pinch, time availability of all sinks decreases from 5-20 to 5-15. Above the pinch, time availability of SK1 increase from 20-35 to 15-35, and time availability of SK2 increase from 20-30 to 15-30, while time availability of sources remains constant. The notable pairings below the pinch are all of CO_2 from SR1 and SR3, which are 36.045 Mt and 20.475 Mt, respectively, are not captured at all. The carbon exchange occurs from SR2 to SK1 and from SR4 to SK2. There is an increase in alternative storage. The alternative storage comes from SR1 to SR4 as 36.045 Mt, 36.21 Mt, 20.475 Mt, and 7.892 Mt respectively. Above the pinch carbon is transferred from SR1 to SK2, SR3 to SK1 and SK2, and SR4 to SK2. The unutilized storage can be obtained from SK1 and SK2 as 61.146 Mt and 22.871 respectively. The total amount of alternative storage and unutilized storage are equal to the result from cascade table.

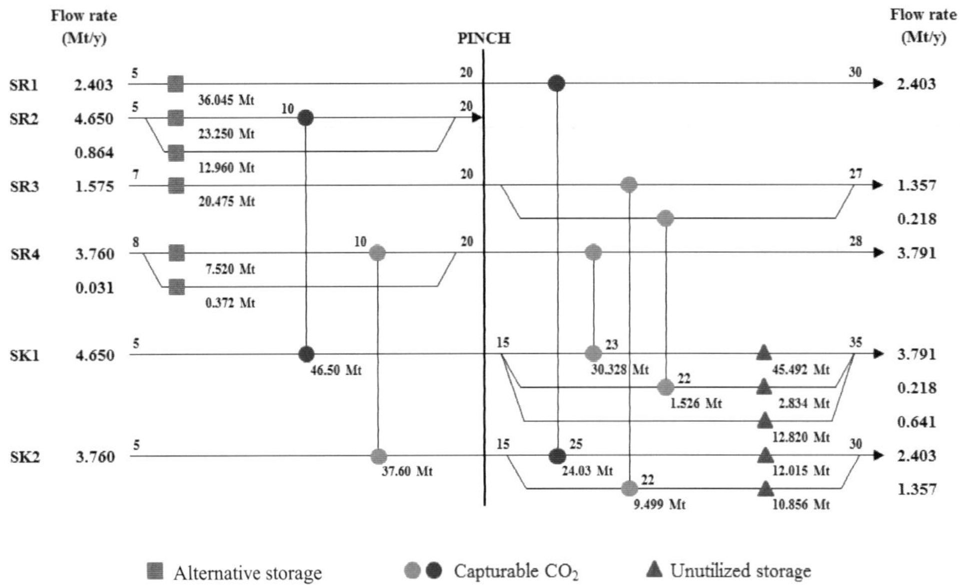

Figure 2. Grid diagram for CCS system with 5 years of time difference

The last cascade table with ten years of time difference is shown in table 8. In this scenario, the operating time of sink is shifted for 10 years. The amount of alternative storage required for the system is 142.672 Mt, almost three times the amount of alternative storage with zero time difference, while unutilized storage is 158.617 Mt. The pinch point for the system remains constant, which is in year 20. However, the sink pinch point is in year 10. The sink pinch point is obtained by subtracting pinch point with time difference as 10 years. There is no positive CO_2 load in this scenario, it means the CO_2 emission larger than feasible CO_2 storage in all time interval.

Figure 3 shows the grid diagram for ten years of time difference. Below the pinch, there are only 5 years of time availability of all sinks, from year 5 to year 10. Above the pinch, time availability of SK1 increases to 25 years for SK1 and 20 years for SK2, while time availability of sources remains constant. The carbon transfer occurs from SR2 to SK1 and from SR4 to SK2, below the pinch. The alternative storage comes from SR1 to SR4 as 36.045 Mt, 59.46 Mt, 20.475 Mt, and 26.692 Mt respectively. Above the pinch carbon is transferred from SR1 to SK2, SR3 to SK1, and SR4 to SK1 and SK2. The unutilized storage can be obtained from SK1 and SK2 as 85.753 Mt and 40.314 respectively.

Table 8. The cascade table for 10 years of time difference

t (y)	Source, $S_{i,t}$ (Mt/y)				Sink, $D_{j,t}$ (Mt/y)		Δt	CO_2 rate (Mt/y)	CO_2 load (Mt)	Cum. load (Mt)	Feasib. cascade (Mt)
5	SR1	SR2								0.000	142.67
	2.403	5.514					2	-7.917	-15.834		
7	↓	↓	SR3							-15.834	126.84
	↓	↓	1.575				1	-9.492	-9.492		
8	↓	↓	↓	SR4						-25.326	117.35
	↓	↓	↓	0.501			7	-13.28	-92.981		
15	↓	↓	↓	↓	SK1	SK2				-118.31	24.365
	↓	↓	↓	↓	4.65	3.76	5	-4.873	-24.365		
20	↓	↓	↓	↓	↓	↓				-142.67	0.0
	↓		↓	↓	↓	↓	7	0.641	4.487		**pinch**
27	↓		↓	↓	↓	↓				-138.18	4.487
	↓			↓	↓	↓	1	2.216	2.216		
28	↓			↓	↓	↓				-135.97	6.703
	↓				↓	↓	2	6.007	12.014		
30	↓				↓	↓				-123.96	18.717
					↓	↓	10	8.410	84.100		
40					↓	↓				-39.855	102.82
					↓		5	4.650	23.250		
45					↓					-16.605	126.07

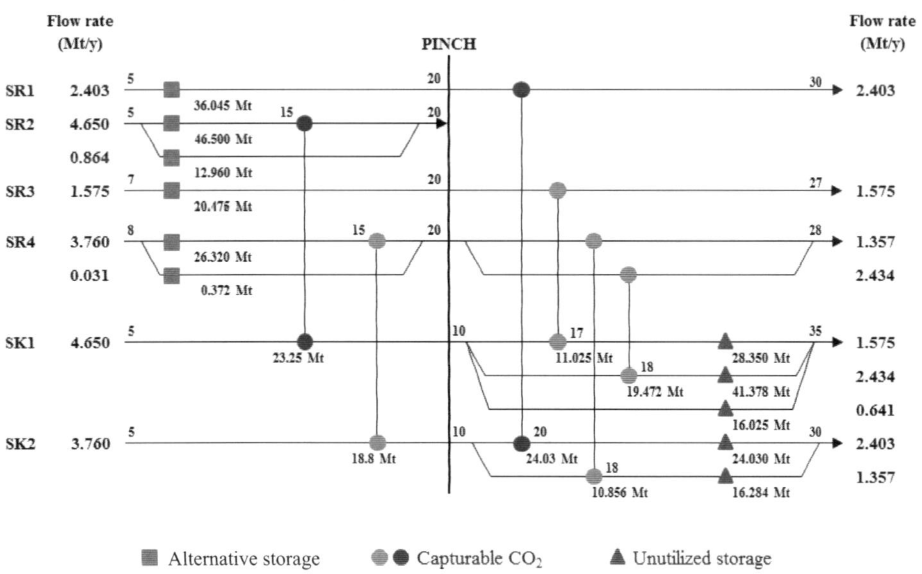

Figure 3. Grid diagram for CCS system with 10 years of time difference

The result of CCS cost estimation for the central part of Indonesia is tabulated in table 9. Total annual cost (TAC) of three CCS network designs are US$556.48 million for zero time difference, US$478.52 million for 5 years of time difference, and US$476.33 million for 10 years of time difference. Since there is only slight difference in TAC between 5 years and 10 years, it means the cost per unit mass of CO_2 is too high, so the design is not effective. The carbon incentive and carbon tax are used to assess the cost sensitivity of the CCS network design implementation. Several European countries give carbon incentive through certified emission reduction (CER) with fluctuating Price. As noted, the CER prices fluctuate from US$20 per ton CO_2 before financial crisis to less than US$3 per ton. The World bank reported that a half of current emission covered by carbon pricing initiative are priced at less than US$10 per ton CO_2e. We use the initial carbon prices of US$8 per ton CO_2 for carbon trading and incentive to assess the sensitivity design. A tax rate on carbon emission is set at price US$20 per ton CO_2.

Table 9. The CCS costs and carbon pricing for central part of Indonesia

	0 year	5 year	10 year
TAC (US$)	556.48	478.52	476.33
Carbon tax (US$)	1191.16	2012.44	2853.44
Carbón incentive (US$)	1886.13	1675.76	859.464
Surplus/déficit	138.49	(815.20)	(2470.30)

When the carbon tax and incentive are implemented, then the two scenarios, namely 5 years and 10 years of time difference will deficit in the financing, while zero time difference scenario will surplus. But, as noted before, the transportation costs are not taken into account. This increasingly shows that time delay in CCS Project will be very costly. As shown in table 9, the carbon tax will increase due to time difference. The longer the time difference, the greater carbon tax due to the large amount of uncaptured carbon emission is and the less carbon incentive due to the less carbon stored.

4. Conclusion

Our results suggest that it is possible to use the combination of pinch analysis with carbon pricing mechanism for CCS planning and design. Different CCS costs for time difference scenarios suggest that management strategies for CCS should be done properly. The reason is time delay in CCS project will affect the CCS supply chain and increase in cost. In this study a CCS network design only considers the costs related to CO_2 capture and storage. However, the combination with carbon incentive and carbon tax can provide a brief description of overall CCS system costs. To obtain more accurate cost calculation, the further studies on CCS transportation costs are needed. The method is currently applied to planning and design CCS network in central part of Indonesia, which represents a multi-region and multi-period CCS network system. In general, the method can also be applied to both single region and multi-region CCS network systems. Variations of carbon trading and carbon tax can be applied by this method so that the most attractive carbon price for the industry can be obtained.

Acknowledgment: We thank to LPPM ITS for the financial support through Penelitian LAB Based Education (LBE) under contract number 911/PKS/ITS/2019

References

[1] Yusuf A A and Resosudarmo B P 2014 On the distributional impact of a carbon tax in developing countries : the case of Indonesia *Environ. Econ. Policy Stud.*

[2] Leung D Y C, Caramanna G and Maroto-Valer M M 2014 An overview of current status of carbon dioxide capture and storage technologies *Renew. Sustain. Energy Rev.* **39** 426–43

[3] Diamante J A R, Tan R R, Foo D C Y, Ng D K S, Aviso K B and Bandyopadhyay S 2014 Unified pinch approach for targeting of carbon capture and storage (CCS) systems with multiple time

periods and regions *J. Clean. Prod.* **71** 67–74

[4] Tan R R and Foo D C Y 2007 Pinch analysis approach to carbon-constrained energy sector planning *Energy* **32** 1422–9

[5] Tan R R, Ng D K S, Foo D C Y and Aviso K B 2009 Pinch analysis approach to carbon constrained planning for sustainable power generation *J. Clean. Prod.* **17** 940–4

[6] Tjan W, Tan R R and Foo D C Y 2010 A graphical representation of carbon footprint reduction for chemical processes *J. Clean. Prod.* **18** 848–56

[7] Putra A A, Juwari and Handogo R 2018 Multi Region Carbon Capture and Storage Network in Indonesia Using Pinch Design Method *Process Integr. Optim. Sustain.* **2** 321–41

[8] Rubin E S, Short C, Booras G, Davison J, Ekstrom C, Matuszewski M and McCoy S 2013 A proposed methodology for CO2 capture and storage cost estimates *Int. J. Greenh. Gas Control* **17** 488–503

[9] Wang X and Zhang H 2018 Optimal design of carbon tax to stimulate CCS investment in China's coal-fired power plants : A real *Greenh. Gases Sci. Technol.* **13** 1–13

[10] Iskandar U P, Usman U and Sofyan S 2013 Ranking of Indonesia sedimentary basin and storage capacity estimates for CO2 geological storage *Energy Procedia* **37** 5172–80

[11] Indonesia CCS Study Working Group 2009 *Understanding Carbon Capture and Storage Potential in Indonesia*

[12] Diamante J A R, Tan R R, Foo D C Y, Ng D K S, Aviso K B and Bandyopadhyay S 2014 Unified pinch approach for targeting of carbon capture and storage (CCS) systems with multiple time periods and regions *J. Clean. Prod.* **71** 67–74

[13] Thengane S K, Tan R R, Foo D C Y and Bandyopadhyay S 2019 *A Pinch-Based Approach for Targeting Carbon Capture, Utilization, and Storage Systems* vol 58

[14] Hasan M M F, Boukouvala F, First E L and Floudas C A 2014 Nationwide, regional, and statewide CO2 Capture, Utilization, and Sequestration supply chain network optimization *Ind. Eng. Chem. Res.* **53** 7489–506

[15] IPCC 2005 *IPCC Special Report on Carbon Dioxide Capture and Storage. Prepared by Working Group III of the Intergovernmental Panel on Climate Change* ed B Metz, O Davidson, H De Coninck, M Loss and L A Meyer (New York: Cambridge University Press)

[16] Zhang X, Wang X, Chen J, Xie X, Wang K and Wei Y 2014 A novel modeling based real option approach for CCS investment evaluation under multiple uncertainties *Appl. Energy* **113** 1059–67

Modelling of Saltwater Evaporation using WAIV

Henrietta[1], H Santoso[1] and J R Witono[1]

[1]Department of Chemical Engineering, Parahyangan Catholic University, Jalan Ciumbuleuit 94, Bandung 40141, INDONESIA

E-mail: hsantoso@unpar.ac.id

Abstract. Indonesia as a country, which has the second longest coastline in the world, has a big opportunity to improve its salt production. The productivity of salt production in Indonesia from 2011 to 2014 was around 67, 92, 40, and 90 tons/ha/year, respectively. This low productivity was mainly due to the method of salt production used, which was very traditional and depended mainly on the availability of solar energy for evaporation. The longer the dry season was, the higher the salt production would be in that year, and vice versa. Since the performance of evaporation process holds the key to the improvement of productivity of salt production, it is important to study a method that can speed up the evaporation process in the salt production. One of the recent methods that can be used to achieve that purpose is the Wind Aided Intensified Evaporation (WAIV) process. In this study, a model for saltwater evaporation in a WAIV unit using Harbeck equation is proposed. Two set of data for saltwater evaporation in a WAIV unit was extracted from literature. The first set of data was used to obtain the parameters of Harbeck equation by minimizing the sum of error squares of the cumulative evaporation rate in a WAIV unit compared to its model. The second set of data was used to validate the model. It is found that the proposed model can estimate the evaporation rate in a WAIV unit based on the surrounding climate data reasonably well.

1. Introduction

Indonesia has the second longest coastline in the world. With this geographical resource, Indonesia is expected to be able to fulfil its domestic demands for salts. However, until now Indonesia still imports salts from overseas. The main reason for this problem is the low productivity of salt production in Indonesia. Table 1 shows the salt production data from 2011 to 2014 [1].

Table 1. Salt production in Indonesia from 2011 to 2014.

	Year			
	2011	2012	2013	2014
Production (tons/year)	1,623,786	2,473,716	1,163,608	2,502,891
Land areas (ha)	20,066	22,632	25,098	23,411
Productivity (tons/ha/year)	67.27	91.70	39.62	89.72
Dry season (months)	4	5	2	5
Demands (tons)	3,228,750	3,270,086	3,573,954	3,611,990

From Table 1, the domestic demands for salts in Indonesia increased every year while the productivity of salt production was low and unstable. This low and unstable productivity was due to the very traditional method used in the salt production, which relied heavily on the weather conditions for evaporating the seawater. Hence, the longer the dry season was in a particular year, the higher the salt production would be, and vice versa.

In general, the evaporation rate of saltwater is affected by the climatic conditions (i.e. temperature, humidity, solar radiation and wind speed), the properties of saltwater (i.e., concentration and composition of saltwater) and the size of saltwater body [2]. The climatic conditions and the properties of saltwater depend on the geographical location and the nature of saltwater available in that location. These factors are practically given by nature and cannot be modified easily. On the other hand, the size of saltwater body is a factor that can be modified using a certain method which in turn can increase the evaporation rate significantly. One of the recent methods that can be used to achieve that purpose is the Wind Aided Intensified Evaporation (WAIV) process.

The Wind Aided Intensified Evaporation (WAIV) process is a process that uses wind energy to increase the evaporation rate of liquid. The WAIV unit consists of a structure that holds several vertically hung hydrophilic sheets that serve as an enhanced surface area for evaporation. As the liquid flows down and spreads slowly on the hydrophilic sheets, it contacts with the wind blowing pass the sheets and evaporates with a significantly higher evaporation rate per footprint area compare to the traditional evaporation pond [3].

In this study, a model for saltwater evaporation in a WAIV unit using Harbeck equation is proposed. Two set of data for saltwater evaporation in a WAIV unit was extracted from literature. The first set of data was used to obtain the parameters of Harbeck equation by minimizing the sum of error squares of the cumulative evaporation rate in a WAIV unit compared to its model. The second set of data was used to validate the model.

2. Methodology

In this study the saltwater evaporation in a WAIV unit was modelled using the mass transfer model. The mass transfer model is one of the oldest models that are commonly used to model water evaporation. This model has a simple form. The parameters of model can be easily determined. The model also can provide a reasonably good result. The mass transfer model for free water surface can be written as follows [4]:

$$E = C\,(e_s - e_a) \tag{1}$$

where E is the free water surface evaporation rate (mm day^{-1}), e_s is the saturated vapour pressure at the temperature of the water surface (mbar), e_a is the vapour pressure in the air (mbar), and C is a coefficient that is dependent on the wind speed, the size of water body, the roughness of water surface, the thermally induces turbulence, the atmospheric stability, the barometric pressure, the density and the kinematic viscosity of the air [4, 5, 6].

The mass transfer driving force for evaporation is due to the higher vapour pressure of the water surface compared to the vapour pressure of the surrounding air [7]. If the evaporation rate can be assumed to be proportional to the wind speed u (mm day^{-1}) [5], Equation 1 can be expressed as:

$$E = N\,u\,(e_s - e_a) \tag{2}$$

where coefficient N is given by the following correlation:

$$N = 3.367 \times 10^{-9} \times A^{-0.05} \tag{3}$$

where A is the size of water body (m^2). Equation 2 is known as Harbeck equation. The correlation given in Equation 3 holds for water body of the size between 4000 m^2 and 120 km^2 [7].

Saturated vapour pressure e^* (mbar) at the temperature of the water surface and at the temperature of the air can be calculated using the modified Clausius-Clapeyron equation where temperature T are expressed in degree Kelvin [8]:

$$ln\, e^* = 53{,}67957 - \frac{6743{,}769}{T} - 4{,}8451\, ln\,(T) \tag{4}$$

For evaporation of saltwater in the WAIV process, the mass transfer driving force for evaporation is the difference between the saturated vapour pressure of the saltwater at the wetted surface and the vapour pressure of the surrounding air blowing pass the wetted surface. The temperature at the surface of the falling saltwater can be estimated using the wet bulb temperature of the air moving through the WAIV array. The mass transfer driving force for evaporation $(e_s - e_a)$ then can be calculated using the following equations [8]:

$$(e_s - e_a) = \left[\left(e^*_{TWB}\right)\left(\frac{e}{e^*}\right)_{TWB} - \left(e^*_{Ta}\right)(RH)\right] \tag{5}$$

$$\left(\frac{e}{e^*}\right) = \mathrm{a_W} \tag{6}$$

where T_a, RH, and T_{WB} are the temperature, the relative humidity, and the wet bulb temperature of the air moving through the WAIV array, respectively while a_W is the water activity of the saltwater.

The WAIV data used in this study was taken from the demonstration trial conducted on the WAIV module at Queensland, Australia [3]. The WAIV module uses around 2850 m^2 sheets, thus has approximately 5700 m^2 wetted surface for evaporation. The dimensions of the full scale WAIV module is illustrated in Figure 1.

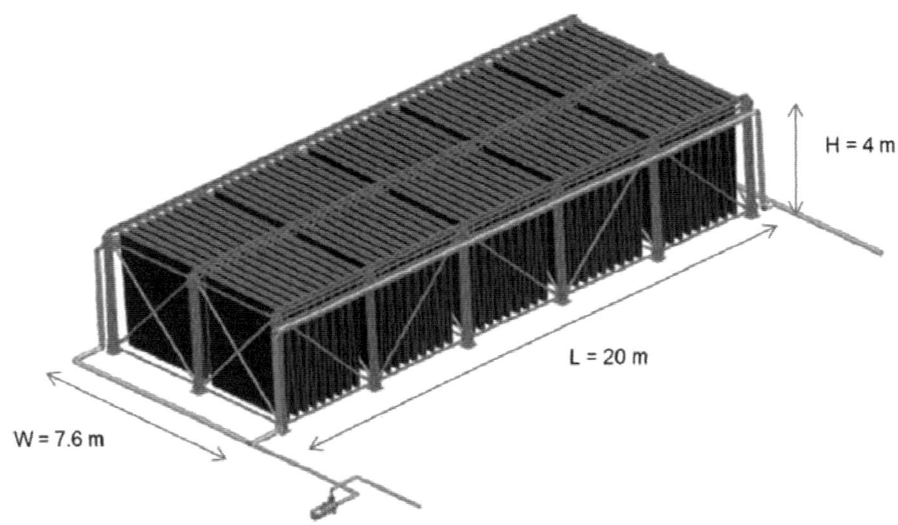

Figure 1. The WAIV module [3].

The feed water for the WAIV unit was taken from the reverse osmosis reject saltwater of desalination plant. The composition of the feed water was shown on Table 2. The evaporation performance of the WAIV unit were observed for 24 hours. The results from February 21st, 2013 and and November 20th, 2012 were plotted in the graphs shown in Figure 2 and Figure 3 [3].

Table 2. The feed water composition for WAIV unit [3].

Parameter	Unit	Min	Max	Average
pH		9.37	9.62	9.49
Temperature	°C	18.8	30.9	25.1
Total Alkalinity (as $CaCO_3$)	mg/L	2160	3110	2593
Calcium (Ca)	mg/L	3	13	7
Magnesium (Mg)	mg/L	2	6	4
Sodium (Na)	mg/L	2820	4430	3513
Potassium (K)	mg/L	50	89	65
Chloride (Cl)	mg/L	2970	5260	3995
Sulfate (SO_4^{2-})	mg/L	85	524	275
Boron (B)	mg/L	0.75	1.02	0.87
Fluoride (F)	mg/L	4.9	9.7	8.6
Total Dissolved Solid (TDS)	mg/L	8230	11200	9648
Electrical Conductivity (EC)	mS/cm	13000	18500	15300

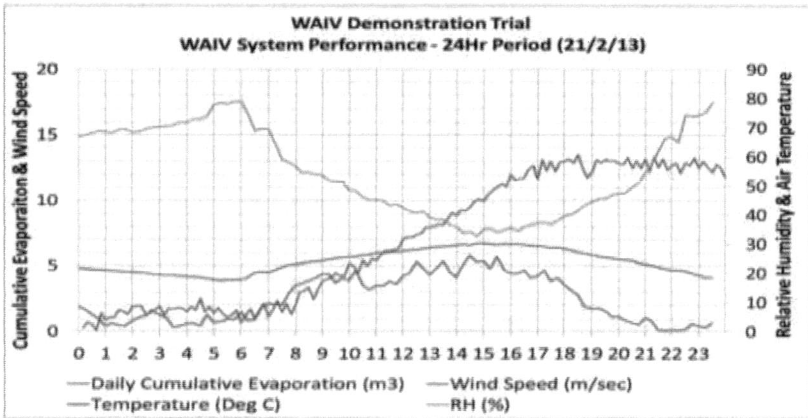

Figure 2. Evaporation performance of the WAIV unit on February 21st, 2013 [3].

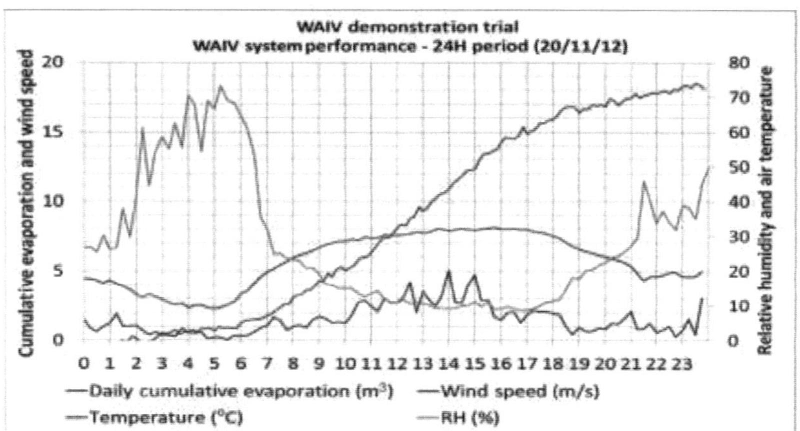

Figure 3. Evaporation performance of the WAIV unit on November 20th, 2012 [3].

The evaporation performance data of the WAIV unit on February 21st, 2013 was used to obtain the parameters of Harbeck equation by minimizing the sum of error squares of the cumulative evaporation rate in a WAIV unit compared to its model. The evaporation performance data of the WAIV unit on November 20th, 2012 was used to validate the model.

3. Results and discussion

The parameters of Harbeck equation obtained by minimizing the sum of error squares of the cumulative evaporation rate of the WAIV unit on February 21st, 2013 is given in Equation 7.

$$N = 3.719 \times 10^{-9} \times A^{-0.0459} \tag{7}$$

The coefficient of determination (R^2) for the model is 0.9074.

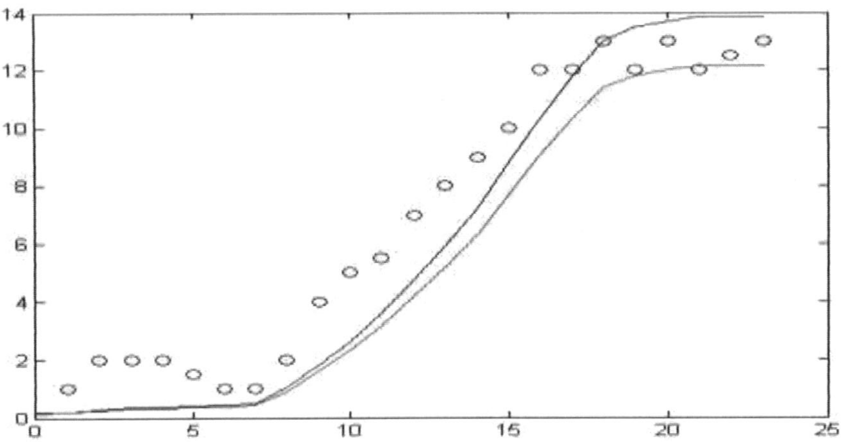

Figure 4. The comparison between the actual cumulative evaporation rate on February 21st, 2013 (blue dot) and Harbeck model calculated using the parameters given by Equation 3 (red line) the parameters given by Equation 7 (black line).

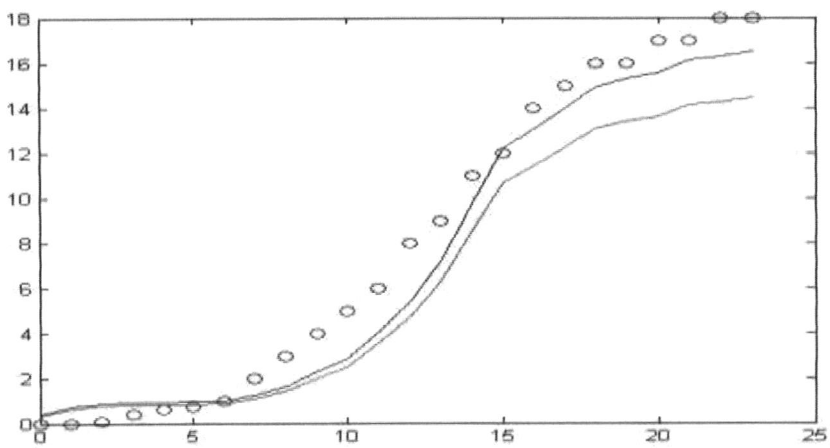

Figure 5. The comparison between the actual cumulative evaporation rate on November 20th, 2012 (blue dot) and Harbeck model calculated using the parameters given by Equation 3 (red line) the parameters given by Equation 7 (black line).

Figure 4 shows the comparison between the actual cumulative evaporation rate in the WAIV unit on February 21st, 2013 shown as blue dot with the predicted cumulative evaporation rate calculated using Harbeck model. The red line is the cumulative evaporation rate calculated using Harbeck equation with the parameters given by Equation 3 while the black line is the cumulative evaporation rate calculated using Harbeck equation with the parameters given by Equation 7. It can be seen that the Harbeck model calculated using the parameters given by Equation 7 is slightly better that the Harbeck model calculated using the parameters given by Equation 3.

The parameters of Harbeck model given in Equation 3 and 7 are then tested to predict the cumulative evaporation rate in the WAIV unit on November 20th, 2012. Figure 5 shows the comparison between the actual cumulative evaporation rate in the WAIV unit on November 20th, 2012 shown as blue dot with the predicted cumulative evaporation rate calculated using Harbeck model with the red line represents the cumulative evaporation rate calculated using the parameters given by Equation 3 and the black line represents the cumulative evaporation rate calculated using the parameters given by Equation 7. The Harbeck model calculated using the parameters given by Equation 7 again is slightly better that the Harbeck model calculated using the parameters given by Equation 3.

The Harbeck model calculated using the parameters given by Equation 3 tends to give an underestimate prediction of the cumulative evaporation rate in the WAIV unit. In the WAIV unit, the evaporating saltwater is not static but continuously flowing. Therefore, the evaporation rate is slightly higher than the evaporation rate for conventional pond with a similar size of evaporation surface.

4. Conclusion

A model for saltwater evaporation in a WAIV unit has been developed. This model is based on Harbeck equation. The parameter for this model is obtained by minimizing the sum of error squares of the cumulative evaporation rate in a WAIV unit. When tested using the actual cumulative evaporation rate in the WAIV unit, the proposed model can estimate the evaporation rate in the WAIV unit reasonably well using the surrounding climate data.

References

[1] *Produksi Garam Indonesia* [Internet] [cited: 20th November 2019] Available from: https://bulelengkab.go.id/assets/instansikab/126/bankdata/data-statistik-produksi-garam-indonesia-tahun-2014-56.pdf

[2] Dama-Fakir P and Toerien A 2009 Evaporation rates on brine produced during membrane treatment of mine water *Int. Mine Water Conf.* 19th - 23rd October 2009 Pretoria - South Africa pp 666-670

[3] Murray B McMinn D and Gilron J 2015 Waiv technology: An alternative solution for brine management: Results of a full-scale demonstration trial conducted at a location near Roma in Queensland *Journal of the Australian Water Association* vol 42 no 5 pp 77-83

[4] Singh V P and Xu C Y 1997 Evaluation and Generalization of 13 Mass-Transfer Equations for Determining Free Water Evaporation *Hydrological Processes* vol 11 pp 311-323

[5] Pallavi K and Rajeev S 2016 Evaluation of evaporation models Based on climate factors *International Journal of Civil Engineering and Technology* vol 7 no 4 pp 111–120

[6] Harbeck G E Jr. 1962 *A Practical Field Technique for Measuring Reservoir Evaporation Utilizing Mass-Transfer Theory* United States Government Printing Office Washington DC

[7] Gilron J Folkman Y Savliev R Waisman M and Kedem O 2003 WAIV-wind aided intensified evaporation for reduction of desalination brine volume. *Desalination*, vol 158 pp 205-214

[8] Katzir L Volkmann Y Daltrophe N Korngold E Mesalem R Oren Y and Gilron J 2010 WAIV-Wind aided intensified evaporation for brine volume reduction and generating mineral byproducts *Desalination and Water Treatment* vol 13 no 1-3 pp 63-73

IOP Conf. Series: Materials Science and Engineering **742** (2020) 012014 doi:10.1088/1757-899X/742/1/012014

Improvement of Biogas Production by Sequential Pretreatment of Rice Straw and Coconut Shell for Power Generation Applications

A Magomnang[1] and S Capareda[2]

[1]Mechanical Engineering Department, University of Science and Technology of Southern Philippines, Cagayan de Oro City, Philippines
[2]Biological and Agricultural Engineering Department, Texas A&M University, College Station, Texas, USA

E-mail : a_magomnang@yahoo.com

Abstract. Different pretreatment methods (milling, ultrasonication, liquid hot water, and alkali hydrolysis) were evaluated on rice straw and coconut shell biomass as a substrate for biogas production for vehicle fuel, provide heat and generate electricity. This was achieved by the sequential combination of ultrasonication, hot water, and alkali hydrolysis treatment at (4% and 5% w/v) concentration was evaluated. Additionally, Fourier Transform Infrared (FT-IR) spectroscopy was used in identifying structural differences and identification of compounds in the treated biomass. The results show that the combination of ultrasonication-alkali hydrolysis, hot liquid water using 3% of NaOH improved methane yield of 150% for rice straw and 290% for coconut shell compared with the original treatment. Moreover, the Fourier-transform infrared spectroscopy (FT-IR) methodology identified modifications in the biomass structure after different types of pretreatment and conditions. The results of this research are the initial steps for the development of new processes using the Ultrasonication-NaOH LHW pretreatment, for the production of biogas from agricultural biomass waste.

1. Introduction

Renewable energy produced from lignocellulosic materials (i.e. agriculture and forest residues), non-food crops (i.e. algae and grasses) or industrial waste and residue streams have considerably lower greenhouse gas emissions (GHGs) due to its closed carbon cycle. The use of second-generation feedstocks will diminish the concern over competition between food versus fuel and will lessen the potential impact on food supply.

At present, feedstocks from the municipal solid waste, industry, food processing plants, and wastewater treatment plant, are limited; thus, new renewable sources are sought after. With the abundant availability of agricultural biomass in the Philippines makes it a highly interesting feedstock sourced from a variety of crops like palay, corn, sugarcane, cassava, and coconut; and livestock such as carabao, cattle, hog, goat, and dairy, that can be utilized for biogas production. These sources can serve as fuel for power generation or as feedstock for advanced biofuels production. Given the abundant supply of biomass raw materials in the country, it would be advantageous for the Philippines to exploit these resources to address the country's energy dependence, mitigate climate change and eventually achieve economic growth and prosperity.

Biogas is a renewable, high-quality fuel which can be produced from various organic raw materials and used for various energy services. Biogas technology has been developed and widely used over the world because it has several advantages – reduction of the dependence on non-renewable resources, high energy-efficiency, environmental benefits, available and cheap resources to feedstock, relatively easy and cheap technology for production, and extra values of digestate as a fertilizer. But the current status of biogas production and utilization varies largely among continents.

To overcome the issues related to the utilization of lignocellulosic materials like rice straw and coconut shells for biogas production, new pretreatment strategies have been evaluated and developed. One of these strategies is the combination of sodium hydroxide hydrolysis, ultrasonication, and liquid hot water pretreatment. Typically, the ultrasonication and liquid hot water pretreatment were used to modify the lignin and cellulose structure of the biomass to generate ethanol but for this case, said pretreatments were done to improve biogas yield. In a study by sodium hydroxide hydrolysis was found to substantially increase the biogas yield of rice straw. Currently, there are type of research on the pretreatment of biomass wastes similar to paddy straw using sodium hydroxide-microwave pretreatment. However, a combination of these pretreatments (ultra-sonication, liquid hot water, and NaOH hydrolysis), has not been tested in rice straw (RS) and coconut shell (CS) biomass wastes. Thus, the synergic effect of these technologies may raise de-lignification, cellulose conversion, and biogas yield from RS & CS.

Anaerobic co-digestion experiments were performed in this study to provide information on the long-term effects, such as eventual inhibition during the digestion of a lignocellulosic substrate. A co-digestion is often beneficial, since it supplies the system with more nutrients, leading to a better balance in the C/N ratio and the pH. It also improves the stability of the process and increases methane yield due to positive synergistic effects thereby increasing the economic value of the biogas. This study aims to determine the effects of various pretreatment methods (alkalinity using NaOH, ultrasonication, liquid hot water using autoclave) on the biomass properties and biogas yield in the anaerobic co-digestion of agricultural biomass waste utilizing cattle (cow) manure co-digested with different biomass feedstocks [1] such as rice straw, coconut shell, and sewage sludge as inoculum for power generation applications.

This study investigated the production of biogas from the anaerobic co-digestion of cattle manure with various feedstocks such as rice straw, coconut shell. Due to ammonia inhibition, different biomass pretreatment methods were employed to increase biogas yield. The feedstocks were subjected to liquid hot water pretreatments with NaOH via ultrasonication. It was noted that methane yields increased by two or threefold from the pretreatments of rice straw and coconut shells, respectively. The results of this study can contribute to the development of new processes using the ultrasonication-NaOH hydrolysis and LHW pretreatment for the production of biogas from agricultural biomass waste.

This study aimed to develop a zero-waste technology that optimizes resource recovery of agricultural wastes to produce bioenergy efficiently and in an environment-friendly approach. This could address the issues of power shortage being observed in Mindanao, Philippines by utilizing as a fuel for the stationary non-road engines and engine-generator units.

2. Materials and Methods

2.1. Biomass waste pretreatment for anaerobic co-digestion process

In this research, three different pretreatments (alkaline-ultrasonication, hot water) and their combinations were evaluated to obtain the best biogas production from rice straw and coconut shell. The experiment followed the selected sequence of pretreatment such as ultrasonication, liquid hot water. However, the ultrasonication step was modified to simultaneously perform basic hydrolysis using the different concentration of NaOH. The experimental procedure in this study was shown in Figure 1 and the experimental design was completely randomized with the NaOH concentration as a factor with two levels (3 and 4% w/v) control of unpretreated rice straw and coconut shell. All the experiments were developed in two replicates, using yield as response variables. The ultrasonicator (Hielscher Ultrasonic Processors, Ringwood, NJ, USA) was set at the highest value of amplitude (100%) and cycle (1). The biomass was not washed before the hot water treatment, thus remaining particles of NaOH can be present for the hot water treatment. The liquid hot water pretreatment used Erlenmeyer flasks with 10% solution solids at 121°C, 1.02 atm for 1 h in an autoclave.

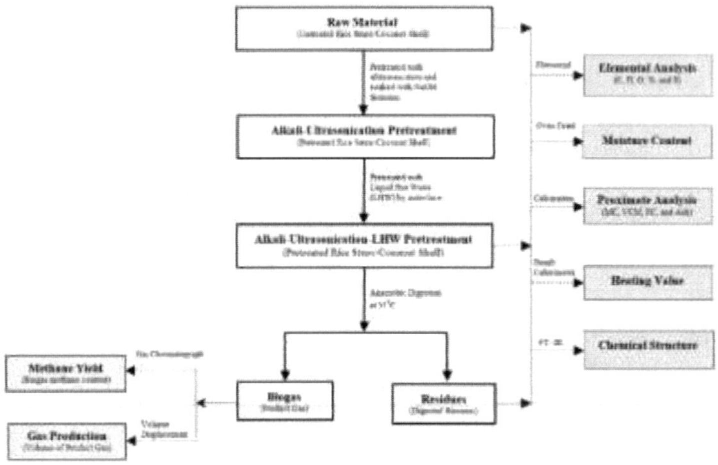

Figure 1. Schematic diagram of the biomass pretreatment experimental procedure conducted in this study

2.2. *Sequential pretreatment process*

The experiment follows a sequence of pretreatments (ultrasonication, liquid hot water, and alkaline treatment) through NaOH [2]-[7]. However, for this experiment, the ultrasonication step was modified to simultaneously perform basic hydrolysis using different concentrations of NaOH. Moreover, the experiment was completely randomized with the NaOH concentration as the factor of 3 and 4% w/v and control of untreated (pretreated) rice straw and coconut shell. All the experiments were developed in two replicates and biogas yield considered as response variables.

As shown in Table 1, rice straw and coconut shell samples were subjected to ultrasonication process with a solution of 10% biomass solids. It was set at the highest value of 100% amplitude and full cycle as the waves create cavitation in the liquid medium (NaOH) which contains 250 grams of rice straw sample that was placed in 3000 ml Erlenmeyer flask manually stirred, then sonicated for 60 min. After sonication, it was then transferred to the autoclave and subjected it for 121°C, 15 psi for 1 h. After cooling from the autoclave, it soaked the biomass-NaOH solution for three days at a controlled temperature of 25°C. The liquid solution was drained off and the biomass was filtered and washed with deionized water until the pH becomes neutral. Then, the washed rice straw was dried overnight in an oven dryer at 105°C. Pretreated and dried rice straw and coconut shell samples were stored in polyethylene bags and stored for the subsequent test such as proximate analysis (total solids, volatile solids, and ash) and compound determination using FT-IR. The control biomass (untreated rice straw and coconut shell) was analyzed simultaneously to determine the extent of degradation.

Table 1. Pretreatment evaluated in the anaerobic co-digestion experiments

Biomass Substrate	NaOH Conc.	Vol. of NaOH Solution	Weight of Biomass	Ultrasound Treatment	Liquid Hot Water Treatment
Rice Straw T1	3% w/v	2,000 mL	250 g	1 hour, 100%, cycle (1)	121 °C,15 psi, 1 hour
Rice Straw T2	4% w/v	2,000 mL	250 g	1 hour, 100%, cycle (1)	121 °C,15 psi, 1 hour
Coconut Shell T1	3% w/v	2,000 mL	250 g	1 hour, 100%, cycle (1)	121 °C,15 psi, 1 hour
Coconut Shell T2	4% w/v	2,000 mL	250 g	1 hour, 100%, cycle (1)	121 °C,15 psi, 1 hour

2.3. Anaerobic co-digestion process

Anaerobic co-digestion experiments were performed in this study to provide information on the long-term effects, such as eventual inhibition during the digestion of a lignocellulosic substrate. Furthermore, a co-digestion is often beneficial, since it supplies the system with more nutrients, leading to a better balance in the C/N ratio and the pH; it also improves the stability of the process and maximizing the methane due to positive synergistic effects, which can further increase the economic value. To get a more realistic picture of the biogas production from rice straw and coconut shell substrate that are subject to pretreatments, an anaerobic co-digestion should be employed. In the present study, nine different lab-scale reactors were used to determine the methane yield in a batch process and also to evaluate the effects of the co-digestion processes using the pretreated rice straw and coconut as feedstock (substrates).

2.4. Biomass analysis using Fourier Transform InfraRed (FTIR) Spectroscopy

The structural composition of the rice straw biomass before and after the pretreatments be determined using the analytical protocols NREL LAP was followed. The Fourier Transform InfraRed (FT-IR) spectroscopy was then used to evaluate the structural properties of the rice straw and coconut shell with and without pretreatments. Further, the infrared spectra collected range was 4000 to 700 cm^{-1} at a resolution of 4 cm^{-1}.

3. Results and Discussions

The rice straw and coconut shell were subjected to a sequence of pretreatments such as ultrasonication with alkaline treatment through NaOH simultaneously; liquid hot water through the autoclave. However, the ultrasonication step for this experiment was modified to simultaneously perform basic hydrolysis using different concentrations of NaOH (3 and 4% w/v) and control of untreated rice straw as shown in Table 2. The pretreatment of rice straw and coconut shell with 3% w/v NaOH concentration, yielded the highest heating value, followed by the 4% NaOH.

Table 2. 45-days digestion results of pre-treated agricultural residues with cow manure.

Parameter	Cattle Manure & Rice Straw		Cattle Manure & Coconut Shell	
	3% w/v	4% w/v	3% w/v	4% w/v
C/N ratio	27.05	27.03	24.79	24.79
Biogas Yield (L)	18.18	12.54	23.12	8.44
Methane Yield (L)	9.88	6.96	13.84	4.77
Average Methane content (%)	53.80	53.00	55.90	54.50

Furthermore, the results are shown in Figure 2 that the pretreated rice straw produces higher biogas production than other pretreatments and the same effect was achieved in the pretreatment of the coconut shell. This implies that the fermentation/anaerobic digestion process properly converts the waste organic matter into a mixture of carbon dioxide and methane gas. Hence, the co-digestion of pretreated biomass helps to overcome the deficiencies of mono-digestion, which typically becomes the rate-limiting step for the AD process [8]-[9]. Thus, anaerobic co-digestion of different organic materials may enhance the stability of the anaerobic processes because of better carbon-to-nitrogen balance. Besides, the co-digestion of rice straw and coconut shell at 4% w/v reduce the methane yield due to the inhibitory effect of high ammonia and sulfide concentrations [10] and exhibits a more stable biogas production.

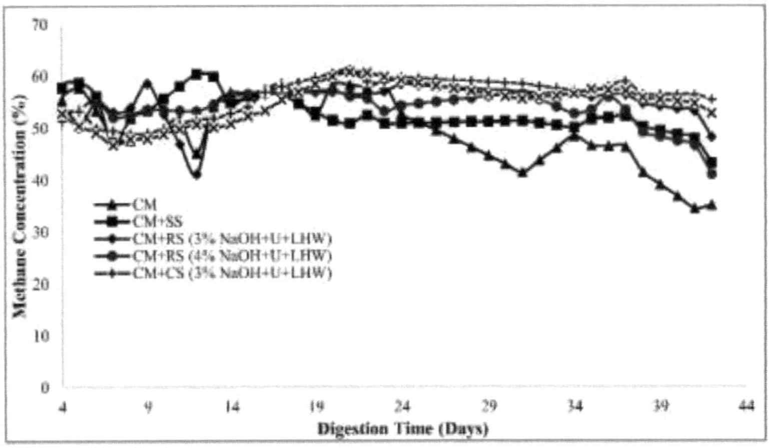

Figure 2. Methane concentration after 45-day anaerobic digestion of cattle manure, treated rice straw & coconut shell.

3.1. Biomass waste pretreatment

As shown in Tables 3 and 4, the rice straw and coconut shell biomass were subjected to varying concentrations of NaOH (3 and 4% w/v) and control of untreated rice straw. These were selected to prevent a waste chemical solution and also to reduce the pretreatment cost [14]. Furthermore, 3 to 4% w/v concentration is the optimal dose for five NaOH samples (2, 4, 6, 8, and 10%) as it was used to pretreat rice straw digestibility and biogas production presented in various publications [7], [11].

Table 3. Data for the Rice Straw Pretreatment Experiments

Treatment No.	Conc.	Vol of NaOH Solution	Wt of Rice Straw	Ultrasound Treatment	Liquid Hot Water Treatment
1	3% w/v	2,000 mL	250 g	1 hour, 100%, cycle (1)	121 C, 15 psi, 1 hour
2	4% w/v				
3	3% w/v				
4	4% w/v				

Table 4. Data for the Coconut Shell Pretreatment Experiments

Treatment No.	Conc.	Vol of NaOH Solution	Wt of CS	Ultrasound Treatment	Liquid Hot Water Treatment
1	3% w/v	2,000 mL	250 g	1 hour, 100%, cycle (1)	121 C, 15 psi, 1 hour
2	4% w/v				

Table 5. Characteristics of the untreated biomass materials used in the anaerobic co-digestion experiments

Property	Cattle Manure	Sewage Sludge	Rice Straw	Coconut Shell
Moisture (% FW)	77.57	96.43	7.20	22.29
TS (% FW)	54.61	1.92	96.69	81.71
VCM (%TS)	48.54	70.38	66.56	92.42
Ash (%TS)	5.99	3.35	8.49	5.10
TKN (%TS)	5.23	39.48	5.03	5.38
C (%TS)	21.64	36.17	33.71	45.21
H (%TS)	2.83	5.13	4.73	5.93
O (%TS)	73.51	51.36	60.54	47.83
N (%TS)	0.84	6.32	0.81	0.86
S (%TS)	1.18	1.02	0.20	0.17
C:N Ratio	25.86	5.73	41.88	52.57
Heating Value (MJ/kg)	16.99	18.81	13.59	18.51

Analysis and tests were conducted form the untreated and treated rice strew and coconut shell biomass. The test employed is the percentage of total solids and moisture content determination through proximate and ultimate analysis as the results shown in Table 5. Furthermore, the volatile combustible matter (VCM) content in rice straw is significantly higher than the untreated ones. The untreated rice straw increased VCM content from 18-20 percentage points for the 3% NaOH+U+LHW and 4% NaOH+U+LHW rice straw pretreatments. Moreover, the carbon content of the pretreated coconut shell is higher than the pretreated rice straw by 7 - 8 percentage points for 3% and 4% NaOH+U+LHW pretreatment respectively.

Table 6. Proximate Analysis data for the Rice Straw Pretreatment Experiments

Sample	Volatile Combustible Matter (%)	Ash Content (%)	Fixed Carbon (%)	Heating Value (MJ/kg)
Pretreated – RS 1 (3% NaOH)	84.73	11.26	4.01	15.62
Pretreated – RS 2 (4% NaOH)	88.18	4.79	7.03	10.32
Pretreated – RS 3 (3% NaOH)	83.00	9.62	7.38	15.90
Pretreated – RS 4 (4% NaOH)	92.42	0.51	7.08	13.08
Pretreated – CS 1 (3% NaOH)	78.85	18.78	2.37	19.28
Pretreated – CS 2 (4% NaOH)	78.19	19.55	2.26	19.33

As shown in Table 6, the alkaline-ultrasonication and Liquid Hot water using autoclave pretreatment of rice straw and coconut shell biomass with 3% w/v NaOH concentration, yielded the highest heating value, followed by the 4% NaOH. Moreover, the biomass samples are subjected to combustion and it was shown that it was combustible compared with the untreated rice straw. Consequently, the amount of the combustible matter had increased with 30 percentage points more than the untreated ones. Then, the said pretreatments can be interpreted that the lignin component of the rice straw is degraded during the pretreatments.

3.2. Analysis of the pretreated biomass using FTIR

As shown in Figures 3 and 4, the analyzed peaks for the pretreated samples of rice straw and coconut shell spectra's. In all of the samples, the highest peaks are the ones related to celluloses for pretreated rice straw biomass, which also has the highest variation among the pretreatments. The principal variation in the cellulose peaks is around 1300 to 1700 cm^{-1} for rice straw and similar peaks trends for the pretreated coconut shell biomass. This pretreatment leads to the increased strength of the signal (peaks) as compared with the control of rice straw and coconut shell. The NaOH-ultrasonication and liquid hot water pretreatments at varying concentrations showed the highest differences in signals (peaks) as compared with the biomass that was not subjected to any pretreatment.

The increase in improving biogas production due to the biodegradability of pretreated biomass can partly be explained by a decrease in the crystallinity of the cellulose fibers. These characteristics were measured by FTIR, having a wavenumber of 900 cm^{-1} for rice straw and 1200 to 1700 cm^{-1} for coconut shell (Figure 4). Further, the analyzed compounds and their IR wavenumbers are enumerated, identified and presented in Tables 7 and 8.

It was observed in the 1600 cm^{-1} spectrum, the lignin signals of pretreated rice straw were reduced principally. This reduction is related to the modification of the lignin aromatic skeletal structure generated in the pretreatments mixture.

This increase of biogas production due to sequential pretreatments is desirable due to the breakdown of the crystalline structure of the lignocelluloses, and consequently due to the higher accessibility of the bacteria to the cellulose and hemicellulose. Likewise, the hydrolysis of lignocellulosic biomass is the rate-limiting step in anaerobic digestion and pretreatment improves digestion efficiency and biogas production [12].

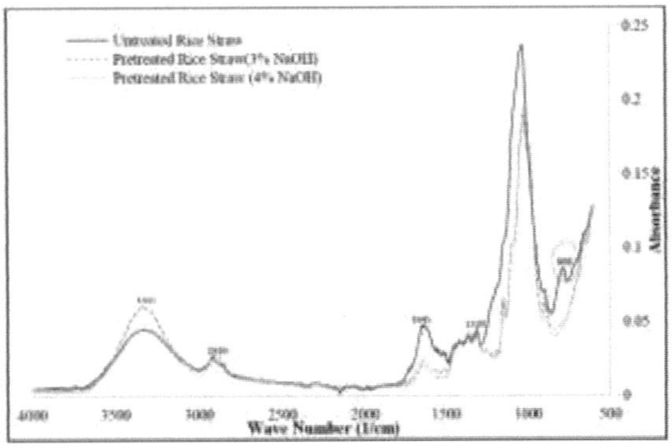

Figure 3. FT-IR spectra of the rice straw

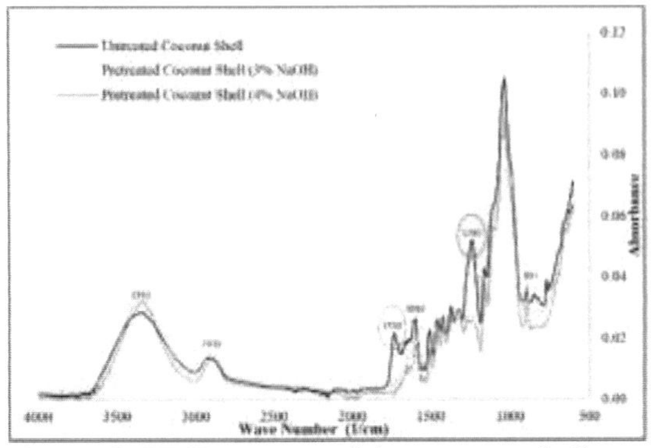

Figure 4. FT-IR spectra of the coconut shell

3.3. Anaerobic co-digestion process: methane quality

The biogas production from rice straw and coconut shell subjected for sequential pretreatment were shown in Figure 6, where the pretreated rice straw produces higher biogas production than other pretreatments and the same effect was achieved in the pretreatment of the coconut shell, whereas twenty-three (23) liters of biogas was produced (Table 2). The co-digestion of rice straw increased the biogas by 1.90 times, and 1.50 times higher for the coconut shell as compared with the untreated ones.

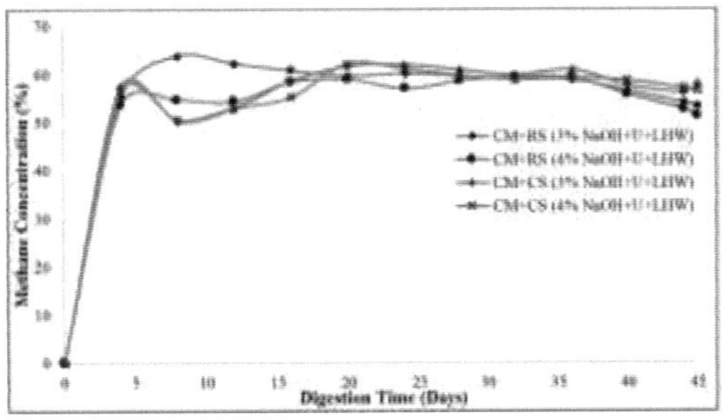

Figure 5. Average Methane Concentration

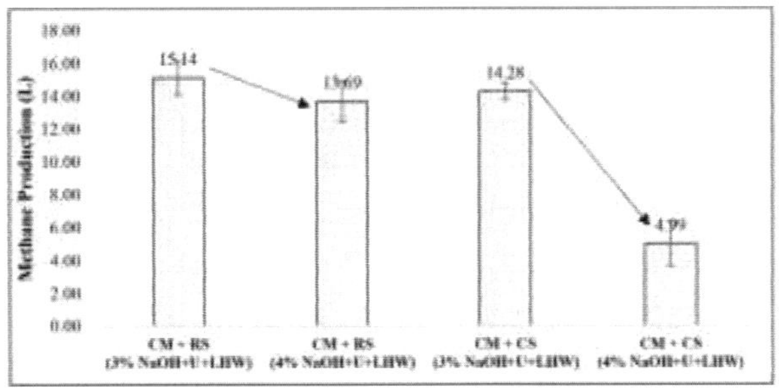

Figure 6. Cumulative methane production

As shown in Figure 5, the biogas produces methane components higher than the 50% concentration mark except for the cow manure as observed on the 10th to the 12th day. Likewise, parallel reports were reported [13] in the literature that the anaerobic co-digestion of cow manure at the various time resulted in a methane concentration of around 58% from the co-digestion of cow manure with rice straw. Moreover, the average percent methane concentration for these treatments shown in Table 8 are all in the 50% mark as compared to the untreated rice straw and coconut shell that it took 25 days to reach the 50% methane concentration, which was way slower than the pretreated ones.

Table 7. Wave numbers of IR vibration frequencies used for rice straw characterization

Wave Number (cm^{-1})	Compound	Functional Groups
3335	α-cellulose	Hydrogen bond hydroxyl (Alcohol/Phenol), O-H bond stretch
2920	Lignin	Alkane (alkyl), C-H stretch
1645	Lignin	Conjugated carbonyl (amide), C=O Stretch
1320	Cellulose	C-H plane bending of cellulose I and cellulose II
898	Cellulose	β-D- cellulose

Table 8. Wave numbers of IR vibration frequencies used for coconut shell characterization

Wave Number (cm^{-1})	Compound	Functional Groups
3340	α-cellulose	Hydrogen bond hydroxyl (Alcohol/Phenol), O-H bond stretch
2930	Lignin	Alkane (Alkyl), C-H stretch
1730	Hemicellulose	Ester Carbonyls, C=O stretch
1592	Lignin	Aromatic, C-H deformation
1240	Hemicellulose	Ester Carbonyls, C=O stretch
897	Cellulose	β-D- cellulose

On all the biogas treatments, the methane concentration level reaches above 50% on its fourth day and readings maintain with slight variations and reported the same levels with other publications [9], [12] presented. This implied the conversion of the waste organic matter through an anaerobic digestion process into a mixture of carbon dioxide and methane gas [9].

The biogas production trends of the biomass (pretreated and untreated) as shown in Figure 5 follows similar production trends for all the digesters. In just five (5) days, the methane production rates using treated biomass increases over the initial production rates (Day 4) as compared with the untreated ones for more than thrice than the untreated ones. Hence, the co-digestion of pretreated biomass helps to overcome the deficiencies of mono-digestion, which typically becomes the rate-limiting step for the AD process [11], [13].

3.4. Scanning electron microscopy (SEM) of the pretreated biomass samples

The rice straw and coconut shell samples (Figures 7 and 8) subjected to sodium hydroxide ultrasonication and liquid hot water treatment were analyzed at the Material Characterization Facility (MCF) at Texas A&M University, College Station, Texas for SEM evaluation. SEM specimens were prepared by spreading particles of each sample on carbon double-stick tabs and subsequently coating it with PtPd (80/20) ~ 10 nm thickness. The carbon tape and film were used for the fixation of particles and removal of accumulated charges. Micrographs were taken using a JEOL JSM 7500F field emission scanning electron microscope with a semi-in-lens and equipped with a Gentle Beam to deliver high-resolution. It was operated at a 5 kV acceleration voltage with a working distance of 1-10 mm (coconut shell) and 10-100mm (rice straw).

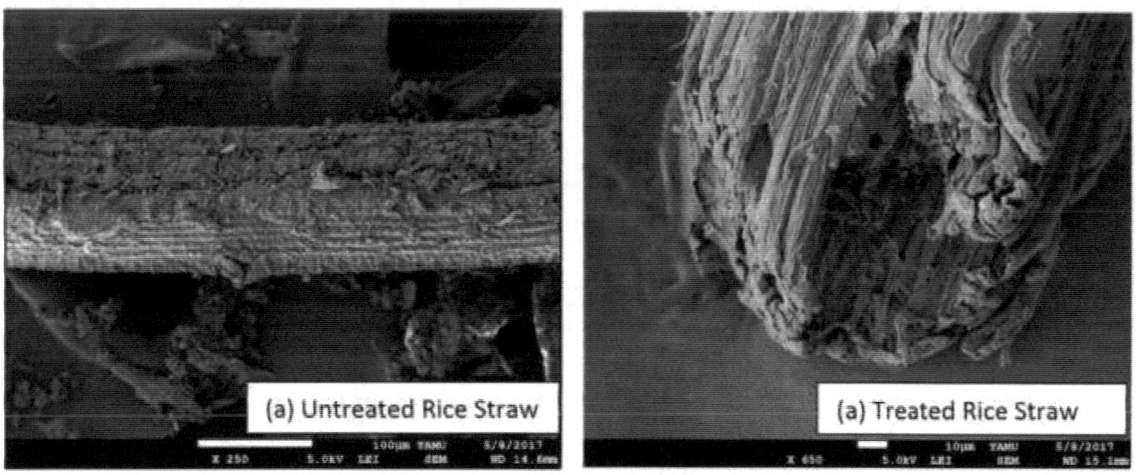

Figure 7. Scanning Electron Microscopy image of Rice Straw

IOP Conf. Series: Materials Science and Engineering **742** (2020) 012014 doi:10.1088/1757-899X/742/1/012014

Figure 8. Scanning Electron Microscopy image of Coconut Shell

3.5. Statistical analysis of various pretreatment

In this research, considering the possibility of the material characteristics limitation was observed on the pretreatment of coconut shell and rice straw. In fact, by comparing the means of the pretreated coconut shell and rice straw using the paired samples t-test at 95% confidence interval shows that the effect of NaOH concentration is not significant for the rice straw biomass, but it was significant for the coconut shell, and it implies that the results of the said experiments are repeatable at the same time, it tells us that the varying the NaOH concentration (3% and 4%) on the coconut shell. Furthermore, under the alkaline environment of the biomass (pH 11 for RS and pH 13 for CS), it helps to reduce the acidity of the anaerobic digester by increasing the pH into near alkaline range (pH 6.83 – 7.00). Aside from that, the dominant peaks are shown in the FT-IR spectra of the pretreated coconut shell and rice straw qualifies that the cellulose and hemicellulose component is broken down and been reduced due to the presence of NaOH in the mixture since it was carbon-bearing and it greatly helps the digestibility of the biomass material.

Thus, treating the coconut shell biomass will generally increase the biogas production volume at the same time, it will produce high-concentration methane gas, but there is another composition of gases present as by-products during the anaerobic co-digestion process. shown in Tables 3 and 4, the rice straw and coconut shell biomass were subjected to varying concentrations of NaOH (3 and 4% w/v) and control of untreated rice straw. These were selected to prevent a waste chemical solution and also to reduce the pretreatment cost [14]. Furthermore, 3 to 4% w/v concentration is the optimal dose for five NaOH samples (2, 4, 6, 8, and 10%) as it was used to pretreated rice straw digestibility and biogas production presented in various publications [11],[12].

4. Conclusion

The sequential pretreatment of agricultural biomass helps to increase the cellulose and lignocellulose conversion and resulted in a significant increase in biogas (methane) yields from the co-digestion of rice straw and coconut shell with cow manure and sewage sludge as inoculum as compared with untreated material. The use of FT-IR helps to identify the variations in the signal of the cellulose, hemicellulose, and lignin from rice straw and coconut shell after subjecting into different pretreatments. The effect of pretreatment on biomass produces more methane gas due to the increase of biogas volume, and thus the digestibility of the biomass feedstocks. Thus, the study shows how substrate composition, pretreatment methods, and operational parameters during anaerobic digestion affect the microbial consortia working in the digester.

References

[1] Philippine Statistics Authority. https://psa.gov.ph/. Date accessed: 18/07/2018.
[2] Karray R, Hamza M, Sayadi S. *Bioresour technol* **187** 205-213.
[3] Plácido J, Imam T, and Capareda S 2013 *Bioresour technol* **139** 203-08.
[4] Oz NA, Uzun AC. *Ultrason sonochem* **22** 565-572.
[5] Jiang D, Ge X, Zhang Q, Li Y. *Bioresour technol* **216** 60-68.
[6] Kaur K, Phutela UG. *Ren Energ* **92** 178-184.
[7] Yang CY, Fang TJ. *Bioresour technol* **164** 198-202.
[8] Zhang W *et al* 2014 *Appl Energ* **128** pp 175-83.
[9] Lianhua L *et al* 2010 *Jour hydro Energ* **35(13)** pp 7261-66.
[10] Zhang C, Xiao G, Peng L, Su H, and Tan T 2013 *Bioresour technol* **129** 170-76.
[11] He Y, Pang Y, Liu Y, Li X, Wang K. *Energ & Fuel*. **22(4)** 2775-2781.
[12] Silverstein RA, Chen Y, Sharma-Shivappa RR, Boyette MD, Osborne J. Bioresour technol **98(16)** 3000-3011.
[13] Sterling Jr MC, Lacey RE, Engler CR, Ricke SC. *Bioresour technol* **77 (1)** 9-18.
[14] Teghammar A, Chandra R, Saddler JN, Taherzadeh MJ, Horváth IS. *BioResour* **7(3)** 3921-34.

IOP Conf. Series: Materials Science and Engineering **742** (2020) 012015 doi:10.1088/1757-899X/742/1/012015

Alternative Method of Radiochemical Purity Testing for Technetium-99m Tetrofosmin

Widyastuti W.[1], Puji W.[1], A. Ariyanto[1], E. Lestari[1] and N.E. Rosilawati[2]
[1]Center for Radioisotope and Radiopharmaceutical Technology, Batan, INDONESIA.
[2]Dept. Nuclear Medicine and Molecular Imaging, Hasan Sadikin General Hospital, INDONESIA.

E-mail: widyast36@gmail.com

Abstract. Tc-99m tetrofosmin is a radiopharmaceutical preparation used in nuclear medicine. Radiochemical purity as main parameter of quality should be more than 90%, and was determined using thin layer chromatography with ITLC-SG as stationary phase and mixture of acetone and dichloromethane as mobile phase. Since ITLC-SG is expensive and the solvent used is rather noxious, an alternative method is expected to replace it for routine QC use. Besides, the radiolabeling capability of Tc-99m in low radioactivity is expected to reduce the utilization of Tc-99m in QC of kits manufacturing. Experiment using SepPak C18 column and TLC system with TLC-SG and ITLC-SG were carried out, then the results were compared. Radiochemical purity measurement using TLC-SG and ITLC-SG showed 99.49 ± 0.78% and 96.17 ± 2.37% respectively with the peaks observed at Rf 0.0 and ~ 0.5 respectively, whereas that using SepPak C-18 showed RCP of 98.51 ± 0.78%. Various radioactivity of Tc-99m from 2 to 100 mCi have been used to label tetrofosmin kits and the radiolabeling quality were almost the same. It can be concluded that SepPak C18 as well as TLC method using TLC-SG can be used as alternative methods for radiochemical purity measurement of Tc-99m tetrofosmin, and low radioactivity of Tc-99m is sufficient to label tetrofosmin with high radiochemical purity.

1. Introduction

The most successful cationic, lipophilic Tc-99m radiopharmaceuticals for myocardial perfusion agents are the well-known compounds Tc-99m sestamibi (Tc-99m MIBI) and Tc-99m tetrofosmin which were developed since almost 3 decades [1]. To date Tc-99m tetrofosmin and Tc-99m MIBI have been widely used as myocardial perfusion imaging agent, but in the last decade was discovered that both Tc-99m radiopharmaceuticals display uptake in the tumor tissues, such as in breast cancer, small cell lung cancer (SCLC), parathyroid adenoma and brain tumors [2-5]. The uptake of Tc-99m MIBI and Tc-99m Tetrofosmin in this kind of tumors is supposed to be related with the over expression of tumor's multidrug resistance phenotype namely P-glycoprotein (Pgp) and multidrug resistance related proteins (MRPs) in which Tc-99m tetrofosmin is superior than Tc-99m MIBI [5]. The high uptake of Tc-99m tetrofosmin is comparable, even more useful than brain MRI which is accepted as gold standard for brain tumor imaging modality [6]. Besides, the multidrug resistance gene 1 encoding human P-glycoprotein (Pgp) is thought to play an important role in the multidrug resistance of lung cancer. Tc-99m tetrofosmin lung SPECT can accurately predict the chemotherapy response, and the same phenomena was also found in patients with malignant bone and soft tissue tumors [7-8].

Tetrofosmin kits as a pharmaceutical product must conform the requirements including the physical characteristics, biologic purity, and radiochemical purity. Biological purity includes tests for sterility and apyrogenicity, while radiochemical purity (RCP) is the percent of total radioactivity present in the desired chemical form which is a main quality parameter of radiopharmaceutical products. Unacceptable RCP can lead to localizing of the radiopharmaceutical in an unintended target, which may result in unclear image of target organ that may lead to misdiagnosing the disease. Three types of radiochemical impurities to be determined are as follows: free technetium Tc-99m pertechnetate (Tc-99mO_4^-), hydrolyzed-reduced Tc-99m (insoluble Tc-99m dioxide and/or Tc-99m tin colloid), and bound Tc-99m to the ligand of interest (desirable radiochemical form). These radiochemical impurities can hinder clinical interpretation by decreasing the target to non-target ratio. In clinical practice, the RCP analysis must be quick, accurate, and economical whereas in a research environment greater emphasis is placed upon accuracy, while the time and economical factors are sometimes less important [9].

All of impurities can be determined using a chromatography system, consisting of a stationary phase and a mobile (solvent) phase. The components of the mixture traveling the farthest are those that are most soluble in the solvent and least attracted to the stationary medium. The most frequently encountered procedures involve solid-phase extraction or instant thin-layer and paper chromatography. Mobile phase (solvent) moves the solute through the stationary medium affecting a separation from other solutes that are eluted earlier or later. Thin layer chromatography (TLC) is frequently used in radiochemical purity testing of radiopharmaceuticals, which uses TLC-SG as stationary phase and suitable eluants as mobile phase. The newer product of TLC-SG sheets namely ITLC-SG is widely used in research and clinical practice since it gives better performance and much faster than using TLC-SG. ITLC-SG provides superb resolution for nonpolar compounds, unfortunately, the manufacturer of ITLC-SG strips once discontinued their production, although now another manufacturer continues the production but the price is very expensive [10, 11].

According to the manufacturer's leaflet of Myoview®, QC protocol involves TLC system which uses ITLC-SG as stationary phase and mixture of acetone and dichloromethane as mobile phase [12-14]. Optimization of using shorter size of ITLC-SG strips has been reported to be as effective as that of regular size, which makes it more efficient in using this expensive consumable material [15].

Better and faster QC method was reported using SepPak cartridge as an alternative method for radiochemical purity testing of radiopharmaceuticals. Sep-Pak® cartridges are useful for chromatography since they operate on the principle of solid-phase extraction or column chromatography. The manufacturer's package inserts for Tc-99m mertiatide4 and In-111 pentetreotide recommend the Sep-Pak® C18 chromatography cartridge for the determination of radiochemical purity. This cartridge is composed of a solid, nonpolar material (sorbent) that enables the separation of the various radiochemical species of interest [9]. Another kind of SepPak cartridge, namely SepPak silica was reported to give good performance in radiochemical purity measurement of some radiopharmaceuticals including Tc-99m tetrofosmin, and can be used as an alternative for TLC method [11]. SepPak C18 cartridge was used to measure radiochemical purity of Tc-99m tetrofosmin (from Myoview® and local produced tetrofosmin kit), and found to give the same result as that of using TLC, so it can be used as an alternative method for TLC in case if there is a shortage of ITLC-SG in the market [16]. Moreover, RCP measurement using SepPak C18 is safer than using TLC, since solvents used for SepPak C18 are saline (0.9% sodium chloride solution) and ethanol, whereas the TLC system are using dichlorometane which can be classified as noxious chemicals. Labeling study of tetrofosmin kits using various radioactivity of Tc-99m of 150 mCi to 400 mCi has been reported and showed good results, this will be helpful to send a high radioactivity of Tc-99m tetrofosmin to hospitals in a ready-to-inject dosage form, to anticipate the decay during transportation since the halflife of Tc-99m is quite short (6.02 hours) [17].

Tc-99m tetrofosmin of local product (developed by PTRR-BATAN) has been previously compared with commercial product (Myoview®, GE Healthcare) in the quality of images using gamma camera (SPECT) and the results showed no significant difference between both products [18].

IOP Conf. Series: Materials Science and Engineering **742** (2020) 012015 doi:10.1088/1757-899X/742/1/012015

Labeling of tetrofosmin kits using low radioactivity of Tc-99m will be carried out, with the aim to ensure that it can perform the same quality as the one using normal and high radioactivity of Tc-99m. This information will be beneficial particularly for the manufacturer of tetrofosmin kits that needs to do quality control (QC) for each production batch. Radiochemical purity testing of final product has to be accomplished for every production batch as a requirement of Good Manufacturing Practice (GMP) for all pharmaceutical products. Therefore, the use of small amount of Tc-99m is preferable and will be more efficient.

The aim of this study is to obtain a more practical method for radiochemical purity testing of Tc-99m tetrofosmin as a part of Quality Control process in tetrofosmin kits manufacturing. Besides, the aim is to get known the minimum radioactivity of Tc-99m that can be used to label tetrofosmin kit to form Tc-99m tetrofosmin with RCP of more than 90% as stated in Pharmacopoeia as a radiopharmaceutical requirement. The use of low radioactivity of Tc-99m for routine QC activity will reduce the risk of radiation exposure to the QC personnels and will also be efficient.

2. Methods

Tetrofosmin kits of local product (PTRR BATAN)) were used to be labeled with technetium Tc-99m to form Tc-99m tetrofosmin. Tc-99m was obtained from in-house Mo-99/Tc-99m generator (PTRR-BATAN). To ensure that tetrofosmin kit was completely labeled with Tc-99m, the radiolabeled products were analysed for radiochemical purity using thin layer chromatography (TLC). The standard method of radiochemical purity testing was TLC method using ITLC-SG (Agilent) as stationary phase and mixture of acetone and dichloromethane (35:65) (both are from Merck) as mobile phase (eluant). Two alternative methods were studied, i.e. column separation using SepPak C18 cartridge (Waters) and TLC method using TLC-SG (Merck).

2.1. RCP measurement using TLC method

TLC-SG strip (1.5 cm x 10 cm) was used as stationary phase, on which 5 ul of sample was dropped at the baseline or ~ 2 cm from the strip edge, dip into mixture of acetone-dichloromethane (35:65) in a glass chamber and eluted until the solvent reached the upper edge of the strip (~1 cm from the top edge). TLC using ITLC-SG strips were used as a standard. The strips were then taken out from the chamber, air dried and measured for radioactivity by means of TLC scanner (Comecer TLC 204).

2.2. RCP measurement using SepPak C18 column/cartridge. Other separation method using SepPak C18 cartridge was studied to analyse radiochemical purity of Tc-99m tetrofosmin, in which ~ 100 ul of sample was injected into the column (cartridge) which has been previously activated with 5 ml of ethanol, water and air consecutively. The polar impurity (TcO_4^-) was eluted out of the column with 10 ml of saline, the main complex was eluted with 10 ml of ethanol-saline (4:1), and the colloid impurity (TcO_2) remained in the column. By measuring the radioactivity of the eluted fractions and the column, % radiochemical purity of Tc-99m tetrofosmin can be calculated.

2.3. Labeling capability of Tc-99m tetrofosmin with various radioactivity of Tc-99m. Labeling of Tc-99m tetrofosmin using various radioactivity of Tc-99m with the range of 2 mCi to 100 mCi was carried out and the RCP was measured using TLC (ITLC-SG as stationary phase) and SepPak C18.

3. Results and Discussion

Radiochemical purity measurement of several batches of Tc-99m tetrofosmin has been carried out using TLC system and column separation method (SepPak C18). The results were compared to the one using ITLC-SG as standard method. TLC method using TLC-SG strips performed higher radiochemical purity (RCP) than that using standard method, but since the Tc-99m complex could not migrate so the peak might be overlapped with Tc-99m colloid impurity, as can be seen at Table 1.

Table 1. Comparison between radiochemical purity analysis by TLC using TLC-SG and ITLC-SG as stationary phase

Stationary phase	Radiochemical purity (%)	Rf	
		Tc-99m Tetrofosmin	Free pertechnetate
TLC-SG strips	99.49 ± 0.78	0.0-0.1	0.9-1.0
ITLC-SG strips	96.17 ± 2.37	0.5-0.7	0.9-1.0

The identification of a radiochemical species is due to its transfer through a stationary phase via a mobile phase until it is finally separated from the other radiochemical species that are eluted earlier or later. The movement of various radiochemical species is hindered by the stationary phase's electrostatic forces (adsorption) as the solvent phase moves them forward [10].

The spotted sample volume used for all experiments was 5 μl since the volume sample spotted on TLC strips can affect the migration of the complex along the chromatogram, leading to overestimate the impurities level, this was the uniqueness of Tc-99m tetrofosmin compared to other radiopharmaceuticals [15].

The results of radiochemical purity analysis using SepPak C18 column was also similar with the one using TLC method. The advantage of using SepPak C18 column was that the separation process of the main product from its impurities was running fast, and to some experiences, the column can be reused up to 3-5 times. The series of these experiments were replicated 12 times (n=12). The results of the 3 methods can be seen in Table 2.

Table 2. Radiochemical purity measured by TLC and SepPak C18 methods

Methods	Radiochemical purity (%)
SepPak C-18	98.51 ± 0.78
TLC using TLC-SG strips	99.49 ± 0.78
TLC using ITLC-SG strips	96.17 ± 2.37

The chromatograms which used TLC-SG as stationary phase showed different peaks of Tc-99m tetrofosmin and free pertechnetate, i.e at Rf 0.0-0.1 and Rf 0.9-1.0 respectively (Figure 1), whereas the one using ITLC-SG showed the peaks at ~Rf 0.5-0.7 and Rf 0.9-1.0 respectively (Figure 2). In TLC system using TLC-SG as stationary phase the labeled compound seem to remain at the origin, and the time used to complete the elution was a bit longer, compared to the one using ITLC-SG. On the other hand, when using ITLC-SG as stationary phase the labeled tetrofosmin can move to the middle of the strip length, so it can be distinguished from the impurities which might be present at Rf 0.0 (Tc-99m colloid) and Rf 0.9-1.0 (free pertechnetate), and it conformed with the specification [12]. The advantage of using ITLC-SG compared to TLC-SG is that on ITLC-SG strip the migration of species is much faster and show better separation, but unfortunately, ITLC-SG is much expensive and made of fragile material which tends to be easily broken.

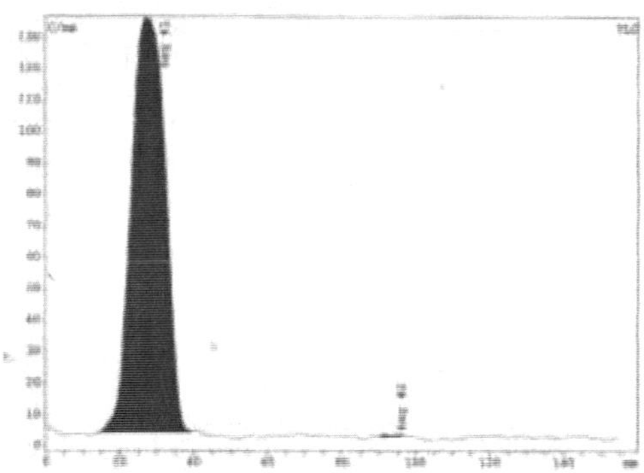

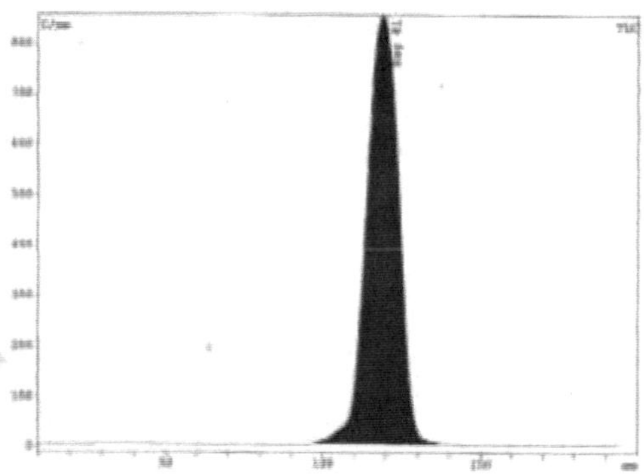

Fig 1(a) Fig 1(b)

Figure 1. Position of peaks on TLC-SG chromatogram, Fig 1(a) peak of Tc-99m tetrofosmin and Fig 1(b) peak of free pertechnetate Tc-99m

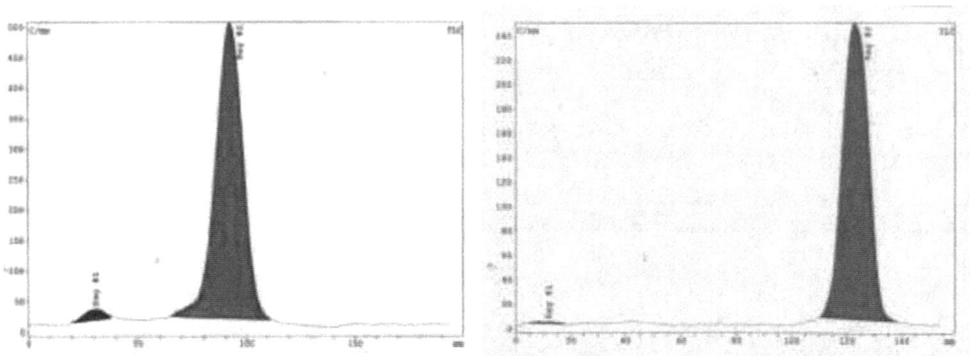

Fig 2(a) Fig 2(b)

Figure 2. Position of peaks on ITLC-SG chromatogram, Fig 2(a) peak of Tc-99m tetrofosmin and Fig 2(b) peak of free pertechnetate Tc-99m

The use of various radioactivity of Tc-99m from 2 mCi to 100 mCi in the labeling of tetrofosmin kits did not give significant difference in RCP, all of them showed RCP of more than 90% which means that the labeled compound meets the requirement of tetrofosmin kits (Figure 3). The capability of tetrofosmin kits to be labeled with low radioactivity of Tc-99m is preferable for the manufacturer of tetrofosmin kits since they do not need to consume much Tc-99m for routine QC use for every production batch.

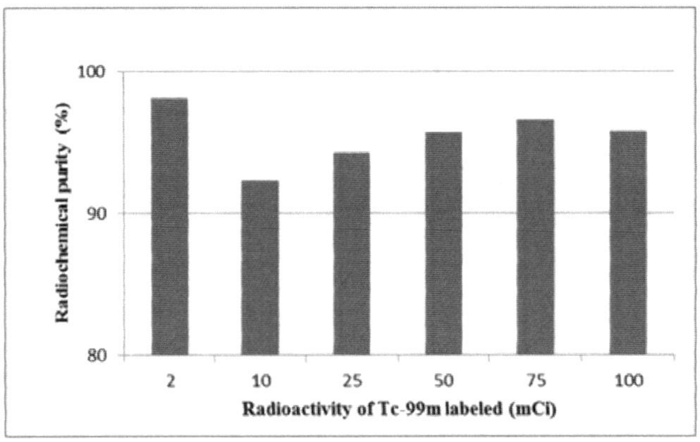

Figure 3. Labeling efficiency of Tc-99m tetrofosmin using various radioactivity of Tc-99m

4. Conclusion

RCP measurement using SepPak C18 column showed an average of 98.51 ± 0.78%, whereas using TLC method with TLC-SG as stationary phase showed RCP of 99.49 ± 0.78%, which were slightly higher than that using standard TLC method which RCP was 96.17 ± 2.37%. It is concluded that SepPak C18 can be used as an alternative method for radiochemical purity measurement of Tc-99m tetrofosmin which gives the same performance but much faster and efficient compared to standard TLC method. However, when TLC method is still preferred, an alternative TLC-SG strips can be used to substitute a much expensive ITLC-SG strips although its performance is moderate. The capability of tetrofosmin kits to be labelled with Tc-99m of only 2 mCi has been proven effective, and this will

IOP Conf. Series: Materials Science and Engineering **742** (2020) 012015 doi:10.1088/1757-899X/742/1/012015

be beneficial for the kit manufacturer since it can efficiently reduce the use of Tc-99m in implementing QC process.

Acknowledgment
This work was partly funded by The Ministry of Research and Technology through PPTI program 2018. The author would like to thank to CRRT BATAN and Nuclear Medicine Department, Hasan Sadikin Hospital staffs who are involved in this work for the collaboration.

References
[1] Verbruggen, 2007, Technetium Chemistry and Radiopharmaceuticals Session 1. In Trends in Radiopharmaceuticals (ISTR-2005), Proceedings of an international symposium organized by the International Atomic Energy Agency, Vienna, 14–18 November 2005.Vienna, Austria, Vol. 1, pp. 3-18.
[2] E Abo-Salem and M Gerson, 2014, *J Nucl Cardiol*, 21, pp.1258–59.
[3] G Silov et al, 2014, *Hell J Nucl Med*, 17 (2), pp. 87-89.
[4] K Romanidis et al, 2014, *BMC Research Notes*, 7, pp.335-39.
[5] G A Alexiou et al, 2014, *Int J Mol Imaging*, Article ID 471032, 5 pages.
[6] T Birbilis et al, 2013, *MAEDICA – a Journal of Clinical Medicine,* 8(4), pp. 347-50.
[7] Yeh J J et al, 2003, *Tumor Biol,* 24, pp. 151–55.
[8] Z Yapar et al, 2003, *Annals of Nuclear Medicine*, 17 (6), pp. 443–49.
[9] V S Loveless, 2009, *Quality Control of Compounded Radiopharmaceuticals, Continuing Education for Nuclear Pharmacists and Nuclear Medicine Professionals*, vol. 15, lesson 3, pp. 7-17.
[10] C Decristoforo and I. Zolle, 2007, Quality Control Methods of Tc-99m Pharmaceuticals. In *Technetium-99m Pharmaceuticals : Preparation and Quality Control in Nuclear Medicine,* edited by I. Zolle, Springer, pp. 123-37.
[11] R Hammes et al, 2004, J Nucl Med Technol, 32, pp. 72-78.
[12] GE Health Care, 2006, Myoview package insert.
[13] US Pharmacopoeia 32, Monograph of Tc-99m tetrofosmin.
[14] I Zolle and PO Bremer, 2007, Monographs of Tc-99m Pharmaceuticals. In: *Technetium-99m Pharmaceuticals : Preparation and Quality Control in Nuclear Medicine*, edited by I. Zolle, Springer, pp. 245-51.
[15] M C Geyer et al, 1995, *Nuclear Pharmacy*, 23, pp. 186-89, downloaded from tech.snmjournals.org July 18, 2018.
[16] W Widyastuti, S Setiyowati, Cecep TR, Yunilda A, 2013, *Proceedings of National Seminar in Chemical Technology Kejuangan*, Yogyakarta, pp. J10 1-7 (in Indonesia).
[17] Lindawati, W Widyastuti, A Ariyanto et al, 2017, *Proceedings of Seminar on Utilization of Nuclear Technology* 2017, BATAN, South Tangerang, pp. 53-62 (in Indonesia).
[18] N Elly Rosilawati, AHS Kartamihardja, A Mutalib, 2014, IJAS, 4 (1), pp. 36-40.

Quantitative Determination of Vanillin in Coated Paper Using Gas Chromatography

Katherine Kho*[1], Solmaz Aslanzadeh[1]

[1]Biotechnology Study Program, School of Life Sciences, Indonesia International Institute for Life Sciences, Indonesia

E-mail: katherine.k@i3l.ac.id

Abstract. One of the strategies to improve the food shelf – life is to utilize functional packaging. The functional packaging could contain antimicrobial agents to inhibit pathogenic microorganisms or antioxidants to prevent oxidative degradation. Vanillin, known as one of the most common food flavoring ingredients, could serve as a potent antioxidant. This paper focuses on the development of a methodology to determine vanillin content in the coated paper using gas chromatography. The vanillin was extracted from vanillin-coated paper using water as the aqueous phase and dichloromethane (DCM) as the organic phase. The effect of filtration on vanillin extraction and water to DCM ratio used for extraction were investigated. It was concluded that the extraction without filtration with water to DCM ratio of 1:2 was the best setting for the vanillin extraction from vanillin-coated paper.

1. Introduction

Paper is commonly used for packaging material of food products [1]. To prevent microbial growth on the product surface and food spoilage and increase the shelf life of food products, many antimicrobial substances are coated either directly on food surfaces or into the food packaging materials [17]. However, sometimes these substances immobilized on the surface of the packaging material could migrate into the food products, thus leading to unacceptable product quality [2-4]. Therefore, the antimicrobial compounds used for food packaging or even coming into human contact should be safe and correspond to the regulatory concern. As a result, the industry is searching for a natural compound that has antimicrobial with abilities. There are some studies in the literature showing that polyphenols have antimicrobial activity. Vanillin is a familiar flavour and has shown to be a potential antimicrobial agent due to the effect of the phenolic compound that prevents bacteria, yeasts, and molds to grow [1, 5]. The antimicrobial property of vanillin is the effect of the phenolic group in chemical structure. One of the most promising possibilities of stabilizing aromatics such as vanillin is their inclusion complex formation (molecular encapsulation) with β-cyclodextrin as a powder-like crystalline, which made its industrial utilization possible. However, the main applications of the microencapsulated vanillin powder are as food additives [6].

One other potential application is as a coating agent for the production of aroma paper [7, 8]. Despite vanillin's advantages as an antimicrobial agent in the packaging material for the enhancement of food's shelf-life, no study has reported the use of vanillin in the paper packaging or materials [1, 9].

IOP Conf. Series: Materials Science and Engineering **742** (2020) 012016 doi:10.1088/1757-899X/742/1/012016

Furthermore, methods for extraction and identification of vanillin from the coated paper is absent. Consequently, there are still industrial challenges with standardized methods for the extraction and determination of vanillin from coated paper [1, 10]. Thus, this paper focuses on the methodology developments for quantifying vanillin content in vanillin coated paper using Gas Chromatography (GC).

2. Experimental Section

2.1. Liquid – liquid extraction

For liquid-liquid extraction, 1 g of commercial vanillin (obtained from PT Mane Indonesia) and 0.05 g of ethyl vanillin (obtained from PT Mane Indonesia) were completely dissolved in 15 mL of deionized water at 100 °C and 300 rpm during 10 minutes. In the single-step extraction, the solution were cooled down, and mixed with 10 mL of dichloromethane (DCM). and the mixture was stirred for 10 minutes at 300 rpm. The aqueous phase (water) and the organic phase (DCM) were separated using the decantation method. The organic phase was then centrifuged at 4000 rpm for 5 minutes and filtered before injection to GC.

In the multi-step extraction, the solution was cooled down, then 5 mL of DCM was added to the solution, and the mixture was stirred for 10 minutes. The aqueous phase (water) and the organic phase (DCM) were separated using the decantation method, while the organic phase was centrifuged at 4000 rpm for 5 minutes. The supernatant was filtered to remove solid impurities and 0.5 mL of the aliquot was then mixed with ethyl vanillin solution. In the next step of extraction, another 5 mL of DCM was added to the aqueous phase and the extraction, recovery, and sample preparation for analysis as described previously were repeated. The extraction steps were repeated five times.

2.2. Vanillin coated paper sample preparation

An A4 size paper was soaked in 3.5 % commercial vanillin solution for one minute. Excess water was removed by using blotting paper. The paper was then dried at 60°C for two minutes.

2.3. Effect of filtration on vanillin extraction method from coated paper

Approximately 1 g of coated paper and 0.05 g of ethyl vanillin were added into 15 mL of deionized water. The sample was heated at 100 °C and stirred at 300 rpm for 10 minutes. The sample was then left to cool down. After cooling, both vanillin and ethyl vanillin were extracted from the sample using 10 mL of DCM by manually stirring the mixture for 2 minutes. The organic phase was then recovered according to the method described in section 2.1.

In another experiment, a different approach was taken, in which after the sample was cooled down, the aqueous phase was recovered by filtration first before adding 10 mL of DCM to the aqueous phase. The mixture was then stirred for 2 min and the DCM was recovered according to the method described in section 2.1.

2.4 Effect of water to DCM ratio on vanillin extraction method from coated paper

The extraction was carried out using the protocol described in section 2.3 without filtration. To obtain water to DCM ratio of 1:2, the amount of DCM added was 30 mL instead of 10 mL.

2.5 Analysis of vanillin by GC – FID

Before injection, all samples were filtered using a Minisart membrane filter with 0.2 μm pore size. The analysis was performed using gas chromatography (GC) (Thermo Scientific™ TRACE™ 1300) with an FID detector. The capillary column TG-5MS column (Thermo Scientific) with the dimension of 30 m x 0.25 mm and 0.25 μm film thickness was used. The oven temperature started at 150°C kept for 1 minute, and then increased to 210 °C at the rate of 15°C/min for 2 min. The injector temperature were maintained at 290 °C and the detector temperature at 325 °C. 0.8 μl of the sample was injected in a split mode with the split ratio of 75. Nitrogen gas was maintained at 40 mL/min, while air and hydrogen were

maintained at 350 mL/min and 35 mL/min, respectively. Helium gas was used as carrier gas at a flow rate of 1 mL/min.
Pure vanillin (analytical grade, Merck) was used for calibration. The pure vanillin was dissolved in DCM (analytical grade) at the concentration ranging from 0.05% - 0.25% w/v. In addition, the sample solution contained 0.25% w/v ethyl vanillin (analytical grade) as an internal standard.

2.6. Statistical analysis
In this study, all experiments were carried out in replicates. The data were analysed using the statistical tool of Excel 2016, MS Office. The graphs were plotted using mean values and the standard deviation is shown as error bars.

3. Result and Discussion

3.1. Extraction of vanillin from commercial vanillin powder
As one of the most common flavouring agent in the food industry, vanillin is commonly used in food packaging in a small amount. Vanillin is usually supplied to the market in the form of powder, containing carriers such as maltodextrin. The carrier is used to improve the property of the powder such as flowability, and to increase the yield during spray drying. Although maltodextrin is highly soluble in water, it has minimal solubility in DCM. Therefore, for quantification of vanillin using GC, it was necessary to release the entrapped vanillin from the powder by dissolving the powder in water. Afterward, the released vanillin was extracted from the aqueous phase using DCM. Preliminary experiments were conducted to determine the optimum time for extraction and the total content of the vanillin in the powder. It was concluded that 10 minutes of mixing was enough for vanillin to diffuse to DCM.

Figure 1 shows the vanillin content detected using single-step extraction and multiple-step extraction. In the single-step extraction, 8.23±0.06% vanillin in commercial powder was reported. Meanwhile, a higher vanillin content of 8.57±1.11% was reported in multi-step extraction. The result of a one-way analysis of variance indicated that both methods gave a similar result with a 95% confidence level. Although multi-step extraction could extract more vanillin, the standard deviation was relatively larger due to the volatility of the solvent and difference in solvent recovery in each stage of extraction.

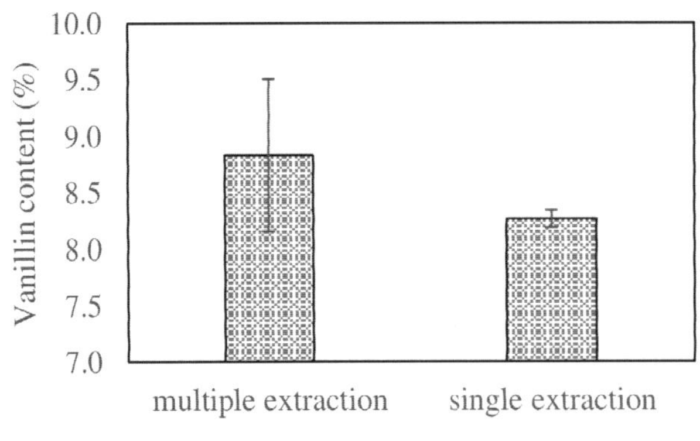

Figure 1. Comparison of single- step and 5- steps vanillin extraction from commercial powder

3.2. Extraction of vanillin from coated paper

As demonstrated in section 3.1. using commercial vanillin for the vanillin extraction, the single-step vanillin extraction gave a smaller variability. Hence, the single-step extraction was adapted for the extraction of vanillin from coated paper. The result is presented in Figure 2. The vanillin content detected was 0.17±0.01%. A challenge occurred during the dispersion of paper in water, in which the paper disintegrated into fibers. The dispersion of fibers in water modified the consistency of the suspension such that the suspension had a slurry-like consistency. The consequence of this was an increase in energy required for mixing to ensure that the vanillin was completely extracted. In addition, the difficulty of organic phase recovery at a later stage was higher. To solve this problem, the fibers were separated from the slurry before DCM was added. This gave rise to a lower vanillin recovery, with vanillin content of 0.14 ± 0.00% (Figure 2). Furthermore, two other methods were also investigated. First, increasing the extraction step (data not shown), and second by increasing the water to DCM ratio, while maintaining the volume of the water constant. The result showed that changing the water to DCM ratio from 3:2 to 1:2 while maintaining the water volume constant gave the best result in terms of the vanillin extracted and the lowest variability, with the vanillin content from 1:2 ratio was 0.30 ± 0.00% (Figure 3).

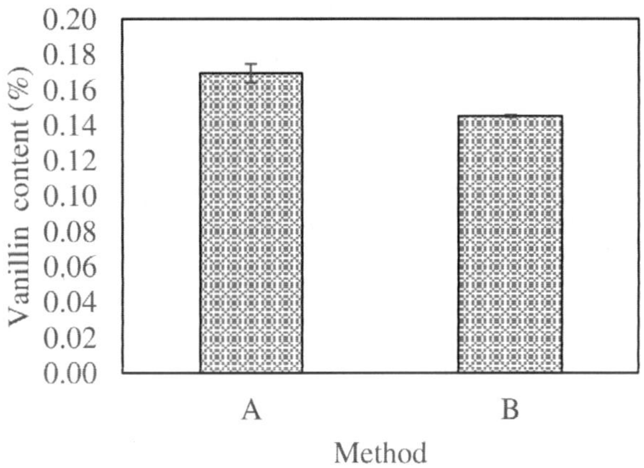

Figure 2. Effect of adding filtration step in single-step vanillin extraction. Method A: the paper was dispersed in the water before DCM addition. Method B: the recovered filtrate after dispersing the paper in the water mixed with DCM.

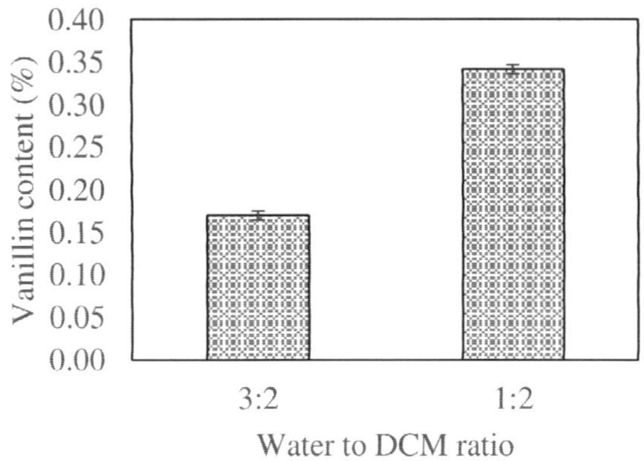

Figure 3. Effect of water to DCM ratio.

A higher amount of organic solvent is needed in the extraction from coated paper compared to that of commercial vanillin powder. It was speculated that the fibers hindered the diffusivity of vanillin to both water and DCM. The presence of hydroxyl groups in cellulose fiber renders the paper hydrophilic. As such, the paper is easily wetted and the fiber will disperse in water. However, the same functional group is responsible for inter and intra-chain bonds in the fiber. As such, the paper will only dissolve in certain solvents, of which water and DCM do not belong to [11]. The implication to the vanillin extraction process from the coated paper was a three-phase extraction process. Solute recovery using liquid-liquid extraction is an already established method with the effect of most factors known. However, when the solute is dispersed in a heterogeneous matrix-like fiber in water, the complexity for modeling and prediction rises exponentially, as both the structure of the solid phase and the interaction between the solid phase and the solute affect the diffusivity of the solute to the solvent [12-14].

4. Conclusion

The goal of the experiment was to establish the protocol for extraction and quantification of vanillin from the coated paper. A modification was needed in terms of water to DCM ratio in which the ratio of 1:2 was concluded to be the best for the vanillin extraction process from paper. The protocol could be useful to consistently recover and quantify vanillin from coated paper. Thus, the method would be appropriate for sampling and quality control. Future work includes modification of protocol for faster analysis.

5. Acknowledgement

The authors acknowledge PT Bukit Muria Jaya, Indonesia for funding this Project. The authors would like to thank Kezia Soeweno for the help in GC optimization, and Adam Jaya for conducting the main part of the experiment. Last but not least, the authors are grateful to Ester Yusmarina for the help in vanillin calibration and in making sure all experiments ran smoothly.

6. References

[1] Rakchoy S, Suppakul P and Jinkarn T 2009 Antimicrobial effects of vanillin coated solution for coating paperboard intended for packaging bakery product *Asian Journal of Food and Agro-Industry* **2** 138-47
[2] Han J H 2000 Antimicrobial Food Packaging *Food Technol.* **54** 56-65
[3] Vermeiren L, Devlieghere F, van Beest M, de Kruijf N and Debevere J 1999 Developments in the active packaging of foods *Trends Food Sci. Technol.* **10** 77-86
[4] Quintavalla S and Vicini L 2002 Antimicrobial food packaging in meat industry *Meat Science* **62** 373-80
[5] Sinha A, Sharma U and Sharma N 2007 A Comprehensive Review on Vanilla Flavor: Extraction, Isolation and Quantification of Vanillin and Others Constituents *Int. J. Food Sci. Nutr.* **59** 299-326
[6] Szejtli J, Lindner and Szente L 1981 Food flavoring with β-cyclodextrin-complexed flavour substances *Acta Alimentaria* **10** 175-86
[7] Kant A, Linforth R, Hort J and Taylor A 2004 Effect of β-Cyclodextrin on Aroma Release and Flavor Perception *Journal of agricultural and food chemistry* **52** 2028-35
[8] Sharma N and Baldi A 2014 Exploring versatile applications of cyclodextrins: An overview *Drug Deliv.* **23** 1-19
[9] Mourtzinos I, Konteles S, Kalogeropoulos N and Karathanos V 2009 Thermal oxidation of vanillin affects its antioxidant and antimicrobial properties *Food Chem.* **114** 791-7
[10] Erdohan Ö and Turhan K 2012 pp 1094-101
[11] Bochek A M 2003 Effect of Hydrogen Bonding on Cellulose Solubility in Aqueous and Nonaqueous Solvents *Russ. J. Appl. Chem.* **76** 1711-9
[12] Kosto K B and Deen W M 2004 Diffusivities of macromolecules in composite hydrogels *AICHE*

J. **50** 2648-58

[13] Phillips R J, Deen W M and Brady J F 1990 Hindered transport in fibrous membranes and gels: Effect of solute size and fiber configuration *Journal of Colloid and Interface Science* **139** 363-73

[14] Leonidakis K A, Bhattacharya P, Patterson J, Vos B E, Koenderink G H, Vermant J, Lambrechts D, Roeffaers M and Van Oosterwyck H 2017 Fibrin structural and diffusional analysis suggests that fibers are permeable to solute transport *Acta Biomater.* **47** 25-39

IOP Conf. Series: Materials Science and Engineering **742** (2020) 012017 doi:10.1088/1757-899X/742/1/012017

An Evaluation of the Effect of Takakura and Effective Microorganisms (EM) as Bio Activators on the Final Compost Quality

Solmaz Aslanzadeh[1]*, Katherine Kho[1], Irnayuli Sitepu[2]

[1]Biotechnology Study Program, School of Life Sciences, Indonesia International Institute for Life Sciences, Indonesia
Email: *solmaz.aslanzadeh@i3l.ac.id, katherine.k@i3l.ac.id
[2]Phaff Yeast Culture Collection, Department of Food Science and Technology, University of California Davis, One Shields Ave, Davis, USA

E-mail : irsitepu@ucdavis.edu

Abstract. Composting is an organic waste management option, where suitable microorganisms are used to accelerate the rate and the decomposition process and improve the final compost quality. In this study, a commercial bio-activator EM4 (EM) and a traditional bio-activator prepared according to the Takakura method (Tak) were investigated. The bio-activators were mixed with a fermenting bed (fb) material consisting of rice husk and rice bran mixed in a ratio of 1:1. The inoculated fb was then used in concentrations of 0%, 20%, and 40% in a composting process of food waste. The composting process was monitored daily for 48 days, and the quality of the final compost based on its physical and chemical characteristics was measured and compared. The result revealed that all samples containing 40% fb material with both Tak and EM bio- activator contained lower moisture content in the final compost compared to samples with 20% fb material. The C/N ratio of the samples with 20% fb material for all samples decreased to 20:1, while 40% fb material increased above 20:1, compared to its initial level. Furthermore, the nitrogen level for all samples with 20% fb material increased while samples with 40% fb material decreased from the initial level. The phosphorus and potassium content also increased for all samples with both 20% and 40% fb material from the initial level. Additionally, no significant difference could be found for the final compost quality between the two bio-activators.

1. Introduction

Composting is a degradation of organic matter under a controlled environment, which is widely recognized as a mean of organic waste management today[1]. However, the process of composting in nature is slow and it is not considered an efficient process in an ever-rising organic waste in our modern society [2]. Many methods have been developed to increase efficiency and ultimately the rate of the composting process while keeping high compost quality. Numerous attempts have been done to control parameters such as temperature, pH, and moisture content. to create a pleasant environment for the composting microorganisms [3, 4]. Others have tried to add beneficial microbes to the composting pile. These microbes that aid the composting process could come from the natural environment and found often locally, in manure or from organic material such as fruit and vegetable commonly used in the Takakura composting method, which was developed at the Institute for Global Environmental Strategies

(IGES) of Japan [5-8]. The method is based on two fermented solutions, cultured from locally available materials and inoculated with rice bran and rice husk as fermenting bed (fb) material. The inoculated fermenting bed is then used as a seed compost or starter culture for composting organic fraction of municipal solid waste [9, 10]. Usually, the exact microbial community in Takakura in the cultured solutions is unknown and can vary depending on the source of the materials used. Studies that using Takakura culture helped compost food waste into good-quality compost with a C/N ratio of 15:1, which is an indicator of compost maturity within six weeks. Abushammala et al [11] also found that the Takakura method was superior in terms of shorter composting process and produced high-quality compost, to other microorganisms produced using 60% water and 40% fruit waste only. Similarly, Saad et al [12] came to the same conclusion using Takakura culture and fruit waste cultivated microorganisms for 7 weeks. According to the study, the C/N ratio of the compost reached 20:1, which is suitable to be used for agriculture.

Lately, commercial bio-activators, containing microorganisms with specific strains known as effective microorganisms (EM) have found their way into the market. The microbial community in these products is well defined and it contains a mixture of specific species of beneficial fermentative microorganisms [13]. Husen et al [14] reported that bio-activators or commercial bio-fertilizers at the market (agricultural shops) in Bandung District, West Java, contain various strain of bacteria from genus *Rhizobium, Azotobacter, Bacillus, Lactobacillus Azospirillum, Aspergillus, Actinomycetes, Lactobacillus, or Pseudomonas that has been mixed together with Actinomycetes or fungi such as Streptomyces or Aspergillus.* According to a study conducted by Ihsanullah [15], the addition of commercial culture (EM1) to a mixture of cow, poultry and kitchen manure in a ratio of 2:1:1 enhanced the final compost quality. Furthermore, the study also indicated that compost C/N stabilization occurred faster (6 weeks) compared with samples without the addition of the EM (18 weeks).

Although the benefits of commercial and Takakura culture has been proven in terms of processing time and final compost quality, few studies that have investigated the difference between EM and those prepared by cultivating local material containing unknown community in a composting process [6]. Therefore, this project aims is to investigate the difference between final qualities of compost produced with commercial effective microorganism to that of using the Takakura method.

2. Materials and methods

2.1. Materials

Tempeh, *oncom,* and the food waste, consisting of leftover food, fruits, and vegetables, were obtained from the traditional wet market in Jakarta, Indonesia. The garden waste was gathered from around the campus area and a local flower shop. The paper waste, consisting mostly of unused office paper and tissue paper were gathered from within the campus. The rice husk and the rice bran used as fermenting bed materials were obtained from a local traditional market in Jakarta. The commercial effective microorganisms (EM4, agriculture organic fermentation bacteria) starting culture was purchased from a local shop containing proprietary microbes of fermenting bacteria, from the genus *Lactobacillus*, fermented mushrooms, *Actinomycetes,* photosynthetic bacteria, phosphate solubilizing bacteria, and yeast.

2.1.1. Takakura culture preparation

The starter culture consisted of two different fermenting solutions; a sugar solution and a salt solution that were fermented separately and then mixed. For the sugar solution, 6 L of water were mixed with 400g brown sugar (carbon source), 600 g tempeh and 600 g *oncom* (fermenting microorganisms). For the salt solution, 6 L of water were mixed with 120 g salt and 1200 g of shredded vegetable waste. Both fermentation solutions were loosely capped to allow better aeration and to prevent possible overpressure due to microbial activities and were incubated for two weeks inside a shaker incubator at 25 °C and 100 rpm.

2.1.2. Fermenting bed (fb) preparation
Three different fermenting beds were prepared, which were Tak (Takakura solution inoculated fermentation bed), EM (commercial EM4 inoculated fermenting bed material) and Blk (fermenting bed with no inoculation). For the Tak preparation with a moisture content of 50%, the salt and sugar solutions prepared in section 2.1.1 were mixed and 924 ml of the mixed solution was added to 2.6 kg rice bran and 210 ml solution was mixed with 0.5 kg rice husk. The inoculated rice bran and rice husk fermenting beds were then mixed in a ratio of 1:1. Similarly for EM, the rice bran and rice husk were mixed with EM4 commercial culture first and then mixed together in a ratio of 1:1 to reach a final moisture content of approximately 50%. For the Blk samples, only pure water type III was used to mix with the fermenting bed materials which were then mixed to gather in a 1:1 ratio for adjusting to a final moisture content of 50 %. These inoculated fermenting beds were incubated for 1 week at 25°C before use. At the end of the incubation period, the TS (total solid content) content of each fermenting bed materials were checked and the moisture content were adjusted before setting up the experiment.

2.1.3. The organic waste mixture preparation
The food waste was sorted, and the substrate was further treated in the lab by shredding it into of 1-2 cm size. The garden waste, which was a mixture of green leaves and branches were cut down in size using a laboratory garden scissor to a size between 2-5 cm. The office paper waste was also shredded in smaller pieces. The TS of all the substrates used in this study was measured and the data is presented in Table 1. The food waste, garden waste, and paper waste were mixed based on individual C/N ratios 25:1, 30:1, and (175:1) of food scraps, garden waste, and paper, respectively described by [16]. The final waste mixture contained approximately 45% food scrap, 50% garden waste and 5% paper with a final C/N ratio of approximately 35:1.

2.2. Experimental setup
The experiment was carried out in 8 L home-made reactors with aeration holes at the sides with filter space for leachate removal at the bottom. Each experiment contained a total volume of 5 L substrate containing organic waste mixed with 20% and 40% Tak (Tak20 and Tak40), EM (EM20 and EM40) or Blk (Blk20 and Blk40). Organic waste without any fermenting bed material (FB0) was also set up as a negative control. The initial moisture content of all samples was adjusted to approximately 70 ± 7% in this study. All samples were aerated by turning the pile every day in the first two weeks in the composting house outdoors and after that only once a week for the rest of the experimental period. During the composting period, the change in temperature and pH were monitored every day. The initial and the final moisture content, C/N ratio, nitrogen, phosphorus, and potassium (NPK) content for the experiments were measured. The experiment was carried out in triplicates.

2.3. Analytical methods
The moisture content of the substrates was measured according to a protocol described by NREL procedures [17]. The solid fraction was dried at 105°C until the stable weight was recorded. Nitrogen, phosphorus, potassium, and carbon were analyzed at a service laboratory at the Bogor Agriculture Institute (IPB). The carbon analysis used ash gravimetric method [18]. The nitrogen content was analyzed using the Kjeldahl method based on block digestion and steam distillation [19]. The phosphorus analysis used spectrophotometer in the form of P_2O_5 molecules. The potassium analysis used F-AAS in form of K_2O molecules [20].

3. Results and Discussion

3.1. Temperature and pH
The daily temperature and pH of all samples are presented in Figure 1. The temperature in all the experiments started to increase slowly in the first day after the set-up and reached the mesophilic range

at a peak between 26.5°C and 30°C after two weeks. However, the temperature never reached a thermophilic range as expected before it started to decrease for all samples at week three, which then fluctuated between 26°C and 28°C, and stayed constant at this temperature until the end of composting process. The increase in temperature in the initial stage of the composting process for all experiments is an indication of microbial activity, even though the increase was only by 2°C to 5°C at maximum. This slight temperature increase could be due to a combination of daily manual mixing for optimum aeration of the waste and small scale reactors which increases the heat loss, causing lower temperatures [21, 22].

The pH at the start of the experiments ranged between 6 and 7. The pH of all samples showed a decreasing trend during the first three weeks of the composting process. All samples reached the lowest pH (4.5 – 6) in the earlier stages of the process. This is a clear indication that the decomposition of organic waste had started due to the accumulation of organic acid [23, 24]. After week three, the pH started to slowly rise again for all samples, until the end of the process with a final pH between 6 and 7. The increase in pH was also an indication that organic acids starting to become neutralized. Moreover, no specific difference was observed between the samples inoculated with Tak culture and those of EM culture in their average pH trend during the entire process for either fb compositions used in this study and the pH of all samples was within recommended pH range (between 6 – 7) for mature compost [3, 25].

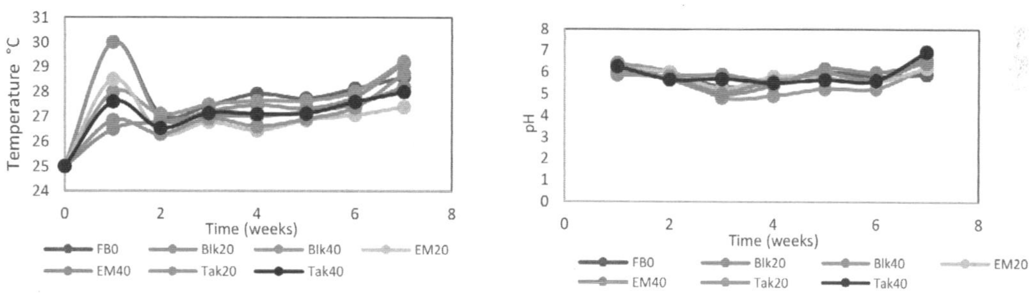

Figure 1. The temperature (left) and pH (right) of the experiments during 7 weeks (48 days) of composting process.

3.2. *Moisture content*

Moisture is important to support microbial activity because it increases the rate of metabolism [18]. The moisture content of the final compost is presented in Table 1. The moisture content reduced during the composting process for all samples studied. The reduction in the moisture content at the end of the composting process is a positive sign of decomposition and it is an important parameter and indicator for mature compost [26]. Furthermore, the samples with 40% fb material contained lower moisture content compared to the samples with 20% fb material at the end of the experimental period (Table 1). The moisture content of EM40, Tak40, and Blk40 were 13%, 27% and 11%, lower than EM20 and Tak20 and Blk40, respectively. This could be because samples with higher fb concentration contain proportionally lower organic matter. The fb material acts as a bulking agent causing the lower bulk density in the compost. Results of this study is in accordance with a previous study carried out by Monson and Murugappan [27], who confirmed that the decrease of bulk density will lead to higher porosity in the compost, causing moisture loss especially with frequent aeration rate at the initial stages of the composting process [28].

In general, the EM and Tak inoculated samples in both 20% and 40% fb material concentrations was lower compared to that of the Blk samples. However, no significant difference was observed in moister content between the EM and Tak (Table 1). The highest moisture content of the final compost was observed for the FB0 sample, with 66.3±1.5%. The addition of 20% fb material decreased the final moister content to 63.5±2.0 % for Blk, 45.0±1.5% for EM and 55.3±2.1% for Tak40. Increasing the fb material to 40% further decreased the final moister content to 56.9±1.8%, 45.0±1.5%, and 43.4±1.6%

for the Blk, EM, and Tak, respectively. The higher water loss observed in all Tak and EM inoculated samples could be due to a higher concentration of microbial population causing higher microbial activity and water loss during the composting process, particularly for samples containing a higher concentration of fermenting bed material.

Table 1. The final moisture content (%), the initial and final C/N ratio, nitrogen (%), phosphorus (%) and potassium (%) content of the composting experiments with and without Tak and EM bio-activators.

Samples[a]	Moisture Content (%)	C/N Ratio		Nitrogen (%)		Phosphorus (%)		Potassium (%)	
		Initial	Final	Initial	Final	Initial	Final	Initial	Final
FB0	66.3±1.5	25:1	26:1	0.33±0.03	0.36±0.05	0.07±0.12	0.06±0.10	0.53±0.1	0.05±0.03
Blk20	63.5±2.0	22:1	20:1	0.38±0.02	0.56 ±0.03	0.14±0.04	0.39±0.01	0.5±0.03	0.62± 0.05
Blk40	56.9±1.8	16:1	25:1	0.6±0.01	0.48±0.08	0.24±0.01	0.55±0.02	0.44±0.02	0.57±0.02
EM20	52.3±2.0	23:1	20:1	0.52±0.02	0.57±0.02	0.15±0.02	0.31±0.04	0.51±0.04	0.68±0.02
EM40	45.0±1.5	20:1	26:1	0.55±0.05	0.39±0.01	0.18± 0.04	0.49±0.02	0.46±0.02	0.62±0.04
Tak20	55.3±2.1	25:1	21:1	0.46±0.03	0.54±0.04	0.18± 0.02	0.35±0.02	0.49±0.02	0.66±0.05
Tak40	43.4±1.6	16:1	19:1	0.68±0.02	0.7±0.02	0.29±0.01	0.52 ±0.01	0.50±0.05	0.61±0.04

[a] FB0 (Organic waste mixture with no fermenting bed material), Blk20 and Blk40 (Organic waste mixed with 20% and 40% fermenting bed material without culture inoculation, [Blk=Blank]), EM20 and EM40 (Organic waste mixed with 20% and 40% fermenting bed material inoculated with commercial EM4 culture), Tak20 and Tak40 (Organic waste mixture with 20% and 40% fermenting bed material inoculated with Takakura culture).

3.3. Carbon to Nitrogen (C/N) ratio

The results of the C/N ratio before and after the composting process were measured and the results are presented in Table 1. In general, the C/N ratio of all the samples containing 20% fb material slightly decreased after the composting process, which was stabilized at a C/N ratio between 20:1 and 21:1 at weeks 7 composting process. The decrease in C/N ratio in the final compost during the process is not uncommon, as C/N ratio decreases for the compost with increasing maturity [29] and it is caused by the release of carbon as CO_2 as a result of microbial activity [30]. The C/N ratio below 20:1 for Municipal solid waste (MSW) has been suggested to be an indication of whether the composts have reached maturity during the conventional composting process[12]. The decrease in C/N ratio for samples containing 20 fb was not enough for any of the samples to be considered mature compost. However, some studies argued that the C/N ratio is not an appropriate parameter for assessing compost maturity when inoculants are added [31, 32].

In contrast, the C/N ratio of all samples containing 40 % fb material increased during the composting process. Increasing the fb material resulted in the samples having a somewhat higher final C/N ratio compared to their initial C/N level, which was measured to be 16:1 for the Blk and Tak samples and slightly higher around 20:1 for the EM sample. After the composting process, the C/N ratio was increased 25:1, 26:1 and 19:1 for the Blk, EM, Tak samples, respectively. The reason could be due to the low C/N ratio in the starting material (16:1-17:1) which is lower than the recommend optimum of 20:1 to 40:1 recommended for an effective composting process. A low initial C/N ratio would cause higher NH_3 formation and emission leading to higher nitrogen loss especially at higher bulking agent concentrations and as a result the C/N ratio of the final compost increases [33, 34]. When the C/N ratio is less than 18.1, the microorganisms transform the nitrogen in a higher degree, favoring its volatilization[35]. Furthermore, no significant difference in final C/N was observed between the EM cultures. On the other hand, the FB0 sample did not significantly change from its initial value and the C/N ratio of 25:1 which increased to 26:1 during the entire composting period which could be due to low microbial community activity mostly due to limitation of microbial respiration [36] and these samples were not inoculated.

3.4. The nitrogen, phosphorus and potassium content

The nitrogen, phosphorus and potassium content of the experiments before and after the composting process were measured, and the results are presented in Table 1. An increase in the final nitrogen content was observed for all samples containing 20 % fb material during the composting process as compared to that of the initial nitrogen level. The increase was almost by 17% for Blk and the Tak sample, while for EM sample the increase was 9.6%. In contrast, increasing the ratio of the fb material to 40% decreased the final nitrogen content as compared to its initial value. The Blk and EM sample decreased by almost 20% and 29%, respectively, while no difference in the initial and final nitrogen content was observed for the Tak and FB0 samples. This contradictory result in nitrogen content between samples containing 20% and 40% fb material can be explained by the difference in the ratio of organic waste to fb materials in the samples. The fb material is made up of rice husk and rice bran consisting of lignocellulose as the major component [37] with low availability and digestibility of the carbon content. Thus, having a higher ratio of fb material would affect the C/N ratio of the sample and cause higher nitrogen loss in the final compost. Some studies showed similar compost C/N ratios, a higher content of biodegradable carbon with reduced nitrogen emission, while less biodegradable carbon such as lignocellulose material would increase the nitrogen loss in the form of NH_3 emission [35]. Another reason for lower nitrogen content in the samples with higher fb material could be due to low moisture content, which affects also microbial activity. Lower moisture content in the compost samples with higher fb material could have caused limitation of the microbial activity to fix nitrogen as NH_3 emission increases at low moisture content. Furthermore, there was no difference in the initial and final nitrogen for the Tak40 and FB0 samples. For the Tak40 the low moisture content (43%) in the sample might have lowered the microbial activity at some point during the composting process where the nitrogen content neither increased nor decreased in the sample [23]. While for the control sample FB0, the lack of fermenting bed material with high moisture content causing oxygen transfer limitations in the compost, thus inhibiting the microbial activity [23, 38, 39].

The phosphorus content was also increased for all experiments during the composting process compared to their initial content. The initial phosphorous content for the experiments containing 20% fb material was 0.14±0.04%, 0.15±0.02% and 0.18±0.02% for the Blk, EM, and Tak, respectively. After the composting process, the phosphorus content of the final compost more than doubled for all samples and ranged between 0.31±0.04% to 0.39±0.01%. The same trend was also observed for the samples containing 40% fb material with the initial phosphorus content ranging from 0.18 ± 0.04% for the EM sample to 0.29 ±0. 01 % for the Tak sample, which after the composting process more than doubled to 0.49±0.02 and 0.52±0.01%, respectively. The highest final phosphorus content was observed for the Blk sample with 0.55±0.02%. Moreover, no change in phosphorus content was observed for FB0 which remained at the same value as its initial phosphorous content of 0.07 ± 0.12 %. The final phosphorus content for Em40, Tak40, and Blk40 was 58%, and 48% and 41% higher than EM20, Tak20, and Blk20, respectively.

The potassium content also increased during the composting process for all samples containing fb material. The initial potassium content ranged from the lowest for Blk40 0.44 ± 0.02 % to the highest for EM20 with 0.51% ± 0.04%. After the incubation period, the final compost increased 1.3 times higher compared to the initial level for all samples studied. Moreover, the final potassium content for the EM20, Tak20, and Blk20 contained approximately 10% and 8% and 9% higher than EM40, EM20 and Blk20, respectively. The lowest potassium content in the final compost was observed for Blk40 which was about 0.57 ± 0.02 % and the highest potassium content was observed for EM20 with 0.68 ± 0.02%. Furthermore, the final potassium content for FB0 measured to be significantly decreased from 0.53% to only 0.05 ± 0.03 % in the final compost.

The increase in the phosphorus and potassium content in the final compost in both 40% and 20% fb material could be due to the concentration effect. According to Lin [29], the concentration effect arises from higher carbon loss rate in form CO_2 or methane during decomposition process as compared to the loss rate of phosphorus and potassium, thus leading to a higher content of phosphorus and potassium content in the compost as a result of this concentration effect. The difference in phosphorus and potassium levels in 20% and 40% fb material could be due to microbial activity and the environmental condition. Irshad et al [40] observed the potassium and phosphorus at various stages of composting and compared with the initial level. It was found that both phosphorus and potassium concentration decreased with the time of composting and concluded that the fluctuation observed in the release of P nutrient could be due to the variation in the microbial activity and C/N ratio. This indicates that the 48 days of composting time was not enough time to mature and stabilize in this study. Moreover, there was no significant difference in the final potassium content in the FB0 as compared to its initial level, which is probably due to the lack of fb material causing none or low microbial activity. The lowering of the final potassium observed in the FB0 is probably due to the loss of potassium content during the leaching process [40].

4. Conclusion

In this study, composting with 20% and 40% fermenting bed inoculated with culture fermented according to Takakura method and commercial bio-activator (EM4) mixed with organic waste was evaluated during the 48 days and compared to that of the blank sample with no inoculation. The results showed the concentration of fermenting bed material was effective in the final compost quality. On the other hand, there was no significant difference in the final compost quality between EM4 culture and Takakura culture compared to that of the blank samples based on the C/N ratio, moisture content and NPK content.

Acknowledgment
The authors would like to acknowledge the Indonesia International Institute for Life Sciences Waste Management Program for funding this project. Moreover, the authors would like to thank Muhammad Imam Kurniawan for the practical part of this work and Givens Pratiwi Marpaung for the laboratory support of this project. The authors are also grateful to Sugeng Roh Muryanto for designing and building the composting reactors used in this study.

Reference.

[1] Argun Y, Karacalı Tunç A, Çalışır U and Kılınç N 2017 Composting as a Waste Management Method *Journal of International Environmental Application & Science* **12** 244-55
[2] Silpa Kaza, Lisa Yao, Perinaz Bhada-Tata and Frank Van Woerden 2018 WHAT A WASTE 2.0- A Global Snapshot of Solid Waste Management to 2050 *World bank group- Urban development series*
[3] Khairul, Binti B, Hamid A, Huzairi M, Mohd Zainudin M H, Armi M, Samah A and Yunus K 2016 Improving Composting Process By Controlling Aeration Time *International Journal of Applied Chemistry* **12** 130-5
[4] Gigliotti G, Valentini F, G Erriquens F and Said-Pullicino D 2005 Evaluating the efficiency of the composting process: a comparison of different parameters *Geophys. Res. Abs.* **7**
[5] Manullang R R and Dan Daryono R 2018 Combination microorganism as local bio activator compost kirinyuh *International Journal of Scientific and Technology Research* **7** 1-5
[6] Hartono D, Kristanto G and Kusuma A 2011 Evaluating the Role of Bioactivators in Home Made Composting *International Journal of Applied Engineering Reserach* **6** 1437-44
[7] Sondang Y, Anty K and Alfina R 2014 The Influence of Bioactivator Cattle Feces Against The Length of Composting and C/N Ratio from Three Kind of Organic Material *International Journal on Advanced Science, Engineering and Information Technology* **4** 278
[8] Jiménez Antillón J, Calleja-Amador C E and Romero L 2018 Food Waste Recovery with Takakura Portable Compost Boxes in Offices and Working Places *Resources* **7** 84

[9] Maeda T 2009, Retrieved from http://pub.iges.or.jp/modules/envirolib/ view.php?docid=2661 Reducing waste through the promotion of composting and active involvement of various stakeholders: Replicating Surabaya's solid waste management model. In: *IGES Policy Brief.* : Institute for Global Environmental Strategies)

[10] Nuzir F, Shiko Hayashi, and Koji Takakura 2019 Takakura Composting Method (TCM) As An Appropriate Environmental Technology for Urban Waste Management *International Journal of Building, Urban, Interior and Landscape Technology (BUILT)* **13** 67-82

[11] Abushammala M, Ezlin Ahmad Basri N, Md Zain S, Mat Saad N F and Afida Zainudin N 2015 Green Biological Transformation of Food and Yard Waste *Jurnal Teknologi* **73** 21-6

[12] Mat Saad N F, Nadrah Ma'min N, Md Zain S, Ezlin Ahmad Basri N and Sofia Md Zaini N 2013 Composting of Mixed Yard and Food Wastes with Effective Microbes *Jurnal Teknologi* **65**

[13] Daly M J and Stewart D P C 1999 Influence of "Effective Microorganisms" (EM) on Vegetable Production and Carbon Mineralization–A Preliminary Investigation *J. Sustainable Agric.* **14** 15-25

[14] Edi Husen R D M S, Rasti Saraswati and Irawan 2007 Characterization and quality assesment of Indonesian commercial biofertilizers *Indonesian Journal of Agricultural Science* **8** 31-8

[15] Daur I 2016 Study of commercial effective microorganism on composting and dynamics of plant essential metal micronutrients *J. Environ. Biol.* **37** 937-41

[16] Hirrel S S, Riley T and Andersen C R Retrieved 2018 Composting. (USA: United States Department of Agriculture, and County Governments Cooperating)

[17] A. Sluiter, R. Ruiz, C. Scarlata, J. Sluiter and Templeton D 2008 Determination of total solids in biomass andtotal dissolved solids in liquid Process samples In: *National renewable energy laboratory (NREL),* (Golden, CO., USA

[18] Miyazawa M, Pavan M A, de O, Ionashiro M and Silva A K 2000 Gravimetric determination of soil organic matter *Brazilian Archives of Biology and Technology* **43**

[19] Boström C Å, Cedergren® A, Johansson G and Pettersson I 1974 Rapid coulometric method for the Kjeldahl determination of nitrogen *Talanta* **21** 1123-7

[20] Motsara M R and Roy R N 2008 Guide to laboratory establishment for plant nutrient analysis In: *FAO fertilizer and plant nutrition bulletin no 19,* (Rom: Food and Agricultural Organisation of United Nations (FAO))

[21] Zhao R F, Gao W and Guo H Q 2017 Comprehensive review of models and methods used for heat recovery from composting process *International Journal of Agricultural and Biological Engineering* **10** 1-12

[22] Misra R V, Roy R N and Hiraoka H 2003 On-Farm Composting Methods. Food and Agriculture organization of united nations)

[23] Jamaludin S, Abdul Kadir K and Wahidah N 2017 Study on NPK Performance in Food Waste Composting by Using Agricultural Fermentation *MATEC Web of Conferences* **103** 05015

[24] Sundberg C, Yu D, Franke-Whittle I, Kauppi S, Smårs S, Insam H, Romantschuk M and Jönsson H 2013 Effects of pH and microbial composition on odour in food waste composting *Waste Management* **33** 204-11

[25] Atalia KR, Buha DM, Bhavsar KA and NK S 2015 A Review on Composting of Municipal Solid Waste *IOSR Journal of Environmental Science, Toxicology and Food Technology (IOSR-JESTFT)* **9** 20-9

[26] Ameen A, Ahmad J and Raza S 2016 Effect of pH and moisture content on composting of Municipal solid waste *International Journal of Scientific & research publication* **6** 35-2250

[27] Monson C C and Murugappan A 2010 Developing Optimal Combination of Bulking Agents in an In-Vessel Composting of Vegetable Waste *E-Journal of Chemistry* **7**

[28] Sundberg C and Jönsson H 2008 Higher pH and faster decomposition in biowaste composting by increased aeration *Waste Management* **28** 518-26

[29] Lin C 2008 A Negative-Pressure Aeration System for Composting Food Wastes *Bioresour. Technol.* **99** 7651-6
[30] Azim K, Khalid O, Aomar A, Périssol C, Thami-Alami I and Soudi B 2014 *Dynamic composting optimization through CN ratio variation as a startup parameter*
[31] Sugahara K and Inoko A 1981 Composition analysis of humus and characterization of humic acid obtained from city refuse compost *Soil Sci. Plant Nutr.* **27** 213-24
[32] Chukwujindu I, A.C E, F.N E and N.O I 2006 Compost Maturity Evaluation and its Significance to Agriculture *Pak. J. Biol. Sci.* **9**
[33] Garge A and Tothill I E 2009 A review of solid waste composting process- The UK perspective *Dynamic soil, Dynamic plant* **3** 57-63
[34] Sun X-p, Lu P, Jiang T, Schuchardt F and Li G-x 2014 Influence of bulking agents on CH_4, N_2O, and NH_3 emissions during rapid composting of pig manure from the Chinese Ganqinfen system *J Zhejiang Univ Sci B* **15** 353-64
[35] Sánchez Ó J, Ospina D A and Montoya S 2017 Compost supplementation with nutrients and microorganisms in composting process *Waste Management* **69** 136-53
[36] Eiland F, Klamer M, Lind A M, Leth M and Bååth E 2001 Influence of Initial C/N Ratio on Chemical and Microbial Composition during Long Term Composting of Straw *Microb. Ecol.* **41** 272-80
[37] Kodali B and Ravindra P 2006 Pretreatment studies of rice bran for the effective production of cellulose *Elect. J. Environ. Agric. Food. Chem* **5** 1253-64
[38] Sarkar S, Pal S and Chanda S 2016 Optimization of a Vegetable Waste Composting Process with a Significant Thermophilic Phase *Procedia Environmental Sciences* **35** 435-40
[39] Hao X and Benke M B 2008 Nitrogen transformation and losses during composting and mitigation strategies *Dynamic Soil, Dynamic Plant; Hao, X., Ed.; Global Science Books: London, UK* **2** 10-8
[40] Irshad M, Inoue M, Shezadi M, Khan T and Faridullah F 2011 Ammonium, phosphorus and potassium release from animal manure during composting *Journal of Food, Agriculture and Environment* **9** 629-31

IOP Conf. Series: Materials Science and Engineering **742** (2020) 012018 doi:10.1088/1757-899X/742/1/012018

Antioxidant Activity of Blended of Robusta Coffee (Coffea canephora L.) with The White Turmeric (Curcuma zedoaria (Berg.) Roscoe) and Wild Ginger (Curcuma xantorrhiza, Roxb.)

P A Gafar[1), S Agustini[1)

[1] Palembang Institute of Research and Standardization For Industry

E-mail: patoni@kemenperin.go.id

Abstract. This research aims to study the manufacture of herbal mixed coffee by blending the ground roasted coffee with the white turmeric extract and wild Javanese turmeric extract as a natural additive. The treatment was the ratio between ground roasted coffee and white turmeric and wild Javanese turmeric extract, which consisted of three levels namely 80:20%, 60:40%, 40:60%. The measurement of antioxidant activity was carried out using the DPPH method as % inhibition. The results showed that the antioxidant activity of coffee mixed with white turmeric was 18%, 19%, 48% respectively for each ratio and IC_{50} of 12.98%; while coffee mixed with Javanese turmeric extract showed the antioxidant activity of 20%; 32%; 48% with IC_{50} 7.71%.

1. Introduction

Coffee (*Coffea* sp) is economically one of most important commodity and the most popular beverages in the world. Coffee known as a mayor contributor for dietary antioxidant intake, since it has antioxidant activity and radical scavenging activity, anti inflammatory, anti aging, anti bacterial, antiviral, and suppressive activity of metalloproteinase expression [1,2,3,4,5].

Coffee is known to be rich in biologically active substances such as chlorogenic acid, nicotinic acid, trigoelline, quinolic acid, tannic acid, pyrogallic acid and, caffeine. It has antioxidant activity and radical scavenging activities [Ratz]. Recent study by Liang showed that bioactive component in coffee can trigger tissue anti oxidant gene expression and protect against gastrointestinal oxidative stress. Kawamura and Muraoki stated that antioxidant play important roles in delaying or preventing oxidation of intracellular and extracellular bio molecules [7]

Herbal coffee is a coffee product that has been enriched with herbs or spices to improve the taste, and nutritional value as well as health benefit, due to its antioxidant properties. Herbs and spices function as a flavouring agents as well as a source of bioactive components that have health benefit [ariefandie, Durak] such as ginger, panax (from the genus of *Panax* L.), black cumin/black seed/*Nigella sativa*, cinnamon, cardamon and others [febrianto et al]. Hoffman et al stated that the combination of coffee and herbal more effective for enhancing metabolic rates than consuming caffeine alone. Some spices and herb have been used traditionally to treat diseases and ailment and reported to be excellent source of phenolic compounds with good antioxidant activities [Otunola].

White turmeric (CZ) or commonly known as Zedoary (Curcuma zedoaria), and temulawak or Javanese turmeric (CX/*Curcuma xantorrhiza*, Roxb.) are type of plant which known as herbal medicine. CZ rhizome contains bioactive compounds such as flavonoids, polyphenols, curcuminoid, terpenoids and essential oils so it has antioxidant, anti-inflammatory, anti-cancer and anti-microbial activity [Dosoky, Marliani]. The effectiveness of anti oxidant activity (EC_{50}/effective concentration at 50%) of aqueous extract of CZ by DETBA method was 7.1±0.11 μg/mL more effective than ascorbic acid (66.15 μg/mL), and similar to BHA (6.62 μg/mL) and α–tocopherol (6.71 μg/mL). EC_{50} by TBARS (8.9±0.08 mg/mL), Scavenging abilty by DPPH (6.2±0.76 mg/mL) and chelating ability by FRAP method was 20.4±0.19 mg/mL (Huang et al).

CX (*Curcuma xantorrhiza*, Roxb.) has been shown to exhibit anticancer, anti bacterial, antifungal anti-hepatoxic, anti-inflammatory, antidysmenorrheal, antispasmodic cholesterol reducing activities and anti oxidant (Dosoky, Rosidi et al 2016, Kawiji, Nurcholis, Devaraj). It also known as a traditional medicine to treat hepatitis, liver complaints, diabetes, rheumatism, cancer, hypertension, and heart disorders. Water extract of CX contain bioactive component such as curcuminoids (0.319 %), phenols (3.258 %) [Kawiji], and flavonoids (10.60 mg/g) [Devaraj]. The main substances of C starch (48.18-59.64%), volatile oils (phelandren, camphor, tumerol, sineol, borneol) 3-12%, and xanthorrhizol (1.48-1.63%), sesquiterpenes (β-curcumene, ar-curcumene, bisabolane, lactone germacone), flavonoids (catechin, epicatechin, quercetin, myricetin, kaemferol, apigenin, luteolin, naringenin), and also curcuminoids. Aqueous extract of CX had a total phenolic content accounting of 24.3 mg/g. At the concentration of 1 mg/mL aqueous extract of CX had 62.3 % antioxidant activity by using DPPH and FRAP value 358 mmol (Rohman et al).

In addition there are several factors contribute to the oxidative stress, including pollution and radiation, diet and lifestyle. Oxidative stress is an imbalance between oxidant particularly reactive oxygen species (ROS/ free radicals) and antioxidants in the body which results in pro-oxidative state, which can lead to cell and tissue damage. Oxidative stress occurs naturally and plays a role in the aging process. The body's cells produce free radicals during normal metabolic processes. However, cells also produce antioxidants that neutralize these free radicals. In general, the body is able to maintain a balance between antioxidants and free radicals. Uncontrolled oxidative stress can accelerate the aging process and may contribute to the development of a number of disease conditions. Recently exposure to environmental pollutant has become mayor health concern, and thus fostered interest for natural antioxidant which can be consumed as a food while having health benefits.

A combination of bioactive component from coffee and herbs, such as white turmeric and wild ginger is believed to have health benefit and be able to inhibit oxidative stress to prevent degenerative disease. Therefore, the purpose of this study was to evaluate the antioxidant activity of coffee mixed with herbs such as turmeric extract and wild ginger extract.

2. Materials and Methods

The Robusta coffee bean used in this research obtained from public farming in South Sumatera, while white turmeric and temulawak (Javanese turmeric) extracts were Suri Brand from small scale industry. The equipments used consisted of roasting machine, grinder, hexagonal blender, analytical balances, oven, sealer, plastic pack.

The ground roasted coffee was prepared by roasting the Robusta coffee bean in a roaster which is a drum rotary machine that allow the coffee bean heated thoroughly, and equipped with cooling fan to cool the roasted coffee. The color of coffee bean can be monitored by using checking window to determine the desired roasting level. Prior to the roasting, coffee bean (10% moisture) were preheated in the roaster until the roaster temperature was 200°C. Soon after the coffee bean was fed into the roaster, and heated until the colour of roasted coffee was as desired. Normally roasting time was 40 min. The roasted coffee then finally air-cooled until the temperature 40°C to stop the roasting process completely and resting for degassing. The roasted coffee was ground by using grinding machine up to 80 mesh in size. The ground roasted coffee was then stored in dry place until used.

The herbal coffee powder were prepared as follows : Simple mixture design was used to determine the proportion of the factors (C: ground roasted coffee, T: white turmeric extract and G Javanese turmeric extract). The levels were in the range of 0% to 100% (80:20; 60:40 and 20:80), with the total mixture was 100% (C+T/G=100). The ground coffee and white turmeric/Javanese turmeric extract was mixed until evenly distributed powder was obtained. For hedonic test, the powder was brewed using hot water prior to testing.

2.1. Antioxidant Activity

Antioxidant activity determined by using scavenging effect on 2,2-diphenyl-1-picryl hydrazyl radical (DPPH). The radical scavenging ability was determined as described by Liyana-Pathiranan and Shahidi (2005) with slight modification. Briefly, 1.0 g of coffee mixed was added with 10 ml and heated at 100°C. It was allowed to cool and filtered. 100 µl coffee mixed extract in a test tube was added with 2.9 ml of 0.5 mM methanolic solution of DPPH, shaken vigorously and incubated at room temperature in the dark for 30 min. The absorbance was read at 517 nm. DPPH free radical scavenging activity as calculated as :

$$\% \text{ DPPH inhibition} = (Abs_{(control)} - Abs_{(sample)})/Abs_{(control)} \times 100$$

Where $Abs_{(control)}$ is the absorbance of DPPH radical + methanol and $Abs_{(sample)}$ is the absorbance of DPPH radical + sample extract.

2.2. Statistical Analysis

The result was expressed as an average and standard deviation of three replication. Data were analyzed using statistical program SAS 9.1.3.

3. Result and Discussion

The results of antioxidant test on coffee mixed showed in Table 1.

Table 1. The average of antioxidant activity (% inhibition) and Total Phenol Content (TPC)

Sample Code	(%) inhibition	Sample Code	TPC mg GAE /g
CT20	54.75±1.60	CT20	7.30±0.67
CT40	59.04±1.03	CT40	10.27±0.50
CT60	64.66±1.20	CT60	12.62±1.52
CG20	51.59±1.90	CG20	7.22±0.72
CG40	60.27±1.78	CG40	8.82±0.56
CG60	69.04±1.39	CG60	12.28±0.48

Data in Table 1 and Figure 1 showed that the higher of natural additives in a coffee mixed, the higher the antioxidant activity. It showed that the antioxidant activity of coffee mixed significantly different ($p<0.05$) for each treatment for both white turmeric and Javanese turmeric. The highest inhibition for coffee mixed is obtained at 60% white turmeric. This result also applies for coffee mixed with Javanese turmeric. These results indicate that the antioxidant activity of white turmeric and Javanese turmeric is higher than the antioxidant activity of coffee. This results is in good agreement with Mulato and Suharyanto (2012), who reported that the antioxidant activity of light, medium, and dark roasted coffee was about 11%, 10 % 14% respectively.

Plant food contains complex array of phenolic compound that may help increase antioxidant ability, and are the main source antioxidant in the diet. Test result on the total phenol content of herbal coffee was range from 7.30±0.67 up to 12.62±1.52 mg GAE/g. White turmeric known to contain bioactive component such as flavonoids, polyphenols, curcuminoids, terpenoids, and essential oils. According to Marliani, et al (2017), the phenolic content of the water extract of white turmeric was 19.9456±6.831 up to 23.463±2.010 mg GAE/g. Devaraj et al (2014) stated that the water extract of Javanese turmeric only contained total phenol of 0.752±0.015 mgGAE/mg. substance which have antioxidant properties as well as white turmeric and Javanese turmeric. This explains why the total phenol of herbal coffee from coffee and ginger mixture is lower than herbal coffee from coffee with white turmeric.

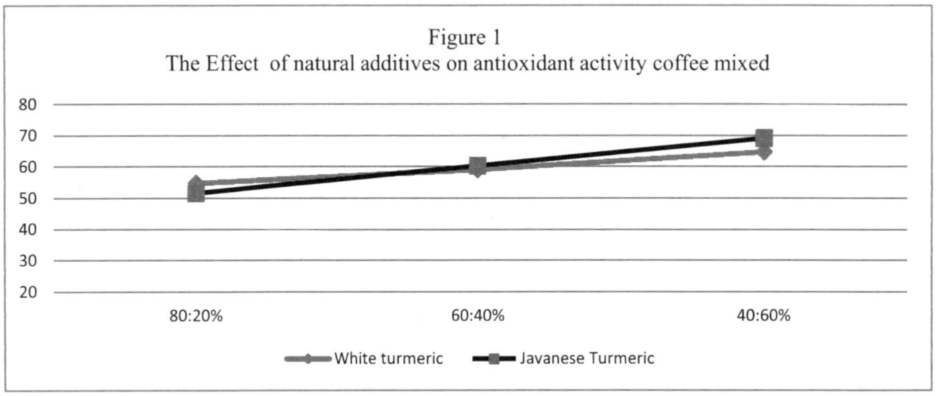

4. Conclusion

The herbal coffee mixed product and formulation has been successfully developed. The addition of white turmeric and Javanese turmeric instant powder significantly increased the antioxidant activity of herbal coffee. The formulation of 60% white turmeric and Javanese turmeric instant powder provide the highest antioxidant activity. Both herbal coffee showed abilities to scavenge free radical activity.

Acknowledgement
The author research programs are supported by Palembang Institute of Research and Standardization for Industry, Ministry of Industry, Republic of Indonesia.

References

[1] Febrianto AN, Riki VM, and Djumiarti 2015 *Pelita Perkebunan* 31 (1);49-58.
[2] Moreira DP, Monteiro MC, Ribeiro-Alves M, Donangelo CM, and Trugo LC 2005 *journal of Agricultural and Food Chemistry* 53(5) 1399-1402
[3] Almeida AAP, Farah A, Silva DAM, Nunan EA, and Gloria MBA 2006 *Journal of Agricultural and Food Chemistry* 54(23) 8738-8743
[4] Utsunomiya H, Ichinose M, Uozaki M, Tsujimoto K, Yamasaki, H and Koyama AH, 2008 *Food and Chemical Toxicology* 46(6) 1919-1924
[5] Dogasaki C, Shindo T, Furuhata K, and Fukuyama M 2002. *Yakugaku Zassi* 122:487
[6] Ratz-Yko A, Arct J, Majewski S and Pytkowska K 2015 *Phytotherapy Research* 4 (2) 509-517.
[7] Affonso RCL, Voytena APL, Fanan S, Pitz H, Coelho DS, Hortmann A, Pereira A, Uarrota VG, Hillmann MC, Varela LAC, Ribeiro-do-Valle RM and Maraschin M 2016 *Oxidative Medicine and Cellular Longevity* ID 1923754.
[8] Kawamura T, and Muraoka I 2015 *Antioxidants* 7, 119; doi:10.3390/antiox7090119
[9] Hofman JR, Kang J, Ratamess NA, Jennings PF, Mangine G, Faigenbaum 2009 AD *J Int Soc Sports Nutrs* 3(1);35-41
[9] Otunola GA, and Afolayan AJ 2013 *Journal of Applied Botany and Food Quality* 86 : 66 – 70 DOI:10.5073/JABFQ.2013.086.010
[10] Durak A, Gawlik-Dziki U, Pecio L 2014 *Food Chemistry* 162 : 81-88
[11] Liyana-Phatiranan CM, and Shahidi F *2005 J.Agric Food Chem.* 53:2433-2440
[12] Stoilova I, Krastanov AI, Stoyanova A, Denev P, and Gargova S 2007 *Food Chemistry* 102(3):764-770

[13] Mulato S, Widyotomo S, and Suharyanto E 2014 *Indonesian Coffee and Cocoa Research Institute* :94- 98
[15] Devaraj S, Ismail S, Ramanathan S and Fey Yam M 2014 *A scientificWorld Journal ID3553128*
[16] Rosidi A, Khomsan A, Setiawan B, Riyadi H and Briawan D 2016 *Pakistan Journal Nutrition*15(8):556-560.
[17] Nurcholis W, Ambarsari L, Kartikasari NPE, darusman LK 2012 *Proceedings of Unesa's National Seminars on Chemistry* :C284-C292.
[18] Kawiji K, Atmaka W, Otaviana PR 2011 *Journal of Agricultural Product Technology* iv (1):32-40
[19] Huang S, Mau J, Tsai C, Chen C, Mau J, and Tsai S 2015 *Advanced Materials Research* 1120-1121: 920-925
[20] Dosoky NS, and Setzer WN 2018 *Nutriens* 10(9):1196.
[21] Rohman A, Widodo H, Lukitaningsih E, Rafi M, Nurrulhidayah AF, and Windarsih A 2019 *Food Research*

IOP Conf. Series: Materials Science and Engineering **742** (2020) 012019 doi:10.1088/1757-899X/742/1/012019

Antioxidant Stability Testing On Liquid And Powder *Eichhornia Crassipes* Extract

Nandhita Octavia Nugriani[1], Dr. –Ing. Diah I. W., S.T, M.Sc[1], Silvya Yusri, S.Si, M.T [1]

[1)] Department of Chemical Engineering, Faculty of Life Sciences and Technology, Swiss German University, Tangerang Selatan, Banten, Indonesia.

E-mail: nandhitanugriani@gmail.com

Abstract. Water hyacinth or *Eicchornia crassipes* is a free-floating aquatic plant and is easily found in swamps or rivers. The rapid growth of water hyacinth can disrupt the aquatic environment, thus water hyacinth is categorized as one of the world's worst weed. The challenge is not on how to eliminate the number of this weed, but on how to take advantage by the presence of this water plant. Water hyacinth is known to contain high phenolic content and antioxidant activity. Water hyacinth was extracted using 96% ethanol/water with mass to solvent ratio of 1:30 at 3 different temperatures (30°C, 40°C, 50°C). Spray drying was conducted to produce extract powder with the addition of maltodexrin and Arabic gum as the encapsulating agents. The stability of liquid and powder extract was compared based on the Total Phenolic Content (TPC) and antioxidant activity. The highest TPC was obtained from extraction at 30°C whereas the highest antioxidant activity was obtained from extraction at 50°C. In powder extract, the highest TPC and antioxidant activity was obtained from powder with ratio 2:1 of TSS to encapsulating agents. The stability was performed under different storage conditions. The degradation of TPC was faster in room temperature.

1. Introduction

Eichhornia crassipes or water hyacinth (WH) is a floating warm water aquatic plant. It belongs to the world's worst weeds and accordingly is considered as a plant with a very high growth rate that can harm other aquatic life due to its natural behavior in consuming huge quantities of nutrients in water. In Indonesia, there are 15 critical lakes, which are threatened by the growth of this water hyacinth. As in Rawa Pening, East Java, the growth of water hyacinth has covered around 800 hectares of the total surface water in the area [1]. Billions of dollars have been spent on removing water hyacinth from waterways, but in most cases, this removal is almost impossible to be done. Thus this condition presents its own challenge, not on how to eliminate the number of these weeds but on how to take advantage by the presence of this water plant.

Based on the present studies, *Eichhornia crassipes* is found to have antioxidant properties, as antioxidizing agents are detected in the leaves and its ethanol extract showed good DPPH radical scavenging activities [2]. It could be used as a potential source of antioxidant for food ingredients as well as pharmaceutical products. Antioxidants are compounds that inhibit the oxidation reaction rates, in which the oxidation reaction itself is a chemical reaction that will produce free radicals. Free radicals are molecules, ions or atoms containing unpaired electron or more. Due to the free radical's presence, these radicals are unstable and tend to have a degree of reactivity to chemical reaction of other molecules [3]. They act by either accepting electron from or donating electron to other molecules. For this reason, they behave as both reductants and oxidants. Free radicals have the capability to attack the macromolecule in the body, causing damage in cells.

efore being further applied, requirements should be fulfilled includes the stability studies of *Eichhornia crassipes* extract. There has been a research regarding the stability of *Eichhornia crassipes* liquid extract conducted by Givianty [4], but It still need to be improved and developed because not only liquid extract available in markets, there are still many choices of dosage forms and the demand for powder form is still increasing; therefore the stability testing should be conducted for powder extract as well to determine which forms have the higher stability. Thus, this problem leads to this research that aims to determine and compare the stability of *Eichhornia crassipes* liquid and powder extract under different temperature conditions.

2. Research Methods

2.1. Chemicals and Equipment

2.1.1. Chemicals. Eichhornia crassipes or water hyacinth was used as the main material in this research, which is known to contain high antioxidant activity and was analyzed for its stability in different kind of forms during storage. Eichhornia crassipes was taken from Situ Cipondoh Lake, Tangerang. Another materials that used for analysis during the research process were listed below. For extraction, the required chemicals were 96% ethanol (v/v), and distilled water that was purchased from Harum Kimia, a chemical supplier in Jakarta. Folin Ciocalteu reagent (Merck, Germany), Gallic acid (Merck, Germany), and Sodium Carbonate (Na2CO3) (Sigma-Aldrich, UK) were used for the determination of total phenolic content of the extract. For antioxidant activity analysis, the extract of Eichhornia crassipes was being assessed based on the radical scavenging effect against DPPH, as the stable free radical. Thus, 2,2-Diphenyl-1-Picrylhydrazyl hydrate (DPPH) (Sigma-Aldrich, USA) was required. Other materials needed were maltodextrin and arabic gum, as the encapsulating agents during spray-drying process that were purchased from Harum Kimia, a chemical supplier in Jakarta.

2.1.2. Equipments. The equipments listed are equipment that were needed for all processes in this research, including the extraction process, determination of total phenolic content in the extract, analysis of antioxidant activity, encapsulation process and shelf life test. The equipment includes were automatic experimental drying oven (EXDRY-51, Indonesia), digital balance (Mettler, Switzerland), weighing scales (O'haus, USA), miller (Fomac, Indonesia), sieve shaker (CISA, Spain), vacuum pump (GAST, USA), cuvette (Brand GmbH, Germany), micropipette (Eppendorf, Germany), vortex (Heidolph, Germany), centrifuge (Pro Research Centrifuges, UK) and UV-VIS spectrophotometer (PG Instruments, UK), that were available in Laboratory of Swiss German University, whereas other equipment needed for spray-drying method was spray dryer (Buchi 190 Mini Spray Dryer, Germany), that available in Bogor Agricultural University, Bogor.

2.2. Experimental Procedure

2.2.1. Preliminary Treatment of Eichhornia crassipes. Preliminary treatment for Eichhornia crassipes was conducted before extraction, where the collected Eichhornia crassipes taken from Situ Cipondoh Lake was washed with water several times to remove the dirt and other possible contaminants which might interfere with the analysis process. It was then dried in shade and the stems, roots and leaves of Eichhornia crassipes were separated, for the reason that only the leaves were used in the extraction process. The leaves of Eichhornia crassipes was then weighed and dried in oven with a temperature of 55°C until reaching moisture content of below 10%. The dried samples was milled using food-processing miller in order to produce smaller size of particle. It was sieved in mechanical

IOP Conf. Series: Materials Science and Engineering **742** (2020) 012019 doi:10.1088/1757-899X/742/1/012019

sieve shaker; powder that passed the 60 mesh sizes was collected and stored in refrigerator until extraction.

2.2.2. Stage I: Extraction, Total Phenolic Content and Antioxidant Activity Analysis. For the extraction process, 10 gr of dried Eichhornia crassipes leaves was mixed into 300 mL solution of 96% (v/v) ethanol and distilled water (1:1) as the solvent. The sample was prepared for duplicate analysis. It was extracted under three different extraction temperatures (30°C, 40°C, 50°C) to determine which extract have the highest total phenolic content and antioxidant activities. Extraction process was done under the optimum condition of 3 hours. Filtration was then done using vacuum filter and filter paper to separate the floating materials and to get only the supernatant. Centrifuge was also used before vacuum filter, to firstly separate the floating materials from the solution so that the filtration process was done easier.

For total phenolic content analysis, Folin-Ciocalteu method with gallic acid as a standard was applied. The extract samples was mixed with Folin Ciocalteu reagent and Sodium Carbonate (Na2CO3). The standard curve using the standard solution of gallic acid was prepared. For the ethanolic/water extract of *Eichhornia crassipes,* 200 L will be mixed with the same reagents. The absorbance was then measured after 30 minutes incubation to obtain the total phenolic content in the extract. The absorbance was measured at 765 nm wavelengths.

Antioxidant activity of the extract was assessed using 2,2-diphenyl-1-picryhydrazyl or DPPH solution, as a free radical. DPPH solution was mixed with the extract at minimum four different concentrations. The mixture was allowed to stand for 30 minutes under room temperature. The absorbance was measured at 517 nm, and was compared with a control solution without extract in UV-Vis spectrophotometer. The assays were conducted in duplicate.

2.2.3. Stage II : Pulverization of Extract. Spray - drying method was applied in producing powder extract of Eichhornia crassipes, because it is known as the best drying process in the industry. This facilitates quick drying compared to any other drying method. In this process, the liquid extract was sprayed into a stream of hot vapor through a nozzle and being vaporized. The inlet temperature was set at 170°C, according to the common method of spray drying. The pulverization of Eichhornia crassipes extract was conducted using spray-drying method with the addition of encapsulating agents. The encapsulating agents used were maltodextrin and arabic gum, with the ratio 60:40 respectively. The encapsulating agent was added with ratio 1:1, 1:2 and 2:1 of the total soluble solids of extract to the encapsulating agents, in order to see the effect of encapsulating agents' quantity to the product stability.

2.2.4. Stage III : Stability Studies. After extraction, pulverization, analysis of total phenolic content and antioxidant activity, the last step was the shelf life analysis that involving two different storage conditions; shelf life testing at 30°C in room temperature and shelf life testing in refrigerator temperature at 4°C. By interacting the result with kinetic order reaction principles, the product shelf life can be observed in each storage temperatures.

2.3. Analytical Procedure

2.3.1. Total Phenolic Content Analysis

2.3.1.1. Preparation of Folin-Ciocalteu reagent, Sodium carbonate and Gallic Acid Standard Curve. Total Phenolic Content analysis was conducted using Folin - Ciocalteu method with gallic acid as the standard. The folin reagent and sodium carbonate was prepared. The Folin reagent was prepared by diluting the concentrated Folin- Ciocalteu reagent with distilled water with a ratio of 1:10. The sodium carbonate was prepared by diluting 7.5 grams of solid sodium carbonate into 100 mL distilled water to

IOP Conf. Series: Materials Science and Engineering **742** (2020) 012019 doi:10.1088/1757-899X/742/1/012019

get 75% (w/v). For standard curve preparation, 1000 mg/L stock solution was prepared by dissolving 0.1 grams of solid gallic acid into 100 mL distilled water. The stock solution then was diluted to 5, 40, 80, 120, 160 and 200 ppm or mg/L gallic acid solution. From that diluted solution, 200 μL was taken into test tube, mixed with 1 mL Folin reagent and 800μL of sodium carbonate solution (Na2CO3). All mixtures were incubated for 30 minutes at room temperature were then poured into cuvettes and the absorbance was measured using UV-Vis Spectrophotometer at 765 nm. The Gallic Acid standard curve was obtained by plotting a linear regression based on the measured absorbance.

2.3.1.2. Total Phenolic Content of Eichhornia crassipes Extract. For obtaining total phenolic content of the extract, 200 μL of extract was taken into test tube, mixed with 1 mL Folin reagent and 800 μL of sodium carbonate solution (Na2CO3). It was incubated for 30 minutes at room temperature, and was then poured into cuvettes and the absorbance was measured using UV-Vis Spectrophotometer at 765 nm. The total phenolic content was calculated using Gallic Acid Equivalent (GAE) according to the equation below.

$$Total\ phenolic\ content\ \left(\frac{mg\ GAE}{L}\right) = \frac{(Abs - c)x\ DF}{m} \quad (1)$$

where;

Abs : Absorbance (A)

M : Gradient of the Gallic Acid standard curve (A/mg L)

c : Intercept of the Gallic Acid standard curve (A/mg L)

DF : Dilution Factor

2.3.1.3. Total Phenolic Content of Eichhornia crassipes Powder Extract. For obtaining total phenolic content of the powder extract, 0.02 grams of extract powder was dissolved in 10 mL solvent (distilled water and ethanol) in order to obtain 2000 mg/L solution. Then, 200 μL of extract was taken into test tube, mixed with 1 mL Folin reagent and 800 μL of sodium carbonate solution (Na2CO3). It was allowed to stand for 30 minutes for incubation at room temperature, and was then poured into cuvettes and the absorbance was measured using UV-Vis Spectrophotometer at 765 nm. The total phenolic content was calculated using Gallic Acid Equivalent (GAE) according to the mentioned equation.

2.3.2. Antioxidant Activity Analysis by DPPH Assay

2.3.2.1. Preparation of 0.25 mM DPPH Stock Solution. The antioxidant activity of Eichhornia crassipes extract was tested using DPPH Assay method. In order to prepare the 0.25 mM DPPH stock solution, 10 mg of powder DPPH was diluted into 100 mL 96% (v/v) pro-analysis ethanol solution. The stock solution was stored in refrigerator and stored in amber colored bottle that wrapped with aluminium foil.

2.3.2.2. Antioxidant Activity Analysis of Eichhornia crassipes Extract Solution. Antioxidant activity was assessed using 2,2-diphenyl-1-picryhydrazyl or DPPH solution. For control preparation, 0.5 mL of ethanol pro-analysis was taken in a test tube and added with 0.5 mL of 0.25 mM DPPH solution. For obtaining the sample absorbance, 0.5 mL of ethanol was replaced with extract solution at minimum of four different concentrations. The mixture was shaken well, and allowed to stand for 30 minutes in a dark room at room temperature. The absorbance was then measured using UV-Vis Spectrophotometer at 517 nm. From the collected absorbance data, the DPPH inhibition percentage was calculated using the following equation. IC_{50} value is expressed as the concentration of the sample to be able to inhibit 50% of free radical (DPPH).

$$Radical\ scavenging\ activity\ (\%) = \frac{Ac - As}{Ac}\ x\ 100\% \qquad (2)$$

Where,

Ac : Absorbance of control (A)

As : Absorbance of sample (A)

3. Results and Discussions

3.1. Preliminary Treatment and Extraction of Eichhornia crassipes

Preliminary treatment was done in order to separate the raw plant into its stems, roots and leaves as only the leaves were taken for analysis. The leaves were washed with tap water several times to remove dirt and any possible contaminants that might interfere the analysis process, and was dried in shade before cut. The leaves were cut into smaller pieces to facilitate further drying process in oven. The leaves that used for analysis were the leaves with the size ranging from 13cm to 17cm diameter, the other were excluded during the separation of this plants part. The cut leaves were dried in oven at 55°C for 24-48 hours until the weight was constant and moisture content to below 10% were obtained. After drying, the dried sample was milled using food-processing miller for 15 minutes to reduce the size and sieved using sieve shaker to obtain powder, powder that passed the 250 micrometer pores or 60 mesh size was collected in plastic zipper bag and stored in refrigerator until extraction in order to inhibit or slow down the growth of microorganism so that the product has a longer shelf life [5].

The next step after pre-treatment of *Eichhornia crassipes* was solid-liquid extraction. The extraction process was done and partially adopted from the previous extraction process done by Jimmy [6] and Givianty [4], where the optimum condition for extracting *Eichhornia crassipes* within 3 hours was already selected. In this research, extraction of *Eichhornia crassipes* was carried out in 3 different temperatures within 3 hours, means that the dried sample of *Eichhornia crassipes* was extracted at 30°C, 40°C and 50°C with the combination of distilled water and 96% (v/v) ethanol as the solvent. It was extracted with 1:30 sample to solvent ratio. Vacuum filtration and centrifuge was used in order to filtrate the extract solution and to remove the insoluble impurities to get a clear solution. The filtrate was then centrifuged with speed of 6000 rpm for 15 minutes to precipitate solids followed, so that the supernatant is obtained. Because in general, one of the requirements for using UV-Vis spectrometry is the sample must be in the form of a clear solution to avoid the misleading in data interpretation [7].

3.2. Pulverization of Eichhornia crassipes Extract

In order to produce powder extract, the liquid extract from extraction condition 1 (30°C), 2 (40°C) and 3 (50°C) was spray-dried. It was firstly added with encapsulating agents (maltodextrin and arabic gum) with ratio 1:1 of the total soluble solids to total encapsulating agents and other variation ratios (1:2 and 2:1) was then conducted to analyze the effect of encapsulating agents to the TPC and IC_{50} value of the extract. The inlet temperature of spray drying was 170 °C and the outlet was 70°C, based on the default setting parameter in mini spray drier in IPB, Bogor.

Encapsulation is a technique of core material entrapment in certain coatings. The advantage of encapsulation techniques is to protect and control the release of active ingredients [8]. The coating material used must be food grade, biodegradable, and stable during the processing process. Coatings commonly used in encapsulation include carbohydrate-based ingredients, proteins and fats. Maltodextrin is composed of glucose units, and is not effective for stabilizing oil or flavor in a viscosity solution. For this reason, maltodextrin is usually combined with ingredients such as arabic gum or other modified starches for the purposes of material stability. As has been stated by Supriyadi and A. S. Rujita [9], maltodextrin has a good ability to inhibit oxidation reactions. While Arabic gum is soluble in water. In addition to its high solubility, the main characteristics of arabic gum are texture-forming, film forming, binding and also good emulsifier in the presence of protein components in arabic gum. Gum arabic can retain the flavor of food dried by spray drying method because these gum can form layers that can protect from the process of destructive change. Hence, the combination of Arabic gum and maltodextrin was chosen in this research.

3.3. Stability based on Total Phenolic Content

Stability testing for liquid and powder extract was conducted based on Total Phenolic Content. The changes in TPC result from day to day were analyzed.

TPC Stability of Liquid Extract – at Room Temperature Storage

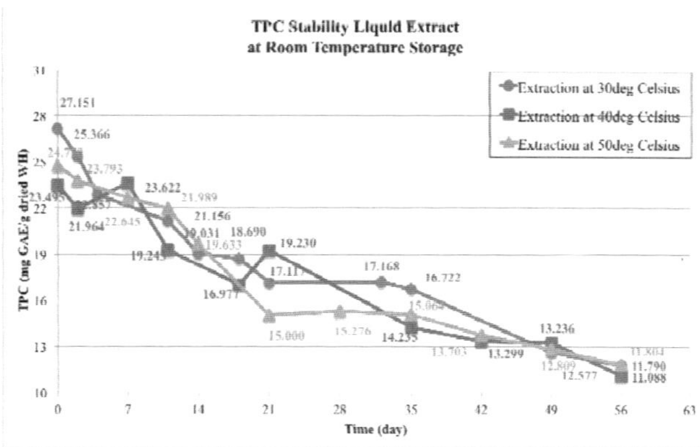

Figure 1. The TPC Stability of Liquid Extract – Room Temperature Storage

IOP Conf. Series: Materials Science and Engineering **742** (2020) 012020 doi:10.1088/1757-899X/742/1/012020

Development of Skin Lotion Containing Antioxidant Extract from Coffee Pulp and Study on Its Stability

D I Widiputri[1], S Wijaya[1], and S P Kusumocahyo[1]

[1]Department of Chemical Engineering, Faculty of Life Sciences and Technology, Swiss German University, The Prominence Tower, Alam Sutera, Tangerang 15143, Indonesia

E-mail: samuel.kusumocahyo@sgu.ac.id

Abstract. Coffee pulp, a major coffee waste in the coffee industry, has been found to contain bioactive compounds that can provide high antioxidant activity and hence, can be utilized as a natural antioxidant source. One of the beneficial utilizations of coffee pulp extract is its application in skin cosmetics, such as skin lotion. The application of antioxidant on the skin is one way to avoid skin damages and diseases from happening, which can be caused by free radicals generated through ultraviolet (UV) radiation from sunlight. The aim of this study is to apply the coffee pulp extract powder into skin lotion formulation, and to study the phenolic content, antioxidant activity and the stability of the formulated lotions. First, the coffee pulp (CP) was extracted at sample to solvent ratio of 1:10 (w/v) using ethanol/water mixtures at various extraction temperatures and times to obtain the optimum extraction condition. The resulting coffee pulp extract at optimum condition (50°C, 120 minutes) was pulverized into coffee pulp extract powder. The yield of the pulverization was 0.217146 g powder/g CP. The coffee pulp extract powder was then added into skin lotion formulation with four different concentrations (0.125 %, 0.25 %, 0.5 % and 1 % by mass). All coffee pulp skin lotions were found to show good physical stability and considerably higher antioxidant activity (IC_{50} : 5,805 ppm – 55,776 ppm) compared to commercial lotions (IC_{50} : 505,018 ppm – 557,218 ppm).

1. Introduction

Indonesia as a tropical country located at both sides of the equator has relatively long daytime duration per year, that causes the people to be often exposed to sunlight and essentially endangered by the amount of ultraviolet (UV) radiation reaches their skin. Exposure to chronic ultraviolet radiation can cause skin damage or serious diseases such as skin cancer [1], since free radicals can be generated through various interactions of UV with the surroundings. The application of antioxidant on the skin is one way to fight free radicals generated from ultraviolet radiation, which can avoid further skin damages from happening. Although human body already retains natural antioxidant coming from different source of external intake, applying antioxidant topically is important to protect the skin from UV radiation, dust, pollution etc. Hence, applying cosmetics products such as skin lotions and skin gels are the possible ways to provide antioxidant to the skin directly.

Nowadays, many cosmetics products use natural plant extract as the natural antioxidant source. Such extracts can be obtained from many plants, among other is coffee. Coffee is one of the most popular beverages consumed by Indonesian people. The coffee plants themselves are spread throughout several provinces of Indonesia, and it has been reported that Indonesia is a top coffee producer after Brazil, Vietnam and Colombia during the crop season in 2016-2017. Moreover, the amount of coffee produced in Indonesia in 2017-2018 is 636,000 tons of coffee cherries and it has been estimated to increase by 30,000 tons in 2018-2019 according to Global Agricultural Information Network. Due to a large coffee cherries production, it also results in a large amount of coffee waste in the coffee industry. A high amount of coffee waste during the coffee processing, known as coffee pulp, has become a major problem in the coffee industry, since it represents 40 % of the whole coffee cherries by weight [2]. In the effort to find solution to this waste problem, several studies have found that coffee pulp still contains hydroxycinnamic acids, which have been identified as a natural antioxidant, such as caffeic acid, chlorogenic acid and ferulic acid. Coffee pulp has been found to be able to provide high antioxidant activity [3], hence it can be considered as natural antioxidant source.

Researches on the utilization of coffee wastes extract, coming from coffee silver skin and coffee pulp, as antioxidant source in cosmetic products have been thoroughly conducted. In previous studies, it was confirmed that the extract of coffee pulp and coffee silver skin show high antioxidant activity, hence is still very beneficial to be utilized [4] [5]. Efforts to apply coffee pulp extract on skin cosmetics have also been made. Coffee pulp extract has been applied to skin gel [6], while coffee silver skin extract was applied to skin lotion by Kusumocahyo [4]. However, the skin lotion was found to be unstable during the storage time, shown by a phase separation after certain period and after going through a mechanical test.

A comparison was also made between coffee silver skin and coffee pulp extract, which led to a result that the antioxidant activity of coffee pulp extract is higher than coffee silver skin [6]. Realizing this fact, this research was aimed at developing a skin lotion with coffee pulp extract as antioxidant source. The stability of the skin lotion is an important parameter to be studied, since it can be affected by the physicochemical properties of the skin lotion. Therefore, a stability study of the different formulations of skin lotion was also performed.

2. Methodology

The coffee pulp underwent a sequence of pre-treatment, consisted of washing with water, drying, milling to reduce the size and washing with n-Hexane by stirring for 1 hour. The extraction process of coffee pulp was divided into 2 phases. The first phase was extraction of coffee pulp at sample to solvent ratio of 1 : 40 and 1 : 10 (w/v) at 60°C for 60 minutes, by using a mixture of ethanol : water (50:50) (v/v) as solvent. In the second phase, coffee pulp was extracted at sample to solvent ratio of 1 : 10 (w/v) at various temperatures (Room temp., 40°C, 50°C and 60°C) and extraction times (40, 60, 90, 120, 180 and 240 minutes). The extract obtained from the second phase was further prepared to undergo spray drying process. The extract solution was filtered using vacuum filtration, then it was dried using spray dryer with inlet temperature of 175°C and outlet temperature of 125°C, at a feed flow rate of 17 mL/min to produce powder extract.

In the next stage of the research, the pulverized CP extract was used for the further process, which was skin lotion formulation. The coffee pulp extract was applied into the skin lotion formulation in the form of powder at various extract concentrations, namely 0.125 %, 0.25 %, 0.50 %, and 1.00 % CP powder extract by mass. During the storage time, the antioxidant activity and the physicochemical properties of skin lotion were analysed, while stability testing according to CTFA [7] was conducted. The physicochemical properties observed in this research include organoleptic, spreadability, pH measurement, homogeneity and viscosity.

3. Results and Discussion

A different sample to solvent ratio could affect the phytochemical characteristic of a sample. In this study, the phytochemical observed was the total phenolic content. Dewi [5] has discovered the optimum extraction condition for CP Extract solution with sample to solvent ratio of 1:40 (w/v). However, according to Tan, et al. [8], a sample to solvent ratio of 1:10 (w/v) could increase the total soluble solid (TSS) of the yielded extract, which would be more suitable for a spray drying process subsequently.

In this experiment, both sample to solvent ration were applied in the extraction of coffee pulp using a mixture of ethanol and water as the solvent. The result has shown that the TPC of CP extract obtained through extraction using sample to solvent ratio of 1:40 (w/v) and 1:10 (w/v) was 53.378 ± 0.974 mg GAE/ g CP and 45.945 ± 0.795 mg GAE/ g CP respectively and the difference was statistically significant ($p<0.05$). Even though the ratio of 1:10 (w/v) was found to be resulting in a lower TPC of the extract, the TSS obtained was much higher and has reached 36.106 %, which is sufficient for spray drying process subsequently without the need for additional filler. This ratio is found to be more beneficial to be used since it requires less solvent and less energy to evaporate the solvent completely prior to spray drying.

Table 1. Total Phenolic Content of CP Extract Solution at Various Condition

Sample to solvent ratio 1:10 (w/v)				
No.	Solvent	Temp. (°C)	Time (min)	TPC (mg GAE / g CP)
1	Ethanol/Water Mixtures (50:50) v/v	Room	60	48.872 ± 3.359
2		40		52.655 ± 0.795
3		50		55.704 ± 1.110
4		60		45.945 ± 0.79
5		50	40	27.728 ± 0.952
6			60	33.463 ± 1.153
7			90	38.655 ± 0.760
8			120	54.427 ± 9.521
9			180	32.631 ± 1.269
10			240	33.993 ± 0.675

Since the ratio of 1:10 (w/v) was chosen, the selection of best extraction temperature and time was conducted. It was concluded from the results that the optimum extraction condition is 50°C and 120 minutes, when using ethanol/water mixtures (50:50) as the solvent. After the CP extract solution was prepared, the sample was spray-dried to change the form of the extract from liquid into powder form. The aim of the spray drying process was to make sure that due to the hot steam, the ethanol from the extract solution can be totally evaporated to make sure that the CP extract powder does not contain any ethanol anymore as it will be applied to the skin lotion formulation. The yield of the pulverization was 0.2171416 g powder/g CP.

Table 2. DPPH Inhibition and TPC of CP Extract Solution and CP Extract Powder

	Extract Solution	Extract Powder	% Reduction
DPPH Inhibition	926.003 mg DPPH/ g solid	734.960 mg DPPH/ g powder	20.63 %
TPC	170.595 mg GAE/ g solid	135.313 mg GAE/ g powder	20.68 %

After CP extract solution was spray-dried, there was a degradation of the TPC in the CP extract powder when compared to its liquid form, due to the high temperature during spray drying process. Table 2 shows the data of DPPH inhibition and total phenolic content before and after spray drying. The reduction of the antioxidant activity and total phenolic content was due to the hot temperature by spray drying, which caused a theviscosityrmal degradation in which a compound who was sensitive to heat such as ferulic acids, can be damaged. The IC_{50} value for extract solution and extract powder are 148.328 ppm and 308.853 ppm respectively.

Table 3. DPPH Inhibition of Coffee Pulp Skin lotion before and after CP extract powder was added

	DPPH Inhibition (%)
Basic lotion (without CP extract powder)	3.678
F2 (with 0.125 % CP Extract powder)	30.036
F3 (with 0.25 % CP extract powder)	69.602

In the formulation stage, four different concentrations of CP extract powder (0.125 %, 0.25 %, 0.50 %, and 1.00 % by mass) was added into the skin lotion formulation. Table 3 shows the %-inhibition of the skin lotion with and without coffee pulp extract powder. The DPPH inhibition was observed before and after adding the antioxidant compound into the skin lotion formulation. Before CP extract powder was added, the DPPH Inhibition was only 3.678 %, however, after adding 0.125 % and 0.25 % of CP extract powder into the formulation, the CP skin lotion could inhibit as much as 30.036 % and 69.602 % DPPH respectively.

The CP skin lotion was stored at room temperature and low temperature for 6 weeks. The antioxidant activity and physicochemical properties were analysed during storage. In both storage temperatures, the IC_{50} of the CP skin lotion was considered stable as it only showed slight increase from time to time, as there is no sudden change on the IC_{50} value. While the IC_{50} between week 0 and week 6 from each concentration increased insignificantly ($p>0.05$), the CP extract powder concentration significantly affect the IC_{50} ($p<0.05$). The higher the IC_{50} value indicates the smaller amount of CP extract powder in the lotion formulation.

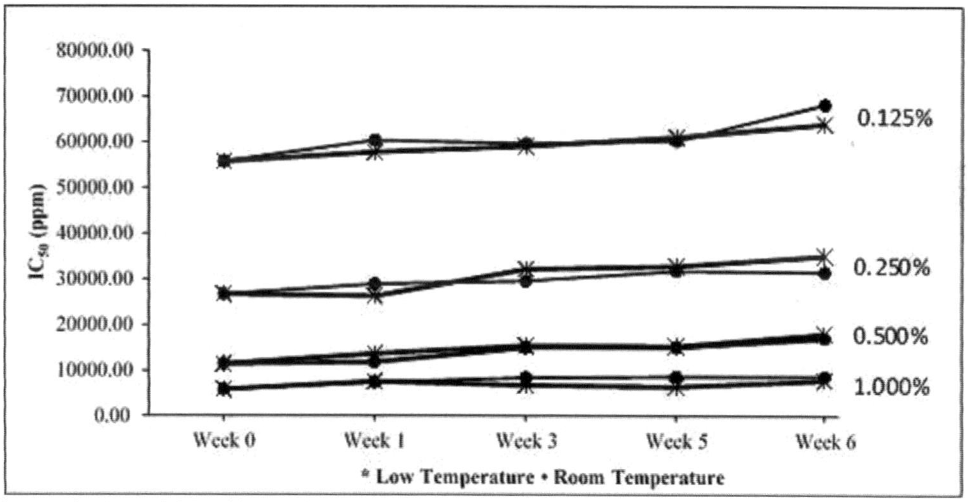

Figure 1. IC_{50} of CP Skin Lotion stored at room temperature and low temperature for 6 weeks

Physicochemical Properties of CP Skin Lotion

The physicochemical properties, including organoleptic, pH, viscosity, spreadability and homogeneity of the CP skin lotion were analysed for 6 weeks to study their stability. Organoleptic is one of the physicochemical properties for semisolid preparations, to describe colour, consistency, and homogeneity. Organoleptic properties of a skin lotion could affect the aesthetics and comfort when applied, and are also related with the acceptability of the skin lotion. Through the observation on the lotions within a study period of 6 week, the colour did not visually show significant change in real condition.

Homogeneity will affect the effectiveness of the skin lotion. If an emulsion is homogeneous, the active ingredients will spread evenly. By applying a CP skin lotion to a certain area, there is an opportunity for the active ingredients to occupy the area applied. All skin lotion formulations from F1 to F5 from this study were examined, and after 6 weeks all of the formulations were still found to be homogenous and no coarse particle was felt when rubbed on skin, nor a phase separation could be observed during this time period.

The CP skin lotion acidity is one of the important physical characteristics in the formulation, because it could affect drug solubility, activity, absorption, stability, and patient comfort [9]. The pH measurement was conducted on CP skin lotion, as it is important to ensure that the lotion is in a tolerable range of pH to be applied topically on skin, to avoid any irritation. According to SNI 16-4399-1996 a good and safe skin lotion should have pH around 4.5 – 8. If it was below 4.5, it would be too acidic and could cause skin irritation, and if it was above 8, it would be too alkali which could cause the skin to become scaly. During the stability testing conducted in 6 weeks period, all skin lotion formulations with different amounts of CP extract added, have shown a stable pH condition in the range of 6.6 – 7.34 when stored at room temperature, and in the range of 5.62 – 7.1 when stored in refrigerator.

Viscosity is one of the parameters of physical properties in semisolid preparations. The viscosity determination could determine the tendency of a semisolid preparation to breaks into oil phase and water phase. According to Stoke Law, the time to reach phase separation of an emulsion is inversely proportional to the viscosity. The higher the viscosity of an emulsion, the stable the emulsion is as the time to reach phase separation is longer. Furthermore, viscosity of an emulsion related to the therapeutic effects and comfort in use. CP skin lotions formulated in this research have a viscosity that is not too thick nor too damp, and is within the range of viscosity regulated by Standar Nasional Indonesia (SNI), which is 2,000-50,000 cP. After six weeks of stability testing, the viscosity of all skin lotion formulation was found to remain within the regulated range, which was 3,733 – 9,933 cP for storage at room temperature, and between 4,333 cP and 8,800 cP for storage in refrigerator.

Table 4. Comparison of skin lotion properties

	SNI	CP Skin Lotion	Commercial Lotion
pH	4.5-8	5.62 – 7.34	7.38 – 8.50
Viscosity (cP)	2,000-50,000	3,733-9,933	11,200-19,800
TPC (mg GAE/g lotion)	-	4.368 – 9.261	1.706 – 1.742
IC_{50} (ppm)	-	5,805-55,776	505,018-557,218
Mechanical Test	Stable	Stable	Stable
Cycling Test	No Phase separation	No Phase separation	-

After conducting all of the stability testing and physicochemical analysis, the CP skin lotion can be concluded to show a good behaviour during 6 weeks storage period. This was confirmed through the observation on increasing viscosity with time, which is indication of a stable lotion; the pH was maintained stable; and through the positive results of mechanical test and cycling test. Table 4 shows the comparison between CP skin lotion, the standardized requirement by SNI and several commercial lotions. The CP skin lotion characteristic was positioned between the SNI standard and the commercial lotions that are widely marketed in Indonesia. In term of pH and viscosity, CP skin lotion are still in the range of SNI's requirement. On the other hand, in term of TPC and IC_{50}, CP skin lotion has much higher TPC value and much lower IC_{50} value than the commercial lotions. This shows an antioxidant activity that is 10 to 100 times higher than the commercial lotion.

4. Conclusion
A development of skin lotion containing coffee pulp extract as antioxidant source was conducted in this research. The research began with the search for the most optimum condition of coffee pulp extraction, which can provide highest yield of coffee pulp powder extract at maintained antioxidant activity level. The optimum extraction method was determined to be applying sample to solvent ratio of 1:10 (w/v) using ethanol/water mixtures (50:50) as solvent, and conducted at 50°C for 120 minutes. After spray drying process, the yield of CP powder obtained was 0.2171416 g powder/g CP.

The application of the CP extract into skin lotion with different formulations (0.125 % - 1.0 % by mass) was successfully performed. All formulations showed good physicochemical stability and comply with SNI requirement as cab seen in Table 4. When compared to commercial lotions in Indonesia, the developed CP lotions have shown antioxidant activity that is 10 to 100 times higher.

References
[1] D'Orazio J, Jarrett S, Amaro-Ortiz A and Scott T 2013 UV radiation and the skin *International Journal of Molecular Sciences*
[2] Ramirez-Coronel M A, Marnet N, Kolli V S K, Roussos S, Guyot S and Augur C 2004 Characterization and estimation of proanthocyanidins and other phenolics in coffee pulp (Coffea arabica) by Thiolysis-High-Performance liquid chromatography *J. Agric. Food Chem.*
[3] Arellano-González M A, Ramírez-Coronel M A, Torres-Mancera M T, Pérez-Morales G G and Saucedo-Castañeda G 2011 Antioxidant activity of fermented and nonfermented coffee (Coffea arabica) pulp extracts *Food Technol. Biotechnol.*
[4] Kusumocahyo S P, Widiputri D I, Tangguh P, Kusumonegoro R, Annelies C D, Verdiana M 2018 Extraction of coffee silverskin and the development of antioxidant-rich products *Proceedings of International Conference on Food, Agriculture and Biotechnology (ICOFAB)* DOI: 10.14457/MSU.res.2018.30
[5] Dewi A A C 2017 *Optimization of Antioxidant Extraction from Coffee Pulp* http://library.sgu.ac.id/index.php?p=show_detail&id=31525
[6] Diamantina C 2018 *Utilization of Coffee Processing Waste in the Form of Antioxidant Skin Gel Containing Coffee Pulp Extract* http://library.sgu.ac.id/index.php?p=show_detail&id=31927
[7] CTFA 2004 Cosmetics Europe: Guidelines on stability testing of cosmetic products," *Guidel. Stab. Test. Cosmet. Prod.*, no. March
[8] Tan S, Kusumocahyo S P, Widiputri D I 2017 Pulverization of Coffee Silverskin Extract as Source of Antioxidant *IOP Conf. Series: Materials Science and Engineering* **162**(2017)012027
[9] Baran R and Maibach H I 2010 *Textbook of Cosmetic Dermatology* (USA: CRC Press)

IOP Conf. Series: Materials Science and Engineering **742** (2020) 012021 doi:10.1088/1757-899X/742/1/012021

Biogas Utilization for supporting the Sustainability of Palm Oil Mill

Irhan Febijanto

Centre for Assessment of Energy and Process Industry- BPPT
Business and Technology Innovation Centre Building, 2nd floor
PUSPIPTEK area, Serpong - Tangerang Selatan 15314

E-mail: irhan.febijanto@gmail.com

Abstract. Indonesia, as the largest palm oil production country, currently facing various criticisms from European countries related to environmental issues from Crude Palm Oil industries. One way to reduce the criticism impacts, is to obtain a certificate related to the sustainability of Crude Palm Oil industries, which issued by a national or even an international body. Based on the sustainability standards, POM (Palm Oil Mill) should prevent the gas emission/methane gas generated from Palm Oil Mill Effluent (POME) which emitted to the atmosphere. By captured methane gas and utilize as fuel for the boiler (biogas co-firing) or fuel for the gas engine for captive power, the emission gas amount emitted to the atmosphere can be reduced. All selected technologies in this study utilize biogas/methane gas. The economic analysis of each selected technology is assessed by using an economic parameter of IRR (Internal Rate of Return) and NPV (Net Present Value). It is proved that the biogas power plant integrated with the pellet factory shows the highest IRR, however, the biogas co-firing is the easiest technology among the selected technology and also the implementation is feasible for supporting the sustainability of POM though the IRR is the lowest one.

1. Introduction

Indonesia is the biggest palm oil producer with a production of 32.5 million tons of CPO (Crude Palm Oil) in 2017[1], and the production increases year by years, rose to 43 million tons in 2018[2], and it is predicted will climb to 48 million tons in 2025[3]. During the production of CPO, a considerable amount of solid by-product of empty fruit bunch (EFB), palm kernel shell (PKS) and mesocarp fruit fibre (MF), and aqueous by-product which is also called POME (Palm Oil Mill Effluent) are discharged.

The aqueous by-product is heavily polluted with biodegradable organic material with COD (Chemical Oxygen Demand) value is over 50,000 mg/ltr and requires pre-treatment before discharged to the river or to the surrounding environment. Usually, the aqueous by-product is treated in open ponds before discharged to the river. The open ponds are a wastewater treatment system mainly used in Indonesian Palm Oil Mill (POM) widely. This system is low cost and consists of anaerobic and aerobic ponds, which is easily amenable to biodegradation due to its high organic content. Based on the national standard, the minimum of COD shall be 350 mg/ltr for releasing to the river [4] and 2000 mg/ltr for land application use[4]. The biogas is produced during an anaerobic decomposition of POME in anaerobic pond and it is emitted to the atmosphere and then contributes to global warming effects with the impact of 22 times larger than CO_2.

The solid by-product of MF and PKS are used for boiler fuel to produce electricity and process steam in POM to produce CPO. The surplus both of MF and PKS may be sold to the market, as an additional revenue of POM. The implementation of EFB for boiler fuel is still limited due to a constraint of the high moisture content [5][6] and high potassium content [7], though it is very potential to be developed. In Indonesian POMs, EFB is usually utilized for a mulch or land application [8]. However, in some POMs, the implementation is not a full success due to the high cost to transport back EFB to the plantation. Moreover, Menon [9] proved that the EFB use for renewable fuel has several times of economic values compared to use for a mulch.

In this study, biogas generated from aqueous by-products in the anaerobic pond is utilized for energy to implement several kinds of technology. The selected technologies have a TRL (Technology Readiness Level) [10] value over a value of 7. Each technology is reviewed based on the sustainability of the palm oil mill, which considered based on economic (financial), environmental, and social aspects.

The selected technologies in this study should meet with the requirement of sustainability in RSPO [11] (Roundtable on Sustainable Palm Oil) or ISPO[12] (The Indonesian Sustainable Palm Oil System).RSPO is the Roundtable on Sustainable Palm Oil (RSPO) was established in 2004 with the objective of promoting the growth and use of sustainable oil palm products through credible global standards and engagement of stakeholders. ISPO (The Indonesian Sustainable Palm Oil System) is the policy adopted by the Government of Indonesia in this case the Ministry of Agriculture with the aim of increasing the competitiveness of Indonesian palm oil on the world market and participating in the framework of fulfilling the President of Indonesia's commitment to reduce greenhouse gases and provide attention to environmental problems. Both RSPO and ISPO ask POM to compliance the requirements of their standards.

One of the criteria of both standards is to identify and mitigate GHGs emissions. The emission sources which generates GHGs during oil palm growing and FFB (Fresh Fruit Bunch) processing in POM are the use of fuel for transportation and machinery in the plantation[13], the use of artificial fertilizers [14] and emission from POME (Palm Oil Mill Effluent) [15]. As we know the fuel used in the POM to produce CPO use solid by product of MF and PKS. Based on the literature review'[16][17], the highest emission source is generated by POME with the emission of 546,85-896,48 CO_2/t-CPO.

The GHG emitted from POME is methane gas, which can be captured and utilized it as a fuel to generate electricity. The technology has been mature and most of the biogas utilization is as a fuel to generate electricity.

In Indonesia, the situation is different compared to the developed countries who provide the premium selling electricity price for power generation based on renewable energy. Since the selling electricity price for renewable energy below 10 MW which determined by Ministerial Decree of ESDM (Energi dan Sumber Daya Mineral) No.: 1772 K/20/MEM/2018 [18] issued, the selling electricity price became lower than the previous selling price [19] and due of this, almost of renewable energy investors from local and foreign became not interested anymore in investing in the renewable energy development. Several references show that 85% of BPP (Biaya Penyediaan Pokok)/main cost electricity supply in each region as its new reference price is relatively lower than the electricity cost from most renewable technologies in Indonesia, except in the Eastern region of Indonesia 20].

1.1 Material Balance

The by-product ratio of CPO production consists of PKS, EFB and MF which indicated below based on several references.

Table 1. By-Product Ratio proposed by Several Researchers

	NTT [21]	Erliza [22]	Agus [23]	CE [24]
PKS	6%	6,4%	6,4%	6,0%
EFB	23%	21%	22,6%	22%

MF	13%	14,4%	7,9%	13,5%
POME	50%	58,3%	-	67%

Based on the conservative consideration, the lowest value of ratio proposed by Erliza[22] is selected and is used to estimate by-product amount in this study.

1.2. Proposed Technologies
The proposed technologies which can be integrated with a biogas power plant are 1) biogas for flaring 2) biogas for fuel 3) biogas for electricity, and 4) biogas for boiler fuel (biogas co-firing).

Each proposed technology is explained below,

1) Biogas for flaring is not economically feasible due to no revenue.
2) Biogas for fuel, the biogas is converted into renewable fuel such as bio-CNG or bio-methane. However, POM is located in the remote area which low household energy demand. It is not economically feasible to develop pipe-line distribution in the remote area. The possibility to utilize as bio-CNG is low due to there is no regulation which supports bio-CNG as a transportation fuel [23].
3) Biogas for electricity, for captive power and for selling to the state-owned electric power company (PT PLN) and for captive power (pellet factory or compost factory).
4) Biogas for co-firing, the biogas is supplied to the existing boiler to substitute biomass fuel of PKS and MF [24][25]. The excess of both biomasses is sold and will be the revenue of this option.

Therefore, the option of 3) or 4) are able to be implemented, and it will be discussed in this study.

1.2.1. Biogas for Electricity. The covered lagoon reactor has the lowest investment cost among the others kind of biodigester technology [26], therefore, the technology is subjected to be discussed in this study.

At the beginning of the biogas power plant technology implementation time, the technology was directly imported from Germany or Malaysia, which had already been implemented the technology widely. However, in line with the many constructions of biogas plants in Indonesia, the developers have many opportunities to improve local content resulted decreased the technology investment cost compared to the early phase, several years ago.

The unit investment cost used in this study is based on the unit investment cost of the biogas power plant in Bangka built recently, 1,987,867 USD/MW[27], and the ratio of main facilities cost is shown in Table 2.

Tabel 2. Ratio of Investment Cost for Biogas Power Plant Components [28]

Komponen		**Ratio**
Biodigester System	:	25%
Biogas Management System	:	16%
Instrumentation & Electricity System	:	10%
Logistic, Insurance, Instalation, Commisioning, Delivery,	:	19%
Project Contigency	:	5%
Biogas to electricty convertion ratio	:	25%

Based on the ratio above, the biogas digester cost is 75% of the total investment cost. Biogas power plant capacity and the energy electricity generated are calculated based on the parameter of COD (Chemical Oxygen Demand) which is assumed by 60.000 mg/ltr.

1.2.2. Selling Electricity Price. One of the selected options for the proposed technology is a biogas power plant for selling electricity to PT PLN (The state-owned electricity company). The selling electricity price is determined by the Ministerial Decree [29]. In this study, the Riau Province area's price of USD 0.097/kWh (=85% x Rp 1655/kWh with a currency exchange of 1 USD = Rp 14,500) is used.

1.2.3. Captive Power. Captive Power is implemented as integration between biogas power plant with a pellet factory or a compost factory. For the implementation at the POM, both factories have been implemented in the commercial base.

1.2.3.1. Pellet Factory. The investment cost of the pellet factory is assumed based on the Feasibility Study conducted by JICA in 2016 [30]. The characteristic of the technology proposed by JICA, a pellet material of EFB is dried and pressed all at once in one machine made by SYNG. Usually, in the conventional method, the EFB is dried and pressed by drying machine and pressuring machine sequentially. Based on the FS, the energy source for shredding, drying and pressing EFB to produce pellet with the production amount of 600 ton/yr is 1,200,000 kWh/yr with use for one unit of SYNG 400E. The electricity generated is utilized maximumly.
The investment and operational cost of pellet factory with a capacity of 14,400-ton pellet/yr is indicated in Table 3.

Table 3. Implementation Cost (USD)

Investment Cost		
Pellet machine	:	55,172
Binder machine	:	275,862
Operational Cost:		
Salary	:	13,241
Maintenance	:	2,069

The pellet price is assumed by 48.28 USD/ton. The investment cost of the biogas power plant depends on the biogas production capacity determined by the calculation, which uses the flow rate and the COD value of POME as a primary variable.

1.2.3.2. Composting. The investment and operational cost of the composting factory including the compost price are assumed based on the feasibility study conducted recently in North Sumatera Province [31] and the value are shown in the table 4.
The biogas power plant investment cost depends on the biogas production capacity. The ratio of POME used to moisten EFB during the composting process (85 days) is 0.3 m^3/t-compost.

Table 4. Implementation Cost

Biogas Power Plant	:	1,987,867 USD/MW
Compost factory	:	1,250,058 USD
Operational	:	19 USD/t-compost
Compost Price	:	61.5 /t-compost

1.2.3.3. Biogas Co-firing. Biogas which captured and generated in the bio-digester is used fuel and burned in the boiler to replace by-product of PKS which usually used as boiler fuel. The excess fuel of the PKS is sold, and the revenue is the profit of this project.
The investment, operational cost of this factory is shown in the table 5.

Table 5. Investment and Operational Cost

Investment		
Biogas reactor is 75% of biogas power plant investment cost		
1-unit of burner	:	USD 137,931
Operational		
3% of biogas reactor investment cost		
Revenue		
PKS price	:	0.042 USD/kg
Fibre price	:	0.014 USD/kg

1.2.4. GHG Reduction. GHG emitted from POME in the anaerobic pond is calculated using the methodology determined by the United Nations Framework Convention on Climate Change [32].

1.2.5. Economic Analysis. The economic analysis for each proposed technology is analyzed using the equation (2) , where the assumptions used are shown below,

POM capacity	:	60 tph
Operational time	:	20 hours in a day
		330 days in a year
Project lifetime	:	15 years
Depreciation time	:	10 years
Debt equity ratio	:	30: 70
Bank rate	:	11,28%

An average of interest rate of rupiah loan of state bank for working capital loan during 2014-2018 is r=11.28% [33].

$$NPV = \sum_{t=0}^{n}\left(\frac{CF_t}{(1+r)^t}\right)\text{..................}(1)$$

t : time period cash flow is received
CF_t : net after-tax cash inflow-outflows during a single period *t*
r : Internal rate of return that could be earned in alternative investments

This paper discussed the utilization of biogas for electricity in POM with a capacity of 60 tph. The biogas is used for an energy source for utilizing by-product of EFB for pellet, composting or biogas co-firing, as the solution to reduce GHG generated from POME without a connection to the PLN's grid and to comply the criterion of the principal oil palm sustainability standards, such as determined in RSPO or ISPO.

The economic analyses and the GHG reduction for each proposed technology are discussed and analysed in this study.

2. Methodology

The by-product amount in POM with a capacity of 60 tph is calculated based on the by-product ratio, and the potential electricity amount in each proposed technology is calculated based on the wastewater amount and COD of wastewater.

The investment cost and operational cost in each proposed technology is determined below

Table 6. Kind of The Proposed Technology and the Component of Investment and Operational Cost

Proposed Technology	Investment and Operational cost
Biogas Power plant	The capacity of biogas power plant (a covered lagoon reactor + a gas engine) which uses all biogas generated from POME to generate electricity
Biogas Power plant + Pellet factory	a) The capacity of biogas power plant which uses all biogas generated from POME to generate electricity b) Pellet factory which supplied electricity from biogas power plant maximumly.
Biogas Power Plant + Composting Facility	a) The capacity of biogas power plant which uses all biogas generated from POME to generate electricity b) Composting facility which use EFB amount maximumly.
Biogas Power Plant + Biogas Co-firing facility (burner unit and piping line)	a) The capacity of covered lagoon reactor which produces biogas to substitute biomass fuel of PKS and MF in the existing boiler. b) Burner unit for biogas co-firing in the existing boiler and piping line from the reactor to the boiler.

Economic analysis is conducted using as an economic parameter of IRR and NPV.GHG emission is calculated based on the methodology determined by UNFCCC.

3. Results and Discussion

3.1 Electricity Generated

The by-product amount of PKS, EFB, and MF produced in POM with a capacity of 60 tph, and also the POME/wastewater are shown in the table 7.

Table 7. By-product Amount

Kind of Biomass	Amount
PKS (ton/yr)	25,344
EFB (ton/yr)	83,160
MF (ton/yr)	57,024
POME (ton/yr)	230,868

Based on the methodology, the potential electricity can be generated is 13.23 x 10^6 kWh annually, with the own consumption is 10% from electricity generated. The gas engine with a capacity of 1.86 MW is able to be installed. The investment and the operational cost are USD 3,68 x 10^6 and USD 184,441/yr, respectively.

Based on the parameters assumed in I.4, the IRR and NPV of the selling electricity to PT PLN are 13.95 % and -1,126,080 USD. Therefore, the project is not feasible due to the NPV indicates a negative value.

3.2. Captive Power for Pellet Factory

Two units of pellet machine are installed with a capacity of 14,400-ton pellet/yr. The electricity demand of the machine is 4.8 x 10^6 kWh/yr, and it will be supplied by electricity generated from biogas plant with a capacity of 750 kW. The investment and the operational cost for two-units of pellet machines are USD 2.15 x 10^6 and USD 0.11 x 10^6, respectively. EFB amount used in pellet production is 63,202 ton/yr, the EFB is dried until the water content of 8%. The remaining EFB is used as a mulch in the plantation as usual.

The revenue from selling pellet of 28,800 t-pellet/yr is USD 1.39 x 10^6.

Based on the parameters assumed in I.4, the IRR and NPV of the selling pellet are 38.06 % and USD 19,5 x 10^6/yr.

3.3. Captive Power for Compost Factory
The EFB amount of 790,200 ton/yr is used to produce compost with the amount of 79,200 ton/yr. The remaining EFB of 3,960 ton/yr is used as a mulch as usual. Using the ratio of 0.3 m^3 for 1 ton of EFB, therefore, 24,948 t-POME/yr is used in this composting process. The remaining POME in the covered lagoon produces biogas is used to generate electricity of 1.04 x 10^6 kWh/yr by using a gas engine with a capacity of 300 kW. The remaining biogas which not used as a fuel for biogas engine is flared in the flaring system. The investment for the composting facility and biogas power plant are USD 3.0 x 10^6 and USD 0.6 x 10^6, respectively. The operational cost of both facilities is USD 0.03 x 10^6 and USD 3,03 x 10^6/yr, respectively.

The revenue from selling compost of 79,200 t/yr is USD 4.88 x 10^6/yr.

Based on the parameters assumed in I.4, the IRR and NPV of the selling electricity to PT PLN are 32.54 % and USD 25,381,595.

3.4. Biogas Co-firing
In this proposed technology, all biogas generated in the covered lagoon is supplied to the boiler through the burner to substitute biomass fuel of PKS and MF.

The investment cost for a covered lagoon and burner are USD 1.2 x 10^6 and USD 0.14 x 10^6, respectively. The amount of PKS and MF which substituted by biogas supply are 11,179 t/yr and 2,837, respectively. The selling price of PKS and MF are USD 0.04/kg and USD 0.013/kg, respectively. The total revenue earned from the selling of both biomasses is USD 0.50 x 10^6/yr.
Based on the parameters assumed in I.4, the IRR and NPV of the selling pellet are 17.44 % and USD 1.14 x 10^6/yr.

3.5. GHG Reduction
The proposed technologies burn all biogas generated from the covered lagoon reactor certainly through the gas engine system or flaring system or boiler system and through these systems, methane gas is converted to carbon dioxide. Therefore, the amount of GHG which reduced in each proposed technology is the same, 37,374.4 t-CO_2/yr.

Based on the economic analysis for each technology, the pellet factory integrated with a biogas plant has the highest IRR, however, the pellet business is required the standard international or the standard of buyer country. Therefore, the pellet standard should be well known, before doing the pellet business. Biogas co-firing technology though shows the lowest IRR compared to other proposed technology, the implementation of biogas co-firing does not involve third/outside parties, moreover the PKS market for domestic and abroad are quite mature. Therefore, the biogas co-firing technology is the easiest technology to be implemented for supporting the sustainability of POM although the IRR is the lowest one.

4. Conclusion
To build a sustainable palm oil industry the GHG mitigation and manage is most required by several criteria stated in RSPO, ISPO and also government regulation. The highest GHG sources in the POM is methane gas emitted from POME treated in the anaerobic pond. The covered lagoon technology has the lowest investment cost and the easiest handling, this biodigester technology is selected in the simulation for several proposed technologies which implemented in POM with a capacity of 60 tph. In each technology is economically analysed.

Biogas factory integrated with pellet factory has the highest IRR, however, the market for pellet should be assessed more detail, because the international standard is implemented in the pellet business. The market study is required to be a next topic study to complete study on the biogas power plant for supporting the sustainability of the palm oil mill.

Biogas co-firing is the easiest technology to be implemented for supporting the sustainability of POM and to mitigate the GHG in POM, although the IRR is the lowest one.

References

[1] N., Tampubolon and M. Pasaribu, 2017, The role of oil palm companies in Indonesia as a nation's competitive advantage, IOP Conf. Series: *Materials Science and Engineering*, pp.237, doi:10.1088/1757-899X/237/1/012036

[2] https://www.rambuenergy.com/2019/02/indonesias-cpo-production-up-by-12-65-exports-increases/ [accessed on August 27, 2019]

[3] http://www.palmoilindonesia.com/indonesia-palm-oil/palm-oil-production/ [accessed on August 27, 2019]

[4] The Decree of Minister Environment No: 51/1995

[5] Husain, Z., Zainal, Z. A. & Abdullah, M. Z., 2003, Analysis of biomass-residue-based cogeneration system in palm oil mills, *Biomass Bioenergy*, **24**, pp.117–124.

[6] Mohammed MA, Salmiaton A, Wan Azlina WA, Mohamad Amran MS,2012, Gasification of oil palm empty fruit bunches: a characterization and kinetic study, *Bioresource Technology*, Vol. pp. 628-636

[7] Kunio YOSHIKAWA,2017, Potassium-free Solid Fuel Production from Palm Empty Fruit Bunch by Hydrothermal Treatment, 2017, *International Conference on Energy, Environment and Sustainable Development* (EESD 2017), pp.18-22.

[8] Brury Marco Silalahi dan Supijatno, 2017, Pengelolaan Limbah Kelapa Sawit (Elaeis guineensis Jacq.) di Angsana Estate, Kalimantan Selatan, *Bul. Agrohorti*, **5**(3), pp. 373-383

[9] N. Ravi Menon, Zulkifli Ab Rahman and Nasrin Abu Bakar, 2003, Empty Fruit Bunches Evaluation: Mulch in Plantation vs. Fuel for Electricity Generation, *Oil Palm Industry Economic Journal*, Vol. **382**, pp.15-20.

[10] Indikator Tingkat Kesiapan Teknology (TKT) Technology Readiness Level/TRL), Kementerian Ristekdikti, https://docplayer.info/29978752-Indikator-tingkat-kesiapan-teknologi-tkt-technology-readiness-level-trl-kementerian-riset-teknologi-dan-pendidikan-tinggi.html

[11] Faris Salman, Mukhamad Najib, Setiadi Djoha Faris Salman, Mukhamad Najib, Setiadi Djoha,*2017,* Cost and Benefit Analysis of RSPO Certification (Case Study in PT BCA Oil Palm Plantation in Papua*), Indonesian Journal of Business & Entrepreneurship*, Vol.**3** (3), pp. 219-228.

[12] Sakti Hutabarat, 2017, Sertifikasi Ispo dan Daya Saing Kelapa Sawit Indonesia di Pasar Global Tantangan Perkebunan Rakyat Menghadapi Sertifikasi ISPO, *Agro Ekonomi* Vol. **28**/No. 2, pp. 170-188.

[13] Wood B.J. and R.H.V. Corley,1993, The energy balance of oil palm cultivation, in: *Proceedings 1991 PORIM International Palm oil Conference- Agriculture*, Palm Oil Research Institute Malaysia, Kuala Lumpur, pp 130-143.

[14] Kaewmai, R., A. H-Kittikun, et al. 2013, Alternative technologies for the reduction of greenhouse gases emission from palm oil mills in Thailand. *Environ. Sci. Technol.* **47**(21), pp.12417–12425.

[15] Seyed Ehsan Hisseini and Mazlan Abdul Wahid, 2015, Pollutant in palm oil production process, *Journal of the Air and Waste Management Association*, **65**(7), pp. 773-781.

[16] S. Vijaya, A.N Ma and Y.M. Choo, 2010, Capturing Biogas: A Means to Reduce Green House Gas Emissions for the Production of Crude Palm Oil, *American Journal of Geoscience* 181, pp. 1-6.

[17] Annamari Enstrom, Timo Haatainen and et al, 2018, Introducing a new GHG emission calculation approach for alternative methane reduction measures in the wastewater treatment of a palm oil mill, *Environ Dev Sustain*, https://doi.org/10.1007/s10668-018-0181-4

[18] Ministerial Decree of Energy and Mineral Resources/ESDM (Energi dan Sumber Daya Mineral) No.: 1772 K/20/MEM/2018

IOP Conf. Series: Materials Science and Engineering **742** (2020) 012021 doi:10.1088/1757-899X/742/1/012021

[19] http://www.aseanenergy.org/blog/renewable-energy-power-pricing-in-indonesia/ [accessed on August 27, 2019]

[20] Febijanto, I. (2017). Tinjauan Komponen Harga Jual Tenaga Listrik dari Pembangkit Listrik Tenaga Biogas dengan Teknologi Covered lagoon, *Prosiding Seminar Nasional Kimia,* 13 Mei 2017, Hotel Grand Quality, Yogyakarta, pp.72-73.

[21] Preliminary Feasibility Study on The Palm Oil Mill Wastes-fired Power Generation Systems and CDM Project for Rural Electrification Sumatera, Indonesia, NTT GP-ECO Communication, Inc., 2009.

[22] Erliza H., 2010, *Role of agro-industry process technology in the development of oil palm downstream industry* (Bogor: PT Penerbit IPB Press)

[23] Agus Suandi et al, Analisa Pengolahan Kelapa Sawit dengan kapasitas Olah 30 ton/jam di PT BIO Nusantara Teknologi, *Teknosia,* Vol. II, No.17, Tahun X, September2014, pp.12-19.

[24] Capital Energy, http://capitalenergy.biz/?p=11781 [accessed on 29 August 2019]

[25] Febijanto, I, 2019, Harga Jual Bio-CNG dari POME (Palm Oil Mill Effluent) sebagai Sumber Energi Alternatif, *Serambi Engineering* Vol.**IV**, Edisi Khusus Oktober 2019, pp.603-615.

[26] Nasrin Abu Bakar, Lim Weng Soon and et al, 2016, Co-firing of Biogas in Palm Oil Mill Biomass Boilers, P*alm Oil Engineering Bulletin*, No.120, pp.23-26.

[27] Bambang Sucahyo, Dwi Lukman and et al, 2019, Study of the Technology of Utilizing Biogas from Palm Oil Mill Effluent (POME) to Boiler, *M.I.P.I*, Vol.13, No.1, pp.43-54.

[28] Rahayu, A.S, et al, 2015, *Buku Panduan Konversi POME menjadi Biogas Pengembangan Proyek di Indonesia 2nd edition*, Winrock International, pp. 56-58.

[29] Reza P.A, et al, 2018, Kajian Teknis dan Keekonomian Pembangkit Listrik Tenaga Biogas dari Limbah Cair Sawit (Studi Kasus di PLTB Biogas Bangka), *Prosiding Seminar Nasional-Peneitian dan Pengabdian pada Masyarakat 2018*, Pangkalpinang, 2 Oktober 2018, ISBN:978-602-61545-0-7, pp. 36-40.

[30] Rahayu, A.S, et al, 2015, *Buku Panduan Konversi POME menjadi Biogas Pengembangan Proyek di Indonesia 2nd edition*, Winrock International, pp. 56-58.

[31] Ministerial Decree of Energy and Mineral Resources/ESDM (Energi dan Sumber Daya Mineral) No.: 55 K/20/MEM/2019

[32] Feasibility Survey for the Utilization of Empty Fruit Bunch (EFB) Oil Palm Waste and making valuable as Biomass Energy by SYNG's Extruder in Republic of Indonesia, 2017, JICA's Study team.

[33] Muhammad Ansori Nasution et al, 2017, Selection of Palm Oil Mill Effluent Treatment for Biogas Generation or Compost Production using an Analytic Hierarchy Process, *J. Mater Cycle Waste Management*, DOI 10.1007/s10163-017-0638-9

[34] UNFCCC (*United Nations Framework Convention on Climate Change), (2017) concerning Methane recovery in wastewater treatment*, III.H./Version 14, Sectoral Scope: 13 EB 53. https://cdm.unfccc.int/filestorage/A/N/F/ANF0MTK4BHZC9O7IEY68P5DJ2VRQ3X/EB53_repan17_Revision%20of%20AMS-III.H_ver14.pdf?t=bVZ8b3FuNnM5fDCh-HkC6dg9RQbKR6oj1KlY [accessed on Nov., 27, 2018]

[35] https://www.bi.go.id/seki/tabel/TABEL1_26.pdf [accessed on August 29, 2019]

IOP Conf. Series: Materials Science and Engineering **742** (2020) 012022 doi:10.1088/1757-899X/742/1/012022

Application of Quantitative Structure-Property Relationship (QSPR) Models for the Predictions of Critical Micelle Concentration of Gemini Imidazolium Surfactants

Ely Setiawan[1], Mudasir[2] and Karna Wijaya[3]

[1,2,3]Department of Chemistry, Faculty of Mathematics and Natural Sciences, Universitas Gadjah Mada, Indonesia
[1]Department of Chemistry, Faculty of Mathematics and Natural Sciences, Universitas Jenderal Soedirman, Indonesia
[1]Austrian-Indonesian Centre for Computational Chemistry (AIC), Department of Chemistry, Faculty of Mathematics and Natural Sciences, Universitas Gadjah Mada, Indonesia

E-mail: ely.setiawan@mail.ugm.ac.id

Abstract. The critical micelle concentration (cmc) is important indexes to determine the performance of surfactants quantitatively. In this study, based on the molecular descriptors, which calculated, by Dragon software, were proposed a quantitative structure-property relationship analysis for prediction of cmc of gemini imidazolium surfactants. In the present study, the associative neural networks (ASNN) and multiple linear regression analysis (MLRA) technique were used to interpret the chemical structural functionality (molecular descriptors) against the cmc of gemini imidazolium surfactants. The models were validated rigorously through 5-fold cross-validation. The ASNN method showed to be better than the MLRA method in terms of the internal and the external prediction accuracy with high statistical quality r^2 = 0.96, q^2 = 0.95, RMSE = 0.08 and MAE = 0.04 respectively. The developed QSPR models are publicly available on the web site at https://ochem.eu/model/11583111 for ASNN model and https://ochem.eu/model/43858671 for MLRA model. These models can be applied to predict the cmc of new gemini imidazolium surfactants.

1. Introduction

Gemini surfactants, as a new class of surface-active agent molecules, have attracted the increasing interest of academic research groups over the past decade. These compounds are composed of two surfactants molecules which have two hydrophilic head groups and two hydrophobic groups, linked by a spacer. In case of cationic surfactants, hydrophilic head group of these compound is positively charged nitrogen atom with two hydrophobic groups (long alkyl chain), linked by a flexible or rigid spacer. In comparison with conventional surfactants, gemini surfactants appear to be better physicochemical properties such as lower critical micelle concentration (cmc), higher efficient to reducing surface tension of water, higher adsorption efficiency, and better solubilizing, wetting, and foaming properties [1]. Currently, imidazolium ring bearing cationic surfactants have been widely investigated because of their unique properties and possibility of use in several applications, including gene delivery agent [2], corrosion inhibitors [3], drug delivery agent [4], antimicrobial and antifungal agent [5]. Due to the existence of the imidazolium headgroup, gemini imidazolium surfactants has several advantages over other gemini cationic surfactants, for example; they exhibit a stronger tendency toward self-aggregation [6].

One of the most common reasons for the development of gemini surfactants is that their cmc values are much lower than the corresponding monomeric surfactants. As is well known, the cmc of the surfactant is related to its molecular structure [7]. Usually, the cmc of surfactants are determined by the experimental techniques, such as conductometry, tensiometry, NMR spectroscopy, calorimetry, and fluorescence emission spectroscopy [8]. Recently, QSPR were also used to predict the cmc of surfactants [9]. QSPR is effective technique in relating the structural (descriptors) to physicochemical properties. This method shows it can be a predictor tool for various surfactants properties [10]. In this work, quantitative structure-property relationships (QSPR) modeling was used to predict the cmc of gemini imidazolium surfactants. Various statistical methods are used to build QSPR models like multi-linear regressions or artificial neural networks [11].

Although many QSPR models have been proposed for cmc, even for gemini surfactants [12], to the best of author's knowledge, there are no earlier reports of QSPR models for cmc prediction for gemini imidazolium surfactants. In present study, we developed new QSPR models for prediction of cmc of gemini imidazolium surfactants. The QSPR models were developed using associative neural networks (ASNN)[13] and multilinear regression analysis (MLRA) methods. The quality of QSPR models was evaluated using the 5-fold cross-validation method [14].

2. Materials and methods

2.1. Data set and descriptors calculation

In this study, the initial set of 70 gemini imidazolium surfactants with known cmc were obtained from several published papers [15, 16, 17, 18, 19]. These compounds were randomly divided into 57 compounds for a training set and 13 compounds for a test set (validation set). Structural formulas of gemini imidazolium surfactants were drawn using MarvinSketch, converted in SMILES (Simplified Molecular Input Line Entry System) format and stored with their corresponding cmc (-log cmc or pCMC) in the OCHEM platform [20]. All gemini imidazolium surfactants structures were cleaned by the Chemaxon standardizer and structural coordinates were optimized using CORINA [21].

The descriptors were calculated using Dragon v6 software [22], which covers 4885 molecular descriptors including topological indices, information indices, drug-like indices, ring descriptors, functional group counts, walk and path counts, atom-centered fragments, molecular properties. In order to reduce the interference of multicollinearity, all of the descriptors were filtered before the development of QSPR models. In this study, we used the pairwise decorrelation method for descriptor selection and the duplicated descriptors with the pairwise correlation of more than 0.95 were eliminated.

2.2. Development of QSPR models

In this study, we used the Online CHEmical Modeling environment (OCHEM, htpp://www.ochem.eu) to build QSPR models for predicting cmc of Gemini imidazolium surfactants. OCHEM is a user-friendly web-based platform that aims to reduce the amount of work a QSAR/QSPR researcher should perform to obtain a predictive model. OCHEM provides tools for analysis and modification of the QSAR/QSPR models. Furthermore, models can be published on the Web and publicly used by others.

We developed the QSPR models with associative neural networks (ASNN) and multiple linear regression analysis (MLRA) technique. ASNN is the combination of an ensemble of feed-forward neural networks and the kNN (k-nearest neighbors) method. The ASNN corrects the bias in the ensemble by utilizing the correlation between ensemble responses as a measure of distance. In the present study, the ASNN method employed three-layered back propagation neural network with the input layer was equal to the number of molecular descriptors calculated by Dragon software, the number of neuron in the hidden layer was five and one neurons in the output layer (pCMC). The

IOP Conf. Series: Materials Science and Engineering **742** (2020) 012022 doi:10.1088/1757-899X/742/1/012022

network is trained using the SuperSAB algorithm with 1000 iterations and a set of 64 networks in the ensemble.

Table 1. SMILES notation and experimental and calculated pCMC value of gemini imidazolium surfactants by using the result of QSPR model obtained in this study

Compound	SMILES notation	pCMC		
		Experimental	ASNN	MLRA
GIS_1	CN1C=C[N+](COCCOC[N+]2=CN(C)C=C2)=C1.[Cl-].[Cl-]	2.15	2.18	2.24
GIS_2	CCN1C=C[N+](COCCOC[N+]2=CN(CC)C=C2)=C1.[Cl-].[Cl-]	2.23	2.30	2.19
GIS_3	CCCCN1C=C[N+](COCCOC[N+]2=CN(CCCC)C=C2)=C1.[Cl-].[Cl-]	2.41	2.38	2.42
GIS_4	CCCCCCN1C=C[N+](COCCOC[N+]2=CN(CCCCCC)C=C2)=C1.[Cl-].[Cl-]	2.58	2.57	2.53
GIS_5	CCCCCCCN1C=C[N+](COCCOC[N+]2=CN(CCCCCCC)C=C2)=C1.[Cl-].[Cl-]	2.65	2.67	2.65
GIS_6	CCCCCCCCN1C=C[N+](COCCOC[N+]2=CN(CCCCCCCC)C=C2)=C1.[Cl-].[Cl-]	2.72	2.72	2.74
GIS_7	CCCCCCCCCN1C=C[N+](COCCOC[N+]2=CN(CCCCCCCCC)C=C2)=C1.[Cl-].[Cl-]	2.81	2.83	2.86
GIS_8	CCCCCCCCCCN1C=C[N+](COCCOC[N+]2=CN(CCCCCCCCCC)C=C2)=C1.[Cl-].[Cl-]	2.92	2.93	2.87
GIS_9	CCCCCCCCCCCN1C=C[N+](COCCOC[N+]2=CN(CCCCCCCCCCC)C=C2)=C1.[Cl-].[Cl-]	3.04	3.03	3.02
GIS_10	CCCCCCCCCCCCN1C=C[N+](COCCOC[N+]2=CN(CCCCCCCCCCCC)C=C2)=C1.[Cl-].[Cl-]	3.15	3.16	3.15
GIS_11	CCCCCCCCCCCCCCN1C=C[N+](COCCOC[N+]2=CN(CCCCCCCCCCCCCC)C=C2)=C1.[Cl-].[Cl-]	3.34	3.34	3.31
GIS_12	CCCCCCCCCCCCCCCCN1C=C[N+](COCCOC[N+]2=CN(CCCCCCCCCCCCCCCC)C=C2)=C1.[Cl-].[Cl-]	3.52	3.5	3.43
GIS_13	CN1C=C[N+](COCCCOC[N+]2=CN(C)C=C2)=C1.[Cl-].[Cl-]	2.18	2.29	1.68
GIS_14	CCN1C=C[N+](COCCCOC[N+]2=CN(CC)C=C2)=C1.[Cl-].[Cl-]	2.26	2.26	2.30
GIS_15	CCCCCCN1C=C[N+](COCCCOC[N+]2=CN(CCCCCC)C=C2)=C1.[Cl-].[Cl-]	2.61	2.58	2.57
GIS_16	CCCCCCCN1C=C[N+](COCCCOC[N+]2=CN(CCCCCCC)C=C2)=C1.[Cl-].[Cl-]	2.68	2.66	2.67
GIS_17	CCCCCCCCN1C=C[N+](COCCCOC[N+]2=CN(CCCCCCCC)C=C2)=C1.[Cl-].[Cl-]	2.75	2.73	2.81
GIS_18	CCCCCCCCCN1C=C[N+](COCCCOC[N+]2=CN(CCCCCCCCC)C=C2)=C1.[Cl-].[Cl-]	2.84	2.83	2.86
GIS_19	CCCCCCCCCCN1C=C[N+](COCCCOC[N+]2=CN(CCCCCCCCCC)C=C2)=C1.[Cl-].[Cl-]	2.95	2.95	2.87

GIS_20	CCCCCCCCCCCN1C=C[N+](COCCCOC[N+]2=CN(CCCC CCCCCCC)C=C2)=C1.[Cl-].[Cl-]	3.07	3.05	3.1
GIS_21	CCCCCCCCCCCCN1C=C[N+](COCCCOC[N+]2=CN(CCC CCCCCCCCC)C=C2)=C1.[Cl-].[Cl-]	3.18	3.19	3.15
GIS_22	CCN1C=C[N+](COCCCCOC[N+]2=CN(CC)C=C2)=C1.[Cl-].[Cl-]	2.30	2.34	2.18
GIS_23	CCCCN1C=C[N+](COCCCCOC[N+]2=CN(CCCC)C=C2)= C1.[Cl-].[Cl-]	2.45	2.40	2.47
GIS_24	CCCCCN1C=C[N+](COCCCCOC[N+]2=CN(CCCCC)C=C 2)=C1.[Cl-].[Cl-]	2.53	2.55	2.45
GIS_25	CCCCCCN1C=C[N+](COCCCCOC[N+]2=CN(CCCCCC)C =C2)=C1.[Cl-].[Cl-]	2.64	2.63	2.61
GIS_26	CCCCCCCN1C=C[N+](COCCCCOC[N+]2=CN(CCCCCCC)C=C2)=C1.[Cl-].[Cl-]	2.71	2.68	2.76
GIS_27	CCCCCCCCN1C=C[N+](COCCCCOC[N+]2=CN(CCCCCC CC)C=C2)=C1.[Cl-].[Cl-]	2.79	2.8	2.85
GIS_28	CCCCCCCCCN1C=C[N+](COCCCCOC[N+]2=CN(CCCCC CCCC)C=C2)=C1.[Cl-].[Cl-]	2.88	2.9	2.84
GIS_29	CCCCCCCCCCN1C=C[N+](COCCCCOC[N+]2=CN(CCCC CCCCCC)C=C2)=C1.[Cl-].[Cl-]	3.00	2.98	3.00
GIS_30	CCCCCCCCCCCN1C=C[N+](COCCCCOC[N+]2=CN(CCC CCCCCCCC)C=C2)=C1.[Cl-].[Cl-]	3.13	3.1	3.16
GIS_31	CCCCCCCCCCCCN1C=C[N+](COCCCCOC[N+]2=CN(CC CCCCCCCCCC)C=C2)=C1.[Cl-].[Cl-]	3.22	3.24	3.19
GIS_32	CCCCCCCCCCCCCCN1C=C[N+](COCCCCOC[N+]2=CN(CCCCCCCCCCCCCC)C=C2)=C1.[Cl-].[Cl-]	3.40	3.42	3.38
GIS_33	CCCCCCCCCCCCCCCCN1C=C[N+](COCCCCOC[N+]2= CN(CCCCCCCCCCCCCCCC)C=C2)=C1.[Cl-].[Cl-]	3.58	3.52	3.66
GIS_34	CN1C=C[N+](COCCCCCOC[N+]2=CN(C)C=C2)=C1.[Cl-].[Cl-]	2.25	2.21	2.34
GIS_35	CCN1C=C[N+](COCCCCCOC[N+]2=CN(CC)C=C2)=C1.[Cl-].[Cl-]	2.33	2.35	2.34
GIS_36	CCCN1C=C[N+](COCCCCCOC[N+]2=CN(CCC)C=C2)=C 1.[Cl-].[Cl-]	2.41	2.42	2.36
GIS_37	CCCCN1C=C[N+](COCCCCCOC[N+]2=CN(CCCC)C=C2) =C1.[Cl-].[Cl-]	2.49	2.49	2.59
GIS_38	CCCCCN1C=C[N+](COCCCCCOC[N+]2=CN(CCCCC)C= C2)=C1.[Cl-].[Cl-]	2.58	2.57	2.56
GIS_39	CCCCCCN1C=C[N+](COCCCCCOC[N+]2=CN(CCCCCC) C=C2)=C1.[Cl-].[Cl-]	2.67	2.67	2.65
GIS_40	CCCCCCCN1C=C[N+](COCCCCCOC[N+]2=CN(CCCCCC C)C=C2)=C1.[Cl-].[Cl-]	2.75	2.74	2.78
GIS_41	CCCCCCCCN1C=C[N+](COCCCCCOC[N+]2=CN(CCCCC CCC)C=C2)=C1.[Cl-].[Cl-]	2.83	2.82	2.83
GIS_42	CCCCCCCCCN1C=C[N+](COCCCCCOC[N+]2=CN(CCCC	2.91	2.95	2.93

	CCCCC)C=C2)=C1.[Cl-].[Cl-]			
GIS_43	CCCCCCCCCCN1C=C[N+](COCCCCCOC[N+]2=CN(CCCCCCCCCC)C=C2)=C1.[Cl-].[Cl-]	3.07	3.01	2.99
GIS_44	CCCCCCCCCCCN1C=C[N+](COCCCCCOC[N+]2=CN(CCCCCCCCCCC)C=C2)=C1.[Cl-].[Cl-]	3.16	3.16	3.08
GIS_45	CCCCCCCCCCCCN1C=C[N+](COCCCCCOC[N+]2=CN(CCCCCCCCCCCC)C=C2)=C1.[Cl-].[Cl-]	3.25	3.25	3.22
GIS_46	CCN1C=C[N+](COCCCCCCOC[N+]2=CN(CC)C=C2)=C1.[Cl-].[Cl-]	2.37	2.41	2.36
GIS_47	CCCN1C=C[N+](COCCCCCCOC[N+]2=CN(CCC)C=C2)=C1.[Cl-].[Cl-]	2.45	2.44	2.44
GIS_48	CCCCN1C=C[N+](COCCCCCCOC[N+]2=CN(CCCC)C=C2)=C1.[Cl-].[Cl-]	2.54	2.55	2.59
GIS_49	CCCCCN1C=C[N+](COCCCCCCOC[N+]2=CN(CCCCC)C=C2)=C1.[Cl-].[Cl-]	2.62	2.63	2.60
GIS_50	CCCCCCN1C=C[N+](COCCCCCCOC[N+]2=CN(CCCCCC)C=C2)=C1.[Cl-].[Cl-]	2.71	2.7	2.72
GIS_51	CCCCCCCN1C=C[N+](COCCCCCCOC[N+]2=CN(CCCCCCC)C=C2)=C1.[Cl-].[Cl-]	2.79	2.78	2.82
GIS_52	CCCCCCCCN1C=C[N+](COCCCCCCOC[N+]2=CN(CCCCCCCC)C=C2)=C1.[Cl-].[Cl-]	2.87	2.87	2.87
GIS_53	CCCCCCCCCN1C=C[N+](COCCCCCCOC[N+]2=CN(CCCCCCCCC)C=C2)=C1.[Cl-].[Cl-]	2.95	2.99	2.97
GIS_54	CCCCCCCCCCN1C=C[N+](COCCCCCCOC[N+]2=CN(CCCCCCCCCC)C=C2)=C1.[Cl-].[Cl-]	3.07	3.08	3.10
GIS_55	CCCCCCCCCCCN1C=C[N+](COCCCCCCOC[N+]2=CN(CCCCCCCCCCC)C=C2)=C1.[Cl-].[Cl-]	3.20	3.18	3.20
GIS_56	CCCCCCCCCCCCN1C=C[N+](COCCCCCCOC[N+]2=CN(CCCCCCCCCCCC)C=C2)=C1.[Cl-].[Cl-]	3.29	3.32	3.25
GIS_57	CCCCCCCCCCCCCCN1C=C[N+](COCCCCCCOC[N+]2=CN(CCCCCCCCCCCCCC)C–C2)=C1.[Cl-].[Cl-]	3.47	3.54	3.46
GIS_58	CCCCCCCCCCCCCCCCN1C=C[N+](COCCCCCCOC[N+]2=CN(CCCCCCCCCCCCCCCC)C=C2)=C1.[Cl-].[Cl-]	3.65	3.53	3.67
GIS_59	CCCCCCCCCCCCN1C=C[N+](CC[N+]2=CN(CCCCCCCCCCCC)C=C2)=C1.[Br-].[Br-]	3.19	3.01	2.93
GIS_60	CCCCCCCCCCCCN1C=C[N+](CCCC[N+]2=CN(CCCCCCCCCCCC)C=C2)=C1.[Br-].[Br-]	3.14	3.14	3.16
GIS_61	CCCCCCCCCCCCN1C=C[N+](CCCCCC[N+]2=CN(CCCCCCCCCCCC)C=C2)=C1.[Br-].[Br-]	3.13	3.18	3.27
GIS_62	CCCCCCCCCCCCCCN1C=C[N+](CCCC[N+]2=CN(CCCCCCCCCCCCCC)C=C2)=C1.[Br-].[Br-]	3.92	3.32	3.33
GIS_63	CCCCCCCCN1C=C[N+](CC(O)C[N+]2=CN(CCCCCCCC)C=C2)=C1.[Cl-].[Cl-]	2.36	2.65	2.65
GIS_64	CCCCCCCCCCN1C=C[N+](CC(O)C[N+]2=CN(CCCCCCCCCC)C=C2)=C1.[Cl-].[Cl-]	2.60	2.79	2.88

GIS_65	CCCCCCCCCCC(O)CN1C=C[N+](CCC[N+]2=CN(CC(O)CCCCCCCCCC)C=C2)=C1.[Br-].[Br-]	2.86	2.71	2.85
GIS_66	CCCCCCCCCCC(O)CN1C=C[N+](CCCC[N+]2=CN(CC(O)CCCCCCCCCC)C=C2)=C1.[Br-].[Br-]	2.85	2.81	2.75
GIS_67	CCCCCCCCCCC(O)CN1C=C[N+](CCCCC[N+]2=CN(CC(O)CCCCCCCCCC)C=C2)=C1.[Br-].[Br-]	2.83	2.89	2.75
GIS_68	CCCCCCCCCCC(O)CN1C=C[N+](CCCCCC[N+]2=CN(CC(O)CCCCCCCCCC)C=C2)=C1.[Br-].[Br-]	2.80	2.85	2.93
GIS_69	CCCCCCCCCCC(O)CN1C=C[N+](CCCCCCCC[N+]2=CN(CC(O)CCCCCCCCCC)C=C2)=C1.[Br-].[Br-]	2.80	2.92	2.86
GIS_70	CCCCCCCCCCCCCCCCOCC(O)CN1C=C[N+](CC[N+]2=CN(CC(O)COCCCCCCCCCCCCCCCC)C=C2)=C1.[Br-].[Br-]	3.38	3.23	3.77

Multiple linear regression (MLR) is one of the earliest methods used for constructing QSPR/QSAR models, but it is still one of the most commonly used one to date, because of simplicity and easily interpretable model. The selected descriptors were employed with MLR to develop the linear model of the property of interest, which takes the form:

$$Y= b_0 + b_1x_1 + b_2 x_2 +\ldots+ b_nx_n \qquad (1)$$

in this equation Y is the predicted property value of the dependent variable, x_1 to x_n represent the specific descriptors, while b_1 to b_n represent the coefficient of those descriptors, b_0 is the intercept of this equation.

2.3. Model validation and evaluation

The models developed in this paper were evaluated by 5-fold cross-validation and externally validated by an external validation set independent from the model training. All QSPR models were evaluated with the squared correlation coefficient, r^2, cross-validation coefficient, q^2, root mean squared error (RMSE) and mean absolute error (MAE). QSPR models are considered effective at the value of $r^2>0.6$, $q^2>0.5$, with low RMSE and MAE value, can be used to assess the cmc of new gemini imidazolium surfactants [23].

3. Results and Discussion

In this study, we developed QSPR models for cmc of gemini imidazolium surfactants prediction were generated by ASNN and MLRA along with molecular descriptors which calculated by Dragon v6 packages implemented at the OCHEM platform. As a result, there were two models generated by combination; ASNN model and MLRA model.

Table 2. Statistical quality of the QSPR model

Model	Set	n	r^2	q^2	RMSE	MAE
ASNN	Training set	57	0.96 ± 0.02	0.95 ± 0.02	0.08 ± 0.02	0.04 ± 0.01
	Test set	13	0.91 ± 0.04	0.85 ± 0.10	0.20 ± 0.10	0.07 ± 0.04
MLRA	Training set	57	0.91 ± 0.02	0.89 ± 0.04	0.12 ± 0.02	0.07 ± 0.01
	Test set	13	0.91 ± 0.05	0.85 ± 0.09	0.17 ± 0.10	0.08 ± 0.04

Table 2 clearly shows the better statistical quality performance of the ASNN model compared to MLRA model, with the value of r^2, q^2, RMSE and MAE were 0.96, 0.95, 0.08 and 0.04, respectively. The ASNN method represents an innovative method to calculate nonlinear models between structure and property (cmc) of gemini imidazolium surfactants. The correlation of the experimental pCMC of gemini imidazolium surfactants with the calculated of ASNN model is shown in figure 1.

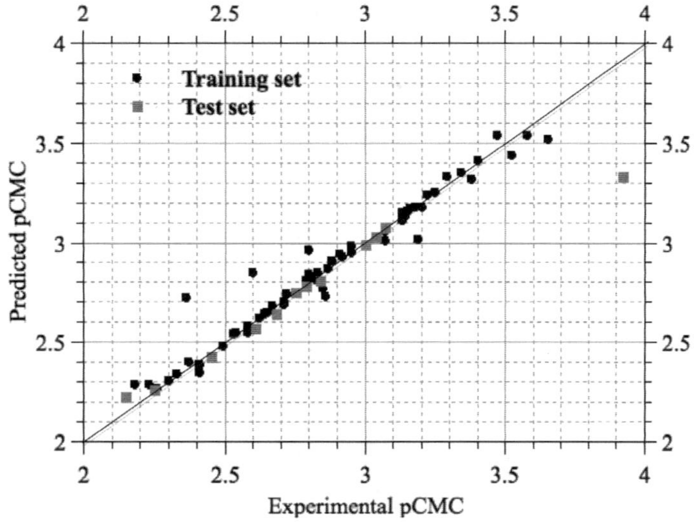

Figure 1. Plot of experimental versus predicted pCMC values obtain from ASNN model

The MLRA model was built with the same number of descriptors as ASNN model. Equation 2 is obtained by MLRA method. The five molecular descriptors with the highest significant contribution to the pCMC of gemini imidazolium surfactants are MW (molecular weight), SpMax_AEA(dm) (leading eigenvalue from augmented edge adjacency mat. weighted by dipole moment), GATS1e (Geary autocorrelation of lag 1 weighted by Sanderson electronegativity), MATS6v (Moran autocorrelation of lag 6 weighted by van der Waals volume), and Eig04_AEA(dm) (eigenvalue n. 4 from augmented edge adjacency mat. weighted by dipole moment). The model equation indicates that the pCMC of gemini imidazolium surfactants has a direct relationship with MW, and has an inverse relationship with SpMax_AEA(dm), GATS1e, MATS6v, and Eig04_AEA(dm).

$$pCMC = 16.3 + 0.00356*MW - 3.83*SpMax\_AEA(dm) - 1.6*GATS1e - 0.483*MATS6v - 1.05*Eig04\_AEA(dm) \quad (2)$$

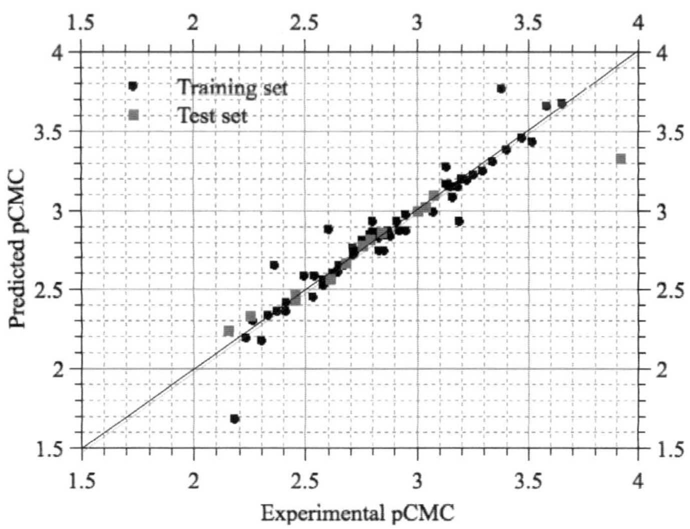

Figure 2. Plot of experimental versus predicted pCMC values obtain from MLRA model

According to the statistical quality of MLRA model (table 2), the squared correlation coefficient (r^2) and cross-validation coefficient, q^2 are 0.91 and 0.89 for the training set, respectively. The RMSE and MAE value (0.12 and 0.07, respectively) are lower enough to indicate successful predictions of the QSPR model developed by MLRA. The graph of the experimental versus predicted cmc of gemini imidazolium surfactants of MLRA model is shown in figure 2.

4. Conclusion

ASNN and MLRA methods were used to build QSPR model of the cmc of gemini imidazolium surfactants using molecular descriptors which calculated by Dragon v6 packages implemented at the OCHEM platform for the first time The results showed that the predicted values from both techniques are close to the true values. High correlation coefficients for ASNN and MLRA model along with low prediction errors demonstrated the ability of the models to predict cmc of gemini imidazolium surfactants. The predictive ability of the ASNN and MLR models was tested by 5-fold cross-validation method. The results show that the ASNN model predicts cmc better with the value of r^2, q^2, RMSE and MAE were 0.96, 0.95, 0.08 and 0.04, respectively. The ASNN model produces high statistical quality and low prediction error model compared with MLRA model. The developed QSPR models are publicly available on the web site at https://ochem.eu/model/11583111 for ASNN model and https://ochem.eu/model/43858671 for MLRA model. These models can be applied to predict the cmc of new gemini imidazolium surfactants.

References

[1] Menger F M Keiper J S 2000 *Angew. Chem., Int. Ed.* **39** 1906–20
[2] Sunitha S Reddy P S Prasad R B N Kanjilal S 2011 *Eur. J. Lipid Sci. Technol.* **113** 756-762
[3] Zhou T Yuan J Zhang Z Xin X Xu G 2016 *Colloid Surf., A* **575** 57–65
[4] Rodrigues M Calpena A C Amabilino D B Garduno-Ramirez M L Perez-Garcia L 2014 *J. Mater. Chem. B* **2** 5419–29
[5] Wang L Qin H Ding L Huo S Deng Q Zhao B Meng L Yan T 2014 *J. Surfactants Deterg.* **17** 1099–1106
[6] Kamboj R Singh S Bhadani A Kataria H Kaur G. 2012 **28** 11969–78
[7] Goronja J Pejic, N Lezăic, A J 2016 *J. Chem. Educ.* **93** 1277–81
[8] Lin C E Wang T Z Chiu T C Hsueh C C 1999 *J. High Resolut. Chromatogr.* **22** 265–270
[9] Roy K Kabir H 2013 *Chem. Eng. Sci.* **87** 141–151
[10] Hu J Zhang X Wang Z 2010 *Int. J. Mol. Sci.* **11** 1020–47
[11] Abrougui K Gabsi K Mercatoris B Khemis CAmami R Chehaibi S 2019 *Soil Tillage Res.* **190** 202–208
[12] Mozrzymas A 2017 *Colloid Polym. Sci.* **295** 75–87
[13] Tetko I V 2002 *J. Chem. Inf. Comput. Sci.* **42**, 717–728
[14] Tropsha A Gramatic P Gombar V K 2003 *QSAR Comb. Sci.* **22** 69–77
[15] Bhadani A Singh S Sakai H Abe M 2016 Synthesis and properties of heterocyclic cationic gemini surfactants. In *Encyclopedia of Biocolloid and Biointerface Science*; Ed Ohshima H (John Wiley Sons Inc.) chapter 43 pp 539–553
[16] Ren Y Liu H Yao X Liu M Hu ZFan B 2006 *J. Colloid Interface Sci.* **302** (2) 669–672
[17] Pałkowski L Błaszczynski J Skrzypczak A Błaszczak J Kozakowska K Wroblewska J Kozuszko S, Gospodarek E Krysinski J Słowinski R 2014 *Chem Biol Drug Des* **83** 278–288
[18] Patial P Shaheen A Ahmad I 2014 *J. Surfact. Deterg.* **17**:253–260
[19] Wang L Liu J Huo S Deng Q Yan T Ding L Zhang C Meng L Lu Q 2014 *J. Surfact. Deterg.* **17** 1107–16
[20] Sushko I Pandey A Novotarskyi S Körner R Rupp M Teetz W Tanchuk V 2011 *J. Comput.-Aided Mol. Des.* **25** (6) 533–554
[21] Kovalishyn V Abramenko N Kopernyk I Charochkina L Metelytsia L Tetko I V Peijnenburg W Kustov L 2018 *Food Chem. Toxicol.* **112** 507–517

IOP Conf. Series: Materials Science and Engineering **742** (2020) 012022 doi:10.1088/1757-899X/742/1/012022

[22] Todeschini R Consonni V 2000 *Handbook of Molecular Descriptors* (Weinheim, WILEY-VCH) p 667
[23] Tropsha, A 2010 *Mol. Inf.* 29 476–488

IOP Conf. Series: Materials Science and Engineering **742** (2020) 012023 doi:10.1088/1757-899X/742/1/012023

Formulation of Curcumin as Natural Dye on Polyester

P Ramadhany[1], J R Witono[1], and O Putri[1]

[1] Department of Chemical Engineering, Faculty of Industrial Technology, Parahyangan Catholic University, Bandung, Indonesia

E-mail: pramadhany@unpar.ac.id

Abstract. Turmeric is perennial herbs that has been used for centuries as natural dye in the Asia. It contains curcumin that generate yellow to orange color. Unfortunately, curcumin is considered as fugitive dyes because its color strength will diminish within times. In order to strengthen the color of curcumin, some additives are added into dye formulation. In this research, turmeric juice was extracted from turmeric without adding any organic solvents. The juice was then mixed with mordant and emulsifier to produce dye solution. Alum and lime juice were used as synthetic and natural mordant, respectively. Tween 80 was used as emulsifier. The polyester was soaked in the dye solution for overnight. The color strength (K/S), the color wash fastness, and the color coordination of the dyed polyester were investigated. The research shows that the ratio of turmeric juice to mordant of 2:1 (without any emulsifiers) gave the best color strength and color fastness.

1. Introduction

Textile industry is one of the fundamental industry, where the products can be easily found in the daily life. Between 2011 and 2015, the development of textile industry expanded by 45% [1]. These increment may be caused by fast-fashion phenomenon, where ready-to-wear clothes are sold in cheaper price in the shorter cycle. However, this phenomenon has negative consequences to ecosystem. The extensive amount of unrenewable products (e.g: synthetic dye and fabrics) are produced, provoking devastating waste pollutions.

These environmental issues have ignite some researchers and industrial practitioner to acquire real solution. Natural dye is one of the solution to replace the synthetic dye. Currently, research about natural dye receives attention due to its sustainability. Natural dye contains natural colorant and is sourced from plants, animals, and minerals. Natural colorant is able to produce color due to its chromophore that can absorb visible light, it has conjugation system and electron resonance to stabilize organic components [2]. Natural dye might also contain color helpers (auxochrome) that defines color gradation and colorant's solubility in the dye.

Currently the market share of natural dye is only 1% of the total market [3]. This might due to several limitations of natural dye, such as tedious extraction of raw materials, weak color fastness thus mordant is required [4], availability of raw materials depending on the season, technology limitation, and limited range of color gradation [5].

Curcumin is the main colorant of turmeric and produces yellow to orange color. It has been used for century as natural dye. Curcumin is classified as diferuloylmethane polyphenol and considered as disperse dye due to its ability to fixate on the fabrics without mordant as fixative. However, like most of natural dyes, curcumin is fugitive dye because it will fade over time. Thus, mordant is added to the dye to enhance curcumin fixation on the fabric.

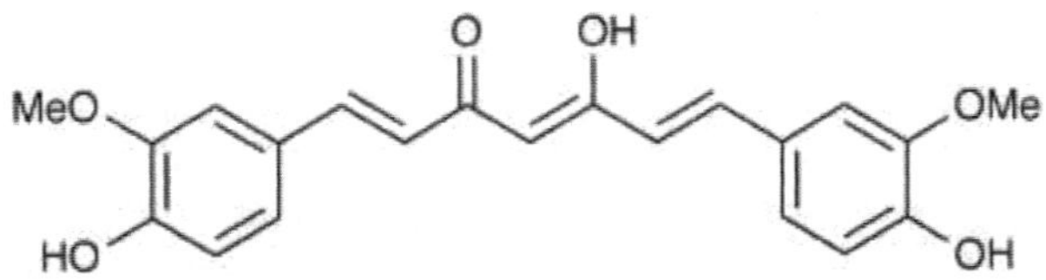

Figure 1. The structure of curcumin

Several studies has been focusing in the development of natural dye. Different mordant will result different color gradation and strength [6]. Dyes for polyester tend to be disperse dye and designed to be hydrophobic in nature [7]. Thus, some researchers have formulate natural dyes with additive such as surfactant to produce emulsion. Water-in-oil (W/O) natural dye emulsion is successfully produced from sappan wood and generates vibrant red color [8]. Sonication during the dyeing process can also improve the dye adherence and fastness properties on the cotton fabric [9].

This research is conducted to understand the influence of concentration of mordant and emulsifier to the color strength, the color fastness, and the color gradation or coordination of curcumin on the polyester.

2. Methodology

Material used in this research was synthetic cotton with 80% Polyester obtained from local textile market. Fresh turmeric (*Curcuma domestica Val*) was obtained from local market. The mordant used were lime juice and alum. Alum was obtained from PT. Brataco Chemika Bandung, while lime juice was obtained from local market. Tween 80 was used as emulsifier and obtained from PT. Brataco Chemika Bandung.

2.1. Curcumin extraction and analysis

The extraction process was conducted with zero used of chemical solvent. Turmeric was washed and cleaned, then cut to smaller pieces and crushed with food processor. Crushed turmeric was filtered and centrifuged at 10,000 rpm at 5 minutes until obtaining turmeric juice. The concentration of curcumin in the turmeric juice was analysed using UV-VIS Spectrophotometry at 420 nm.

2.2. Natural dye formulation and dyeing of polyester

The turmeric juice was formulated with two different type of mordants (alum and lime juice) with volume ratio of turmeric juice (TJ) and mordant 1:0, 2:1, 1:1, and 1:2. The mixture of turmeric juice and mordant was agitated at high speed and room temperature for 30 minutes.

Polyester with the size of 3 x 3 cm was dyed and soaked overnight at room temperature. The dyed polyester must be dried before proceeding into the analysis step.

The formulation with the best color fastness was formulated with emulsifier Tween 80.

2.3. Analysis of dyed polyester

The color strength of dry dyed polyester was analyzed. Color strength analysis was conducted using Spectrophotometer Datacolor 600TM where K/S values follows Kubelka-Munk Theory of Reflectance:

$$\frac{K}{S} = \frac{(1-R)^2}{2R} \quad (1)$$

Where K is absorption coefficient of dyed fabric, S is scattering coefficient, and R is reflectance coefficient. Illuminant D65 and 10° observer were used. The color coordination of dyed polyester was also analyzed based on CIELAB color space.

The color wash fastness was investigated by washing dyed polyester in the water bath (500 ml) for 10 minutes. Each washing process is repeated five times. The washed polyester was dried, afterwards the color strength and the color coordination were measured. The concentration of curcumin in each of the water bath was measured using UV-VIS Spectrophotometry at 420 nm.

The color fastness was analyzed by Grey Scale (GS) method by comparing GS value of dyed polyester before and after washing.

3. Results and Discussion

The concentration of curcumin in the turmeric juice was 7.38 mg/L. The turmeric juice was then formulated with mordant and emulsifier with different volume ratio of turmeric juice, mordant, and emulsifier as follows: (1) 1:0:0, (2) 2:1:0, (3) 2:1:1, (4) 1:1:0, and (5) 1:2:0.

3.1. Evaluation of color coordination

The color coordination of each formulation can be seen on Table 1. The coordination using CIELAB color space with three coordinates letter L*, a*, and b*. L* represents the lightness from black (0) to white (100), a* represents green (-) to red (+), and b* represents blue (-) to yellow (+).

The dyed polyester without using any mordant and emulsifier generated orange color. Adding mordant alum into the formulation produced lighter orange on the polyester. Alum created lower pH in the dye solution. In acidic condition, curcumin emits brighter color gradation [10]. Compared to alum, lime juice is natural mordant because it contains minerals. Additionally lime juice creates acidic condition in the dye solution, thus it emits brighter color. However, in comparison with alum, lime juice generated light yellow color, while alum more into light orange. This might cause by citric acid contained in the lime juice, which is natural anti-oxidant and natural bleach [11].

Adding emulsifier to the formulation will generate dye emulsion. Tween 80 has natural light yellow color, mixing it with curcumin would lighten the dye color.

Table 1. Polyester's color coordination before and after washing for different formulations

Ratio TJ:M:E[a]	Before Washing L*	Before Washing a*	Before Washing b*	After Washing L*	After Washing a*	After Washing b*
Mordant Alum						
(1) 1:0:0	50.611	33.393	23.077	58.561	-0.404	4.132
(2) 2:1:0	62.144	2.88	55.333	60.682	5.964	52.920
(3) 2:1:1	65.896	-9.812	52.223	57.132	8.093	24.408
(4) 1:1:0	50.513	17.938	44.282	48.645	16.012	37.762
(5) 1:2:0	66.672	2.600	56.236	55.886	7.047	35.394
Mordant Lime Juice						
(1) 1:0:0	50.611	33.393	23.077	58.561	-0.404	4.132
(2) 2:1:0	71.874	5.331	62.078	70.451	8.158	60.581
(3) 2:1:1	76.038	0.621	65.543	57.045	-1.847	14.908
(4) 1:1:0	71.040	9.726	61.415	58.906	-2.099	14.140
(5) 1:2:0	63.204	12.021	57.373	61.664	4.687	19.941

[a] TJ (Turmeric Juice), M (Mordant), and E (Emulsifier)

3.2. Evaluation of color strength

The color strength (K/S) of the fabric indicates the depth of the color penetrate in the fabric [12]. Its value can be influenced by the type of mordant used [12]. The results of different formulations can be seen in Figure 2. Dyed polyester without any mordant and emulsifier had the weakest color strength around 8.51 and adding alum as mordant to turmeric juice would increase the color strength of the dyed polyester. The color strength of 15.40 was generated when the alum was increased twice to turmeric juice. The affinity between natural colorant and fibers was considered very limited, thus mordant was applied to improve its affinity. Mordant can form chemical bond between fiber and colorant, hence support the fixation of colorant into the fabric [13].

In comparison to alum, lime juice produced dyed polyester with lower color strength. Increasing lime juice did not have the same effect as alum in enhancing the color strength. One of the parameter that influence the diffusion of dye into fiber is pH value [14]. Lime juice mordant creates medium acidic condition (pH 4 – 5) compared to alum (pH 2 – 3). Curcumin might diffuse better into the fabric in high acidic condition, thus higher color strength when alum was used as mordant.

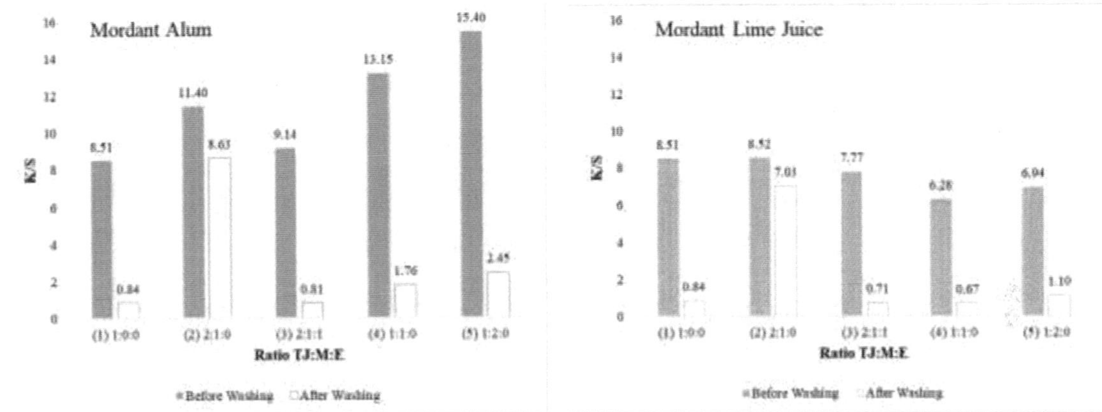

(a) (b)

Figure 2. The color strength (K/S) of curcumin on polyester for different formulations (a) using Mordant Alum, and (b) using Mordant Lime Juice

3.3. Evaluation of color wash fastness

The color strength of curcumin in the fabric after the washing process can be seen in Figure 2. Surely washing would decrease the color strength of curcumin in the fabric for all formulations [12]. The decrease of color strength indicates the weaker fixation rate of the dye in the fabric. In contrast to the rest of formulations, the ratio of turmeric juice to mordant of 2:1 for both alum and lime juice produced the strongest color strength after washing process.

Grey scale is one method to determine the color shading of the fabric. It has different grades with 5 being the highest. It can be used to indicate the color changes after washing process. When the grade drops significantly, it shows low resistance of color changes or weak color fastness of the dye.

Figure 3 displays the grey scale (GS) of the fabric before and after washing. It can be seen that washing process reduces the GS grade of the fabric, which indicating color changes from the initial condition. Excluding the formulation of turmeric juice to mordant 2:1, almost all formulations shows significant drops of GS value. At ratio 2:1 (turmeric juice to mordant), when alum was used as mordant the GS value dropped to 3.50 from 3.60. This change is considered small and indicated that color of fabric after washing still in the same range as initial (Table 1). The similar phenomena also occur when lime juice was used as mordant, the GS value reduced to 3.48 from 3.57. Thus implies the same color range as it is proven in Table 1.

The curcumin concentration in the water bath (washing bath) can be seen in Figure 4. The lesser the concentration of curcumin in the water bath indicates the strong fixative rate of curcumin in the fabric. Figure 4 shows that formulation of turmeric juice to mordant 2:1 diffuses low concentration of curcumin in the water bath. This shows better dye fixation in the fabric and consequently produces stronger color strength and shade of the fabric.

Adding Tween 80 as emulsifier in the formulation would worsen the color fastness in the polyester. Tween 80 can act as detergent and lift the curcumin from the fabric.

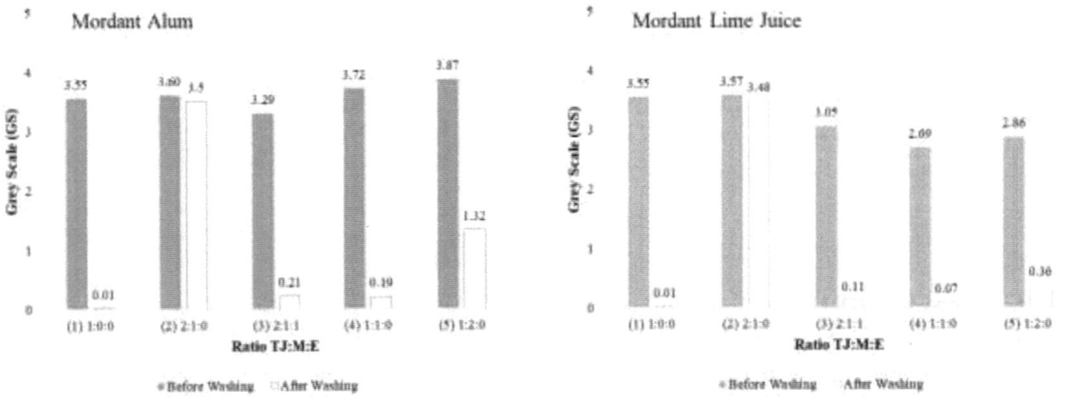

(a) (b)

Figure 3. The Grey Scale (GS) of polyester for different formulations (a) using Mordant Alum, and (b) using Mordant Lime Juice

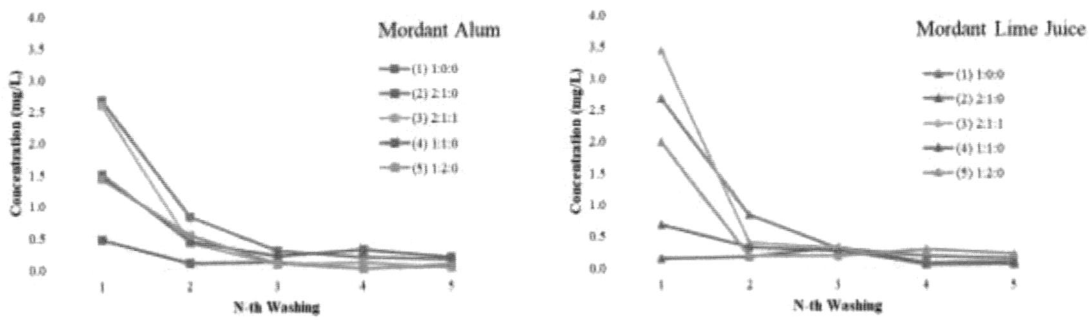

(a) (b)

Figure 4. The concentration of curcumin in the water bath from different formulations (a) using Mordant Alum, and (b) using Mordant Lime Juice

4. Conclusions

This research shows that adding alum and lime juice as mordant would brighten the color of curcumin on the polyester. Increasing alum as mordant strengthened the color strength of curcumin on the polyester, while increasing lime juice gave no similar affect. The color wash fastness of the curcumin on the polyester gave the best result in the ratio of turmeric juice to mordant 2:1 without adding any emulsifier.

Acknowledgement

Authors would like to thanks PT.Idaman Eramandiri for the support given to analyze the products and Lembaga Penelitian dan Pengabdian kepada Masyarakat (LPPM) UNPAR for the financial support to conduct this project with contract number III/LPPM/2019 –01/16–P.

IOP Conf. Series: Materials Science and Engineering **742** (2020) 012023 doi:10.1088/1757-899X/742/1/012023

References

[1] Nugroho R, Ikbal, 2016 Pengolahan Air Limbah Berwarna dengan AOPs *Journal Air Indonesia* **1** (1)

[2] Abrahart E N 1977 Dyes and their intermediates New York: *Chemical Publishing* pp 1–12

[3] Ellen MacArthur Foundation 2017 A new textiles economy: redesigning fashion's future

[4] Vankar P S 2017 Natural dyes for textiles *Chemistry and Applications* Elsevier: UK

[5] Gürses A et al 2016 Dyes and pigments Springer: Turkey

[6] Manhita Manhita A, Ferreira V, Vargas H, Ribeiro I, Candeias A, Teixeira D, Ferreira T, Dias C B, 2011 Enlightening the influence of mordant dyeing technique and photodegradation on the colour hue of textiles dyed with madder - A chromatographic and spectrometric approach *Microchem. J* **98** pp 82-90

[7] International Agency for Research and Cancer (IARC) 2010 Some aromatic amines organic dyes and related exposures

[8] Dong K L 2008 Fabrication of nontoxic natural dye from sappan wood *Korean Journal Chemical Engineering* **25**(2) pp 354 – 358

[9] Vankar P S, Laitonjam W, Tiwari V, Potsangbam L, 2009 Sonicator dyeing of cotton fabric and chemical characterization of the colorant from Melastoma malabathricum *Pigment and resin technology* **38**(1) pp 38-41

[10] Kunnumakkara A B, Anand P, Aggarwal B B, 2008 Curcumin inhibits proliferation, invasion, angiogenesis and metastasis of different cancers through interaction with multiple cell signaling proteins *Cancer Lett.* **269** pp 199-225

[11] Essays UK 2018 Citric Acid to Remove Stains | Experiment. Retrieved from https://www.ukessays.com/essays/biology/citric-acid-in-citrus-fruits-in-removing-stains-biology-essay.php?vref=1

[12] Zhang H 2014 Application of K/S Value In Determination of Fixation Rate *Advance material research* **1048** pp 116-119.

[13] Popoola A V 2000a Dyeability of cellulose fibers using dyestuff from African rosewood (Pterocarpus erinaceous) *Journal of Applied Polymer Science* **77** pp 746-751

[14] Ujhelyiova A, Bolhova E, Oravkinova J, Tino R, Marcincin A, 2007 Kinetics of dyeing process of blend polypropylene/polyester fibers with disperse dye *Dyes and Pigments* **72** pp 212-216

IOP Conf. Series: Materials Science and Engineering **742** (2020) 012024 doi:10.1088/1757-899X/742/1/012024

Effect of Combining Microbial Cellulose Pulp from Nata de Tomato with Tomato Powder on Paper Qualities

D Sugiarto[1], K C Wanta[1] and Y I P A Miryanti[1]

[1]Department of Chemical Engineering, Faculty of Industrial Technology, Parahyangan Catholic University, Bandung, INDONESIA.

E–mail: kcwanta@unpar.ac.id

Abstract. Tomato is a fruit that does not last long (perishable), so it is necessary to find another alternative to increase its economic value. One of those alternatives that can be done is by making a fermented product called nata de tomato. Nata de tomato contains microbial cellulose which is formed by the bacteria *Acetobacter xylinum*. This microbial cellulose has some advantages over the typical cellulose found in common plants. Thus, it potentially could be further developed in the papermaking. This research was conducted to study the effects of combining microbial cellulose pulp from nata de tomato with tomato powder on paper quality, such as water content, grammage and water absorbency. In this research, the tomato filtrate was used as a growth medium for *Acetobacter xylinum* to produce nata de tomato. Then, nata de tomato was shredded into pulp, while its waste was dried into tomato powder and used as additional material for the pulp. This research was designed using a single factor experiment with variation in the addition of tomato powder by 0, 1, 3 and 5 % (w/w). The effects of this variation observed using variance analysis and significant test using the LSD (Least Significant Difference) method.

1. Introduction

Tomato is one of the species of fruit that has the *Lycopersicum* genus and ranks second after potatoes that are interested in the public for consumption[1]. Tomato commonly consumed because it has a lot of benefits and vitamin sources. At first, other than as fresh fruit that can be consumed directly, tomatoes are only used as food and cooking spices. At present, the use of tomatoes has been more developed, such as for flavoring various kinds of dishes and also as industrial materials to be consumed in processed form, such as tomato juice drinks.

However, tomato is a fruit that does not last long (perishable) because it contains much water. So it is necessary to find another alternative in its use to increase its economic value, such as by making a fermented product called nata de tomato[2]. This nata product contains fiber, known as microbial cellulose. Microbial cellulose has several advantages compared to cellulose found in woody plants, which has a better structure, has a degradation ability and has a high level of purity because it does not contain lignin, pectin or hemicellulose. These are the reasons that make nata de tomato products have the potential to be developed further, such as for papermaking[3].

The increasing demand for paper products also increases the need of cellulose which generally comes from woody plants. Other sources of cellulose are needed in order not to increase deforestation rates. Utilization of microbial cellulose is an alternative to overcome weaknesses in the use of cellulose from woody plants in paper production. The use of microbial cellulose in the papermaking does not require the addition of hazardous chemicals that can damage the environment. Besides, the paper products made by microbial cellulose are more easily degraded, so that it will be more eco-friendly. Through this research, it hoped that paper product can be produced with the same quality as a paper product from woody plants and better productivity can be obtained[4].

2. Research Methods

The materials used in this research include tomatoes, *Acetobacter xylinum,* nitrogen sources such as ammonium sulfate (ZA fertilizer), acetic acid (25% vinegar) as an acid regulator and carbon sources such as white sugar. The tools used in this research include a blender, calico as a filter, stainless pan, stoves, stirrer, plastic container and tray dryer. This research consists of three steps, the raw materials preparation, the preliminary step and the main step of research.

The raw materials preparation step was done by washing and cutting tomatoes, smoothing the tomatoes using a blender and filtering using calico to separate the filtrates and the residues. In the preliminary step, nata de tomato and the tomato powder were made. Nata de tomato was made by cooking the tomato filtrates obtained from the raw materials preparation step by the addition of 0,005% (w/v) ZA fertilizer, 0,025% (w/v) white sugar and 0,01% (v/v) of 25% vinegar. Furthermore, the filtrate was put into a plastic container covered with sterile newspaper and given the *Acetobacter xylinum* as much as 0,2% (v/v) after the filtrate was cooled down. This fermentation was carried out for eight days at room temperature.

Tomato powder was made by drying the residues obtained from the raw materials preparation step using a tray dryer at a temperature of 60°C then smoothed with a blender and filtered out with a -80 mesh size. Tomato powder was stored using an aluminum ziplock bag in the refrigerator. In the main step of research, the paper was made from nata de tomato by shredded its cellulose into a pulp using a blender and then filtered out to reduce its water content. Tomato powder produced in the preliminary step was added to the pulp by 0, 1, 3 and 5% (w/w). Furthermore, the pulp will be mold into sheets and dried using a tray dryer at a temperature of 60°C.

Quality analysis conducted on paper products includes water content, grammage and water absorbency. The water content of paper products was analyzed using a moisture analyzer. Paper grammage was calculated by weighing the mass of paper per unit area. Water absorbency of paper products was analyzed by wetting the surface of the paper per time unit and then weighing.

3. Result and Discussion

The paper produced in this research is presented in Figure 1. This research can be said to be successful because physically, the main run and the duplo run produce the same characteristic, which can be seen from the color and texture. The addition of tomato powder affects the physical paper. The more addition of tomato powder into microbial cellulose pulp will darken the paper color and enhance its surface roughness. This phenomenon is according to the aim of tomato powder addition, which is to give color for paper products.

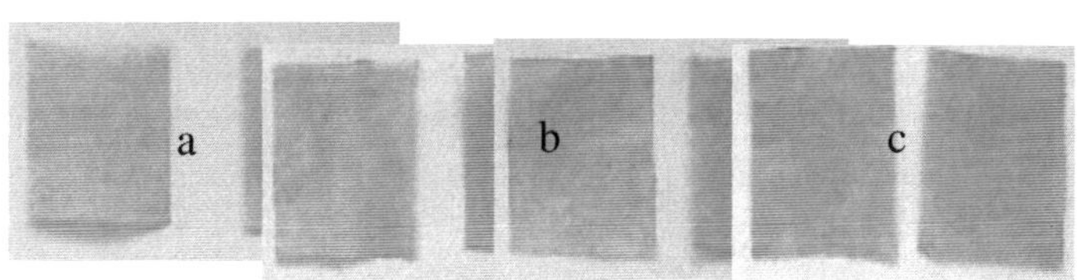

Figure 1. Paper Product (a) Without Tomato Powder Addition (b) Addition of 1% (w/w) Tomato Powder (c) Addition of 3% (w/w) Tomato Powder (d) Addition of 5% (w/w) Tomato Powder

Figure 2 presents the appearance of paper products through a microscope with a magnification of 40×. The figure shows paper that produced without the addition of tomato powder has a denser fiber structure than the paper that produced with the addition of tomato powder. The more addition of tomato powder into microbial cellulose pulp, the higher the strain that occurs between fiber bonds. It can be evidenced by the microscope light that can penetrate the paper products.

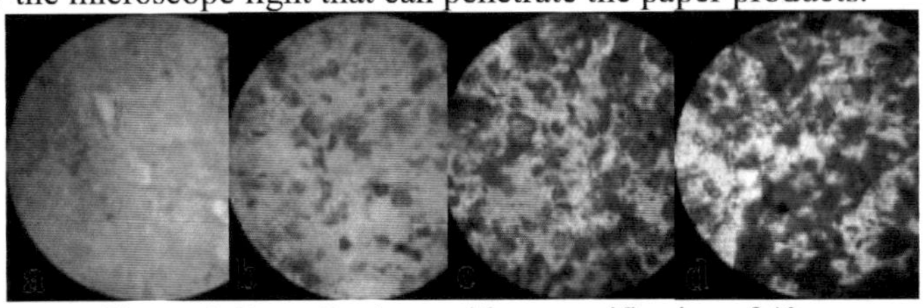

Figure 2. Paper Products with a magnification of 40×

Table 1 shows the result of this research on the value of water content, grammage and water absorbency of paper products with the addition of tomato powder by 0, 1, 3 and 5% (w/w). It can be used as a multifunctional paper which in SNI 6601-2011 has a water content between 3,5-6,5%. This paper product can also be used as an art paper that has a grammage between 100-150 g/m^2. Art paper can be used to print magazines, brochures, or certain documents. Overall, the water absorbency of paper produced in this study is lower than the water absorbency on standard wrap paper, which is 119,73 g/m^2. This results show that paper produced by microbial cellulose from nata de tomato has a high resistance to water penetration [5].

Table 1. Effect of Tomato Powder Addition to Paper Quality

Tomato Powder Addition (w/w)	Water Content (%)	Grammage (g/m^2)	Water Absorbency (g/m^2)	Function
0%	4,65	97,892	38,091	multifunctional & art paper
1%	5,435	113,057	53,257	multifunctional & art paper
3%	5,535	119,918	57,313	multifunctional & art paper
5%	6,505	132,761	78,475	multifunctional & art paper

3.1. Water Content

Water content analysis obtained from this research ranged from 4,65-6,505% where these paper products can be used as a multifunctional paper, which in SNI 6601-2-11 has a water content between 3,5-6,5%. Tomato powder addition into microbial cellulose pulp produced paper with higher water content. The increase of water content can occur because paper products without the addition of tomato powder have denser fibers so that the number of free OH^- groups becomes less and its water content becomes lower [5]. The results obtained from the water content analysis of the paper products are shown in Figure 3.

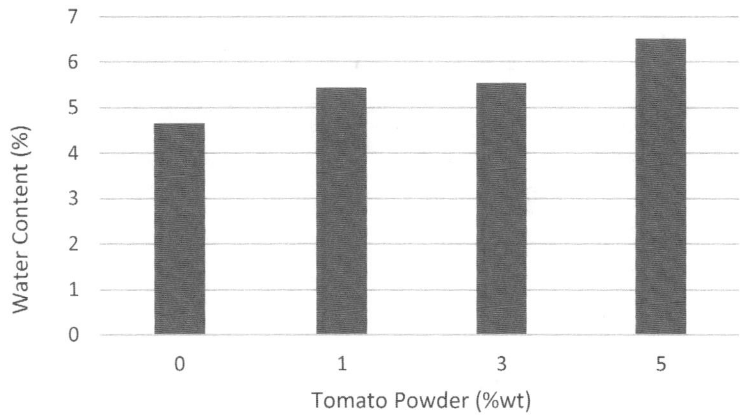

Figure 3. Effect of Addition of Tomato Powder to Water Content

Calculations using the single factor experimental design presented in Table 2, which provide the conclusion that the addition of tomato powder affects the paper quality.

Table 2. Experimental Design Calculation for Water Content

	SS	DOF	MS	Fo		Fo, table	Conclusion
Treatment	3,468138	3	1,156046	147,5019	>	6,59	affects
Error	0,03135	4	0,007837				
Totals	3,499487	7					

3.2. Grammage

Paper grammage analysis obtained from this research ranged between 97,892-132,761 g/m^2. This paper product can be used as an art paper that has a grammage between 100-150 g/m^2. Art paper can be used to print magazines, brochures and certain documents. The more addition of tomato powder into microbial cellulose pulp, the higher the grammage of paper products. The increase of paper grammage can occur because the tomato powder addition will accumulate and increase its weight. The results obtained from the grammage analysis of the paper products are shown in Figure 4.

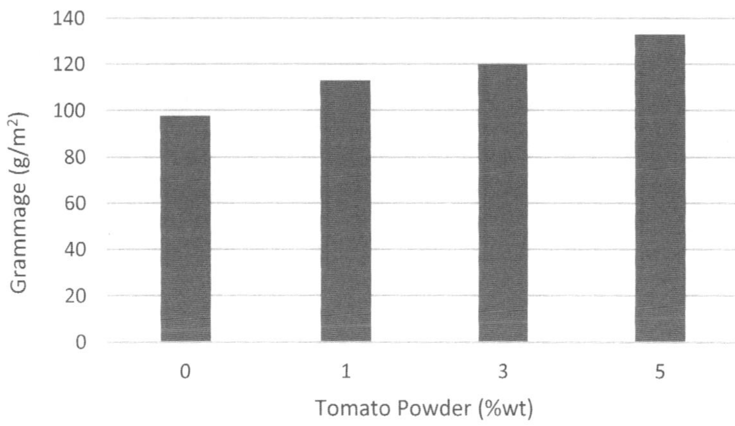

Figure 4. Effect of Addition of Tomato Powder to Grammage

To determine the effect of tomato powder addition on paper grammage, the analysis was done using a single factor experimental design. Based on the calculation presented in Table 3, it can be concluded that the addition of tomato powder into microbial cellulose pulp affects the paper grammage.

Table 3. Experimental Design Calculation for Paper Grammage

	SS	DOF	MS	Fo		Fo, table	Conclusion
Treatment	1265,667	3	421,889	35,59518	>	6,59	affects
Error	47,40968	4	11,85242				
Totals	1313,077	7					

3.3. Water Absorbency

Water absorbency analysis was conducted to study the ability of paper products to absorb water. The result of water absorbency analysis in this research ranged from 38,091-78,475 g/m^2. This result lower than the water absorbency of standard wrap paper, which is 119,73 g/m^2. This shows that paper produced from microbial cellulose has a fairly high resistance to water penetration [4].

The addition of tomato powder into microbial cellulose pulp affects the increase in water absorbency of paper products. The increase of water absorbency can occur because tomato powder is a solid material that cannot absorb water and it can reduce the effectiveness of bonds between fibers in the paper. As a result, there are many air cavities in the contact area between fibers so that water absorbency by paper products can be more easily happen [4]. The results obtained from the water absorbency analysis of the paper products are shown in Figure 5.

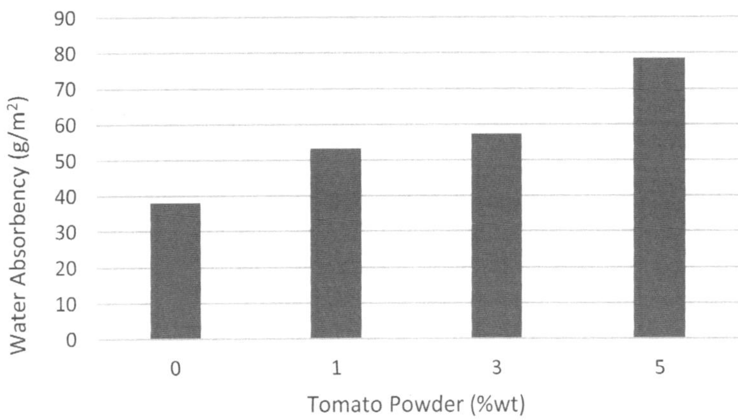

Figure 5. Effect of Addition of Tomato Powder to Water Absorbency

Calculations using a single factor experimental design presented in Table 4 shows that the addition of tomato powder into microbial cellulose pulp affects the water absorbency of paper products.

Table 4. Experimental Design Calculation for Water Absorbency

	SS	DOF	MS	Fo		Fo, table	Conclusion
Treatment	1665,287	3	555,0957	41,99843	>	6,59	affects
Error	52,86823	4	13,21706				
Totals	1718,155	7					

IOP Conf. Series: Materials Science and Engineering **742** (2020) 012024 doi:10.1088/1757-899X/742/1/012024

4. Conclusions

Multifunctional paper and art paper can be made by microbial cellulose pulp from nata de tomato which is formed by the bacteria *Acetobacter xylinum*. The addition of tomato powder into microbial cellulose pulp has a significant effect marked by the increase of water content, grammage and water absorbency of paper products.

References

[1] Pracaya, I. (2009). *Bertanam Tomat*: Kanisius.

[2] Gould, W. A. (1992). CHAPTER 1 – Introduction & History of the Tomato Industry. In W. A. Gould (Ed.), *Tomato Production, Processing and Technology* (pp. 3-17): Woodhead Publishing.

[3] Darmansyah. (2010). *Evaluasi Sifat Fisik dan Sifat Mekanik Material Komposit Serat/Resin Berbahan Dasar Serat Nata de Coco dengan Penambahan Nanofilter*. Tesis Fakultas Teknik. Universitas Indonesia, Depok.

[4] Syamsu, K., Haditjaroko, L., Pradikta, G. I., and Roliadi, H. (2014). Campuran Pulp Tandan Kosong Kelapa Sawit dan Selulosa Mikrobial Nata de Cassava dalam Pembuatan Kertas. *Jurnal Ilmu Pertanian Indonesia, 19(1),* 14-21.

[5] Syamsu, K., Pespitasari, R., and Roliadi, H. (2012). Penggunaan Selulosa Mikrobial dari Nata de Cassava dan Sabut Kelapa sebagai Pensubstitusi Kayu dalam Pembuatan Kertas. *Jurnal Agro Industri Indonesia, 1(2).*

IOP Conf. Series: Materials Science and Engineering **742** (2020) 012025 doi:10.1088/1757-899X/742/1/012025

Subcritical Water as a Solvent for Extraction of Nickel and Aluminum Ions from Reforming Spent Catalysts

K C Wanta[1*], W T Gunawan[1], R F Susanti[1], G P Gemilar[2], H T B M Petrus[3] and W Astuti[4]

[1]Department of Chemical Engineering, Faculty of Industrial Technology, Parahyangan Catholic University, Bandung, INDONESIA
[2] PT. Petrokimia Gresik, Gresik, INDONESIA
[3]Department of Chemical Engineering, Faculty of Engineering, Universitas Gadjah Mada, Yogyakarta, INDONESIA
[4] Research Unit for Mineral Technology, Indonesian Institute of Science, Tanjung Bintang, INDONESIA

Email: kcwanta@unpar.ac.id

Abstract. Spent catalysts are categorized as hazardous waste because they still contain heavy metal elements. It indicates that spent catalysts need to be treated before they discharge to the environment. In this study, spent catalysts treatment was done using subcritical water extraction method. The extraction process was conducted by varying temperature and extraction times. Temperature was varied at 100–250°C, whereas time extraction was varied at 15–45 minutes. Operating pressure and particle size were kept at 70 bars and –60+70 mesh, respectively. The content of aluminum (III) (Al^{3+}) ions and nickel (II) (Ni^{2+}) ions were analyzed using UV–vis spectrophotometer. The experimental results showed that the highest content of both metal ions was achieved at 250°C and 45 minutes. The content of extracted Al^{3+} ions and extracted Ni^{2+} ions at those conditions was 623.39 and 35.03 ppm, respectively.

1. Introduction

Catalysts are the material used in various chemical processes and aim to increase the rate of reaction. Catalysts usually use in chemical industries, such as fertilizer and oil industries. However, the use of catalysts for a long time causes it to become saturated. These catalysts are usually called spent catalysts. Spent catalysts can be classified as hazardous waste because it still contains heavy metal elements, such as aluminum (Al), nickel (Ni), iron (Fe), cobalt (Co), zinc (Zn), and magnesium (Mg) [1]. Thus, in order to avoid contaminating the environment and harming humans, spent catalysts have to be processed first before it is removed to the environment.

One of the spent catalyst processing is using the hydrometallurgy method which heavy metal ions are extracted from the spent catalysts. Most studies related to this method have been conducted using an acid solution in certain operating conditions [2–4]. Based on previous studies, the percentage recovery of heavy metal ions was quite high (more than 90%) depending on operating conditions [5–6]. It indicates that this process has been well developed. However, to maximize the extraction process, it is necessary to use a high concentration of an acid solution. Although for better results, the use of those solutions can pollute the environment. In other words, another green solvent is needed as an alternative to extract those heavy metal ions, for example, subcritical water.

The replacement of solvents from acidic solutions into water has a large advantage because water is cheap, easy to found, easy to be handled and not harmful to the environment. Subcritical water is defined as water which is at 100–374°C and moderate pressure so that it keeps water in the liquid phase [7]. This type of water has specific physical properties, such as the viscosity and density of subcritical water is smaller than water at room pressure and temperature, but the ability of water to diffuse is higher [7–8]. These characteristics are used in various chemical processes, especially in the extraction process. In addition, water in subcritical conditions tends to be more acidic than its initial condition [8–9]. In the extraction process of metal ions, the acidic condition has to be created in the system so that the chemical reaction between solvents and solids can occur as shown in this following reaction:

$$2y(H^+) + M_xO_y \rightarrow xM^{y+} + yH_2O \quad (1)$$

where M is a metal element.

In this study, the main focus was to apply a subcritical water solvent for extracting spent catalyst samples from PT. Petrokimia Gresik. The experimental results from this study are expected to overcome the waste problem.

2. Materials and methods

The spent catalysts used in this study were from the reforming unit, PT. Petrokimia Gresik. These samples were analyzed for metal content in those samples using Energy Disperse X–ray (Shimadzu, EDX–8000). The results showed that the main metal elements of this sample were aluminum and nickel which were 16.67 and 6.50%wt. Thus, the two metal elements were the main focus of this study.

The extraction process of the spent catalyst was conducted using a series of subcritical extraction equipment as presented in Figure 1. Demineralized water was used as a solvent. The temperature was varied at 100, 150, 200 and 250°C while the extraction time was varied at 15, 30 and 45 minutes. Other operating conditions, such as sample mass, sample particle size and operating pressure were maintained at 5 grams, –60+70 mesh and 70 bars, respectively. After the extraction process time was reached, the operating conditions were returned to the atmospheric conditions. The experimental sample was taken and separated between the solid–liquid phase using a centrifuge. The liquid phase was analyzed for aluminum and nickel ions content using a UV–vis spectrophotometer using dimethylglyoxime and eriochrome cyanine R as complexing agent.

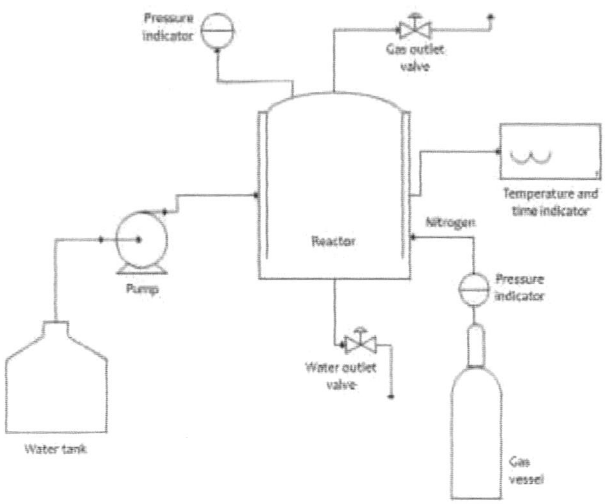

Figure 1. The series of extraction process equipment

3. Results and discussion

3.1. The effect of temperature in the subcritical water extraction process

Temperature is an important parameter in the extraction process because it will affect the diffusion rate and reaction rate. The experimental results of this study are presented in Figure 2 and 3.

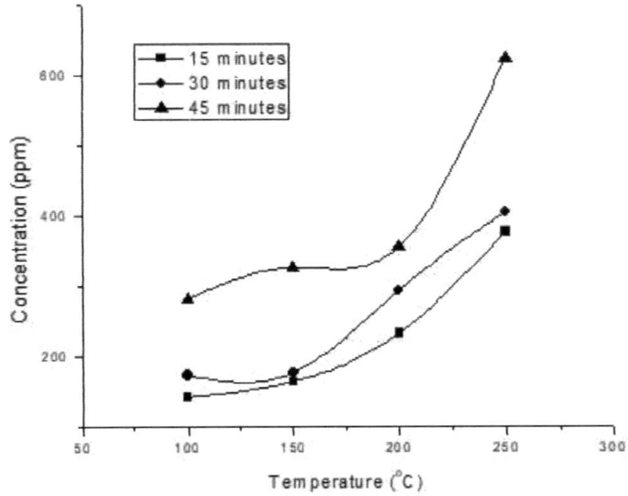

Figure 2. The effect of temperature in the extraction process of Al^{3+} ions

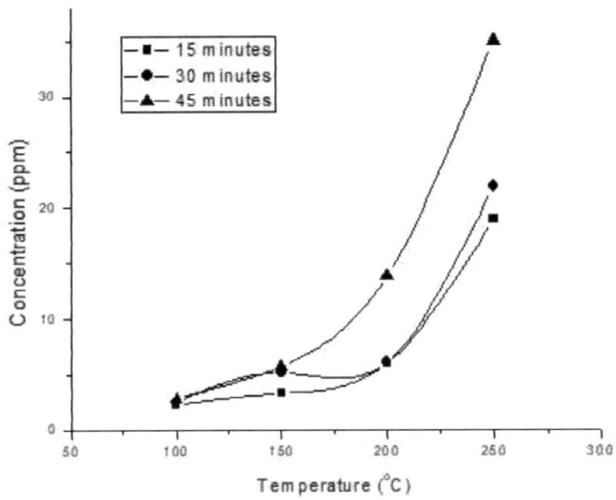

Figure 3. The effect of temperature in the extraction process of Ni^{2+} ions

Based on Figure 2 and 3, the maximum concentration of Al^{3+} and Ni^{2+} ions that can be obtained during the extraction process were 623.39 and 35.03 ppm, respectively. This result was achieved at the same operating conditions, namely at 250°C and 45 minutes. Figure 2 and 3 showed that the trends of aluminum and nickel ions concentration in the liquid phase increase when the operating temperature was raised. It means that the extraction process using subcritical water as the solvent can be activated by temperature. This phenomenon can happen because each molecule moves faster during the diffusion steps and the frequency collision between reactant molecules become bigger during the chemical reaction steps.

The role of temperature in this process is not only related to diffusion and reaction rate but also pH scale in the system [8–9]. pH indicates the amount of hydrogen ions (H^+) present in the system and these ions will react with metal compounds contained in the raw samples. This reaction causes the metal bond in solid phase break and form water–soluble metal ions (Al^{3+} and Ni^{2+} ions) [10]. To obtain the H^+ ions, the breakdown of the hydrogen bonds present in water needs to occur. The amount of hydrogen bond makes H^+ ions difficult to obtain. However, the strength of hydrogen bonds will decrease by increasing temperature [11–12]. In this study, the use of subcritical water conditions was carried out in order to produce acidic conditions in the system so that the chemical reaction steps could occur. Also, the change of lower pH values also affects the solubility of aluminum (III) ions and nickel (II) ions in water [13].

3.2. The effect of extraction time in subcritical water extraction process

Another parameter that affects the extraction process using subcritical water solvents is the extraction time. The experimental results are presented in Figure 4 and 5.

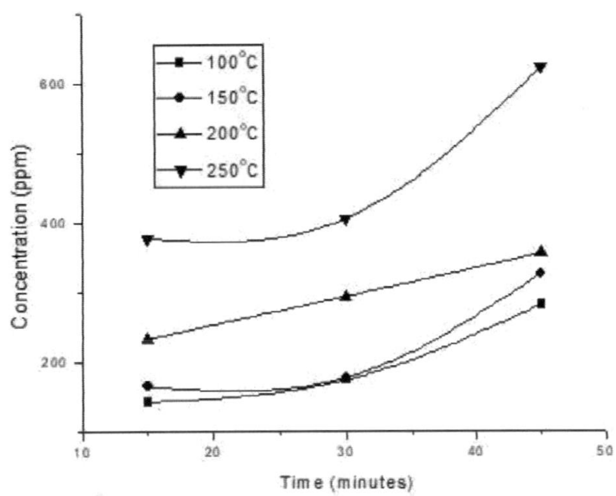

Figure 4. The effect of extraction time in the extraction process of Al^{3+} ions

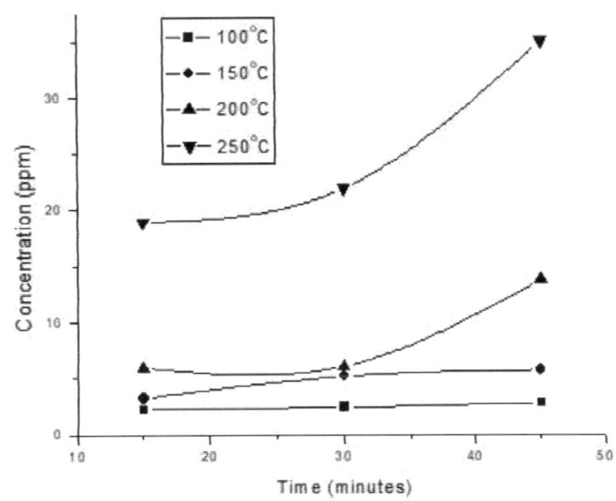

Figure 5. The effect of extraction time in the extraction process of Ni^{2+} ions

Figure 4 and 5 show that the longer the extraction process was carried out, the higher the aluminum and nickel ions would be extracted. The long extraction time will increase the frequency of contact between water (in subcritical condition) and metal compounds (in solids) so that the extraction process will run more maximum and be able to extract more metal ions.

As previously explained, the effect of temperature will affect the quality of the extraction process using subcritical water. Based on Figure 4 and 5, when the operating temperature at 100 and 150°C, the extracted aluminum and nickel ions concentration had different tendencies where the increase in extracted nickel ion concentration was not significant while for aluminum ions, the extracted aluminum ions increases. This phenomenon is caused by the small amount of H^+ ions formed at these two temperatures so that there is a "seizure" of H^+ ions for the reaction of aluminum and nickel compounds. Because the aluminum content in the spent catalyst (raw materials) is more than the nickel content in the spent catalyst, the H^+ ion will attack and react more with aluminum compounds.

Besides being related to the frequency of intermolecular contact reactant (chemical reaction step), the length of the extraction process will also affect the amount of solvent entering the particle. The longer processing time used to extract, the amount of solvent that can enter into solids will continue to increase [13]. This phenomenon will cause lower subcritical water viscosity so that solvents will be easy to enter the sample pore of the spent catalyst.

4. Conclusion

Temperature and time are important parameters in the extraction process include this extraction process using subcritical water. The experimental results showed that the higher the temperature and the longer the extraction process were carried out, the higher the aluminum and nickel ions would be extracted. During the extraction process, the maximum concentration of Al^{3+} and Ni^{2+} ions that can be obtained were 623.39 and 35.03 ppm, respectively. This result was achieved at the same operating conditions, namely at 250°C and 45 minutes.

Acknowledgment

To conduct this study, the authors acknowledge financial support to Lembaga Penelitian dan Pengabdian kepada Masyarakat (LPPM) UNPAR with contract number III/LPPM/2019–01/14–P.

References

[1] Parhi P K, Park K H dan Senanayake G 2013 A kinetic study on hydrochloric acid leaching of nickel from Ni–Al_2O_3 spent catalyst *Journal of Industrial and Engineering Chemistry* **19** pp 589–594.

[2] Petrus H T B M, Wijaya A, Iskandar Y, Bratakusuma D, Setiawan H, Wiratni and Astuti W 2018 Lanthanum and Nickel Recovery from Spent Catalyst using Citric Acid: Quantitative Performance Assessment using Response Surface Method *Metalurgi* **2** pp 91–100.

[3] Oza R, Shah N and Patel S 2011 Recovery of nickel from spent catalysts using ultrasonication–assisted leaching *J Chem Technol Biotechnol* **86** pp 1276–1281.

[4] Barakat M A, Mahmoud M H H and Mahrous Y S 2006 Recovery and separation of palladium from spent catalyst *Applied Catalysis A: General* **301** pp 182–186.

[5] Al–Mansi N M and Monem N M A 2002 Recovery of nickel oxide from spent catalyst *Waste Management* **22** pp 85 – 90.

[6] Miazga B and Mulak W 2008 Leaching of nickel from spent catalysts in hydrochloric acid solutions *Physicochem. Probl. Miner. Process* **41** pp 177–184.

[7] Susanti R F, Garini S, Renaldo I J, Anada R and Stenny A 2013 Ekstraksi Batang *Physalis angulate* dengan Air Subkritik *Research Report* Parahyangan Catholic University.

[8] Weingärtner H and Franck W U 2005 Supercritical Water as a Solvent *Angew. Chem. Int. Ed.* **44** pp 2672–2692.

[9] Marcus Y 2012 *Supercritical Water – A Green Solvent: Properties and Uses* (New York: John Wiley & Sons).

[10] Tremaine P R and Leblanc J C 2980 The solubility of nickel oxide and hydrolysis of Ni^{2+} in water to 573 K *J. Chem. Thermodynamics* **12** pp 521–538.
[11] Ikushima Y, Hatakeda K, Saito N and Arai M 1998 Raman spectroscopy study of subcritical and supercritical water: The peculiarity of hydrogen bonding near the critical point *Journal of Chemical Pysic* **108**(14) pp 5855–5860.
[12] Kalinichev A G and Bass J D 1997 Hydrogen Bonding in Supercritical Water *J. Phys. Chem. A* **101** pp 9720–9727.
[13] Hallet P D, Baumgartl T and Young I M 2001 Subcritical Water Repellency of Aggregates from a Range of Soil Management Practices *Soil Sci. Soc. Am. J.* **65** pp 184–190.

IOP Conf. Series: Materials Science and Engineering **742** (2020) 012026 doi:10.1088/1757-899X/742/1/012026

Biosorption of Copper (II) Ions using Living *Chlorella sp.* from Aqueous Solution

K C Wanta[1], C Hidayat[1], Catherine[1], A Miryanti[1] and A P Kristijarti[1]

[1]Department of Chemical Engineering, Faculty of Industrial Technology, Parahyangan Catholic University, Bandung, INDONESIA

Email: prima@unpar.ac.id

Abstract. Liquid waste from the electroplating industry is classified as hazardous and toxic waste. It contains a lot of heavy metal ions which are toxic and even carcinogenic so that this waste has to be treated first. In this present study, heavy metal waste treatment using biosorption method was investigated. Microalgae *Chlorella sp.* was used as biosorbent. The experimental was conducted using copper (II) (Cu^{2+}) ions which the initial concentration was varied at 20 to 80 ppm. pH scale was also varied at 2 to 5. After several hours, sample was taken and analyzed with UV–vis spectrophotometer. The experimental results showed that the best percentage removal of Cu^{2+} ions was 96.10% which was achieved at 40 ppm and pH 5.

1. Introduction

Metal coating industries, usually called electroplating industries, are developing along with the increasing demand for metal–coated products and metal alloys. However, the rise of this industry not only has a positive impact but also causes a negative impact which generates liquid waste during the electroplating process. Although the amount of electroplating waste is small, this waste is classified as hazardous and toxic waste because it contains heavy metal ions, such as copper (Cu), nickel (Ni), chromium (Cr), iron (Fe) and manganese [1]. The content of this heavy metal is toxic; even some elements are carcinogenic. It means this waste has to be processed first until the environmental quality standards before it discharges into the environment.

On the industrial scale, the heavy metal liquid waste treatment can be carried out physically, chemically or biologically. Physical and chemical waste treatment, such as adsorption or coagulation–flocculation has been widely applied in chemical industries. However, these treatments have several disadvantages when compared with biological treatment, especially in terms of high operational cost, using less environmentally friendly chemicals and inefficient in low concentrations of liquid waste.

Heavy metal liquid waste is biologically treated using biosorption methods and involves microorganisms, such as bacteria, fungi, or (micro)algae as biosorbent [2]. The use of microalgae in biosorption process is fascinating to be further developed. Microalgae are cheap, easy to handle, non-pathogenic and effective biosorbents in removing heavy metals in liquid waste. The effectiveness and efficiency of microalgae in biosorption process can be observed from the biosorption mechanisms that can occur in active uptake (using living microalgae) and passive uptake (using dead microalgae) [3]. In addition, microalgae residues can be reused as raw material for producing biodiesel [4–6].

Several studies related to biosorption process using microalgae have been carried out. Previous studies can provide percentage removal of heavy metal ions more than 70% depending on the operating conditions and the type of microalgae [7–13]. However, most studies have been conducted using dry/dead microalgae (biomass) as biosorbents. In this present study, the biosorption process of using living microalgae was investigated. Copper (II) (Cu^{2+}) ions and microalgae *Chlorella sp.* were chosen as adsorbed metal ions and biosorbent, respectively.

2. Materials and methods

In this present study, biosorption process was conducted using copper (II) sulfate pentahydrate ($CuSO_4.5H_2O$) as mother liquor (source of Cu^{2+} ions) and microalgae *Chlorella sp.* from Ugo Plankton, Central Java as biosorbent. The initial concentration of Cu^{2+} ions was varied at 20, 40, 60 and 80 ppm whereas pH scale was varied at 2, 3, 4 and 5.

900 ml Cu^{2+} ions solution at a certain concentration was prepared and poured into the erlenmeyer flask, then 100 ml microalgae *Chlorella sp.* with cell density of 3,800 cells/ml were also put into that flask. pH scale was adjusted using 1 M hydrochloric acid (HCl) solution until a certain pH scale was reached. When pH scale was reached, biosorption process was started. During the biosorption process, the flask was closed and aerated using air pump. Instead of the source of oxygen, the aeration also acted as agitator in the biosorption system. LED lights with a light intensity of 9,000 lux were used as light sources.

Samples were taken at 0, 1, 2, 3, 4, 5, 6, 24 and 48 hours, then were analyzed the cell density using haemocytometer and the content of remaining Cu^{2+} ions using UV–vis spectrophotometer at 615 nm. During the analysis, the ammonia (NH_3) solution was added as complexing agent so that dark blue $Cu(NH_3)_4^{2+}$ ligand complex compounds were formed. That measured concentration data were calculated as a percentage removal of Cu^{2+} ions using the following equation:

$$\text{Percentage removal of } Cu^{2+}\text{ions} = \frac{C_o - C_t}{C_o} \text{ x } 100\% \quad (1)$$

with C_o is an initial concentration of Cu^{2+} ions; C_t is concentration of Cu^{2+} ions at t hours.

3. Results and discussion

3.1. The effect of pH scale in the biosorption process

The pH scale is one of important factor in biosorption process using microalgae because this parameter also affects the metabolism of microalgae growth. This biosorption process was conducted by varying pH scale at 2–5 while initial concentration of Cu^{2+} ions was 40 ppm. The maximum pH scale at 5 was chosen because in basic conditions (above 5), copper hydroxide [$Cu(OH)_2$] will be formed and this compound will be insoluble in water. The formation of these compounds was avoided because the reduction of Cu^{2+} ions was not only caused by biosorption process but also because of precipitation. The effect of pH scale on this biosorption process is shown in Figure 1.

Based on Figure 1, the higher the pH value used, the better the biosorption process that occurs. It can be characterized by the figure that a higher percentage removal can be reached at higher pH scale at each sampling time. The best percentage removal of Cu^{2+} ions was 96.10% and was achieved at 48 hours and pH 5.

Chlorella sp. can grow optimally at a pH of around 4,9 – 7,3. Although it does not grow optimally, *Chlorella sp.* still grows at pH 2,1 – 4,9 because it has thick cell walls so it can defend itself in acidic pH conditions. In this study, during biosorption process, cell density was also counted to prove that statement, and the experimental data is presented in Figure 2.

Figure 2 showed that cell density increased throughout the sampling time and high pH scale. The more acidic condition caused toxicity level of Cu^{2+} ions to increase and disrupted the metabolism of *Chlorella sp.*. The better growth of *Chlorella sp.* is directly related to the biosorption process. If there are more microalgae in the system, Cu^{2+} ions will be adsorbed well, both actively and passively.

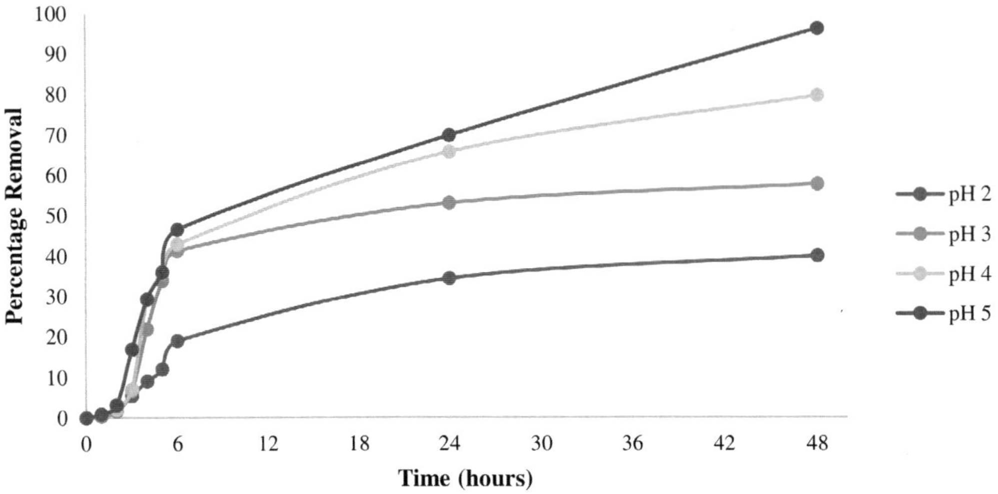

Figure 1. The effect of pH scale in the biosorption process

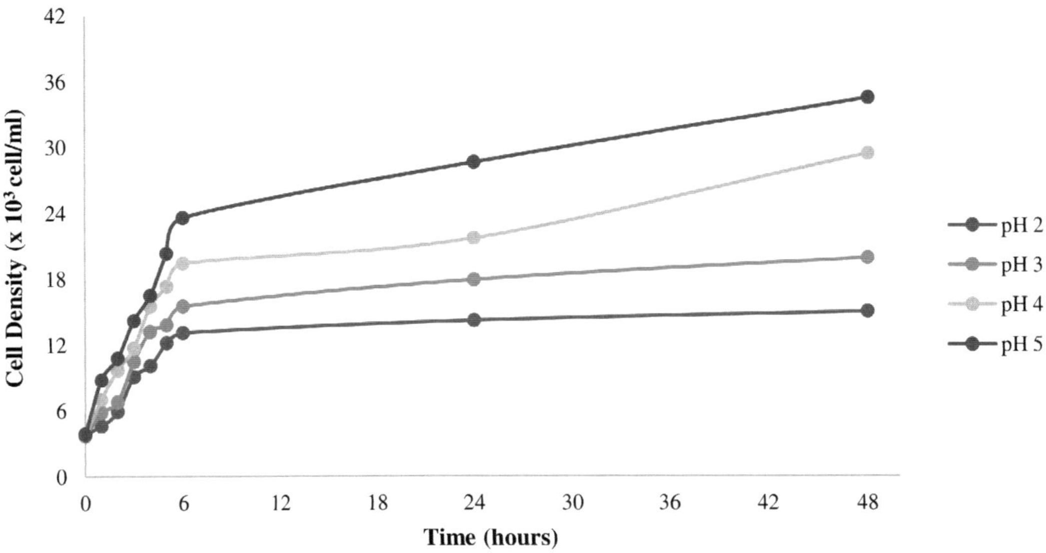

Figure 2. Growth of microalgae *Chlorella sp.* at various pH during the biosorption process.

3.2. The effect of initial concentration Cu^{2+} ions in the biosorption process

In this study, the initial concentration Cu^{2+} ions were varied at 20–80 ppm while pH scale was kept constant at 5. The experimental results can be observed in Figure 3. Figure 3 showed that at a concentration of 40 ppm, the biosorption process run at optimum condition while the performance of biosorption process decreased at concentration of 20 ppm. It occurs because the growth of microalgae was not running optimally. However, the performance of the biosorption process also decreased dramatically at concentration 60 and 80 ppm. In that condition, high Cu^{2+} ions would be "toxic" to microalgae, and the growth of *Chlorella sp.* could not tolerate that condition. It could be proven from the growth of microalgae during the biosorption process and be presented in Figure 4.

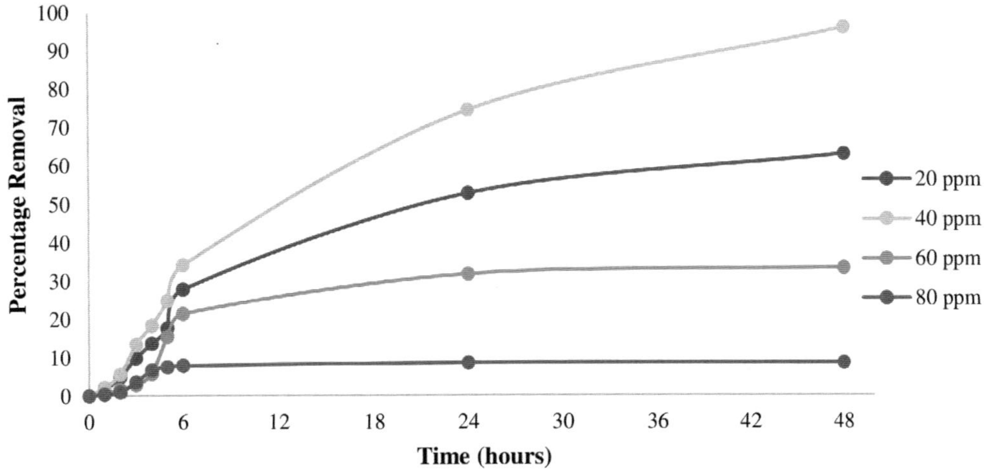

Figure 3. The effect of initial concentration Cu^{2+} ions in the biosorption process

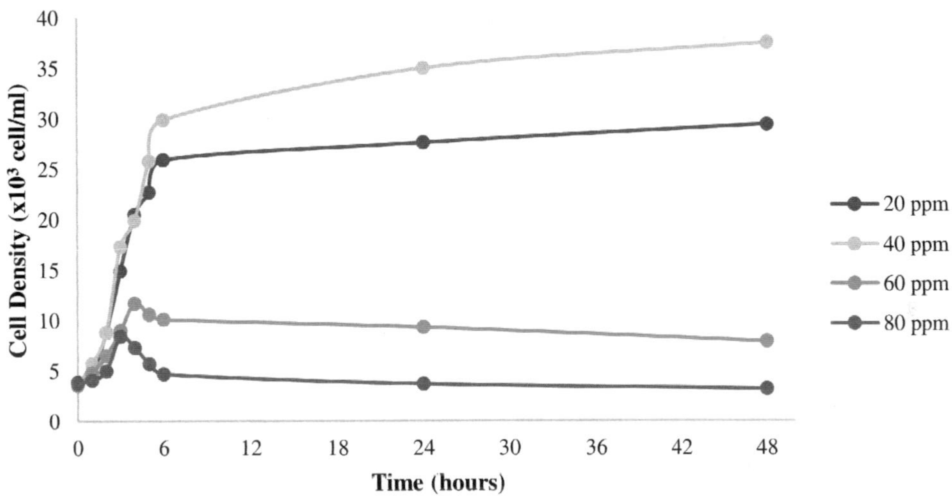

Figure 4. Growth of microalgae *Chlorella sp.* at the various initial concentration of Cu^{2+} ions during biosorption process.

Figure 4 showed that the density cell microalgae decreased at an initial concentration of 60 ppm (after 4 hours) and 80 ppm (after 3 hours). This phenomenon would cause the active uptake mechanism would not occur because microalgae had perished. However, the biosorption could still occur passively but this mechanism ineffectively. It could be seen in Figure 3 were at initial concentration of 60 and 80 ppm, the percentage removal of Cu^{2+} ions changed insignificantly after 6 hours of the biosorption process.

4. Conclusion

The biosorption process of Cu^{2+} ions using microalgae *Chlorella sp.* is influenced by initial concentration Cu^{2+} ions and pH solution. Both parameters are also related to the growth performance of microalgae and affect the biosorption process directly. In this study, the best percentage removal of

Cu^{2+} ions was 96.10% and was achieved at initial concentration of 40 ppm and pH 5 within 48 hours of the biosorption process. At higher initial concentration, microalgae growth was inhibited because high Cu^{2+} ions were toxic for microalgae so that the microalgae will perish and the biosorption process was not optimal.

Acknowledgment
The authors give highly appreciation to Lembaga Penelitian dan Pengabdian kepada Masyarakat (LPPM) UNPAR for the financial support to conduct this project with contract number III/LPPM/2019 –01/21–P.

References

[1] Nurhasni, Salimin Z and Nurifitriyani I 2013 Pengolahan Limbah Industri Elektroplating dengan Proses Koagulasi Flokulasi *Valensi* **3**(1) pp 41–47.
[2] Wang J and Chen C 2009 Biosorbents for heavy metals removal and their future *Biotechnology Advances* **27** pp 195–226.
[3] Suhendrayatna 2001 Heavy Metal Bioremoval by Microorganisms: A Literature Study *Seminar on–Air Bioteknologi untuk Indonesia Abad 21, Sinergy Forum–PPI Tokyo Institute of Technology* 1–14 February 2001.
[4] Chisti Y 2007 Biodiesel from microalgae *Biotechnology Advances* **25** pp 294–306.
[5] Mata T M, Martins A A and Caetano N S 2010 Microalgae for biodiesel production and other applications: A review *Renewable and Sustainable Energy Reviews* **14** pp 217–232.
[6] Huang G, Chen F, Wei D, Zhang X and Chen G 2010 Biodiesel production by microalgal biotechnology *Applied Energy* **87** pp 38–46.
[7] Aung W L, Hlaing N N and Aye K N 2013 Biosorption of Lead (Pb^{2+}) by using *Chlorella vulgaris International Journal of Chemical, Environmental & Biological Sciences* **1**(2) pp 408–412.
[8] Purnamawati F S, Soeprobowati T R, Izzati M 2015 Potensi *Chlorella vulgaris* Beijerinck dalam Remediasi Logam Berat Cd dan Pb Skala Laboratorium *BIOMA* **16**(2) pp 102–113.
[9] Pinem O R B, Sani T F, and Juliastuti S R 2013 Pemisahan Logam Berat Cu dan Cd dari Larutan Logam Sintetis dan Air Limbah Industri dengan Menggunakan Biomassa *Chlorella vulgaris* dan Biomassa *Chlorella vulgaris* yang Terimobilisasi sebagai Adsorben *Jurnal Teknik Pomits* **2**(1) pp 1–5.
[10] Kumar Y P, King P and Prasad V S R K 2006 Removal of copper from aqueous solution using *Ulva fasciata* sp. – A marine green algae *Journal of Hazardous Materials* **B137** pp 367–373.
[11] Chong A M Y, Wong Y S, Tam N F Y 2000 Performance of different microalgal species in removing nickel and zinc from industrial wastewater *Chemosphere* **41** pp 251–257.
[12] Kalyani S, Rao P S and Krishnaiah A 2004 Removal of nickel (II) from aqueous solutions using marine macroalgae as the sorbing biomass *Chemosphere* **57** pp 1225–1229.
[13] Gong R, Ding Y, Liu H, Chen Q, Liu Z 2005 Lead biosorption and desorption by intact and pretreated *Spirulina maxima* biomass *Chemosphere* **58** pp 125–130.

IOP Conf. Series: Materials Science and Engineering **742** (2020) 012027 doi:10.1088/1757-899X/742/1/012027

Effect of Particle Size and Shaking Speed on Enhancing Concentration of Manganese using Shaking Table

Agus Prasetya[1,2], Winayu Mutiara Ahsa[1], Himmah Sekar Eka Ayu Gustiana [1], Widi Astuti [3] and Himawan Tri Bayu Murti Petrus[1,2]

[1]Department of Chemical Engineering (Advanced Materials and Sustainable Mineral Processing Research Group), Universitas Gadjah Mada, Jl. Grafika 2 Yogyakarta, 55281, Indonesia
[2]Unconventional Georesources Research Center, Faculty of Engineering, Universitas Gadjah Mada, Jl. Grafika 2 Yogyakarta 55281, Indonesia
[3]Research Unit for Mineral Technology, Indonesian Institute of Sciences (LIPI), Jl. Ir. Sutami Km. 15, Tanjung Bintang, Lampung Selatan, Indonesia

E-mail: bayupetrus@ugm.ac.id

Abstract. Increased levels of elements or minerals from ore are usually always done in the metallurgical industry. In the present investigation, low grade manganese ores from Kulonprogo, Yogyakarta has investigated by gravity methods using shaking table. In this context, the effect of different process variables such as particle size (-60 + 100 mesh; -100 + 200 mesh and -200 + 270 mesh) and shaking speed (80 rpm, 100 rpm and 120 rpm) was analysed. The table slope and wash water flow rate are a constant variable, which are at 11° and 13 dm^3/minute. The interactional effects between different process variables were analysed using contour plot. It was founded that enhancement of the concentration of Mn is more than 5% with the shaking speed condition of 100-120 rpm and the particle size of 0.17 mm. Second order quadratic equations have developed to predict enhancement of Mn concentration using shaking table.

1. Introduction

Manganese is one of the metals that has an important role in many industrial and other. Manganese is used as food additives, fertilizer additives, battery dry cells, paint materials, iron steel industries and various other chemicals [1]. Global consumption of manganese metal has reached more than 1.3 million tons per year and is predicted to increase in the future [2]. Manganese resources are found in high grade manganese ore (> 45% Mn) and low grade manganese ore. In the past few decades, the availability of high grade ore has dramatically dropped and resulted to explore and develop manganese processes from low grade ore [3,4]. Kulon Progo, one of the regions in Indonesia, has the low grade manganese ore deposits with 5-10% manganese content in which considerable potential to explore.

Ores processing after mining is mineral processing. It is a process of ore preparation, milling and ore beneficiation. Ore beneficiation is a separate process that is based on the physical properties of minerals. Minerals will be more easily separated by gangue if it has a large difference in physical properties [5,6,7,8,9,10,11,12]. Therefore, this study will focus on the method for increasing the manganese fraction in low grade manganese ore by utilizing one of the physical properties of minerals, specifically density through gravitational separation using shaking table.

The purpose of this study was to determine the effect of particle size and shaking speed to increase the concentration of manganese using shaking table.

2. Materials and methods

2.1. Materials
Low grade manganese ore was mined from Kulon Progo, Yogyakarta Province, Indonesia and was associated with limestone. The chemical components of the different size fractions were carried out and given in the Table 1. It can be seen from Table 1, manganese ore from Kulon Progo is classified as low grade manganese ore (<40%) because it only contains about 4-5% manganese.

Table 1. Chemical analysis of low grade manganese ore from Kulon Progo

	Content		
	-60+100 mesh	-100+200 mesh	-200+270 mesh
Ca (%)	32.00	34.20	36.00
Mn (%)	4.92	4.80	3.77
Fe (%)	0.27	0.30	0.27
Cu (ppm)	13.1	14.8	12.30

2.2. Experimental design
Sample puts into the oven at 70-80°C to reduce the water content to facilitate the crushing process. There were three main steps in this research, that are comminution, sieve, and separation using shaking table. In this study, -60+100 mesh; -100+200 mesh and -200+270 mesh particle size were used as feeder of the shaking table. The sample is introduced at the upslope corner of the table (Figure 1). The shaking table was set on a constant table slope and water flowrate, which are at 11° and 13 dm^3/minute. The shaking speed variable has been done in three variations, there are at 80 rpm, 100 rpm and 120 rpm. After the separate process, the results of the shaking table are analyzed using XRF to find out the quantitative amount.

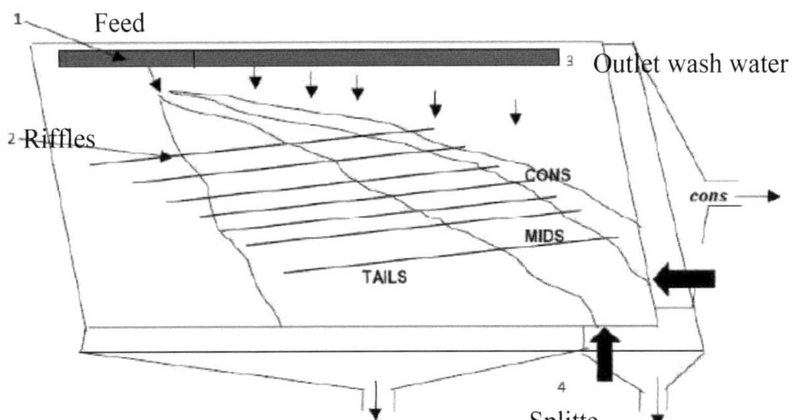

Figure 1. Shaking table scheme

3. Results and discussion
The experiment was conducted using three variations of particle size and three variations of shaking speed to determine the effect of each variable on increasing manganese concentration in ore. The particle size of variations were used at -60 +100 mesh, -100 +200 mesh and -200 +270 mesh. Whereas, the shaking speed of variations carried out at 80 rpm, 100 rpm and 120 rpm. After

conducting the experiment, the results of the shaking table were analyzed using XRF. Increased concentration of manganese in low grade manganese ore can be seen in Table 2.

Table 2. Increased manganese concentration in low grade manganese ore

Particle Size (mesh)	Feed Weight (g)	Shaking Speed (rpm)	Weight Concentrate (g)	Increase of Mn Concentration (%)
-60+100	0.0984	80	0.1720	3.68
		100	0.2000	5.08
		120	0.2140	5.78
-100+200	0.0960	80	0.1572	3.06
		100	0.1968	5.04
		120	0.1744	3.92
-200+270	0.0754	80	0.1432	3.39
		100	0.1778	5.12
		120	0.1386	3.16

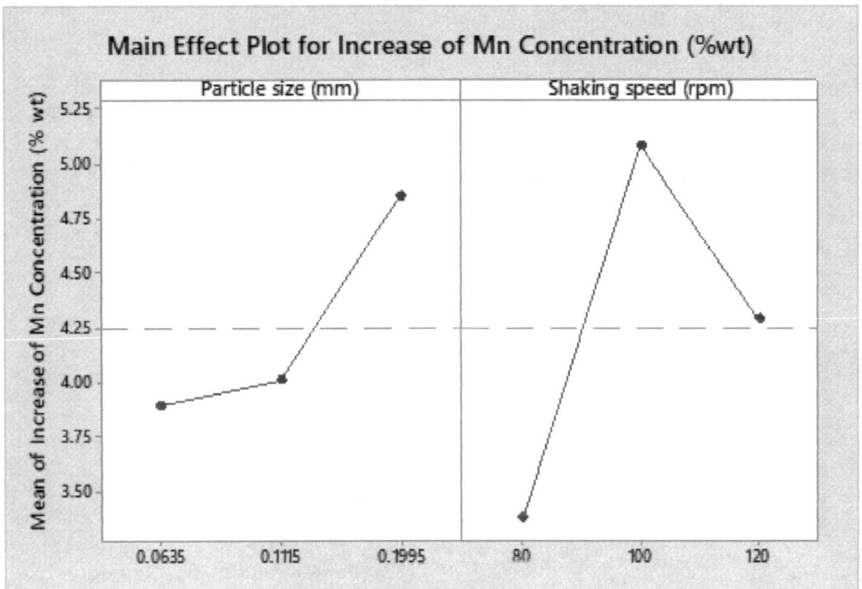

Figure 2. Relation of particle size and shaking speed to increase of Mn concentration

The data in Table 2 are plotted to find out an overview the effect of variables on increasing manganese concentration. Plotting can be seen in Figure 2. From Figure 2, it can be concluded that the larger particle size, the greater manganese concentration is. This is accordance with the existing theory that the larger particle size, the more difficult it carries the flowing water and is restrained by riffles. In addition, motion of the shaking table will push the particle to the concentrate area. Whereas, a small particles will be more easily carried away by the washing water stream which is called tailing. This happens because the concentration in the concentrate is a quite low concentration [13,14]. In Figure 2, it is also seen that the higher shaking speed, the bigger the manganese concentration is resulted at 80 rpm to 100 rpm but the manganese concentration decreased at 120 rpm.

Interaction studies shows the effectiveness of the separation that is indicated by the increasing concentration of Mn. There was influenced by the combination of variables that was seen in Figure 3. Figure 3 shows the contour of the shaking speed vs particle size in increasing the concentration of Mn. In Figure 3 can be seen the area with the increasing concentration of Mn is more than 5.5%, which it

IOP Conf. Series: Materials Science and Engineering **742** (2020) 012027 doi:10.1088/1757-899X/742/1/012027

happens with the condition of the shaking speed is 100-120 rpm and the particle size is more than 0.17 mm.

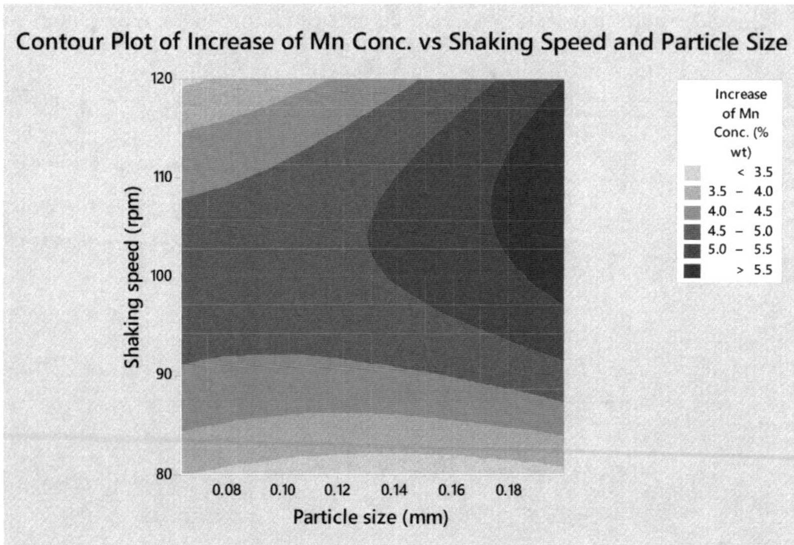

Figure 3. Contour plot of increase of Mn concentration vs shaking speed and particle size

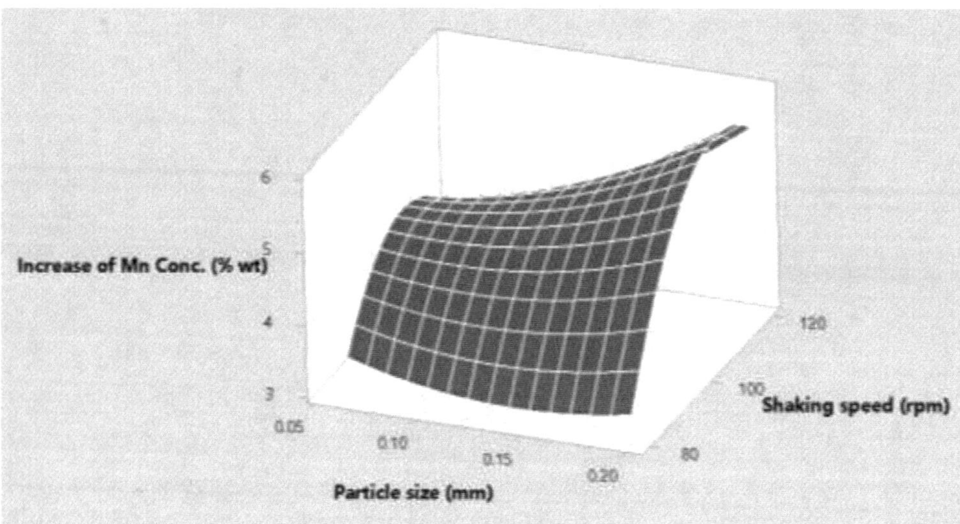

Figure 4. Response surface of increase Mn concentration function of shaking speed and particle size

As can be observed in Figure 4, the highest increase of Mn concentration (5.78%) is achieved at the highest level of shaking speed and particle size. The lower value of Mn concentration (3%) occurs at the lowest shaking speed and particle size of 0.11 mm. The highest of Mn concentration with particle size in 0.2 mm because it has more Mn content than the others (Table 2).

For predicting the increase Mn concentration of shaking table, second order quadratic models have developed and given as,

$$\text{Increase Mn conc.}(\%\ \text{wt.}) = -23.32 - 48.6\,x + 0.595\,y + 52.3\,x^2 + 0.003121\,y^2 + 0.419\,xy$$

Where,

x is particle size ore (mm)
y is shaking speed (rpm)

the predicted increase of Mn concentration derived by using the above equation is in good agreement with experimental data (R^2 value, the accuracy of the model, is 89.51%).

4. Conclusions

Increasing of the concentration of Mn using the shaking table can be reached up to 5% wt. Shaking speed and particle size significantly affected to increase of Mn concentration. From the contour plot, it can be seen that in the range of shaking speed of 100-120 rpm and particle size of more than 0.17 mm, increase Mn concentration more than 5%. The developed second order quadratic equations can be used for predicting the increased of Mn concentration using shaking table, which R^2 value is 89.51%.

References

[1] Su, H., Wen, Y., Wang, F., Sun, Y., and Tong, Z., 2008. Reductive leaching of manganese from low-grade manganese ore in H_2SO_4 using cane molasses as reductant. Hydrometallurgy 93, 136-139.

[2] El Hazek, M.N., T.A. Lasheen., and A.S. Helal. 2006. Reductive leaching of manganese from low grade Sinai ore in HCl using H2O2 as reductant. Hydrometallurgy 84 (3), 187-191.

[3] Li, C., Zhong, H., Wang, S., Xue, J., Wu, F., and Zhang, Z. 2015. Manganese extraction by reduction − acid leaching from low-grade manganese oxide ores using CaS as reductant. Transactions of Nonferrous Metals Society of China, 25(5), 1677–1684.

[4] Anggraeni, R.D., Prasetya, A., Petrus, H.T.B.M., Cahyono, R.B., Sumardi, S., and Astuti W. 2019. Performance of Kulon Progo low grade manganese ore leaching using acetic acid and its selectivity. IOP Conference Series: Materials Science and Engineering. Volume 478, Issue 1, Pages 012008.

[5] Balasubramanian, A. 2015. Overview of Mineral Processing Methods. Centre for Advanced Studies in Earth Science University of Mysore. DOI: 10.13140/RG.2.2.10456.49926

[6] Ferdana, A.D., Petrus, H.T.B.M., Bendiyasa, I.M., Prijambada, I.D., Hamada, F. And Sachiko, T. 2017. Optimization of gold ore Sumbawa separation using gravity method: Shaking table. Proceedings of the 3rd International Conference on Materials and Metallurgical Engineering and Technology (ICOMMET 2017).

[7] Tripathy, S.K., Ramamurthy, Y., Sahu, G.P., Panda, L., Singh, V., and Tathavadkar V. 2010. Proceedings of the XI International Seminar on Mineral Processing Technology (MPT-2010).

[8] Ferdana, A.D., Petrus, H.T.B.M., Bendiyasa, I.M., Prijambada, I.D., Hamada, F., and Sachiko, T. 2018. Study on Sumbawa gold recovery using centrifuge. IOP Conference Series: Materials Science and Engineering. Volume 285, Issue 1, Pages 012027.

[9] Ernawati, R., Idrus, A., and Petrus, H.T.B.M. 2018. Study of the optimization of gold ore concentration using gravity separator (shaking table): case study for LS epithermal gold deposit in Artisanal Small scale Gold Mining (ASGM) Paningkaban, Banyumas, Central Java. IOP Conference Series: Earth and Environmental Science. Volume 212, Issue 1, Pages 012019.

[10] Sandiwan, S.W., Saputra, K.I., Mahdi, M.H., Albarkan, M.S., Prasetya, S., and Petrus, H.T.B.M. 2019. Effect of inlet speed on gravitational air separator for cenospheres accumulation from fly ash: modeling using computational fluid dynamics (CFD). IOP

Conference Series: Materials Science and Engineering. Volume 478, Issue 1, Pages 012028.

[11] Hirajima, T., Petrus, H.T.B.M., Oosako, Y., Nonaka, M., Sasaki, K., and Ando, T. 2010. Recovery of cenospheres from coal fly ash using a dry separation process: Separation estimation and potential application. International Journal of Mineral Processing. Volume 95, Issue 1-4, Pages 18-24.

[12] Petrus, H.T.B.M., Hirajima, T., Oosako, Y., Nonaka, M., Sasaki, K., and Ando, T. 2011. Performance of dry-separation processes in the recovery of cenospheres from fly ash and their implementation in a recovery unit. International Journal of Mineral Processing. Volume 98, Issue 1-2, Pages 15-23.

[13] Burt, R. O. (1984). Gravity Concentration Technology. New York: United States : Elsevier Science Pub.Co.

[14] Taggart, A. F. 1976. Handbook of Mineral Dressing Ores and Industrial Minerals. New York. Chicester. Brisbane. Toronto.

IOP Conf. Series: Materials Science and Engineering **742** (2020) 012028 doi:10.1088/1757-899X/742/1/012028

The Effect of Salt-GDL Coagulant and Temperature on Tofu Quality

Maria Inggrid[1] and Nikolas Hananjaya[2]

[1]Parahyangan Chatolic University, Indonesia.

E-mail : nikolas.hajey@gmail.com

Abstract. Soymilk tofu can be coagulated with salts, acid and lactones. The objectives of this investigation was to study the effect of temperature and different coagulant to the quality of tofu; such as protein content, texture, mass and moisture. Soybean was oaked and grinded to produced soymilk. The soymilk was coagulated with two type of mixed coagulant; $CaSO_4$-Glucono delta-lactone (GDL) and $MgSO_4$-GDL, the variation of composition were 0:4, 3:1, 2:2, 1:3, 0:4 at temperature of 70°C and 80°C. The highest protein content was obtained from the tofu with $CaSO_4$-GDL (2:2) and 70°C. The results showed that the different in coagulation temperature affect the protein content of tofu. Mass and texture of the tofu increased as the composition of salt coagulant increased. General factorial method was used in this experiment, the result supports that the various composition of coagulant increased significantly ($P<0.05$) from 111-135 gram and 241-642 gram, respectively for tofu mass and texture. Temperature has significant result to protein content and mass of the tofu.

Keywords: coagulation, glucono delta-lactone, salt, tofu

1. INTRODUCTION

Malnutrition in Indonesia is still relatively high. Based on Riset Kesehatan Dasar (Riskesdas) and Pemantauan Status Gizi (PSG) from the Indonesian's ministry of Health, it was found that the level of malnutrition continues to increase in Indonesia [1]. According to the result of research by Rizqa Widiasti in 2019, protein deficiency reached 37% in Indonesia [2]. Protein can be classified from its source; vegetable and meat source. In this case, there are tofu and tempe made from soybeans (vegetable based food) that can be improved, especially to their high protein content. Its various nutrition; isoflavones also can reduce the risk of cardiovascular problem. Due to the inexpensive price and its high nutrition, tofu might solve the protein malnutrition in Indonesia.

Tofu is one of vegetable based food that has high nutrition content. The preparation of tofu generally includes soybean soaking and grinding in water, filtering, boiling and coagulation of soymilk, molding and pressing. There are several factors that affect the quality of tofu; such as temperature, type of coagulant, soybean variety, pH, and pressing [3]. Some researchers conducted experiments to compare the quality and the result were diverse, but most of the research stated salt coagulant produces the highest protein content with hard texture. According to Shurtleff and Ayogi (1984) in order to give a good texture, some of the coagulant are mixed with GDL; the texture produced is soft and smooth [4]. There are still few studies that examine the effect of mixed coagulant. This prompted researcher to study the effect of variations in the composition and different type of mixed coagulants (salts and GDL) to the quality of tofu.

2. MATERIAL AND METHOD

2.1 Chemicals

The main materials to conduct this experiments were soybeans that was purchased from the local market (Bandung, Indonesia). Calcium sulphate ($CaSO_4.2H_2O$), Magnesium sulphate ($MgSO_4.xH_2O$), magnesium chloride ($MgCl_2.xH_2O$) and glucono delta lactone (GDL) as coagulant. Sulphuric acid (H_2SO_4), natirum hydroxide (NaOH), boric Acid, hydrochloric acid (HCl), pottasium sulphate (K_2SO_4) and copper sulphate ($CuSO_4.xH_2O$) as the chemicals to support analysis.

2.2 Preparation of soymilk

100 grams soyben were soaked in water with a ratio of 1:10 in a room temperature for 12 hours. The soaked bean which had doubled in weight was mixed and blend with 800 ml of water. Then the soaked bean was grinded by a laboratory blender and raw soymilk was obtained. The slurry in the raw soymilk then was filtered by cotton tack rag cloth to seperate the solid contaminant. The raw soymilk was heated to its boiling point, then cooled into the temperature of the experiment.

2.3 Tofu Processing

Tofu was prepared by coagulating the previous prepared soymilk using salts coagulant for the preliminary experiment and salt – GDL mixed coagulant for the main experiment. The type of salt coagulant used for the preliminary experiment were calcium sulphate, magnesium sulphate and magnesium chloride. The concentration were also varied into 0,3%, 0,5% and 1,0% w/v in a temperature condition of 80°C. The variations of salts and its concentration were chosen based on previous researcher's experiment result. The main experiment was proceeded by using the best coagulant and concentration from the preliminary experiment and mixed it with glucono delta lactone. The mixture coagulant of salt and GDL were varied into 4:0, 3:1, 2:2, 1:3 and 0:4. For each experiment the coagulant was dissolved in 20 mL of water then poured into the soymilk. The soymilk was equipped by an impeller in order to make the coagulants dispersed well and heating plate with temperature of 70°C and 80°C. The curds thus formed after a 10 minutes mixing the soymilk and coagulant. Moreover, the curds was obtained by seperating the whey with cotton rag. The curds was pressed for 30 minutes using a weigth of 1,5 kg in a container, tofu samples were weighed and stored in water at 5-7°C before analysis.

2.4 Protein analysis

The protein content were determined by kjedahl method using a protein conversion factor of 5.75 method. The process consist of three stages; digestion, distillation and titration of the sample.

2.5 Texture analysis

Texture analysis of tofu was carried out using an instrument; texture analyzer. A part from the central of the tofu were cut x2x2 cm and used for texture evaluation. A test speed of 4 mm/s was used with a probe of a cylindrical metal with 6 mm diameter. The sample was compressed to 19 mm of its height. There were two samples carried out for each variation. The hardness of tofu was determined by the peak of the force from the diagram shown by the software.

2.6 Moisture analysis

The water content of tofu were obtained by drying a weighed amount of samples to moisture analyzer. The moisture analyzer operates at 105°C in order to evaporate the water inside the tofu.

2.7 Mass analysis

The mass of samples were obtained by measuring the mass of the tofu in gram. The tofu always came from same amount of soymilk and pressed with the same amount of mass. The measurements were replicated two times and the mean value was obtained.

3. RESULTS AND DISCUSSION

Coagulation is the most important step in the tofu-making process. Coagulation occurs due to the cross-linking of protein molecules in soymilk with salt's cation. There was a preliminary experiment to determine the best salt coagulant ($CaSO_4$, $MgSO_4$, $MgCl_2$) and concentration (0.3%, 0.5%, 1.0% w/v). The result shown in **table 1.1**, tofu sample from 0.5% w/v $CaSO_4$ and $MgSO_4$ produced high protein content with a value of 13,1% and 11,2% respectively. $MgCl_2$ coagulant produced the lowest protein content. Due to their high protein content, $CaSO_4$ and $MgSO_4$ were mixed with GDL as the coagulating agent for further preparation of tofu.

Table 1. Tofu protein content from preliminary experiment

	Protein Content (%)		
Concentration (w/v)	**$CaSO_4$**	**$MgSO_4$**	**$MgCl_2$**
0,3%	12,9	10,9	1,5
0,5%	13,1	11,2	1,5
1,0%	12,6	9,8	1,2

The variation result of protein content is due to the differences of cations and anions which affect the solubility of protein content in the soymilk. Cl^- and SO_4^{2-} ions have the ability to reduce the protein solubility. Based from Hofmeister order, Cl^- can make the protein become more soluble (salting in) rather than SO_4^{2-} [5]. The following figure shows the hofmeister order:

Li^+ Na^+ NH_4^+ Mg^{2+}

Salting out ⟵⟶ Salting in

SO_4^{2-} HPO_4^{3-} $CH3COO^-$ Cl^- Br^- NO_3^-

Figure 1. Hofmeister order on protein solubility [5]

3.1. Composition Variation of Salt and GDL

The purpose of the main experiment was to determine the effect of variation composition in salt and GDL and coagulation temperature on the quality of the tofu produced. The tofu was analyzed based on protein content, mass, moisture and texture of the sample. The protein content was summarized in the following Fig.2

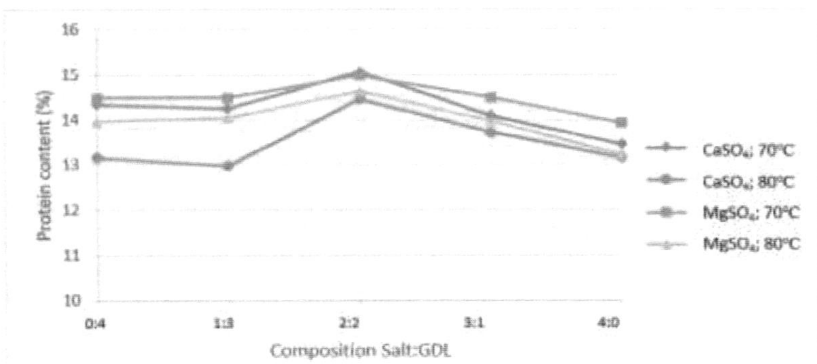

Figure 2. Protein content of the tofu samples profile

The protein content from figure 2 showed that there is a trend in protein content regarding to the salt and GDL composition; most of the data from all variation in the experiment showed the highest protein content at composition 2:2. The result of protein content with $CaSO_4$-GDL as coagulant at temperature of 70°C and 80°C are 14.1-15.4% and 13.1-13.6% respectively. While $MgSO_4$ – GDL relatively have the same protein content at temperatures of 70°C and 80°C; with a range of 14.2-15.3% and 13.1-14.5%. The difference of protein content caused by the composition of coagulant that also manipulated the pH of soymilk. **Table 1.2** showed the pH of soymilk caused by variations on the coagulant composition.

Table 2. pH observed with different composition of salts and GDL.

Composition (Salt : GDL)	pH	
	Ca 70°C-80°C	Mg 70°C-80°C
4 : 0	5,4 - 5,5	5,5 - 5,6
3 : 1	4,9 - 5,2	4,9
2 : 2	4,6 - 4,7	4,4 - 4,5
1 : 3	4,3 - 4,5	4,3
0 : 4	4,3 - 4,4	4,3

Dzikunoo (2015) and Li (2015) stated that soy protein contains two main types of proteins; β-conglycinin (7S) and glycinin (11S) [6][7]. Li (2015) also stated that acidic coagulants coagulate high β-conglycinin (7S) and less glycinin (11S), whereas salt coagulant coagulate higher glycinin (11S) protein than β-conglycinin (7S). Proteins can coagulate themselves at their isoelectric pH (4.5-4.8). GDL is a weak organic acid, so it can reduce the pH of the solution. The highest protein is obtained when the composition of salt-GDL is 2:2. It is predicted due to the mixed salt-GDL coagulant produced solution that reached the isoelectric point. Data from salt:GDL at 3:1 and 4:0 showed the system has a pH above isoelectric point which probably does not coagulate much of the 7S, due to the bonds that occur in the 7S are hydrogen and hydrophobic bonds which does not attracted to salt's ions yet it does attract to other protein when the system reach neutral point (isoelectric point). When the composition of $CaSO_4$: GDL was at 0:4 and 1: 3 the system is below the isoelectric point, the system charge becomes positive and the salt coagulant does not work well when the system below the electrical point.

The experiment showed that variation in composition of coagulant also affects the mass of the tofu. The figure.3 represent that there was an increment of tofu mass along with the increment of salt composition.

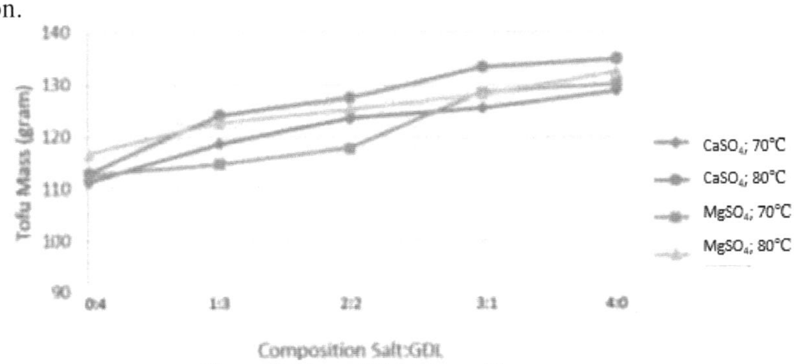

Figure 3. Mass of the tofu profile

11S protein has a greater molecular weight of 350 kDa and 7S has a molecular weight smaller than 11S, which is 170 kDa [4]. Larger molecular weight is proportional to the tofu mass. As it was stated before, salt coagulants coagulate more 11S and acid coagulant coagulate more 7S protein. The result shows the tofu with high salt composition gave higher mass than tofu with high composition of GDL. The highest mass of tofu was obtained from $CaSO_4$:GDL at 4:0 with a value of 135 gram. The data showed corresponding result to the theory which have been stated before.

Most of the data showed an increment of hardness in tofu's texture due to increment of salt coagulant. The following figure 4 showed tofu's hardness data toward the salt and GDL composition

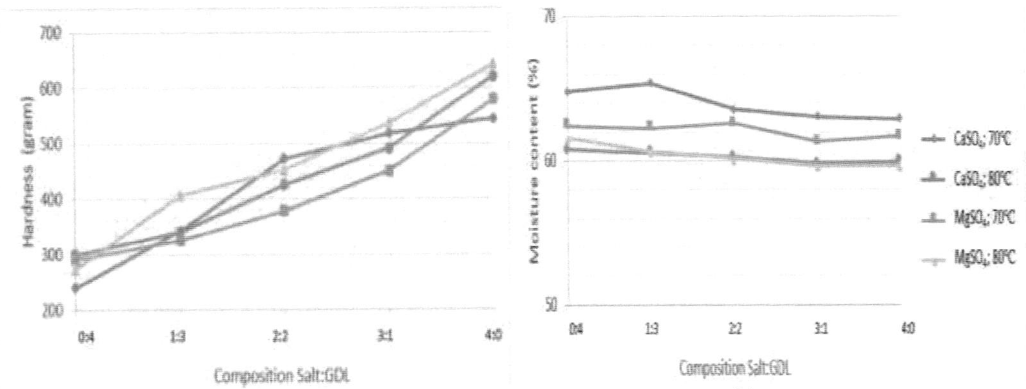

(a) (b)

Figure 4. Quality of tofu's profile diagram for (a) Texture; (b) Moisture content

Figure 4a showed an increment of hardness is corresponding to the salt composition. In variation of $CaSO_4$:GDL at 0:4 showed the lowest hardness (241 gram) and $CaSO_4$:GDL at 4:0 showed the highest hardness (621 gram). $MgSO_4$:GDL showed the same pattern, whereas the hardness at 0:4 was the lowest (275 gram) and at 4:0 was the highest (642 gram. Li (2015) stated tofu with high GDL levels has high water content, it is caused by the increase production of 7S protein [7]. It was suspected that the water content is increasing because 7S protein has many hydrogen bonds that can bind water molecules with it better than 11S can bind with. According to the Syah (2014) and Li (2015), 11S protein has covalent - disulfide bonds between proteins, which are stronger than proteins and water [8][7]. The moisture content from the data were not significance towards the composition variation of coagulants and temperature. According to data from fig.4b, the water content at $CaSO_4$: GDL and $MgSO_4$:GDL at 0:4 has range values of 61-66% and 55-65%.

3.2. Temperature variation of Salt and GDL

The figure 2 observed the effect of different temperature on the protein content. The data shows coagulation at 70°C are higher than 80°C. According general factorial method ($p<0.05$), the deviation were significantly different to coagulant $CaSO_4$:GDL and $MgSO_4$:GDL. The protein can be denatured if the temperature of the system surpasses the temperature of denaturation. Denatured temperature for 7S protein is 70°C and for 11S is 86°C [9].Therefore, the protein content in 70°C relatively higher than 80°C.

The other quality of tofu; mass was lso significantly affected by the temperature. Figure 3 shows that mass at 80°C are relatively higher than 70°C, because the temperature affect the aggregation of protein. Tofu protein has the ability to hold water, fat, polysaccharide compounds and components from other bean milk [10].An increase of coagulation temperature may increase the vibration energy and the rotation of dissolved proteins. These events result in the possibility of proteins colliding and aggregating faster and bind with fat, polysaccharide and other. Therefore, the weight of tofu will be increased [11]. Whereas, the variation in temperature did not affect significantly to the data of moisture content and the hardness of the tofu.

Acknowledgment

This research was supported by the Department of Chemical Engineering, Parahyangan Catholic University.

References

[1] Badan Pusat Statistik, 2018, Rata-Rata Konsumsi per Kapita Seminggu Beberapa Macam Bahan Makanan Penting, 2007-2017, www.bps.go.id, diakses pada Juli 2018
[2] Mia Umi Kartikawati, 2019, 37 Persen Penduduk Indonesia Kurang Protein - gayahidup www.inilah.com, diakses pada bulan Maret 2019.
[3] Y.H.Hui, 2007, Handbook of Food Products Manufacturing, John Wiley & Sons, Inc.. Hoboken,

IOP Conf. Series: Materials Science and Engineering **742** (2020) 012028 doi:10.1088/1757-899X/742/1/012028

New Jersey, hal 1065-1074.
[4] Shurtleff W and Aoyagi A. 1984. Tofu and Soymilk Production, In The Book of Tofu Vol. II. New Age Food Study, Lafayette.
[5] Curtis, R. A., and L. Lue. 2006. A molecular approach to bioseparations: protein-protein and protein-salt intractions. Chemical Engineering Science 61:907-923.
[6] Dzikunoo, J., Ayernor, G. S. & Saalia, F. K., 2015. Effect of Methods of Extraction on Physicochemical Properties of Soy Proteins (Tofu). *American Journal of Food Science and Nutrition Research,*
[7] Li, T. S., 2005. *Functional And Structural Properties of Moleculat Soy Protein Fractions,* Singapore: National University of Singapore.
[8] Dahrul Syah, Azis Boing Sitanggang, RH Fitri Faradilla, Victor Trisna, Yogi Karsono & Dita Adi Septianita., 2014: The influences of coagulation conditions and storage proteins on the textural properties of soy-curd (tofu), CyTA-Journal of Food.
[9] Blazek V. 2008. Chemical and Biochemical Factors that Influence the Gelation of Soybean Protein and the Yield of Tofu. Thesis Faculty of Agriculture, Food and Natural Resources. Univ of Sydney, Sydney, Australia
[10] Milewski S. 2001. Protein structure and physicochemical properties. *In*: Sikorski ZE (ed). *Chemical & Functional Properties of Food Proteins*. Lancaster Pennsylvania: Technomic Publishing Company, Inc.
[11] Zayas J.F. 1997. Functionality of Protein in Food. Springer-Verlag Berlin Heidelberg, Germany

IOP Conf. Series: Materials Science and Engineering **742** (2020) 012029 doi:10.1088/1757-899X/742/1/012029

Fractionation of Turpentine

Danu Ariono, Gigih Aldiyana, Ramadika Adrian, Antonius Indarto,

Department of Chemical Engineering, Institut Teknologi Bandung, Kampus Ganesha, Jl. Ganesha 10 Bandung, Indonesia

antonius.indarto@che.itb.ac.id

Abstract. Turpentine (pine oil) is a liquid phase of the *Pinus mercusii* gum that consists of α-pinene, β-pinene, δ-carene, limonene, and others. These components, if separated, have a higher added value compared to selling crude pine oil. Separation process by distillation is the common method to fractionate all of the components. However, distillation at atmospheric pressure has a potential to decompose some turpentine components. This study aims to obtain the optimum operating conditions that produce compounds of turpentine. The experimental method used consisted of three stages. The first stage is the analysis of components from the main raw material, turpentine. The second step is the turpentine fractionation with the process of party vacuum distillation. The third step is the analysis of the distillate results using Gas Chromatography with FID. The experiment was carried out by varying the operating pressure of the vacuum distillation process which is at 0.2 atm; 0.45 atm and 1 atm. The experimental results show that thermal degradation did not occur at pressures of 0.45 atm and 0.2 atm. The experimental results are in accordance with the simulation results with a column efficiency of 45%. Better results can be obtained by adding stages to 25 stages from 10 stages used and can achieve composition results up to analyst grade (> 99.9%). Distillation of 0.2 atm pressure is the best operating condition judging from the profit generated and its energy consumption.

1. Introduction

Turpentine which is a liquid phase of the *Pinus mercusii* gum product that consists of various cyclic monoterpene hydrocarbon compounds, such as α-pinene, β-pinene, δ-carene and limonene, with general molecular formula of $C_{10}H_6$ [1]. Depending on the type of trees, different composition of species could be found but pinenes showed to be the largest fraction of turpentine (about 80-90% of the total component) [2]. Indonesian turpentine consists mainly of α-pinene 65-85%-wt, β-pinene 1-3%-wt, δ-carene 10-18%-wt, and δ-limonene 1-3%-wt [3].

The above fractionated products of turpentine usually have higher selling value compared to crude turpentine or mixed pine oil. Table 1 shows the comparison of the value added of turpentine derivative products and the world demand for these products [4].

Table 1. Comparison of price and demand of derived turpentine products [4].

No.	Chemicals	Price (USD/ton)	Value added	Demand (ton/year)
1.	Turpentine	3.400	-	-
2.	α-pinene	4.930	45%	300.000
3.	β-pinene	5.916	74%	300.000
4.	Limonene	6.630	95%	45.000

5.	α-terpineol	8.160	141%	330.000
6.	Cineol	12.478	267%	1.100

The quality of the separation of the components of turpentine can be reviewed by comparing the concentrations of the compounds of turpentine before, during and after distillation. This research will study the effect of operating pressure factors on boiling point and the quality of the separation of turpentine compounds, so that it can be known what operating conditions must be pursued to get good separation quality.

2. Experimental Setup

In this study, the experiment was carried out in three stages. The first stage was the initial analysis of raw materials (crude turpentine). The raw material used is turpentine from Perhutani Pine Chemical Industry (PPCI), Pemalang, Indonesia. First, the raw material was analyzed using Gas Chromatography (GC) to determine the composition of species in the feed. The second step was the fractionation process of turpentine with the vacuum distillation process at different operating pressures. The last part was the analysis of distillate obtained from the results of vacuum distillation using GC.

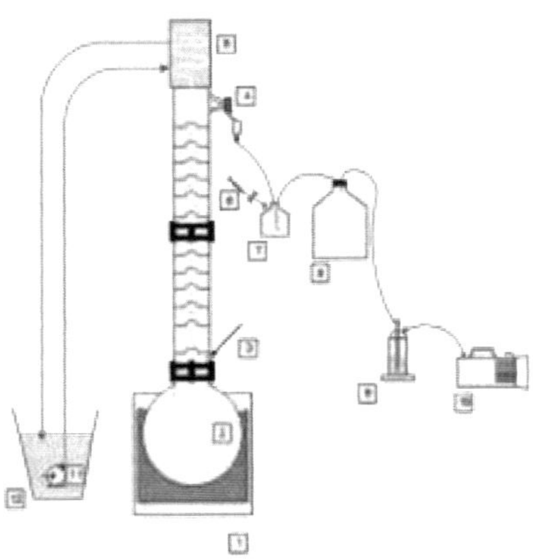

1. *Heater*
2. Distillation feed (10 L)
3. Thermometer
4. Reflux controller
5. Condenser
6. Sample syringe (20 mL)
7. Sample collector
8. Distillated collector
9. Gas wash glass (250 mL)
10. Vacuum pump
11. Submersible pump
12. Basket

Figure 1. Schematic diagram of vacuum distillation unit.

Feed and distillation products were analyzed using Gas Chromatography (GC-2010 Plus type with a Rtx-5 column specifying a length of 30 meters, a diameter of 0.25 mm) with helium as a carrier gas. The analysis was carried out at a Flame Ionized Detector temperature of 240°C, column temperature of 70-190°C with an initial temperature of 70°C for 6 minutes and the temperature rose to 170°C with a rise in temperature at 5°C/minute.

Distillation process was carried out under vacuum with a pressure range of 0.2 to 1 atm. A vacuum condition was required to reduce the operating temperatures that can make the constituent components of turpentine thermally degraded. Vacuum distillation operation was started by charging 5 L of raw turpentine into a 10 L distillation flask. Once the distillation apparatus is ready, the pressure was adjusted according to the experiment variations and feed turpentine was heated by adjusting the heater at certain temperature. Sampling procedure of distillate was carried out after distillate volume was reaching 1500, 3000, and 4000 mL, respectively. Distillation sample was put into vials then analyzed the content using GC.

3. Result and Discussion

This section presents experimental results and interpretations of the results obtained. The influence to be reviewed is the effect of pressure. The performance parameters reviewed in the effect given are the purity of the components of the terpene oil and whether or not a new product is formed due to thermal degradation caused by too high distillation temperatures and the economic analysis of the experiments that have been carried out.

3.1. Turpentine Oxidation at Atmospheric Distillation Condition

Turpentine distillation at atmospheric pressure was reviewed to determine the effect of atmospheric pressure and to compare the purity of its components with turpentine distillation under vacuum pressure. The operating condition of the column with atmospheric pressure produces boiling point of turpentine on the first drop of 152°C. This is quite close to the literature which says that turpentine has a boiling point in the range 154-170°C [4].

Figure 2. Feed color (left) and degraded product (right).

The boiling point of turpentine which is boiled at atmospheric pressure has a high enough temperature, which is 152°C. These operating conditions support the thermal decomposition of turpentine into compounds such as isoprene, benzene, toluene, xylene, naphthalene, phenanthrene, anthracene, and methylanthracene [3]. Limonone which is a constituent of converted turpentine by 30-50% produces isoprene, while α-pinen and β-pinen are converted by 10% each producing isoprene when boiled under atmospheric pressure [3]. Isoprene which is a compound from thermal degradation of turpentine is an undesirable product in the turpentine fractionation process. The formation of isoprene compounds is characterized by changing turpentine from colorless to turbid yellow, which is the color of isoprene compounds. The change in color of turpentine feed after distillation with the atmospheric pressure operating conditions is shown in Figure 2.

Table 2. The composition of the turpentine compound in the feed and the oxidized distillation 1 atm pressure feed of experimental results

Sample Condition	Component	Concentration (%-mol)
Original Feed	α-pinene	79.1
	β-pinene	2.5
	δ-carene	11.6
	Limonene	0.01
Oxidized Feed (after distillation at atmospheric condition)	α-pinene	70.3
	β-pinene	0.2
	δ-carene	24.6
	Limonene	0.05

The color change from turpentine feed to a turbid yellow color indicates the process of thermal decomposition of the compounds of turpentine with the formation of isoprene compounds which reduces the number of compounds making up of turpentine such as α-pinene and β-pinene. The oxidation process of turpentine compounds was also supported by the results of the gas chromatography analysis of the two samples which detected the addition of 5 compounds, from 29 peaks to 34 peaks. In addition, it can be concluded that after turpentine undergoes a process of distillation at atmospheric pressure, the concentrations of turpentine compounds such as α-pinene and β-pinene experience a significant decrease when compared to the initial feed condition with the feed condition after distillation at atmospheric pressure. The α-pinene compound decreased by 8.9%, while the β-pinene compound decreased by 2.3%. Concentrations of α-pinene, β-pinene, δ-carene, and limonene from the results of gas chromatography analysis in the initial feed and the feed after distillation at atmospheric pressure are presented in Table 2.

Decreased concentrations of compounds containing turpentine such as α-pinene and β-pinene indicate these compounds have largely been oxidized to other compounds such as isoprene and other compounds. While the limonene compound was not reviewed because the distillation process carried out had not yet reached the time when the limonene had begun to be fractionated.

3.2. Influence of Pressure to the Distillation Product and Operation

The sampling time of the distillation experiment were done when the distillate volume reached 1500 mL, 3000 mL and 4000 mL in each variation of the experiment. The chromatographic results were quantified using peak area ratios after comparison with standard. The results of the experimental data were compared with the simulation results and shown in Figure 3 and 4.

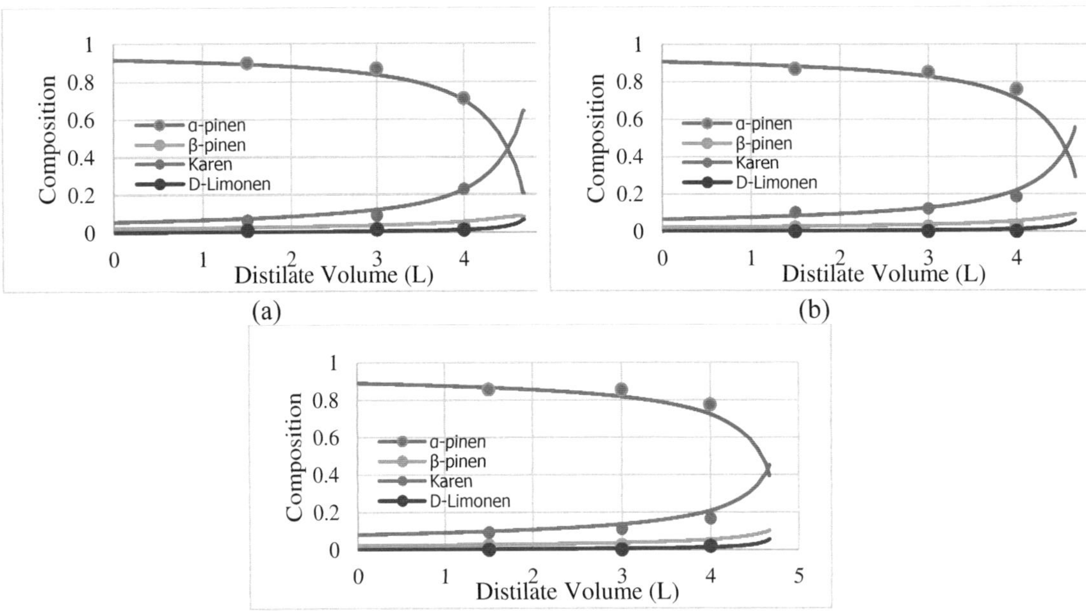

Figure 3. Purity comparison of α-pinene, β-pinene, δ-carene and limonene with respect to pressure: (a) 0.2 atm; (b) 0.45 atm and (c) 1 atm

The results obtained are in accordance with the simulation results with the efficiency of each stage that is not ideal that is equal to 0.45-0.6. Drickamer and Bradford predicted distillation of compounds with viscosity such as turpentine will have an efficiency of 45.5% so that the number of effective stages decreases according to that efficiency. The purity of α-pinene obtained is similar to the simulation results but slightly higher at the end of the experiment. This is because there were variations in the composition of α-pinene at the beginning of the experiment. The δ-carene component trend also has a similar profile.

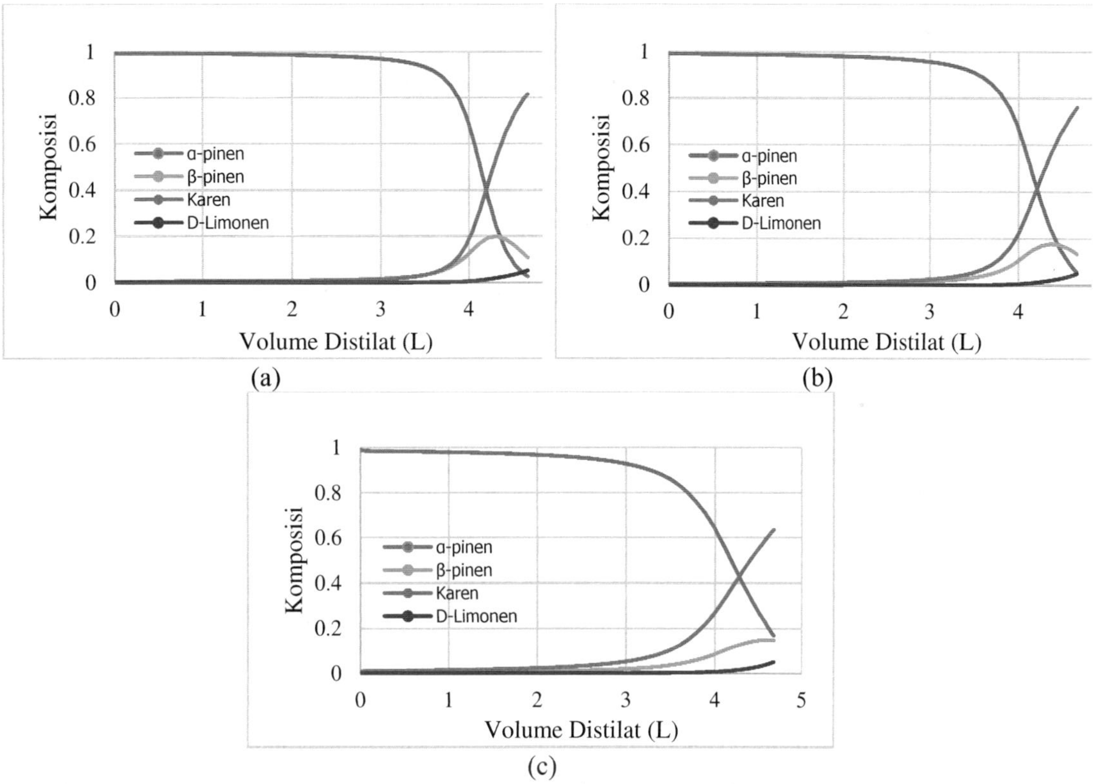

Figure 4. Simulations of the composition of turpentine compounds to the distillate volume under different pressures: (a) 0.2 atm; (b) 0.45 atm and (c) 1 atm

The β-pinene and limonene compounds obtained showed an upward trend, according to the simulation. However, the concentration varies between experiments and does not show a trend, this can be caused by turpentine feed that is used has a number of variations in composition.

Experimental results have shown the suitability of the results with simulations with an efficiency of 0.45, if the efficiency of each stage can be changed to an ideal or the number of stages is changed to 25 stages still with no reflux, better results can be obtained with higher purity. Based on the simulation results, a decrease in pressure from atmospheric pressure to 0.45 atm and 0.2 atm pressure will increase α-pinene recovery by 19.54% and 27.97%, respectively.

3.3. Economic Analysis
The economic aspect is seen from the increase in profits from the addition of the amount of distillate if the distillation can be carried out ideally and changes in operating costs due to the procurement of vacuum conditions. The α-pinene product obtained is assumed to be sold entirely at market prices.

Table 3. Energy consumption comparison

	Energy Consumption		
	Vacuum pump (kJ)	Heater (kJ)	Total (kJ)
0.2 atm	6876	12374	20428
0.45 atm	5157	12008	17545

1 atm	-	8859	8859

The use of energy in distillation is divided into three, for reboilers, condensers and vacuum pumps. The use of vacuum reduces energy consumption in the reboiler and condenser, but there is an increase in energy in the vacuum pump. Based on the calculation results, the energy used by the heater and condenser is almost the same. Following is the energy consumption in each trial with a duration of 5 hours which is presented in Table 3.

Table 4. Cost comparison

	Additional profit (USD)	Additional cost (USD)	Net profit (USD)
0.2 atm	3.55	0.35	3.2
0.45 atm	2.51	0.27	2.24
1 atm	-	-	-

Energy consumption for vacuum pressure distillation was higher than atmospheric pressure by 98% and 130% at pressures of 0.45 atm and 0.2 atm, respectively. The calculation is based on an α-pinene price of 4.93 USD/kg and an electricity price of 0.11 USD/kWh. Comparison among experimental runs were presented in Table 3. The calculation shows that the best results are obtained at an operation pressure of 0.2 atm.

4. Conclusion

The conclusions obtained from the study entitled Turpentine Fractionation with Vacuum Distillation carried out are:

1. Turpentine distillation is carried out at a pressure of 0.45 atm or lower to avoid thermal degradation that can occur.
2. Distillation experimental results in accordance with simulations with column efficiency of 45%.
3. Turpentine distillation operating conditions are best obtained at conditions of 0.2 atm; 0.45 atm and 1 atm respectively.

References

[1] Haneke, K.E. "Turpentine (Turpentine Oil, Wood Turpentine, Sulfate Turpentine, Sulfite Turpentine)", 2002.
[2] Neuenschwander, U., Guignard, F., Hermane, I., ChemSusChem 3, 2010. hal. 75.
[3] Mahood, S. A., "The Thermal Decomposition of Turpentine with Particular Reference to The Production of Toluene and Isoprene", The Journal of Industrial and Engineering Chemistry 12 (1920), 12.
[4] Seader, J. D., dkk. "Separation Proess Principles 3rd Edition". John Wiley and Sons Inc, 2010
[5] Snider, S. R, Ind. Eng. Chem. Anal. Ed. 1945. 17:108-ff.
[6] Soerawidjaja, T. H, "Kuliah Pengantar", Slide perkuliahan TK3210, 2015

IOP Conf. Series: Materials Science and Engineering **742** (2020) 012030 doi:10.1088/1757-899X/742/1/012030

In-vivo Test of Calcium Soap from Palm Fatty Acid Distillate (PFAD) in Three Cow Farms

Antonius Indarto[1], Lienda A. Handojo, Dian Shofinita, Muhammad R. Saadi, Dea Yulistia, Fathinah I. Hasyyati

[1]Department of Chemical Engineering, Institut Teknologi Bandung, Jl. Ganesha 10 Bandung, Indonesia
[2]Departmet of Food Engineering, Institut Teknologi Bandung, Jl. Let. Jend. Purn. Dr. (HC) Mashudi No.1, Kabupaten Sumedang, Indonesia

E-mail: antonius.indarto@che.itb.ac.id

Abstract. Palm Fatty Acid Distillate (PFAD) is a by-product of Crude Palm Oil (CPO) refinery. As the largest CPO producer, Indonesia exported an abundant amount of PFAD without further processing. The use of calcium soap as a supplement for ruminant feed will increase the productivity of dairy cows in Indonesia. Hence, this study reported in-vivo tests that were conducted to determine the effect of calcium soap supplementation on dairy cows. The result showed that feeding calcium soap to dairy cows could increase productivity and extend the lactation period. The milk quality also improved by the increase of fat content and density which lead to an increase in milk prices. Based on the field observations in three animal husbandries, calcium soap improved the health and fertility of cows.

1. Introduction

As the world's largest producer of palm oil, Indonesia covers 58% of total Crude Palm Oil (CPO) production [1]. The process of refining CPO will produce about 4% of Palm Fatty Acid Distillate (PFAD) [2], most of which are exported abroad without further processing. Following the Indonesian Palm Oil Association (GAPKI), Indonesia produced PFAD at *ca.* 1 million tons in 2018. One of the promising applications of PFAD is used as the source of calcium soap that could increase the productivity of dairy cows.

Currently, the supply of fresh milk raw materials from local dairy farmers in Indonesia was only able to meet around 23% of the needs of the domestic milk processing industry (852,000 tons per year) [3]. The average production of cow's milk in Indonesia in 2018 was only 17.2 L of milk /cow/day with a production period of 241 days in 1 year [4].

Early in the breastfeeding period, insufficient feed intake to meet energy needs causes the cows to become malnourished. Therefore, dairy cows are forced to use food reserves from the body to support milk production. This often results in metabolic disorders and also leads to low milk acquisition [5]. Fat is an essential source of energy for cows, but fat intake directly to cows can interfere with the degradation of fibers by microbes. Decomposed fat in the rumen is toxic to certain microbes and reduced the microbe population as well as the digestion of cattle in the rumen [6]. This negative effect can be prevented by giving fat in the form of calcium soap. This is because fat does not directly decompose in the rumen and will only degrade in the abomasum so that the fatty acids are then absorbed by the small intestine. Therefore, calcium soap is often referred to as the intake of protected fat [7].

The fat content in PFAD could reach 81.7% [8] that makes PFAD can potentially be used as a raw material for the production of calcium soap. Research shows that supplementing calcium soap feed from PFAD to dairy cows will increase total SNF (breastfeeding ability) and total milk production [9]. 500 g of rumen-protected fat (Megalac®) intake could increase milk production by 2.3 liters/cow/day. Moreover, based on research work of a team of University of Florida (USA), Holstein cows, with addition of 450 g of Megalac®/day for 120-days from calving, resulted higher conception and pregnancy rates. Therefore, this study aims to determine the effect of calcium fat produced from PFAD on the quantity and quality of Indonesia dairy cows through in vivo testing in three different cow farms.

2. Materials and Method

2.1. Calcium soap production

The production of calcium soap from PFAD refers to the method previously prepared by Handojo et al. [10]. PFAD was reacted with CaO or $Ca(OH)_2$ by the modified fusion method to form calcium soap.

2.2. In-vivo test

In-vivo tests were carried out in several cow husbandries in Java, Indonesia, those of: (1) UPTD Balai Pengembangan Ternak Sapi Perah dan Hijauan Pakan Ternak Cikole, Lembang, West Java; (2) CV. Irs Global Perkasa, a local cow farm in Cihideung, Lembang, West Java; and (3) KUD Karangploso, Malang, East Java.

At CV. Irs Global Perkasa, a cow farm owned by local people in Cihideung, Lembang, an in-vivo test was carried out on first time calving cows by giving 600 grams of calcium soap/cow/day for 120 days. Meanwhile, at UPTD Balai Pengembangan Ternak Sapi Perah dan Hijauan Pakan Ternak Cikole, Lembang, in-vivo tests were carried out on second time calving cows with lactation period between 61 to 120 days by giving calcium soap as much as 600 grams/cow/day. Each group consists of two cows with different treatments: (1) addition of calcium soap supplement during feeding and (2) regular feeding without calcium soap addition. At KUD Karangploso, Malang, East Java, around 440 grams of calcium soap/cow was given to 40 cows for 20 days. In contrast to the in-vivo method in West Java, all cows were given calcium soap supplementation to obtain total milk production. The quantity of milk produced by KUD was measured for each cow per day.

In all cow farms, dairy cows were not given calcium soap for a certain amount of time for the conditioning stage. During the evaluation, all cows were given the same feed and changes in milk quality and production could be assumed due to the influence of calcium soap. Milk from cows was then analyzed using a Lactoscan analyzer to obtain the chemical content of milk, such as fat content and total solids. The amount of milk produced every day was also recorded to obtain the production profile.

3. Results and Discussion

Based on Figure 1 and Figure 2, calcium soap supplementation could increase productivity and extend the lactation period of dairy cows owned by both cow farms in Lembang, West Java.

IOP Conf. Series: Materials Science and Engineering **742** (2020) 012030 doi:10.1088/1757-899X/742/1/012030

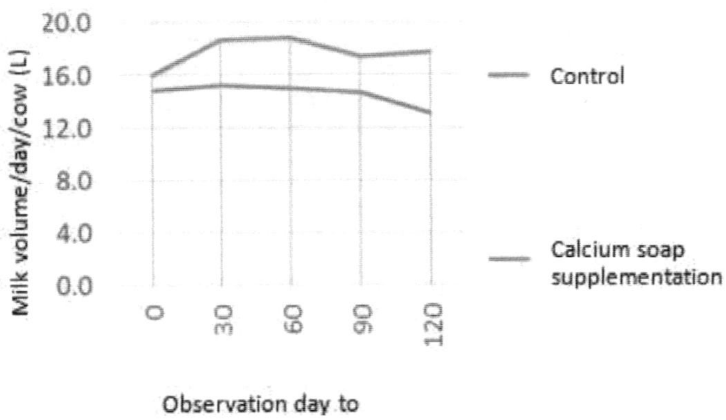

Figure 1. In-vivo test results at cow farm of CV. Irs Global Perkasa in Cihideung, Lembang.

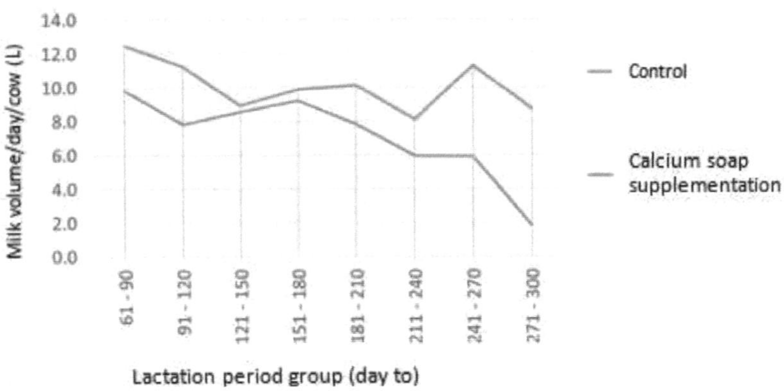

Figure 2. In-vivo test results at Cow Farm of UPTD Balai Pengembangan Ternak Sapi Perah dan Hijauan Pakan Ternak Cikole, Lembang.

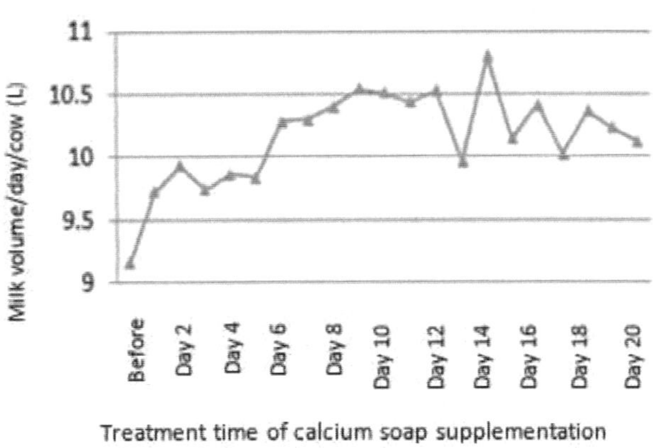

Figure 3. In-vivo test results at Cow Farm of KUD Karangploso, Malang.

Figure 1 shows that calcium soap supplementation for 30 days could increase milk production by about 18%, compared with control cows, which only increased by about 3%. Further supplementation, as shown in the curves up to 120 days, could restrain the rate of decline in milk production compared to controls. Meanwhile, as shown in Figure 2, with the calcium soap supplementation, the rate of decline in milk production was not as sharp as controls. For example, around the dry period, cattle without treatment experienced a decrease in the rate of milk production by 69%, while cattle with calcium soap supplementation only had a decrease in production rate of 3%.

Table 1. The effect of addition of calcium soap on milk quality.

	Average fat content (%)	**Average density (g/cm^3)**	**Average milk price (Rp/L)**
Before	4.03	1.0247	5,386
After 10 days	4.04	1.0246	5,374
After 20 days	4.10	1.0250	5,454

Figure 3 shows that intake of 400 grams of calcium soap/cow/day could increase cow's productivity by around 10-14% for 20 days. Furthermore, milk quality improved as indicated by the increase in fat content and average milk density, which then increased the selling price (from Rp 5.386) as shown in Table 1. Sklan et al. [11] found out that more milk and fat were produced by cows, which were given calcium soap supplementation for 60 days. The bypass fat intake of 300 grams/day/cow significantly increased milk production efficiency (30.22%) compared to controls (27.03%) and increased fat corrected milk by 24.01% over controls [12].

PFAD contains 38-45% palmitic acid (C16:0), 37-42% oleic acid (C18:1), 8-10% linoleic acid (C18:2), and 3-5% stearic acid (C18:0) [13]. Since the high content of saturated fatty acids in PFAD, calcium soap made from PFAD did not have a negative effect on digestion in the rumen, intake of calcium soap actually increases the quantity and quality of milk. Saturated fatty acids increased the percentage of milk fat by 5.1%, while unsaturated fatty acids decreased the percentage of milk fat by 8.0% [14]. Intake of BERGAFAT T-30 produced from palm oil (includes approximately 8% oleic acid which softens milk fat and also includes palmitic acid and stearic acid up to 85% with a melting point of 56-60°C) did not show a decrease in feed intake so that supplementation of BERGAFAT T-30 did not affect fermentation in the rumen of cattle [15]. Feeding a palmitic acid-enriched fat supplement (~ 85% palmitic acid, 2% dry matter) increased the concentration of milk fat from 3.88 to 4.16% and fat yield from 1.23 to 1.32 kg/day compared to cows without supplementation [16].

4. Conclusion

The findings of the current study indicate that supplementation of calcium soap from PFAD increased milk production up to 18% and prolong the milk lactation period of dairy cows. In addition to the quantity of milk, milk quality also increased by the increase of fat content and density which lead to improving milk prices after feeding calcium soap for 20 days.

References

[1] Index Mundi 2019 Palm Oil Production by Country in 1000 MT indexmundi.com
[2] Top A G M 2010 Production and utilization of palm fatty acid distillate (PFAD) *Lipid Technology Journal* **22** (1) pp. 11-3
[3] Panjaitan LY 2017 Indonesia Mengimpor 3,65 Juta Ton Susu Setiap Tahun kemenperin.go.id
[4] BPS 2019 Dairy Cattle Establishment Statistic, 2018
[5] Kronfeld D S 1982 Major metabolic determinants of milk volume, mammary efficiency and spontaneous ketosis in dairy Obs *J. Dairy Sci.* **65** pp.2204-12
[6] Wina and Susana 2013 Manfaat Lemak Terproteksi Untuk Meningkatkan Produksi Dan Reproduksi Ternak Ruminansia (Bogor: Balai Penelitian Ternak Bogor)
[7] Waters J 2010 Fat Supplements for Dairy Cows *Nutrition Files*
[8] Pitoyo 1991 Pemisahan tokoferol dari destilat asam lemak minyak sawit (Yogyakarta: Universitas Gajah Mada)
[9] Alimon A R and Mohamed, W Z. 2012 Recent advances in the utilization of oil palm by-

products as animal feed *International Conference on Livestock Production and Veterinary Technology (ICARD)* 01-04 Okt 2012 Bogor, Indonesia pp. 211-9

[10] Handojo, L A, Indarto, Pramudita D, Shofinita D, Meitha A, Nabila R, Triharyogi H. 2019. Calcium Soap from Palm Fatty Acid Distillate for Ruminant Feed: Reaction Method *International Journal of Engineering and Advanced Technology (IJEAT)* **8** (3S) pp. 422-5

[11] Sklan D, Moallem U, and Folman Y 1991 Effect of feeding calcium soap of fatty acids on production and reproductive responses in high producing lactating cows. *J Dairy Sci* **74** (2) pp. 510-7

[12] Sirohi S, Mohanta R K, Walli T K 2010 Supplementation effect of bypass fat on production performance of lactating crossbred cows *The Indian journal of animal series* **80** pp.733-6

[13] McAskie 1989 Ruminant Feedstuffs, Their Production and Apparatus for Use Therein US Patent: 4,826,694

[14] Bayisa T 2017 Review on Effect of Feeding Dairy Cow with Protected Fat and Protein on Milk Yield and its Composition (Munich: GRIN Verlag)

[15] Sarwar M, Sohaib A, Khan M A, Nisa M 2003 Effect of Feeding Saturated Fat on Milk Production and Composition in Crossbred Dairy Cows *Anim.Sci* **16**(2) pp. 204-10

[16] Lock A L, Preseault C L, Rico J E, DeLand K E, and Allen M S 2013 Feeding a C 16:0-enriched fat supplement

IOP Conf. Series: Materials Science and Engineering **742** (2020) 012031 doi:10.1088/1757-899X/742/1/012031

Thermal Conductivity of Geopolymer with Polypropylene Fiber

Felisitas Aprilia Rastri Utami[1], Andreas Triwiyono[2], Ni Komang Ayu Agustini[3], and Indra Perdana[4]

[1]Magister Student of Civil and Environmental Engineering, Universitas Gadjah Mada, INDONESIA.
[2]Departement of Civil and Environmental Engineering, Universitas Gadjah Mada, INDONESIA.
[3]Ph.D Student of Civil and Environmental Engineering, Universitas Gadjah Mada, INDONESIA.
[4]Departement of Chemical Engineering, Universitas Gadjah Mada, INDONESIA.

E-mail: andreas.triwiyono@ugm.ac.id

Abstract. Geopolymer is widely used to replace cement for fire-resistant products. The benefits of geopolymer based on fly ash are cost-effective and easy process production. Lightweight products are needed to minimize dead load for the existing structure. The addition of polypropylene fiber is intended to produce relative lightweight and strong material applied for fire resistance. This research was conducted to obtain thermal conductivity, density, and compressive strength of the geopolymer. The ratio of activator/fly ash (a/f) was varied 40%, 45%, and 50% with addition of polypropylene fibers of 0.5%, 1%, and 1.5% of the fly ash weight. The experimental test was conducted on the 7th day at ambient temperature while concentration of NaOH used in geopolymer synthesis was kept at 10M with a ratio of alkaline activator (Na_2SiO_3/NaOH) of 2. The test results showed that thermal conductivity (λ) is in a range between 0.19-0.82 W/m.C while the thermal conductivity is less than 1 W/m.C. This indicates that the geopolymer can be applied for thermal insulating material. Geopolymer with polypropylene fiber and a/f 50% can be classified as lightweight material with a density around 1.6283-1.7625 gr/cm^3. The ratio of a/f 50% produces a compressive strength ranging from 18.14 MPa to 28.42 MPa.

1. Introduction

Geopolymers served environmental protection by recycling low calcium-fly ash and slag furnaces. These geopolymers can be used as building material with excellent compressive strength for infrastructure development and structural applications [1]. Fly ash-based geopolymer is also an alternative material for substituting expensive raw materials, such as metakaolin. This procedure uses a low-cost budget and uses available reagents so that it saves material and cost [2]. The use of fly ash with NaOH and Na_2SiO_3 activators dryed at ambient temperature can reduce environmental problems caused by the overproduction of cement. This combination will produced easy-made and cost-effective geopolymer paste [3]. NaOH concentration has the effect on workability and time setting, while the compressive strength depend on Na_2SiO_3 [4] and the molarity of NaOH [5]. According to Jaya [5], low thermal conductivity and high compressive strength were obtained by using NaOH concentration above 10 M. In general, hardened geopolymer paste with polypropylene fiber has a higher strength compared to paste without fiber content [6].

The matrix texture of geopolymer paste was dominated by microporous material while cement paste was dominated by mesoporous material. The microporous material was proven to be thermal-resistance at high temperature indicated by no cracks on the surface, damage or dimensional changes found from laboratory tests. These changes were different from the cement paste which destroyed after high-temperature heating [7]. It was found that additional of carbon and basalt fibers could improve geopolymer thermal properties. Both fibers were shown a good resistance at high temperature applied to wood. These experimental results were shown no fire spread signs, shrinkage and low thermal conductivity [8]. Therefore, fly ash-based geopolymer can be used as eco-friendly fire-resistant building materials, sound and heat insulators, and thermal insulation materials [7,8].

Lightweight products are needed to minimize additional dead load of structures. The addition of polypropylene fiber is intended to produce relatively lightweight and strong geopolymer. The resulting geopolymer is expected suitable for application as fire resistance or thermal insulation. This study examined the properties of fly ash-based geopolymer mixture and polypropylene fiber namely thermal conductivity, density and compressive strength.

2. Materials , Specimens and Test Method

2.1. Materials

2.1.1. Fly ash

Fly ash was obtained from the Paiton power plant which chemical composition is shown in Table 1. Based on XRF test and according to ASTM C618 the fly ash can be classified as type C [9]. The density of fly ash is 1312 kg/m^3 and the specific gravity is 2.46. In this study, the activator/fly ash (a/f) ratio are 40%, 45%, and 50%.

Table 1. Chemical compositions of fly ash from XRF test (% weight).

Chemical Components	SiO_2	Na_2O	Fe_2O_3	CaO	MgO_2	Na_2O	K_2O	TiO_2	Cr_2O_3	SO_3	LOI
Fly ash	38,79	18,21	15,2	14,61	7,18	1,76	1,23	0,95	0,22	0,59	0,69

2.1.2. Alkaline Activator

Alkaline activator was produced through a reaction of sodium silicate (Na_2SiO_3) and sodium hydroxide (NaOH). 10 molarity of NaOH alkaline was made by dissolving NaOH pellets in distilled water. The ratio of Na_2SiO_3/NaOH was kept constant at 2.

2.1.3. Fiber

Figure 1 shows a monofilament polypropylene fiber produced by PT Sika Indonesia, with 0.91 gr/cm^3 density, 12 mm fiber length, 18 micron-nominal fiber diameter, 300-440 MPa tensile strength, and 6,000-9,000 MPa elastic modulus. The ratio of fiber to fly ash weight was varied at 0.5%, 1%, and 1.5%.

Figure 1. Polypropylene fiber.

2.2. Specimens

In this research three series of specimen were tested to determine thermal conductivity, density, and compressive strength. Each series consists of three specimens. All specimens were made of fly ash, alkaline activator, and polypropylene fiber. Fly ash and polypropylene fibers were mixed manually in dry condition. Alkaline activator was added into the mixture and they were mixed for 5 minutes until homogenous, according to the ASTM C 305 [10]. The mixture was poured into a mold and removed from the mold after 24 hours at ambient temperature. All specimens of fly ash-based geopolymer with variation of material combination are shown in Table 2.

Table 2. The proportion of geopolymer paste with polypropylene fiber by weight per m^3.

No	a/f Ratio (%)	Polypropylene Fiber (%)	Fly ash Kg	Na_2SiO_3 kg	NaOH kg	Polypropylene kg
1	40	0.5	1510.71	402.86	201.43	7.55
2	40	1.0	1510.71	402.86	201.43	15.11
3	40	1.5	1510.71	402.86	201.43	22.66
4	45	0.5	1441.20	432.36	216.18	7.21
5	45	1.0	1441.20	432.36	216.18	14.41
6	45	1.5	1441.20	432.36	216.18	21.62
7	50	0.5	1377.80	459.27	229.63	6.89
8	50	1.0	1377.80	459.27	229.63	13.78
9	50	1.5	1377.80	459.27	229.63	20.67

2.3. Test Method

2.3.1. Thermal conductivity test

Thermal conductivity test was carried out with a cylindrical specimen of 4 cm diameter, with variation of 0.2 cm and 0.4 cm thickness, as shown in Figure 2. Thermal conductivity was measured using apparatus machine type OSK 4565-A Tokyo Meter Co.Model HVs - 400000SE according to ASTM E 177 [11]. Temperature investigation in this work was variated at 50^{0}C, 100^{0}C, 150^{0}C, and 200^{0}C.

2.3.2. Density test

A cylindrical specimen of 3 cm diameter with 1 cm thickness, as shown in Figure 3(a), was used for density test. The test was done according to the Archimedes principle according to ASTM D 792 [12]. The specimens were weighed in air and then immersed into a specific liquid to get the volume of specimen. The density was defined as dividing the weight by its volume.

2.3.3. Compressive test

Compressive strength was obtained from compressive test according to ASTM C 109 [13]. Figure 3(b) shows the cube specimen of 5x5x5 cm. The compressive test was executed after 7 day aging using a Compression Testing Machine. The compressive strength was obtained by dividing the maximum load to its cross section area of the original specimen.

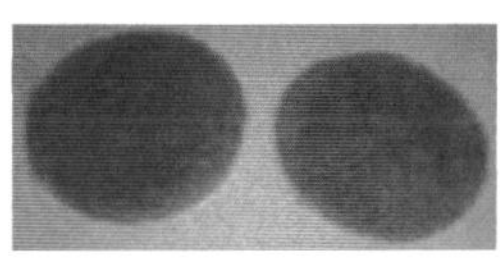

Figure 2. Thermal conductivity specimen.

(a) (b)

Figure 3. Specimen test of (a) density and (b) compressive strength.

3. Results and Discussion

3.1. Thermal Conductivity

Figure 4 to 7 show the thermal conductivity from the measurements. All variance show fluctuated trend of thermal conductivity at low temperature. The thermal conductivity of samples with all a/f ratios and addition of polypropylene fiber shows similar trend. The results indicate that in a temperatures range of 50⁰C-200⁰C, insignificant differences of thermal conductivity were observed. Previous studies showed that, the thermal conductivity at low temperature of 120⁰C was a function of the conductivity of gels and pores caused by water evaporation [14]. It was reported in other study that high water content causes thermal conductivity to increase [15]. The thermal conductivity of samples in the present work was in a range of 0.19-0.82 W/m.C with an average of 0.421 W/m.C. Since the value is less than 1 W/m.C all geothermal synthesized in the present work is applicable and suitable as thermal insulating materials [16].

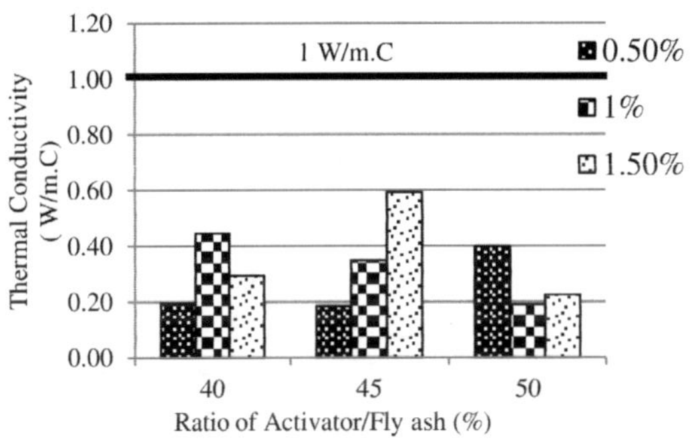

Figure 4. Thermal conductivity of geopolymer with polypropylene fiber (%) at temperature 50⁰C.

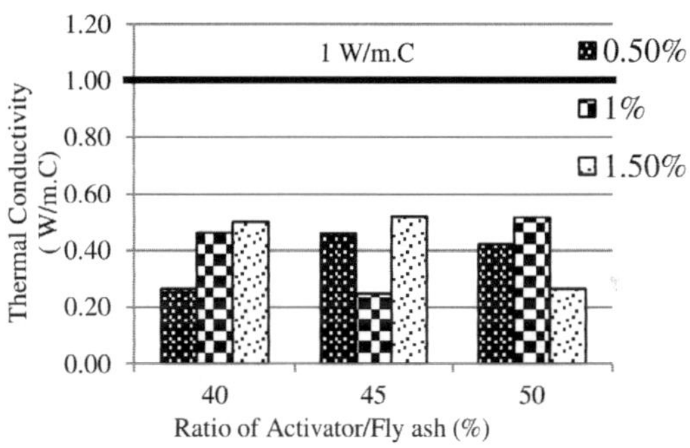

Figure 5. Thermal conductivity of geopolymer with polypropylene fiber (%) at temperature 100⁰C.

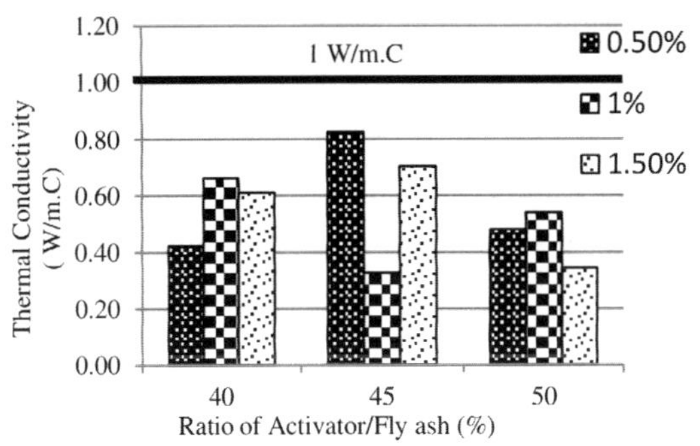

Figure 6. Thermal conductivity of geopolymer with polypropylene fiber (%) at temperature 150⁰C.

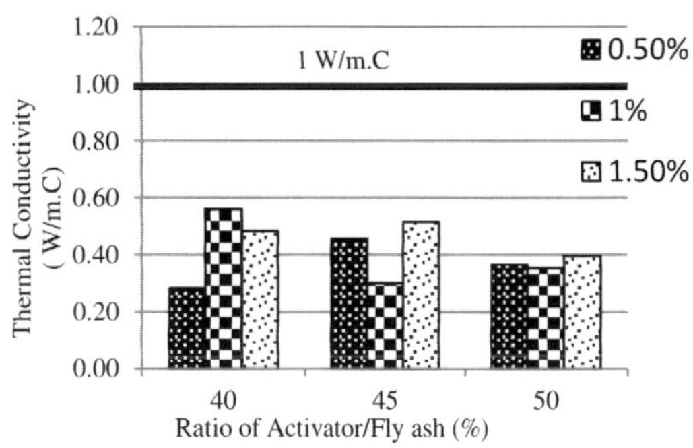

Figure 7. Thermal conductivity of geopolymer with polypropylene fiber (%) at temperature 200⁰C.

3.2. Density

Addition of polypropylene fibers by 0.5-1.5% shows wether the ratio of a/f affects the density of the geopolymer paste. As shown in Figure 8, the density decreases by the increasing a/f ratio. Addition of polypropylene fiber equal or more than 1% lead to formation of geopolymer with a density between 1.6283-1.7625 gr/cm^3. It indicates that the geopolymer paste can be classified as a lightweight product based on ASTM C567 (1140-1840 kg/m^3) [17]. Figure 10 shows that a decrease of density and compressive can be resulted from increasing of a/f ratio. Similar to previous study, compressive strength decreases with decreasing geopolymer density [6,18]. The decrease in density is due to an increase of pore volume in the specimens which consequently decreases the compressive strength [19]. Addition of polypropylene fiber in the paste will replace the amount of fly ash while added number of pores in the material. The pores are trapped between groups of fiber in a fresh matrix which causes a reduction in density in the hardened paste [6]. Moreover, water evaporation from the activator compounds also causes formation of pores which can reduce the density [20].

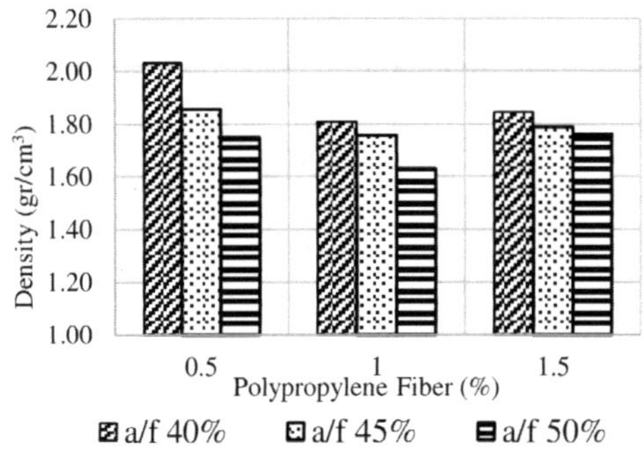

Figure 8. The density of geopolymer with polypropylene fiber.

3.3. Compressive strength

The compressive strength of fly ash-based geopolymers after 7 day aging is shown in Figure 9. The highest compressive strength is produced from the sample with a/f ratio of 40% which strength ranges in 23.63-33.41 MPa. The lowest compressive strength is produced from 50% a/f ratio samples with a strength in a range of 18.14-28.42 MPa. According to previous study, the strength of geopolymers significantly reduced when the a/f ratio is more than 40% [21]. The increasing ratio of a/f will reduce compressive strength of the resulting geopolymer by about 8%-13% with addition of polypropylene fiber by amount of 0.5-1.5%. Increasing a/f ratio will increase the amount of NaOH in the mixture, which causes the increasing number of pores and reduction of compressive strength. The presence of more NaOH will contributed to more OH ions available in the the mixture. OH ions will react with calcium to form $Ca(OH)_2$ and therefore decreases strength of the product [4].

In addition, polypropylene fibers affect significanly on the decrease of the compressive strength by an average of 25%. The decrease of compressive strength is due to shrinkage effect and weak fiber interface contact, which potentially breaks the bonds in geopolymer matrix [6]. Other research reported that the addition of more than 1% fiber will decrease mechanical properties of the products [6,18]. Reduction of fly ash and Na_2SiO_3 in the geopolymer mixture will reduce the compressive strength [4].

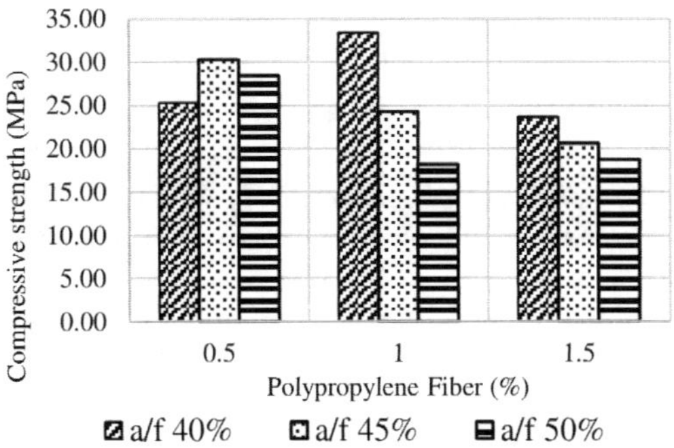

Figure 9. Compressive strength of geopolymer with polypropylene fiber.

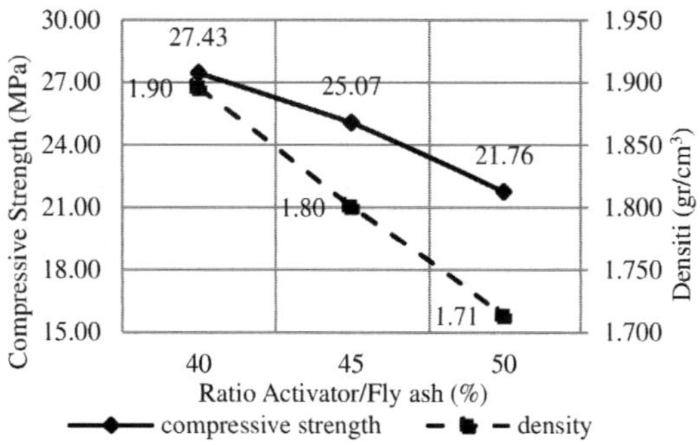

Figure 10. The influence of ratio activator/fly ash on compressive strength and density.

4. Conclusion

In general, geopolymer paste with some variation of polypropylene fiber addition could lower thermal conductivity of the product that make it suitable for thermal insulation application. According to the esperimental results, conclusion can be drawn:

1. The thermal conductivity of all specimens range between 0.19-0.82 W/m.C with an average of 0.421 W/m.C is recommeded as an insulator material.
2. The additional of polypropylene fiber doesn't affect on the thermal conductivity.
3. The ratio a/f 40% is preferable to make high compressive strength (23.63 – 33.41 MPa), but the density is not classified as a lightweight material.
4. The ratio a/f 50% is preferable to produce lightweight material although it has low compressive strength (18.14-28.42 MPa).

Further investigation is needed to identify the strength and behaviour of the product during and after heating at high temperature.

Acknowledgment

This study is a part of the RTA research project financially supported by Universitas Gadjah Mada, Indonesia.

References

[1] Rangan BV. *Geopolymer Concrete for Environmental Protection*. Indian Concr J 2014;88:41–59.

[2] Roviello G, Ricciotti L, Tarallo O, Ferone C, Colangelo F, Roviello V. *Innovative Fly ash Geopolymer-Epoxy Composites: Preparation, Microstructure and Mechanical Properties. Materials* (Basel) 2016;9. doi:10.3390/ma9060461.

[3] Mirza J, Kubba Z, Ismail M, Sam ARM, Fahim Huseien G, Tahir MM. *Impact of Curing Temperatures and Alkaline Activators on Compressive Strength and Porosity of Ternary Blended Geopolymer Mortars*. Case Stud Constr Mater 2018;9:e00205. doi:10.1016/j.cscm.2018.e00205.

[4] Malkawi AB, Nuruddin MF, Fauzi A, Almattarneh H, Mohammed BS. *Effects of Alkaline Solution on Properties of the HCFA Geopolymer Mortars*. Procedia Eng 2016;148:710–7. doi:10.1016/j.proeng.2016.06.581.

[5] Jaya NA, Yun-Ming L, Abdullah MMAB, Cheng-Yong H, Hussin K. *Effect of Sodium Hydroxide Molarity on Physical, Mechanical and Thermal Conductivity of Metakaolin Geopolymers*. IOP Conf Ser Mater Sci Eng 2018;343. doi:10.1088/1757-899X/343/1/012015.

[6] Ranjbar N, Mehrali M, Behnia A, Javadi Pordsari A, Mehrali M, Alengaram UJ. *A*

IOP Conf. Series: Materials Science and Engineering **742** (2020) 012031 doi:10.1088/1757-899X/742/1/012031

Comprehensive Study of the Polypropylene Fiber Reinforced Fly Ash Based Geopolymer. PLoS One 2016;11:e0147546. doi:10.1371/journal.pone.0147546.

[7] Alehyen S, Zerzouri M, El Alouani M, El Achouri M, Taibi M. *Porosity and Fire Resistance of Fly ash Based Geopolymer*. J Mater Environ Sci 2017;8:3676–89.

[8] Uddin Ahmed Shaikh F, Haque S, Sanjayan J. *Behavior of Fly ash Geopolymer as Fire Resistant Coating for Timber.* J Sustain Cem Mater 2018;0:1–16. doi:10.1080/21650373.2018.1537015.

[9] ASTM C618. *Standard Specification for Coal Fly Ash and Raw or Calcined Natural Pozzolan for Use*. ASTM Int 2015:1–5. doi:10.1520/C0618-15.2.

[10] ASTM C305. *Mechanical Mixing of Hydraulic Cement Pastes and Mortars of Plastic Consistency*. ASTM Int 2016:1–3. doi:10.1520/C0305-14.2.

[11] ASTM C177. *Standard Test Method for Steady-State Heat Flux Measurements and Thermal Transmission Properties by Means of the Guarded-Hot-Plate*. ASTM Int 2013:1–23. doi:10.1520/C0177-13.2.

[12] ASTM D729. *Standard Test Methods for Density and Specific Gravity (Relative Density) of Plastics by Displacement*. Am Soc Test Mater 2008;14:6. doi:10.1520/D0792-08.2.

[13] ASTM C109. *Standard Test Method for Compressive Strength of Hydraulic Cement Mortars*. ASTM Int West Conshohocken 2000:1–6. doi:10.1520/C0109.

[14] Jeyalakshmi R, Sivasakthi M, Revathi T, Rajamane NP. F*ire Related Temperature Resistance of Fly Ash Based Geopolymer Mortar*. MATEC Web Conf 2017;97:01034. doi:10.1051/matecconf/20179701034.

[15] Gomes MG, Flores-Colen I, Manga LM, Soares A, de Brito J. *The Influence of Moisture Content on The Thermal Conductivity of External Thermal Mortars*. Constr Build Mater 2017;135:279–86. doi:10.1016/j.conbuildmat.2016.12.166.

[16] Lach M, Korniejenko K, Mikula J. *Thermal Insulation and Thermally Resistant Materials Made of Geopolymer Foams*. Procedia Eng 2016;151:410–6. doi:10.1016/j.proeng.2016.07.350.

[17] ASTM C567. *Standard Test Method for Determining Density of Structural Lightweight Concrete*. ASTM Int 2016:14–7. doi:10.1520/C0567.

[18] Belkharchouche D, Chaker A. *Effects of Moisture on Thermal Conductivity of Che Lightened Construction Material*. Int J Hydrogen Energy 2016;41:7119–25. doi:10.1016/j.ijhydene.2016.01.160.

[19] Liu K, Wang Z, Jin C, Wang F, Lu X. *An Experimental Study on Thermal Conductivity of Iron Ore Sand Cement Mortar*. Constr Build Mater 2015;101:932–41. doi:10.1016/j.conbuildmat.2015.10.108.

[20] Cui Y, Wang D. *Effects of Water on Pore Structure and Thermal Conductivity of Fly Ash-Based Foam Geopolymers*. Adv Mater Sci Eng 2019;2019. doi:10.1155/2019/3202794.

[21] Qingwei Sun, Han Zhu, Haoyu Li, Haiyang Zhu MG. *Application of Response Surface Methodology in The Optimization of Fly ash Geopolymer Concrete* 2018;48:45–52. doi:10.13422/j.cnki.syfjx.2011.19.087.

IOP Conf. Series: Materials Science and Engineering **742** (2020) 012032 doi:10.1088/1757-899X/742/1/012032

Thermal Properties of Geopolymers with Silica Fume

Fanriyanto Sampe[1], Andreas Triwiyono[2], Ni Komang Ayu Agustini[3], Indra Perdana[4]

[1]Magister Student of Department of Civil and Environmental Engineering, UGM, Yogyakarta, Indonesia
[2]Department of Civil and Environmental Engineering, UGM, Yogyakarta, Indonesia
[3]Ph.D Student of Department of Civil and Environmental Engineering, UGM, Yogyakarta, Indonesia
[4]Department of Chemical Engineering, UGM, Yogyakarta, Indonesia

E-mail: fanri.sampe@gmail.com

Abstract. Fly ash-based geopolymer is a renewable technology to minimize the use of cement in the construction field. The development was not only the use of fly ash but also the addition of other materials such as silica fume to geopolymer paste. Insulating materials is one of the examples of fly ash-based geopolymer applications. This paper examines the influence of varied activating solution of fly ash ratio and silica fume on the behavior of thermal properties of the geopolymer paste. The activating solution of fly ash ratio (A/F) was 40%, 45%, and 50%. Silica fume acts as a substitution of fly ash varied from 1%, 2%, and 4%. The activating solution was mixed with NaOH 10 M and Na_2SiO_3 with Na_2SiO_3/NaOH ratio constant at 2. Geopolymer paste was subjected to different tests such as thermal conductivity, bulk density, and compressive strength. The results showed that the geopolymer paste with substitution of silica fume had thermal conductivity, bulk density and compressive strength in the range of 0.051 to 0.932 W/m.C, 1752.457 to 1870.309 kg/m^3 and 25.41 to 38.48 MPa respectively. The thermal conductivity values for all variations are below 1 W/m.C. Results from this work shows that the mixture suitable as a thermal insulation material.

1. Introduction

Geopolymers have recently emerged as an excellent alternative to replacing cement in the construction industry because of its superior mechanical properties and durability. This technology is very promising because CO_2 in the manufacturing process.[1]. Geopolymers are produced from alumino-silicate. Fly ash is commonly used as a geopolymer base. It is one of waste material from coal combustion or electricity generation [2]. The geopolymerization itself is formed between the reaction of alumino silicate source material and the alkali activator. Some studies evaluated the use of different types of alumino silicate and alkali activators on properties of geopolymers. The Alkali activator usually is produced using sodium silicate (Na_2SiO_3) and sodium hydroxide (NaOH) [3,4]. Most research was carried out to study mechanical properties, durability, and thermal conductivity as well as density of geopolymers[5,6]. One of important mechanical properties of geopolymer is compressive strength

In the further development of geopolymers, many studies have been carried out on the addition of silica fume to improve the acid and thermal resistance of the geopolymer [5–7]. Geopolymers with silica fume which have thermal conductivity less than 1 W/m.C is suitable as thermal insulation materials [6,8]. The use of silica fume will also increase the compressive strength [7]. However, there is no reports on mechanical properties of geopolymer for insulation material. This information is important because the stability of the insulation material structure depends on its strength. So, the purpose of this research is to find out thermal conductivity, density and compressive strength of geopolymer paste using silica fume substitution.

2. Materials, Specimens, and Experimental Method

2.1. Materials

In this study, fly ash was obtained from Paiton Power Plant and classified as type C according to ASTM C618 [9] with a density of 2.46 . Table 1 shows the chemical components of the fly ash. The silica fume was obtained from PT. Sika Indonesia, Co with a density of 2.09. Alkali activator was produced from the reaction of sodium silicate (Na_2SiO_3) and sodium hydroxide (NaOH) with a density of 1.59 and 1.52 respectively. 10 molarity of NaOH alkali solution was made of NaOH pellets and Aquades. The ratio of Na_2SiO_3/NaOH was keep constant at 2.0. The activating solution of the fly ash ratio (A/F) were 40%, 45%, 50%. The substitution of silica fume were 1%, 2%, 4%.

Table 1. The Chemical components of *fly ash* (% Weight)

Chemical Components	SiO_2	Na_2O	Fe_2O_3	CaO	MgO_2	Na_2O	K_2O	TiO_2	Cr_2O_3	SO_3	LOI
Fly ash	38.79	18.21	15.2	14.61	7.18	1.76	1.23	0.95	0.22	0.59	0.69

2.2. Specimens

Three types of specimens were used for compressive strength, thermal conductivity, and density test. For compressive test, a cube-shape specimen was used with dimension of 50 x 50 x 50 mm. For thermal conductivity test, the specimen was made of flat cylinder molds with diameter 4 cm and the thick of 0.2, and 0.4 cm. For density test, similar mold was used with diameter of 3 cm and 1 cm thick. Each types of specimen consists of 3 samples.

Geopolymers mixing process refered to ASTM C305 [10]. The process was done in 2 phases, preparation of alkali activator and final mixing with silica fume. First phase, Na_2SiO_3 was mixed with NaOH as activator and then left to cool down for about 30 minutes. Second phase, fly ash and silica fume were stirred in the Hobart mixer for about a minute at 140 rpm before the activator was added. The activator was poured slowly until the mixture became homogeneous for about 4 minutes. The mixture was casted into a mold for one day. The specimen was released and kept in room temperature. Mix proportion for all material per cubic meter is shown in Table 2. From Table 2, the symbols Q, R, and SF show the values of A / F, the ratio of Na_2SiO_3/NaOH, and silica fume substitution. Figure 1 shows a sample for tests of thermal conductivity, density, and compressive strength.

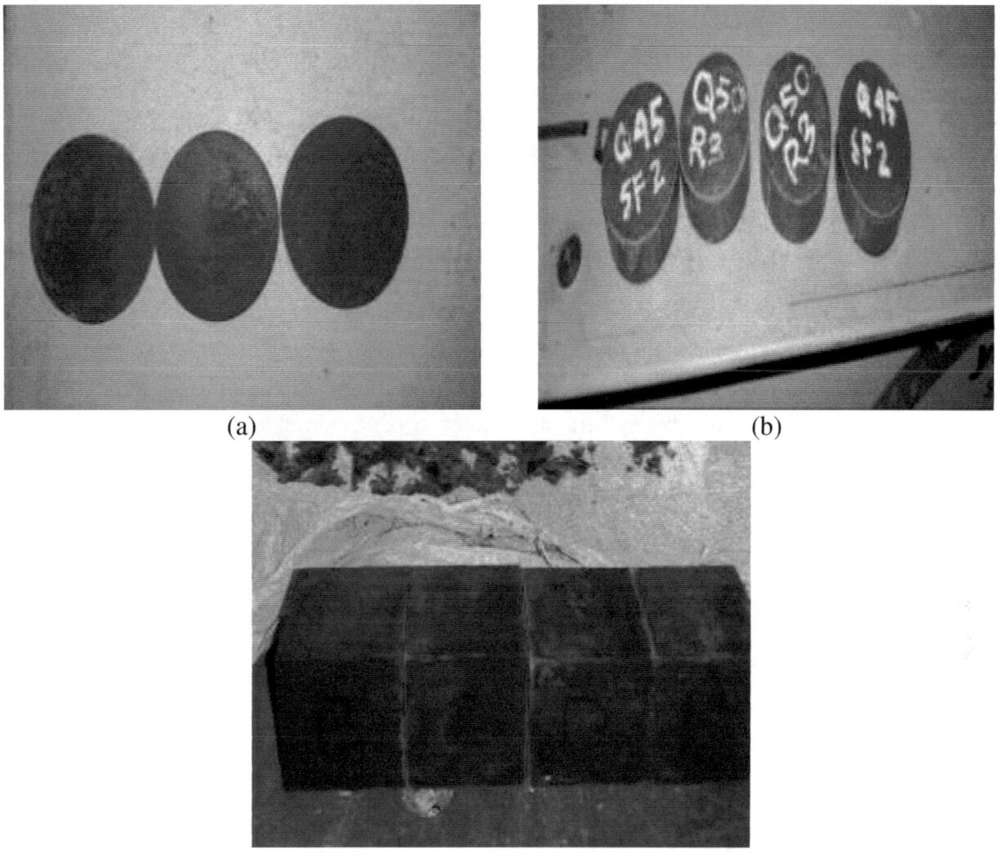

(a) (b)

(c)

Figure 1. The specimen of each testing: (a). Thermal conductivity, (b). Density, (c). Compressive strength

Table 2. Mix proportions of geopolymer paste by weight [per m^3]

No	Specimen	Ratio of Na_2SiO_3/NaOH	A/F Ratio	Fly ash [Wfa] [kg]	Na_2SiO_3 [Wss] [kg]	NaOH [Wsh] [kg]	Silica fume [Wsf] [kg]
1	Q40 R2 SF1	2	0.4	1499.89	399.97	199.99	15.00
2	Q40 R2 SF2	2	0.4	1489.23	397.13	198.56	29.78
3	Q40 R2 SF4	2	0.4	1468.34	391.56	195.78	58.73
4	Q45 R2 SF1	2	0.45	1431.35	429.40	214.70	14.31
5	Q45 R2 SF2	2	0.45	1421.63	426.49	213.24	28.43
6	Q45 R2 SF4	2	0.45	1402.58	420.78	210.39	56.10
7	Q50 R2 SF1	2	0.5	1368.79	456.26	228.13	13.69
8	Q50 R2 SF2	2	0.5	1359.90	453.30	226.65	27.20
9	Q50 R2 SF4	2	0.5	1342.46	447.49	223.74	53.70

IOP Conf. Series: Materials Science and Engineering **742** (2020) 012032 doi:10.1088/1757-899X/742/1/012032

2.3. Experimental Method

Thermal conductivity was measured according to ASTM C177 [11] by using Thermal Conductivity Apparatus machine, Type: OSK 4565-A Tokyo Meter Co. Model: HVs - 400000SE . The temperature was varied of 50^{0}C, 100^{0}C, 150^{0}C, and 200^{0}C. Bulk density test according to ASTM D792 [12] was conducted by dividing the weight and the volume of specimen. Compressive test of geopolymer was carried out at 7 days based on ASTM C109[13] using Avery Dennison 20 kN machine.

3. Result and Discussion

3.1. Thermal conductivity

Figure 2 shows the value of thermal conductivity for each variation. In general, the trend of thermal conductivity values decreases after 100^{0}C. As a whole, the thermal conductivity values range from 0.051 to 0.932 W/m.C. This indicates that the mixture is suitable as a thermal insulation material. That is because all thermal conductivity values are below 1 W/m.C. The changes of A/F has no effect on the value of thermal conductivity. The use of silica fume causes the paste more porous so that the thermal conductivity value decrease. This result is similar to other work of Henon et al. [5], Prud'homme et al. [14], and Dutta et al. [15]. The use of A/F of 40% also causes decrease in thermal conductivity. The use of alkali activators affect the geopolymer process [16]. The higher amount of alkali activator will increase formation of geopolymer. However, if the amount of alkali activator is excessive, the formation of geopolymer will take longer [16].

3.2. Density

Table 3 shows the results of density test for all variation. The density from all variations range 1752.457 to 1870.309 kg/m^3. The changes of A/F variation and silica fume substitution do not affect the density values. It is similar to the results reported in other work [5]. The average value of specific gravity from this research was 1789.351 kg/m^3. This mixture can be classified as a lightweight material based on ASTM C567 [17].

Table 3. Buk density [kg/m^3]

No Specimen	Bulk Density [kg/m3]
Q40 R2 SF1	1835.597
Q40 R2 SF2	1767.668
Q40 R2 SF4	1757.724
Q45 R2 SF1	1819.681
Q45 R2 SF2	1826.080
Q45 R2 SF4	1755.115
Q50 R2 SF1	1752.457
Q50 R2 SF2	1782.136
Q50 R2 SF4	1807.698
Average	1789.351

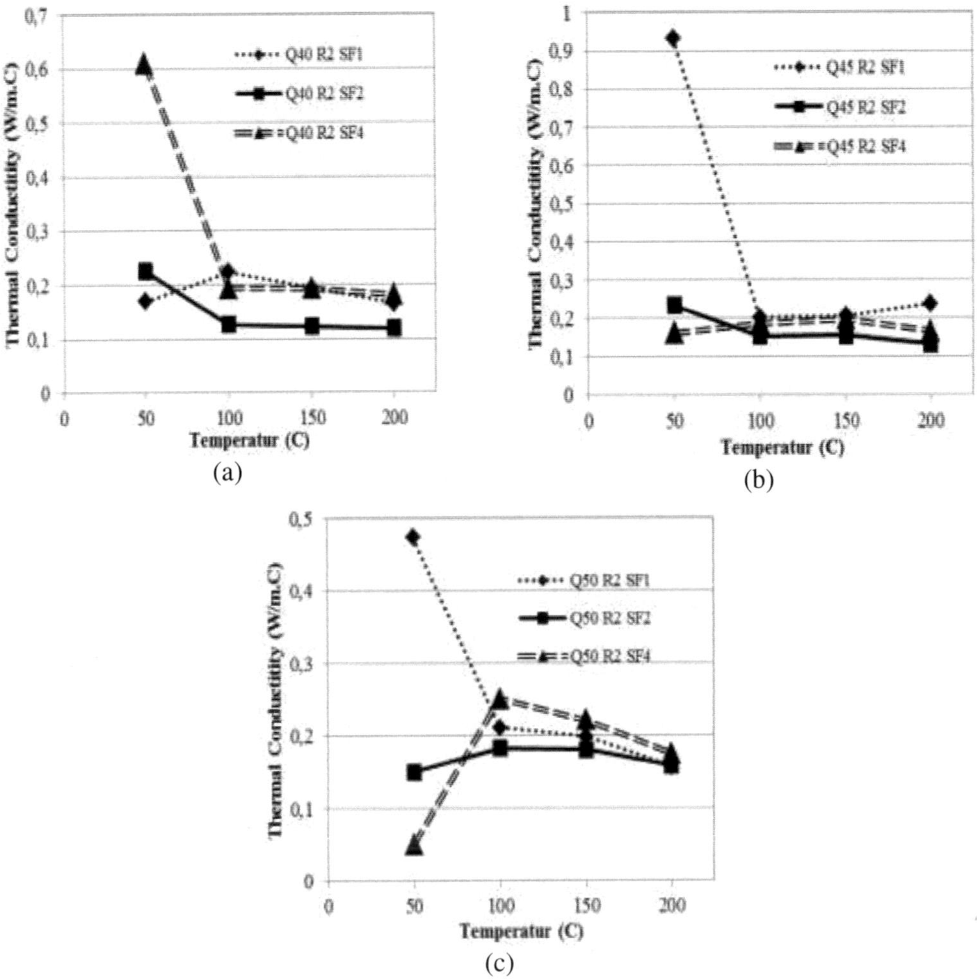

Figure 2. The result of thermal conductivity for variants: (a). Q40 R2 SF1 – Q40 R2 SF4, (b). Q45 R2 SF1 – Q45 R2 SF4, (c). Q50 R2 SF1 – Q50 R2 SF 4

3.3. Compressive strength

Figure 3 shows a comparison of compressive strength for all variations at room temperature. Generally, the compressive strength values range from 25.41 to 38.48 MPa. The use of silica fume can improve the mechanical property of geopolymer so that the compressive strength will increase [18]. The optimal A/F of 40% is found the maximum compressive strength [16,19]. If A/F exceeds 40%, the compressive strength of geopolymer significantly decrease. If A/F is more than 40%, the alkali activator concentration will decrease and therefore relatively retard geopolymerization. This reduces the strength of geopolymer [19].

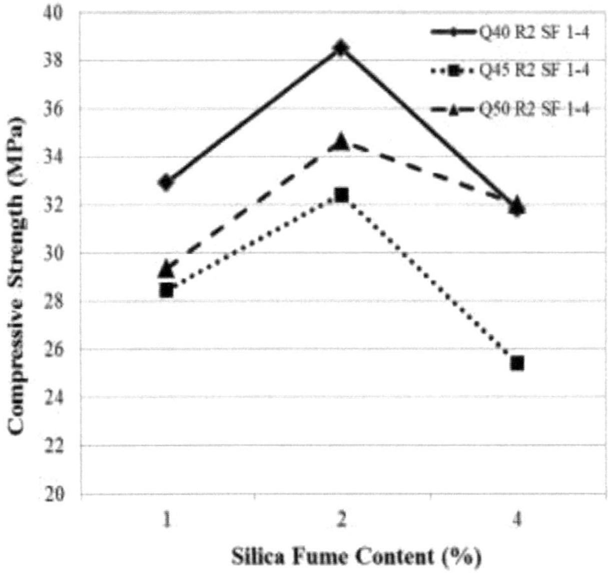

Figure 3. The result of compressive strength for all variants

4. Conclusion

This work investigates the effect of silica fume substitution to fly ash on thermal and mechanical properties of geopolymer. The thermal conductivity of silica fume substituted geopolymer is in the range of 0.051 to 0.932 W/m.C. The use of silica fume causes geopolymer paste becomes more porous so that its thermal resistance increase. The density values obtained in the range of 1752.457 to 1870.309 kg/m^3. The compressive strength in the range of 25.41 to 38.48 MPa. The mixture is suitable as a thermal insulation material, because the termal conductivity value is below 1 W/m.C.

Acknowledgment

The author would like to thank Gadjah Mada University, who provided financial support for this research.

References

[1] Mahendra K, Arunachelam N. Compressive Strength of Flyash Geopolymer Mortar Having Silica Fume as A Partial 2017:33–5.

[2] Saloma, Saggaff A, Hanafiah, Mawarni A. Geopolymer Mortar with Fly Ash. MATEC Web Conf 2016;78:01026. doi:10.1051/matecconf/20167801026.

[3] Ranjbar N, Mehrali M, Behnia A, Alengaram UJ, Jumaat MZ. Compressive strength and microstructural analysis of fly ash/palm oil fuel ash based geopolymer mortar. Mater Des 2014;59:532–9. doi:10.1016/j.matdes.2014.03.037.

[4] Hussin MW, Bhutta MAR, Azreen M, Ramadhansyah PJ, Mirza J. Performance of blended ash geopolymer concrete at elevated temperatures. Mater Struct Constr 2015;48:709–20. doi:10.1617/s11527-014-0251-5.

[5] Henon J, Alzina A, Absi J, Smith DS, Rossignol S. Potassium geopolymer foams made with silica fume pore forming agent for thermal insulation. J Porous Mater 2013;20:37–46. doi:10.1007/s10934-012-9572-3.

[6] Łach M, Korniejenko K, Mikuła J. Thermal Insulation and Thermally Resistant Materials Made of Geopolymer Foams. Procedia Eng 2016;151:410–6. doi:10.1016/j.proeng.2016.07.350.

[7] Haque S. Behaviour of Fly Ash Geopolymer Containing Fillers and Fibres at Elevated

Temperatures and its Application as Fire Resistant Coating for Timber. 2017.

[8] Jaya NA, Yun-Ming L, Abdullah MMAB, Cheng-Yong H, Hussin K. Effect of Sodium Hydroxide Molarity on Physical, Mechanical and Thermal Conductivity of Metakaolin Geopolymers. IOP Conf Ser Mater Sci Eng 2018;343. doi:10.1088/1757-899X/343/1/012015.

[9] Astm. Standard Specification for Coal Fly Ash and Raw or Calcined Natural Pozzolan for Use. Annu B ASTM Stand 2010:3–6. doi:10.1520/C0618.

[10] ASTM. Mechanical Mixing of Hydraulic Cement Pastes and Mortars of Plastic Consistency Astm C 305. 2016. doi:10.1520/C0305-14.2.

[11] ASTM c177. ASTM C177-Standard Test Method for Steady-State Heat Flux Measurements and Thermal Transmission Properties by Means of the Guarded-Hot-Plate. 2001.

[12] ASTM standards. D792 – 13: Standard Test Methods for Density and Specific Gravity (Relative Density) of Plastics by Displacement. vol. 15. 2013. doi:10.1520/D0792-13.2.

[13] Astm. Standard Test Method for Compressive Strength of Hydraulic Cement Mortars (Using 2-in . or [50-mm] Cube Specimens) 1. 2010. doi:10.1520/C0109.

[14] Prud'homme E, Michaud P, Joussein E, Peyratout C, Smith A, Arrii-Clacens S, et al. Silica fume as porogent agent in geo-materials at low temperature. J Eur Ceram Soc 2010;30:1641–8. doi:10.1016/j.jeurceramsoc.2010.01.014.

[15] Dutta D, Thokchom S, Ghosh P, Ghosh S. Effect of silica fume additions on porosity of fly ash geopolymers. J Eng Appl Sci 2010;5:74–9.

[16] Al Bakri Abdullah MM, Kamarudin H, Abdulkareem OAKA, Ghazali CMR, Rafiza AR, Norazian MN. Optimization of Alkaline Activator/Fly ASH Ratio on the Compressive Strength of Manufacturing Fly ASH-BASED Geopolymer. Appl Mech Mater 2011;110–116:734–9. doi:10.4028/www.scientific.net/amm.110-116.734.

[17] ASTM Standard C567. Standard Test Method for Determining Density of Structural Lightweight Concrete. ASTM Int 2004:16–9. doi:10.1520/C0567.

[18] Thokchom S, Dutta D, Ghosh S. Effect of Incorporating Silica Fume in Fly Ash. Concrete 2011;5:243–7.

[19] SUN Q, ZHU H, LI H, ZHU H, GAO M. Application Of Response Surface Methodology In The Optimization Of Fly Ash Geopolymer Concrete. ISA Trans 2018;48:45–52. doi:10.1016/0019-0578(94)90017-5.

IOP Conf. Series: Materials Science and Engineering **742** (2020) 012033 doi:10.1088/1757-899X/742/1/012033

Production Nanocellulose from Raw Materials For Oil Palm Empty Bunches (TKKS) with Hydrolysis and Freeze Drying Methods

Emmanuela Maria W[1], Endang Kusumawati[1], Anggi Regiana[1], and Dian Ratna Suminar[1]

[1]Department of Chemical Engineering, Politeknik Negeri Bandung, Indonesia

E-mail: dian.ratna@polban.ac.id

Abstract. Accumulation Oil Palm Empty Bunches (TKKS) cause environmental problems if not solved. TKKS has a high cellulose composition, so it can be used as nanocellulose which has a higher value. The process of making nanocellulose from TKKS through the isolation stage of α-cellulose with a delignification process using Sodium Hydroxide (NaOH) 17.5% (b / v) at 80°C, bleaching using Hydrogen Peroxide (H_2O_2) 10% (v / v), hydrolysis using sulfuric acid and drying using freeze drying at a heating temperature of 53°C for 7 hours. Cellulose yield produced from the isolation process of α-cellulose from TKKS was 20.8%. The results of drying process will reduce the aggregation of particles in the nanocellulose produced, so that nanocellulose is produced at 160-298 nm. **Keywords:** nanocellulose, hydrolysis, freeze drying.

1. Introduction

Oil palm is one of the plants that thrives in the territory of Indonesia. The main products of oil palm is kernels (palm kernel), side products of oil palm processing is oil palm empty fruit bunches (TKKS) in the range of 20-23% [22]. Components of the TKKS shown in Table 1.

Table 1. Composition of TKKS (Source : [a] [16] and [b] [15])

Composition	Content (%)
Lignin[a]	21,27 - 36.68
Cellulose[a]	35,66 – 57,75
Hemicellulose[a]	6.61 – 15,96
Water content[b]	8.56
Tannin, alkaloids, etc[b]	4,19

This large amount of TKKS waste will cause problems for the environment. High cellulose content in TKKS has the potential to be used as nanocellulose. Cellulose is the main component in lignocellulosic biomass which consists of linear homopolysaccharide from the D-anhydroglucosa unit with β-1,4-glucoside bonds.

Nanocellulose has a higher value than TKKS. The purpose of this research was to make nanocellulose from TKKS and determine the yield and diameter of nanocellulose that produced by the hydrolysis method and freeze drying.

The anhydroglucose unit has a hydroxyl group which forms strong hydrogen bonds with intamolecular glucose units and with intermolecular hydrogen [10]. X-ray analysis proves that cellulose is a long parallel chain that is bound together by hydrogen bonds. Cellulose in the form of long fibers as seen in Figure 1 [3].

Figure 1. Structure Of Cellulose (Source : [14])

Nanocellulose is a new type of cellulose material that has increased crystallinity, aspect ratio, surface area, and increased dispersion ability and biodegradation. With this capability, nanocellulose can be used with good quality for polymer reinforcement fillers, additives for biodegradebale products, membrane reinforcement, dispersion thickener and drug and implant carrier media [5]. Composite cellulose nanofiber is one application of nanocellulose, by adding polymers to strengthen the physical properties of the polymer [11].

Production nanocellulose by isolation process of α-cellulose was carried out from TKKS and acid hydrolysis. The isolation process of α-cellulose is carried out by the alkalization process using NaOH solution to remove the content of hemicellulose and lignin contained in the TKKS which has an alkali soluble nature (delignification process) and followed by H_2O_2 solution to produce purer α-cellulose [20]. The NaOH solution used has a concentration of 17.5% which aims to increase the purity of α-cellulose, which is based on the degree of polymerization and solubility in the compound of sodium hydroxide (NaOH) 17.5%. Cellulose is divided into three types [17]:

- α-cellulose is a long-chain cellulose and does not dissolve in 17.5% NaOH solution or strong base with 600-1500 polymerization degrees.
- β-cellulose is a part of cellulose which dissolves in 17.5% NaOH solution or strong base with a polymerization degree of 15-90.
- γ-Cellulose has a degree of polymerization of less than 15 which easily dissolves in 17.5% NaOH solution and no precipitate will form after the solution is neutralized.

Bleaching using hydrogen peroxide (H_2O_2) for bleach the fiber and eliminate the content of lignin, hemicellulose, and impurity [18]. The use of H_2O_2 was chosen because it was efficient, low cost, and environmentally friendly [12]. The concentration of H_2O_2 used is 10% (b / v) with optimum conditions at temperature 70-80°C for 1 hour which can produce a constant brightness of cellulose [9].

Production nanocellulose is carried out by various methods such as acid hydrolysis, ultrasonic techniques, and enzymatic hydrolysis [23]. The method of acid hydrolysis often used because it produces small size nanocellulose. Strong acids that used such as H_2SO_4, has purpose to break down the glycoside bonds in cellulose [2]. The acid hydrolysis process to remove the amorphous part of the cellulose chain so that cellulose crystal isolation can be carried out [6]. The scheme of acid hydrolysis on cellulose shown in Figure 2.

Figure 2. Acid Hydrolysis can Eliminate Amorphous Parts from Cellulose (Source : [6])

According to Dong et al in Börjesson (2015), hydrolysis process using 64% -weight H_2SO_4 with optimal conditions at temperature 45°C with a ratio of 8.75 ml / gr due to stable nanocellulose obtained [1]. The result of nanocellulose is dried using a freeze drying method through two stages of the process, formation of ice crystals (freeze) and sublimation ice molecules (drying). This sublimation process occurs where solid water is converted into steam without passing through the liquid phase which is carried out under triple water properties in the range 20°C-50°C, to reduce the aggregation of particles in the nanocellulose [21].

Morphology of nanocellulose can be analyzed by SEM (Scanning Electron Microscope). SEM is one type of electron microscope that uses electron beams to describe the surface shape of the material being analyzed. The function of SEM is focus the fine beam of electrons into the sample. This test has a function to determine the morphology, particle size, pore and particle shape material [19].

2. Experimental Details

2.1 Research Procedures

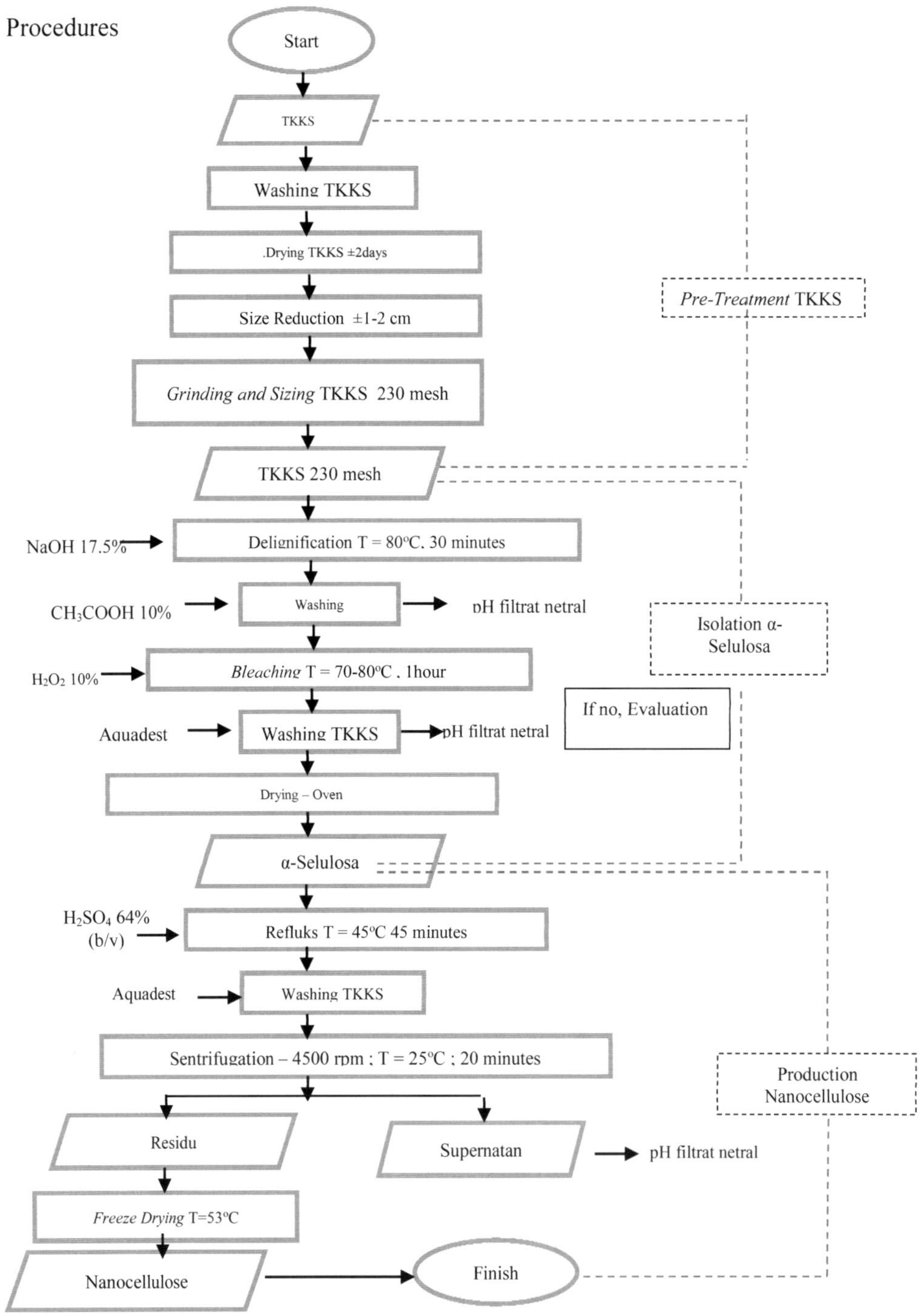

IOP Conf. Series: Materials Science and Engineering **742** (2020) 012033 doi:10.1088/1757-899X/742/1/012033

Figure 3. Production procedure of nanocellulose

2.2 Chemical and Materials
The material that used in this research is Oil Palm Empty Bunches (TKKS) from the Bogor Palm Oil Plantation. Chemicals that used in this research is NaOH 17.5% (b/v), H_2O_2 10% (v/v), CH_3COOH 10% (v/v), aquadest and H_2SO_4 64%.

2.3 Tools
The tools that used in this research is four neck squash, thermometer 0-250°C, stirring motor, stirrer, water bath, hotplate, condenser, water circulator, freeze drying.

3. Result and Discussion

3.1 Pre-Treatment TKKS

In the preparation stage, TKKS are washed using aquadest to remove impurities that still attached at TKKS until clean and dry use sunlight for ± 2 days. Dry TKKS is downsized used grinding and sizing up to 230 mesh in size. TKKS at the pre-treatment stage shown in figure 4.

A B C

Figure 4. (A) Dry TKKS , (B) TKKS Grinding Results, and (C) TKKS Sizing Result 230 mesh

3.2 Isolation of α-Selulosa from TKKS
Isolation of α-cellulose from TKKS is done by alkalizing treatment using NaOH [8] that purpose to eliminate content of hemicellulose, waxes, pectin and some lignin monomers ([7] and [13]). NaOH 17.5% is used to increase the purity of α-cellulose. α-cellulose has an insoluble nature in 17.5% NaOH [17]. In this research, 17.5% (b / v) NaOH solution was used temperature 70-80 °C for 30 minutes. As a result of this process, there is a layer of blackish brown and residue. This blackish brown layer is lignin and hemicellulose which dissolves in NaOH. This blackish brown coating is also produced from previous research, according to Sudiyana, et al (2016) which states that the layer contains lignin, hemicellulose and residual NaOH from the process [28], that's shown in Figure 5.

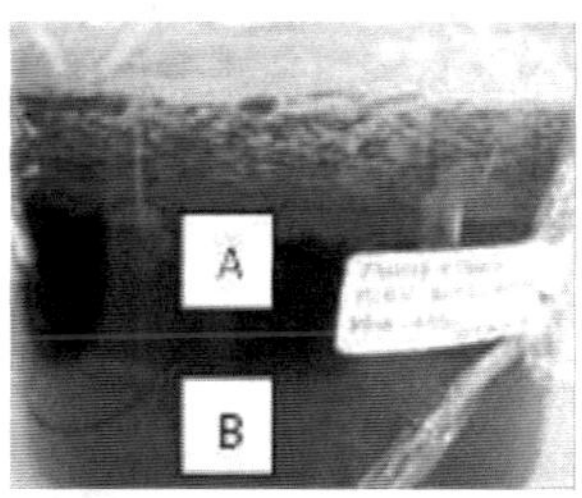

Figure 5. Delignification Result: (A) Lignin and (B) Residu

Bleaching process used hydrogen peroxide (H_2O_2) solution to bleach the fiber by removing content of lignin, hemicellulose, and impurity [18]. Theoretically, perhydroxyl (-OOH) ions are produced from decomposition of hydrogen peroxide into alkaline solutions which can absorb chromophoreic groups from lignin and cellulose (carbonyl and conjugated carbonyl groups and quinones) [18]. Bleaching process using H_2O_2 at concentration 10% has been carried out by Dewanti (2018), which results in a brighter cellulose color which shows that lignin has dissolved, resulting in high purity cellulose [27]. The results of the bleaching process in this research using 10% (v/v) H_2O_2 solution produced brighter colors. Figure 6 shows the results of the bleaching process carried out for three times.

a b c

Figure 6. Bleaching Result one time (A), twice (B), and three times (C)

Bleaching process was carried out three times, because the cellulose results were still dark. Figure 6c shows that the color changes produced after bleaching three times produce brighter cellulose. The isolation process of α-cellulose from this TKKS produced cellulose with a yield of 20.8%. The result has a smaller value compared to the research conducted by Dewanti (2018) of 34%. The use of 12% NaOH carried out by Dewanti (2018) allows the content of β-cellulose and γ-cellulose to be obtained so that the high purity of α-cellulose has not been obtained [27]. Based on the results of SEM analysis, α-cellulose produced was 31.5 μm with α-morphology of cellulose in the form of fibers on its surface. The results of SEM, α-cellulose are shown in Figure 7.

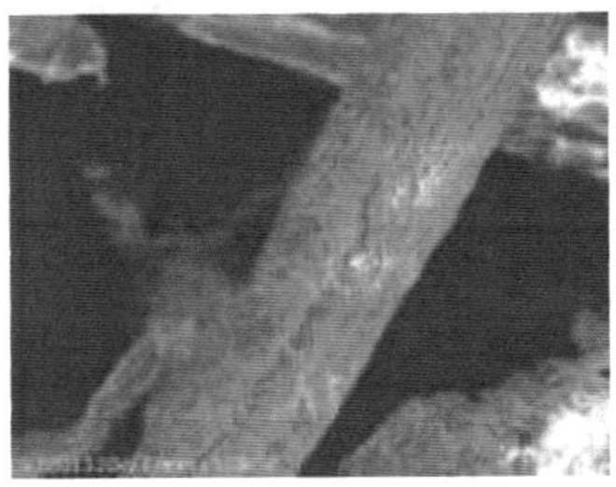

Figure 7. Results of SEM α-cellulose

3.3 Production nanocellulose using Acid Hydrolysis and Freeze Drying Method

Cellulose was hydrolyzed using H_2SO_4 64% for 45 minutes at 45°C by reflux. Hydrolysis of sulfuric acid in cellulose is a heterogeneous process, in which acid diffuses into cellulose and cuts the glycosidic bonds in cellulose polymers. Based on reaction time, hydrolysis can occur in the crystalline region and some hydroxyl groups on the surface of the crystalline area will turn into sulfate groups (-OH groups on cellulose to cellulose $-OSO_3$) [4]. The results of this hydrolysis process have a pH of 2, so that the neutralizing process is needed on nanocellulose formed using distilled water. The results of acid hydrolysis are shown in Figure 8.

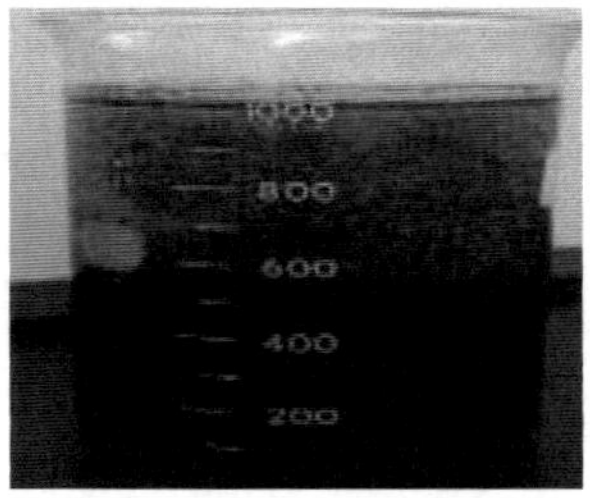

Figure 8. Results of the Hydrolysis Process using H_2SO_4 64%

Nanocellulose was produced by freeze drying method, which was first put into freeze for 24 hours and continued with the freeze drying process at a heating temperature of 53°C for 7 hours. Freeze drying process has purpose to eliminate the water content of nanocellulose which has been frozen with the principle of sublimation [24]. The results of drying process will reduce the aggregation of particles at nanocellulose [21]. The results of the SEM analysis of nanocellulose by hydrolysis and freeze drying method showed that the nanocellulose morphology was still agglomerated, but the diameter produced was small at 160 - 298 nm. According to Othman et al in Lani et al (2014), agglomeration generally occurs due to the presence of Van der Waals forces that interact between nanoparticles [26]. The results of SEM nanocellulose analysis are shown in Figure 9.

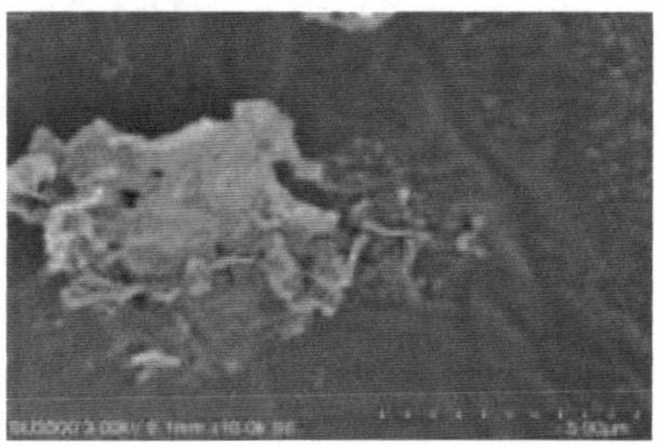

Figure 9. Results of the SEM analysis of nanocellulose

The effect of the freeze drying process on the resulting diameter is quite significant, when compared with the research conducted by Julianto et al (2017) the resulting diamater is 291.4 nm - 8242 μm [25]. This study without the freeze drying process, the freeze drying process is needed to reduce the aggregation of particles in the nanocellulose [21].

4. Conclusions

Production nanocellulose from TKKS by the hydrolysis and freeze drying method has been successfully made with the following yield characteristics: cellulose yield produced from the isolation process of α-cellulose from TKKS was 20.8%, with the size / diameter of nanocellulose produced at 160-298 nm.

References

[1] Borjesson et al. 2015. *Crystalline Nanocellulose-Preparation, Modification, and Properties.* INTECH
[2] Brinchi et al. 2013. Carbohydrate Polymer. 94 154-169
[3] Sumardjo, D. 2009. Pengantar Kimia. Cetakan I. Jakarta: Penerbit Buku Kedokteran EGC
[4] Hamad et al. 2012. *Stucture-Process-Yield Interrelations in Nanocrystalline Cellulose Extraction*. The Canadian Journal of Chemical Engineering. DOI:10.1002/cjce.20298
[5] Ioelovich, M. 2012. *Optimal Conditions for Isolation of Nanocrystalline Cellulose Particles.* Nanocrystals and Nanotechnology. 2012, 2(2), 9-13

[6] IsdinO. 2013. *Nanoscience in Nature: Cellulose Nanocrystals*. Surg, 3(2)

[7] Jayabal et al. 2012. *Effect Soaking Time and Concentration NaOH solution on Mechanical Properties of Coir-Polyester Composite. Bulletin of Material Science*

[8] Kumar et al. 2009. *Method for Pretreatment of lignocellulosic biomass for efficient Hydrolysis and Biofuel Production.* Industrial and Engineering Chemistry Research.

[9] Li et al. 2011. *Hydrogen Peroxide Bleaching of Hardwood Kraft Pulp with Absorbed Birch Xylan and its Effect on Paper Properties.* Bioresources, 721-736

[10] Moon, et al. 2011. *Cellulose Nanomaterials Review : Structure, Properties, and Nanocomposites*, Chem. Soc. Rev. 40 3941-3994.

[11] Muhaimin, Muhammad., dkk. 2014. Fabrikasi Nanofiber Komposit Nanoselulosa/PVA dengan Metode *Electrospinning.* Universitas Gajah Mada: Yogyakarta.

[12] Nazir et al. 2013. *Eco-Friendly Extraction and Characterization of Cellulose from Oil Palm Empty Fruit Bunch.* AIP Conf. Proc 1687, 050016, 1-5.

[13] Ouajai et al. 2005. *Composition, Structure and Thermal Degradation of Hemp Cellulose after Chemical Treatment. Polymer Degradation and Stability.*

[14] Pushpamalar, et al. 2006. *Optimization of Reaction Condition for Preparing Carboxymethyl Cellulose from Sago Waste. Carbohydrate Polymers.* 64:312-318.

[15] Ropiah, Dede. 2010. Pemanfaatan Hidrosilat Tandan kosong Kelapa Sawit (TKKS) untuk Produksi Etanol Dengan Pichia Stipitis. (S). Program Studi Kimia. Fakultas Sains dan Teknologi. Universitas Islam Negeri Syarif Hidayatullah: Jakarta.

[16] Sudiyanti,Yan. 2013. Pemanfaatan Limbah Biomassa Industri Kelapa Sawit untuk Produksi Bioetanol Generasi 2 dan *Co-Products*. Lembaga Ilmu Penelitian Indonesia: Jakarta.

[17] Sumada, dkk. 2011. Kajian Proses Isolasi Alpha-Selulosa dari Limbah Batang Tanaman Manihot *Esculenta Crantz* yang Efisien. Jurnal Teknik Kimia Vol. 5 No.2.

[18] Rayung et al. 2014. *The Effect Of Fiber Bleaching Treatment on The Properties of Poly(Lactic Acid)/ Oil Palm Empty Fruit Bunch Fiber Composites. Int. J. Mol. Sci.*, vol. 15, no. 8, pp. 14728-14742

[19] Triapriani, Yepi. 2016. Pembuatan Nanoselulosa dari Tandan Kosong Kelapa Sawit dengan Metode Hidolisis Asam. FMIPA Universitas Lampung:Lampung.

[20] Tristianti et al. 2017. *Isolation and Characterization of α-cellulose from Blank Bunches of Palm Oil and Dry Jackfruit Leaves with Alkaline Process NaOH. Continued with Bleaching Process* H_2O_2. AIP Conference Proceding.

[21] Voronova et al. 2012. *The Effect of Drying Technique of Nanocellulose Dispersions on Properties of Dried Materials.* Mater Lett 68: 164-167.

[22] Wardani,Aditya dkk. 2014. Pemanfaatan Tandan Kosong Kelapa Sawit Sebagai Material Tekstil dengan Pewarna Alam untuk Produk Kriya. Jurnal Tingkat Sarjana Bidang Seni Rupa dan Desain, No. 1:1-10.

[23] Wulandari et al. 2016. *Nanocellulose Prepared by Acid Hydrolysis of Isolated Cellulose from Sugarcane Bagass*e. IOP Conference Series: Material Science and Engineering.

[24] Zhong et al. 2015. *Drying Cellulose-Based Materials Containing Copper Nanoparticles.* West Virginia University.

[25] Julianto et al. 2017. Ekstraksi Nanoselulosa dengan Metode Hidrolisis Asam sebagai Penguat Komposit Absorpsi Suara. Jurnal Teknik ITS: ITS.

[26] Lani et al . 2014. *Isolation, Characterization, and Application of Nanocellulose from Oil Palm Empty Fruit Bunch Fiber as Nanocomposites.* Journal of Nanomaterial : Hindawi.

[27] Dewanti, Dian Purwitasari.2018. Potensi Selulosa dari Limbah Tandan Kosong Kelapa Sawit untuk Bahan Baku Bioplastik Ramah Lingkungan. Tangerang: Pusat Teknologi Lingkungan

[28] Sudiyana, dkk. 2016. Optimasi Proses Perlakuan Awal NaOH Tandan Kosong Kelapa Sawit untuk Menjadi Bioetanol. Lembaga Ilmu Penelitian Indonesia: Tangerang.

IOP Conf. Series: Materials Science and Engineering **742** (2020) 012034 doi:10.1088/1757-899X/742/1/012034

Preparation of Multi-Walled Carbon Nanotubes Supported Zinc Oxide Catalyst for Transesterification of Kesambi (*Schleichera oleosa*) Oil

N P Asri[1], Y Yuniati[2], H Hindarso[3], Suprapto[4] and R R Yogaswara[5]

[1]Chemical Engineering Department, Faculty of Engineering, Universitas W. R. Supratman, Indonesia.
[2]Department of Chemistry, Faculty of Natural Science, Universitas Ma Chung, Indonesia.
[3]Chemical Engineering Department, Faculty of Engineering, Universitas Katolik Widya Mandala, Indonesia.
[4]Chemical Engineering Department, Faculty of Industrial Technology, Institut Teknologi Sepuluh Nopember, Indonesia.
[5]Chemical Engineering Department, Faculty of Engineering, Universitas Pembangunan Nasional "Veteran" Jawa Timur, Indonesia.

E-mail: nyoman_puspaasri@yahoo.com

Abstract. In this work, multi-walled carbon nanotubes supported zinc oxide (ZnO/MWCNTs) as a heterogeneous catalyst for transesterification had been successfully prepared via the modified Stober-like synthesis method. The effect of calcination temperature during the catalyst preparation towards reaction was investigated. These catalyst was also characterized by X-ray diffraction (XRD) analysis, N_2 adsorption-desorption analysis followed by Brunauer-Emmett-Teller (BET) calculation to determine the total surface area, and scanning electrone microscopy with energy dispersive x-ray spectroscopy (SEM-EDX). Transesterification of kesambi (*Schleichera oleosa*) oil also had been done to test the activity of the prepared catalysts. However, the quantity of dispersed zinc oxide (ZnO) on multi-walled carbon nanotubes (MWCNTs) are quite small that is lower than 5 %wt. It produces low conversion of transesterification after 3 hours reaction that is under 15% of conversion.

1. Introduction

One of the problems of biodiesel development in Indonesia is the higher production cost than that of petroleum diesel. This problem can be happened due to its production process still utilize homogeneous acid or base catalyst [1]. That conventional process has a lot of problem that is side product and alkaline waste formation in huge amount, and also the complicated separation of catalyst from the product [2]. The other problem on biodiesel production in Indonesia is palm oil utilization as a resource. Palm oil is quite expensive and usually used as food sources called an edible oil. Utilizing palm oil as resources of biodiesel synthesis leads higher production cost and competes with food industry [3].

Non edible oil (NEO) has great potential to be used as a feedstock for biodiesel synthesis because of its low price. However, due to its high free fatty acid (FFA) content, synthesis of biodiesel via conventional process from NEO needs two steps, there is esterification and transesterification [4, 5]. These two steps give an impact on rising energy needs so that increasing its production cost [1]. Moreover, utilizing homogeneous catalyst requires further purification step to separate the catalyst from its product [6]. Addition of unit processes in biodiesel production plant surely increases plant investment and production cost. Heterogeneous catalyst development will be the right solution to decrease production cost because it has many advantages, e.g. easy separation from the final product, the catalyst reusability, and environmentally friendly materials [7].

In the previous study, the authors developed several heterogeneous base catalyst, e.g., $CaO/\gamma\text{-}Al_2O_3$ and double promoted $CaO/KI/\gamma\text{-}Al_2O_3$ for transesterification of palm oil [3, 7]. The result of this research shows that both of these catalysts have a high activity towards the transesterification reaction especially $CaO/\gamma\text{-}Al_2O_3$ giving 95 – 97 % yield of biodiesel [3]. However, these catalysts are still sensitive to low grade oil (LGO) which has high FFA content like non edible oil. In the other study, $CaO/KI/\gamma\text{-}Al_2O_3$ was tested to catalyse transesterification of waste cooking oil resulting lower yield of biodiesel at the range of 80 – 83% [5]. The heterogeneous acid catalyst could overcome this problem because it is relatively less sensitive towards FFA than heterogeneous base catalyst [8].

Many researchers had developed solid acid catalysts that can catalyse esterification and transesterification simultaneously for biodiesel synthesis from LGO [4]. This catalyst subsequently named as super acid catalyst because of its strong acid properties [9]. There are many mixed-oxide metal group that had been investigated as super acid catalyst, e.g. titanium oxide (TiO_2) [10], zirconium oxide (ZrO_2) [11], modified zirconium oxide with tungsten oxide (WO_3) [12], and zinc oxide (ZnO) [9]. Moreover, the authors also developed super acid catalyst for transesterification of kesambi (*Schleichera oleosa*) oil namely $ZnO/\gamma\text{-}Al_2O_3$ (ZA) and $ZnO\text{-}CuO/\gamma\text{-}Al_2O_3$ (ZCSA). The result of this study shows that ZCSA catalyst gives low activity in this reaction producing low yield of fatty acid methyl ester (FAME), only 35% at 65 ^oC of transesterification [1]. Otherwise, ZA catalyst gives better result yielding 92.2% of biodiesel at the same condition [1].

The main weakness of this acid catalyst is its active side could be reduced easily by hydration of hydroxyl acid (OH) group which is combination of Bronsted acid sites with the presence of H_2O [9,13]. Carbon based acid catalyst will be a promising heterogenous catalyst for biodiesel production because hydration of OH group can be prevented with carbon sheet hydrophobic characteristics [13,14]. In this study, multi-walled carbon nanotubes (MWCNTs) was used as catalyst support in order to develop carbon based solid catalyst. MWCNTs was chosen as catalyst support due to its high surface area, high purity, and well developed surface morphology and porosity [14]. Furthermore, zinc oxide (ZnO) as acid promoter was doped inside the pore of MWCNTs via Stober-like process followed by precipitation method. This catalyst (ZnO/MWCNTs) subsequently was tested by transesterification of kesambi (*Schleichera oleosa*) oil which is has high FFA content.

2. Material and Methods

2.1. Materials

Multi-walled carbon nanotubes (MWCNTs) was supplied by Advanced Material Esoterica, China. The other analytical grade reagents with no further purification for catalyst preparation including methanol, sodium hydroxide, zinc chloride, ammonia, and ethanol was supplied by Merck, Germany. Kesambi (*Schleichera oleosa*) oil as a low grade oil (LGO) feedstock was provided from the local market. In addition, methanol with commercial grade level as transesterification reactant was purchased from the domestic chemical supplier. Then, all standard reagents for Gas Chromatography (GC) analysis were also provided by Merck, Germany.

2.2. Methods

2.2.1. Preparation of multi-walled carbon nanotubes supported zinc oxide catalyst

Modification of Stober-like process that was proposed by Mukenga, et al. followed by precipitation method was used for the catalyst preparation [15]. This method aims to synthesize ZnO directly inside the pore of carbon nanotubes so its metal-oxide will be an acid active site on the surface of MWCNTs. Zinc methoxide would be used as precursor and was synthesized from zinc chloride ($ZnCl_2$) with sodium hydroxide in alcohol solution. First, 3 grams of sodium hydroxide was mixed in 100 mL of methanol (90% v/v) to form sodium methoxide. Then, this produced sodium methoxide was reacted for 3 hours with 1 M of zinc chloride solution at room temperature. After 3 hours, zinc methoxide solid would be formed and the solution was centrifuged for 5 minutes to separate its solids from the solution. This zinc methoxide solid, then was washed twice by methanol solution and dried at 45 °C.

Two solutions were prepared to synthesize and directly introduce ZnO into the pores of MWCNTs. This alkoxide precursor was dissolved in ethanol solvent as first solution with concentrations according to 15% wt of ZnO loading inside MWCNTs. On the other hand, the second solution was made by dispersing 10 grams of MWCNTs with 59 mL of ammonia as catalyst into 100 mL of water solvent. Then, the first solution was added slowly to the second solution and mixed for 3 hours to form ZnO nanoparticle inside the nanotubes pore. The precipitated solid was respectively separated, washed with alcohol solvent, and dried overnight at 80 °C. After drying process, the solid catalyst was crushed and calcined for 5 hours in muffle furnace at various temperatures. There are several calcination temperatures ranging from 450 – 600 °C used in this research.

The synthesized catalyst was characterized by some analysis to observe its properties. X-Ray diffraction analysis was used for investigation of crystal structure of solid catalyst. The X-ray diffraction analysis of ZnO/MWCNTs catalyst was performed by PANalytical X'Pert Pro MPD. Afterwards, N_2 (nitrogen) adsorption-desorption analysis followed by Brunauer-Emmett-Teller (BET) calculation method was utilized to determine total surface area of catalyst. The morphology of the catalyst surface was captured by scanning electron microscopy with energy dispersive x-ray spectroscopy (SEM-EDX) analysis. The element composition inside the solid catalyst could be analyzed using SEM-EDX so the metal content in the catalyst would be discovered.

2.2.2. Transesterification of kesambi (Schleichera oleosa) oil using ZnO/MWCNTs catalyst

ZnO/MWCNTs catalytic activity was tested into transesterification of kesambi (*Schleichera oleosa*) oil in a batch reactor. Kesambi oil is one of low grade oil that contains gum impurities and has high FFA content. Before esterification and transesterification was conducted simultaneously, kesambi oil must be purified to relieve its impurities called degumming process. Degumming process was carried out by adding phosphate acid (H_3PO_4) solution with amount of 0.1% (v/v of oil) and heating at 70 °C. After this process, the mixture was washed with aquadest to dissolve residual phosphate and left over 48 hours until three layers was formed. Oil was separated from its gum and heated at 110 °C overnight to remove residual water content.

Transesterification of kesambi oil was carried out via batch process using prepared ZnO/MWCNTs solid catalyst. A batch reactor was utilized in this experiment including three necked round bottom flask equipped with reflux condenser. The batch reactor, then was immersed in water bath and this system was heated at 65 °C of methanol boiling point. ZnO/MWCNTs catalyst with amount of 3% (w/w of oil) and 15 %wt loading of ZnO was used in this experiment. Moreover, 1:15 of oil to methanol molar ratio was also utilized as a fixed parameter condition. There are five prepared catalysts from various calcination temperatures was used in this experiment ranging from 450 – 600 °C with 50 °C of increment. These catalysts was coded by ZT4 for 400 °C, ZT4.5 for 450 °C, ZT5 for 500 °C, ZT5.5 for 550 °C, and ZT6 for 600 °C respectively.

The transesterification was conducted under 150 rpm of vigorous stirring for 3 hours at methanol boiling point temperature. After the reaction was finished, the reactor was naturally cooled down to room temperature. Then, the mixture was withdrawn from the reactor and the solid catalyst was separated from the liquid via filtration process. After that, the liquid mixture was left overnight and two layers was formed that is FAME and unreaction oil in the top layer and mixture of glycerol product with excess methanol in the bottom. Finally, the FAME product was separated physically and was dried in oven to relieve the residual methanol.

Fatty acid methyl ester (FAME) content in the product was measured with a gas chromatography flame ionization detector (GC-FID) using a GC HP 5890 equipped with 20 m carbowax column. Nitrogen with a flow of 28 mL min^{-1} was utilized as the carrier gas in this analysis. The operational conditions of analysis consist of the initial temperature and the initial time that was set by 150 °C and 3 minutes respectively. Furthermore, the increment temperature in GC equipment was set by 7,5 °C per minute until 275 °C as a final temperature was reached. The same procedure of GC analysis for biodiesel product had been used in the previous work [1]. The yield of biodiesel and the conversion of transesterification was calculated based on the amount of FAME in the product from GC analysis.

3. Results and Discussion

3.1. Characteristics of ZnO/MWCNTs catalyst

The X-ray diffraction analysis was performed to measure the crystal structure of ZnO/MWCNTs solid catalysts. This analysis was also used in order to detect the synthesized ZnO presence inside the pore of catalyst support. Its detection was shown from diffraction peaks that similar with ZnO diffractogram pattern. The dispersed of ZnO on the carbon nanotubes surface would be an active acid sites and increase the activity of the catalyst [9]. This is 15 %wt loading of ZnO into MWCTNs that applied in this study. From figure 1, it is shown that zinc oxide as an active metal to enhance transesterification kinetic had successfully loaded into MWCNTs. The peak reference of ZnO was given by JCPDS (Joint Comittee of Powder Diffraction Standard) card no. 00-036-1451. This reference has diffraction peak at 2θ angle of 31.4; 33.8; 35.8; 46.8; 55.9; 61.7; 65.5; 66.9; 68.2; 71.5; 75.8; 79.7; 88.04 degree. The same result was shown in previous study with ZnO was dopped into γ-Al_2O_3 as a catalyst support [1]. This diffractogram pattern had indexed to ZnO hexagonal wurtzile structure and seen more crystalline from its high intensity and sharper peak [16].

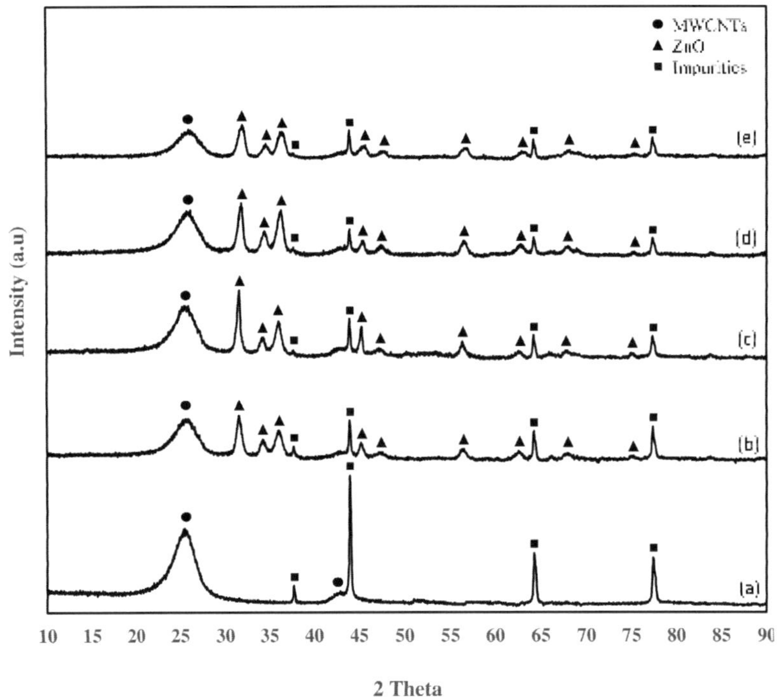

Figure 1. The X-ray diffractogram of solid catalyst comparison, (a) MWCNTs; (b) ZT4.5; (c) ZT5; (d) ZT5.5; (e) ZT6

Meanwhile, diffractogram of MWCNTs shows low and wide peak indicating the amorphous structure and less crystallinity compared to ZnO structure. XRD pattern of MWCNTs shows strong and sharp peak at 27.06° which indicates the graphite form and appears due to the interlayer stacking of graphene sheets [17]. Furthermore, increasing the calcination temperature leads to lower and wider peak of MWCNTs. The high temperature would change the crystal structure of MWCNTs being more amorphous and having less cristallinity. The similar phenonema also appears on the diffractogram of ZT6 that calcined at high temperature (600 °C) resulting wider peaks than ZT5. It can be concluded that the high temperature would transform the catalyst structure more amorphous and reduce its cristallinity.

However, the unknown sharp peaks were seen at 38°; 44°; 64°; 77° meaning some impurities substances was left inside the pores of MWCNTs. These unknown peaks have similar XRD pattern with synthesized aluminum diffractogram pattern given by JCPDS card no. 00-04-0787 [18]. This impurities were also detected by energy dispersive x-ray spectroscopy (EDX) analysis that shown Al_2O_3 impurities inside the carbon nanotubes itself and depicted in table 1. This impurities content is not significant because the MWCNTs was synthesized by chemical vapor deposition (CVD) method. Cabon nanotubes synthesis using CVD is the most efficient method producing high purity of carbon nanotubes and requiring no further purification process [19]. Other impurities like sodium oxide (Na_2O) were left on the surface of MWCNTs after an amount of metal oxide was loaded. This materials might be entrained and formed other compound alongside the desired ZnO particle during catalyst preparation.

Table 1. Solid catalyst composition from EDX analysis

Compound	% mass				
	MWCNTs	ZT4.5	ZT5	ZT5.5	ZT6
C	99,4	89,43	93,45	92,41	93,7
Na_2O	-	2,34	-	2,36	-
Al_2O_3	0,6	0,44	0,49	0,51	0,3
ZnO	-	5,77	4,53	2,43	4,87
Impurities	-	2,02	1,53	2,29	1,13

Table 2 depicts the resulted surface area of catalyst that influenced by calcination temperature. The total surface area was obtained from N_2 adsorption-desorption data followed by Brunauer-Emmet-Teller calculation method. Multi-walled carbon nanotubes will be very promising catalyst support due to its large surface area exceed to 600 m^2/grams. Its total surface area was larger than the other type of catalyst support like γ-Al_2O_3, zeolite, bentonite, and hydrotalcite [2,3,20]. The total surface area of catalyst was decreased with addition of ZnO inside the nanotubes. Otherwise, each of calcination temperature makes different surface area of catalyst with the largest area is obtained when solid catalyst was calcined at 500 ^oC. Generally, increasing calcination temperature until its sintering point will enlarge pore volume and pore diameter due to material decomposition and impurities removal resulting in high surface area [21]. However, the surface area of catalyst decreases gradually at temperature above 500 ^oC signifying the sintering process have occurred.

Table 2. Surface area of ZnO/MWCNTs catalysts with 15% ZnO loading at various calcination temperatures

Catalyst	Surface Area (m^2/gram)
MWCNTs	673,42
ZT4.5	332,111
ZT5	409,043
ZT5.5	303,062
ZT6	271,482

Moreover, the prepared solid catalysts were characterized with scanning electrone microscopy (SEM) analysis. This analysis aims to capture the morphological appearance of the ZnO/MWCNTs catalyst. It is showed that MWCNTs material has randomly nanotubes arrays filling with little amount of Al_2O_3 impurities (figure 2a). This morphological surface was different from other carbon nanotubes synthesizing via chemical vapor deposition (CVD) method. Usually, carbon nanotubes that prepared by CVD has typical well-allignment of nanotubes [19]. Besides that, dispersed white spotting is appeared on the entire surface of the carbon nanotubes which signifies that ZnO particle was successfully filled inside the pores of MWCNTs (figure 2b). It emphasizes the amount of ZnO and other compound that found by EDX analysis shown in table 1. Further phenomena on morphological surface of catalyst was emerged after calcination with high temperature was applied. It is shown that the catalyst became agglomerates and its surface looked broken leaving "crater-like" form broadening its surface area. It was reported that carbon nanotubes would be oxidated while calcined at high temperature above 400 °C releasing a number of its compound and reducing its total mass [21].

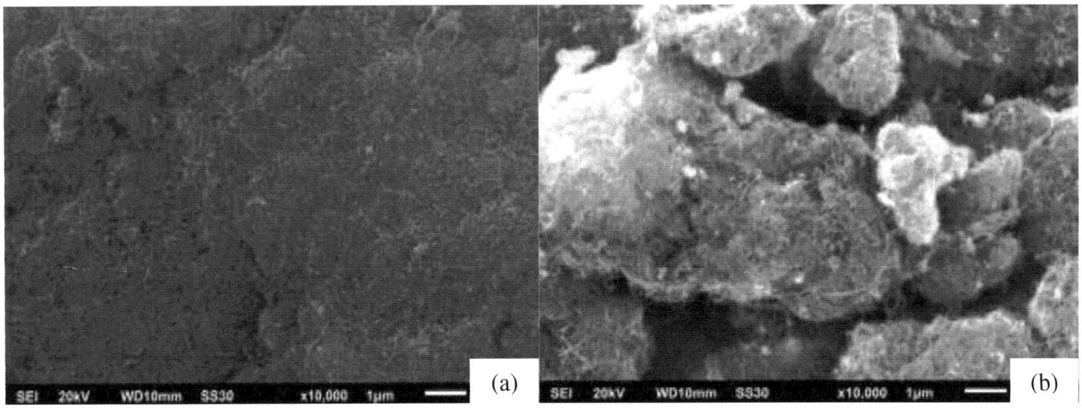

Figure 2. SEM image of the catalyst surface, (a) MWCNTs; (b) ZT5

3.2. Transesterification of kesambi (Schleichera oleosa) oil

The fatty acid methyl ester (FAME) as the product of reaction was measured by Gas Chromatography (GC) analysis. The activity of catalyst is specified by the active site concentration both of base and acid which in this case is zinc oxide (ZnO) loading into MWCNTs. The number of ZnO as an acid active site lead an important role in this transesterification of low grade oil. The effect of calcination temperature on this process was studied during the reaction by varying its temperature from 450 °C until 600 °C. At the same time, the other process conditions were kept constant that is 65 °C of temperature reaction, 3 hours of reaction time, and 1:15 of oil to methanol molar ratio.

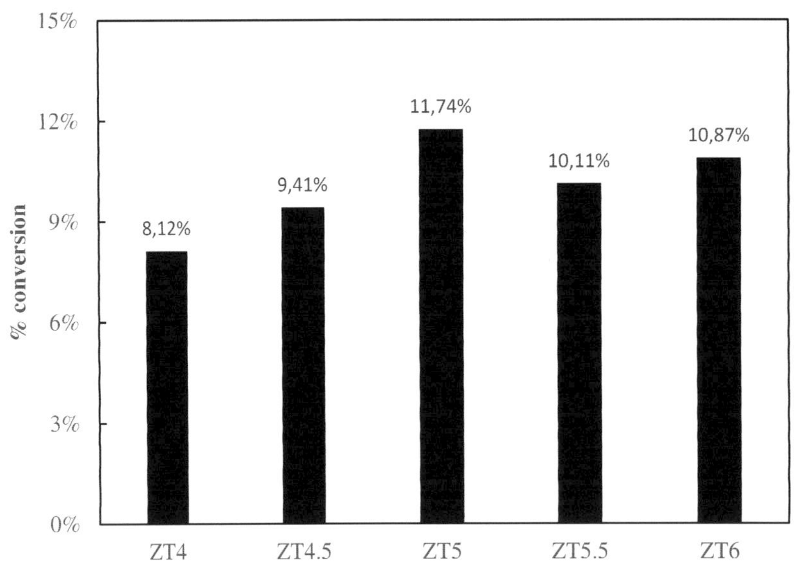

Figure 3. Conversion of transesterification of kesambi oil using ZnO/MWCNTs catalyst

The conversion of transesterification is presented in figure 1 and generally increase while calcination temperature is raised. It is assured that transesterification would not take place without the catalyst presence and FAME might be not formed. The highest conversion of reaction is reached while ZT5 catalyst was utilized in the reaction. Then, increasing calcination temperature exceed to 500 °C properly decrease the conversion of transesterification. But, this conversion of reaction is somehow having low value under 15% (figure 3). Contrarily, the previous study yields conversion of transesterification above 70% during addition of a few amount (1% w/w of oil) ZnO based catalyst. It might be occured because in the previous study ZnO was dispersed over 50 %wt inside the pore of catalyst support that means there is 0,5% (w/w oil) of ZnO in the entire of oil reactant [1]. This huge amount of ZnO active acid site would increase the rate of reaction. It was different with the result of this research that is dispering under 5 %wt of ZnO. With 3% (w/w oil) of solid catalyst in the reaction, it would be only 0,136 %wt of ZnO in the entire of process feedstock.

Theoritically, Zn^{2+} probably has an active Lewis acid sites that gives an effect to trigger both of esterification and transesterification reaction simultaneously [9]. Because of that, ZnO would be appropriate to catalyze these reactions and could process low-grade oil that contain high FFA. But, the result was not satisified yielding low concentration of FAME in the product. This result indicates that preparation of catalyst using modified stober-like process with impregnation and precipitation simultaneously did not give much more dispersed ZnO on the surface of the catalyst support. It was different with the previous study while used modified stober-like synthesis with sol-gel method.

4. Conclusion

The multi-walled carbon nanotubes supported zinc oxide (ZnO/MWCNTs) catalyst for transesterification had been successfully prepared with the modified Stober-like process with precipitation method. Zinc oxide (ZnO) as a metal active was successfully dopped inside the nanotubes of MWCNTs via this modified technique and produced solid catalyst with large total surface area. However, due to only few amount of ZnO was dispersed inside the pores of catalyst, FAME could not be produced with satisfactory results. This result encourages the authors to develop multi-walled carbon nanotubes supported zinc oxide (ZnO/MWCNTs) catalyst using other route of process in order to yield the high conversion of transesterification.

Acknowledgement

The authors would like grateful to Directorate of Research and Community Services, Directorate General of Higher Education, The Ministry of Research Technology and Higher Education, Republic of Indonesia for financial support through The Competence Based Research Grant on fiscal year of 2019 by Contract No. 011/SP2H/LT/MULTI/L7/2019.

References

[1] Asri N P, Soe'eib S, Poedjojono B and Suprapto 2018 *Euro-Mediterr. J. Environ. Integr.* **3** 3
[2] Zabeti M, Daud W M A Wan and Aroua M K 2009 *Fuel Process. Technol.* **90** 770-7
[3] Asri N P, Pujojono B, Puspitasari D A, Suprapto S and Roesyadi A 2015 *Energy Systems and Management (Springer Proceedings in Energy)* (Switzerland : Springer International Publishing) chapter 15 pp 159-169
[4] Lam M K, Lee K T and Mohamed A R 2010 *Biotechnol. Adv.* **28** 500-18
[5] Asri N P, Sari D A P, Poedjojono B and Suprapto 2015 *Mod. Appl. Sci.* **9** 7 99-106
[6] Di Serio M, Tesser R, Pengmei L and Santacesaria E 2008 *Energy & Fuels* **22** 207-17
[7] Asri N P, Machmudah S, Wahyudiono, Suprapto, Budikarjono K, Roesyadi A and Goto M 2013 *Chem. Eng. Process. Process Intensif.* **72** 63-7
[8] de Araujo C D M, de Andrade C C, e Silva E S and Dupas F A 2013 *Renew. and Sust. Energy Rev.* **27** 445-52
[9] Istadi I, Anggoro D D, Buchori L, Rahmawati D A and Intaningrum D 2015 *Procedia Environ. Sci.* **23** 385-93
[10] de Almeida R M, Noda L K, Goncalves N S, Meneghetti S M P and Meneghetti M R 2008 *Appl. Cat.* A: General **347** 100-5
[11] Jitputti J, Kitiyanan B, Rangsunvigit P, Bunyakiat K, Attanatho L and Jenvanitpanjakul P 2006 *Chem. Eng. J.* **116** 61-6
[12] Park Y M, Lee D W, Kim D K, Lee J S and Lee K Y 2008 *Catal. Today* **131** 238-43
[13] Shuit S H and Tan S H 2014 *Bioenerg. Res.* **8** 2 605-17
[14] Shuit S H and Tan S H 2014 *Energy Convers. Manag.* **88** 1283-9

[15] Mukenga M 2012 *Biodiesel production over supported Zinc Oxide nano- particles* (Johannesburg: University of Johannesburg)

[16] Zak A K, Razali R, Majid W H A and Darroudi M 2011 *Int. Journal of Nanomedicine* **6** 1399-1403

[17] Singh K K, Chaudhary S K, Venugopal R and Gaurav A 2017 *Proc. Inst. Mech. Eng. Part N J. Nanomater. Nanoeng. Nanosyst.* **231** 3 141-51

[18] Anaee R A, Ali A H and Hassan A R 2017 Deposition of nicrmo nanowires using anodic aluminum oxide template *Proceeding of 51st IASTEM International Conference* (Berlin Germany, 3rd-4th April 2017)

[19] Andrews R, Jacques D, Qian D and Rantell T 2002 *Acc. Chem. Res.* **35** 1008–17

[20] Gomes J F P, Puna J F B, Goncalves L M and Bordado J C M 2011 *Energy* **36** 6770-78

[21] Bom D, Andrews R, Jacques D, Anthony J, Chen B, Meier M S and Selegue J P 2002 *Nano Lett.* **2** 6 615-19

IOP Conf. Series: Materials Science and Engineering **742** (2020) 012035 doi:10.1088/1757-899X/742/1/012035

Methane Storage Performance in NiO Modified Soybean Straw Based-Activated Carbon

A.R. Nafisah[1], Yuliusman[1*], Nasruddin[2], and Y.W Nugroho.[1]

[1]Chemical Engineering Department, Faculty of Engineering, Universitas Indonesia, Indonesia.
[2]Mechanical Engineering Department, Faculty of Engineering, Universitas Indonesia, Indonesia.

E-mail: usman@che.ui.ac.id

Abstract. Adsorbed natural gas technology can be developed through the development of its adsorbent, especially activated carbon (AC). Biomass-based AC is possible to synthesize, especially by soybean straw which has cellulose content of 44-48 % and hemicellulose content up to 14%. Soybean straw-based AC with surface area 741 m^2/g was used and through modifying process using metal oxide. The NiO-modification was done with a concentration variation of 0.5%, 1%, and 2%. Natural gas adsorption by synthesized AC was executed in the pressure range of 5-10 bars with interval 1 bar and temperature variation of 27°C, 31°C, and 35°C. The best non-modified AC was obtained with $ZnCl_2$ as an activator with S_{BET} of 741 m^2/g and iodine number of 578 mg/g. The best modified AC was obtained with 2%-NiO modification with $ZnCl_2$ as an activator with S_{BET} of 632 m^2/g and iodine number of 535 mg/g. The best adsorption capacity of synthesized AC reached 31.9 mg/g by modified AC on 10 bars and 27°C, and desorption/adsorption efficiency reaches 41.87% on 35°C, showing the effects of pressure and temperature on adsorbed natural gas.

1. Introduction

Utilization of natural gas as the main energy source can be one form of the implementation of the seventh point of Sustainable Development Goals, i.e. the creation of affordable clean energy use. The adsorbed natural gas (ANG) technology is a technology that allows the absorption of natural gas by porous solids which act as adsorbents at relatively low pressures (3.5 MPa). This method is more effective than compressed natural gas (CNG) which requires very high storage pressure, around 3000 psig [1] and liquefied natural gas (LNG) which requires a storage temperature of around -160 ° C and complicated insulation technology [2]. Various factors influence the implementation of ANG, like the characteristics of the adsorbent, the loose capacity, the thermal effect of the adsorption heat, and the storage pressure [3].

Activated carbon has a very large surface area so it is suitable as an adsorbent or chemical reaction [1]. The surface area of commercialized activated carbon ranges from 1,000 to 2,000 m^2/gram. The loading of metal oxides on activated carbon also affects the storage capacity and the release capacity of methane gas. Many studies have used lignocellulose material to produce AC, such as rice husk [4], pam leaf [5], almond shell [6], etc.

Soybean straw (*Glycine max*) waste has potential as an activated carbon raw material. Soybean straw has cellulose content in the range of 44-48%, hemicellulose content in the range of 12-15%, and the low number of ash and lignin [7]. In terms of quantity, according to Kementan RI, the abundance of soybean straw waste is relatively large because soybean production per year in 2015 reached 963,183 tons and growth at 2.35% per year. During this time, this soybean straw waste has not been optimally used and is usually used as animal feed.

Adsorbed natural gas technology can be developed through the development of its adsorbent, especially activated carbon (AC). Biomass-based AC is possible to synthesize, especially by soybean straw. The high cellulose content makes this waste become popular raw material because it can produce good quality AC with a high surface area.

To increase quality, AC can be modified. Modified activated carbon is defined as adding chemical compounds to activated carbon to improve the performance of activated carbon in the adsorption and desorption processes. One of the activated carbon modification methods commonly used is metal oxide insertion. In this study, AC was modified with NiO. This method was expected to improve the adsorption ability of AC and increase methane storage with ANG technology.

2. Materials and Methods

2.1 Activated Carbon Modification

The AC used was produced in the previous study with a surface area of 741 m^2/g. Variation in the amount of loading metal oxide was done by inserting NiO in a portion of activated carbon produced and varied based on the amount of loading with variations of 0.5%, 1%, and 2%. The 10% $Ni(NO_3)_2$ solution was diluted into 0.5%, 1%, and 2% $Ni(NO_3)_2$ concentration solutions. The NiO solution mixed with AC samples and was stirred for 6 hours. This solution then decomposed in the furnace at 360 °C for 1 hour to obtain NiO-AC.

2.2 Activated Carbon Characterization

There are several characteristics of activated carbon that can be done to determine the pore structure, chemical composition, surface area, and absorption capacity of activated carbon made. These methods include iodine number calculation, Scanning Electron Microscopy (SEM), Energy-Dispersive X-ray (EDX), and Brunauer-Emmett-Teller (BET).

2.3 Adsorbed Natural Gas (ANG) Test

AC with the best characteristics was then used as an adsorbent in an adsorbed natural gas (ANG) reactor. The parameter tested was the storage capacity and release of methane gas. The produced AC ability was compared with commercial AC. Storage capacity testing by adsorption mechanism was carried out at various operating pressures in the range of 5 to 10 bar at 1 bar intervals and maintained at isothermal conditions of varying ambient temperatures (27 °C, 31 °C, and 35 °C). Five grams of AC was used for every test.

3. Results and Discussion

3.1 Iodine number test

Analysis of iodine number is a method which states the adsorption capacity of 1 gram of activated carbon to iodine particulates. The relation between the iodine number and specific surface area can be determined by ASTM-4607-94 [8]. Figure 1 shows the iodine number of ACs, $ZnCl_2$ and KOH modified and unmodified, which next called by AC-a and AC-b respectively.

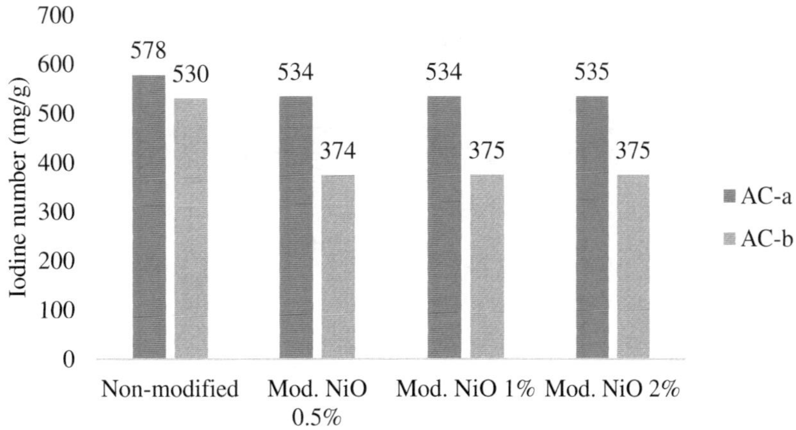

Figure 1. Iodine number calculation

Figure 1 above shows the two highest iodine numbers achieved by AC-a unmodified (578 mg/g) and AC-a NiO 2% modified (535 mg/g). AC-b has a lower iodine number compared to AC-a, defined by the iodine number was 530 mg/g for the unmodified samples. This result indicated that $ZnCl_2$ was more effective as an activating agent to open the pore of soybean straw-based AC. The $ZnCl_2$ induced the decomposition process of lignocellulosic materials and restricted the formation of tar [7].

The number shows that non-modified AC has more free pores than modified AC. This is due to the closure of pores on the surface of AC due to the modified deposit, in this case, NiO. So, it can be assumed that the modified AC has a lower specific surface area than non-modified AC.

3.2 SEM, EDX, and BET Characterization

SEM analysis aims to present an illustration of the surface morphology of the particles. Activated carbon tested was AC-a which unmodified and modified with NiO 2%. The SEM morphological illustrations in this study were arranged at magnifications of 1500 times

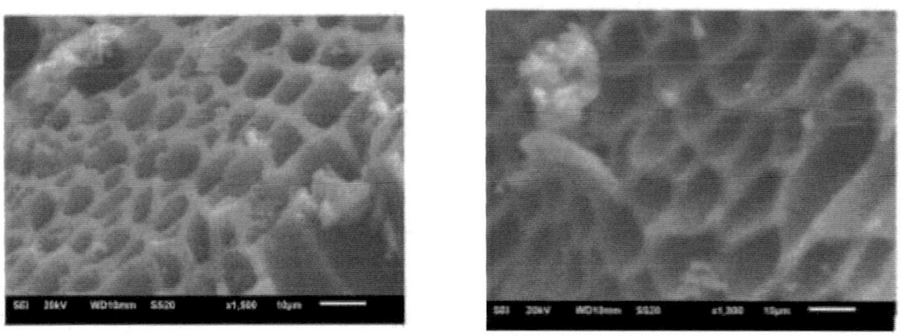

(a) (b)

Figure 2. AC surface morphology, (a) unmodified AC-a; (b) NiO 2% AC-a

The two figures of the morphology of AC analyzed that AC-a had an evenly distributed pore distribution with relatively homogeneous pore sizes. In Figure 2 (b) in the upper left corner, there was a NiO deposit located right on the surface which tended to cover the pore of AC. The NiO seemed not completely mixed with the AC original pore which resulted in the lump closed the surface pore and reduce the active site of AC.

Energy Dispersive X-ray (EDX) is useful for observing the composition of activated carbon elementally. Table 1 presents the composition data of AC-a, both modified and unmodified.

Table 1. EDX characterization results

Components (% wt)	Unmodified AC-a	NiO 2% modified AC-a
C	85,54	86,62
O	10,75	10,2
Zn	3,18	2,03
Cl	0,53	0,97
Ni	-	0,17
Total	100	99,99

The composition of the two types of AC had the main composition of atom C which reaches more than 80% of the AC total mass. $ZnCl_2$ had selectivity to remove atoms in the form of H and O which directed the dehydration of raw materials during carbonization, also prevents the formation of tar and other non-organic carbon substances [7]. The presence of Ni atoms in the modified-AC indicates that the modification process has resulted in a deposit of activated carbon.

BET analysis relies on nitrogen as an adsorbate which adsorbed and desorbed under 77K operating conditions. This analysis will produce pore characterization such as specific surface area (S_{BET}), average pore size, pore-volume, and average particle size. Table 2 states the characterization of the two activated carbons through the BET test.

Table 2. BET characterization results

Parameter	Unmodified AC-a	NiO 2% modified AC-a
S_{BET} (m^2/g)	741	632
Micropore area (m^2/g)	563	468
External surface area (m^2/g)	178	164
Micropore volume (cm^3/g)	0,287	0,237
Pore width (nm)	2,136	2,169

NiO plugging occurred on the mesoporous side of carbon so the total specific surface area possessed by modified AC was lower than unmodified. This phenomenon was supported by other quantitative parameters shown in the table; there was a decrease in the value of the micropore area and the micropore volume.

In order, the unmodified and modified AC pore width values were 2.136 nm and 2.169 nm, categorized as particles with mesoporous. Mesopore has a pore size between 2 and 50 nm. For better performance, AC should have a micropore structure that has a pore size of less than 2 nm [4]. This affected the performance of AC to absorb gas particulates in the pore because the gases had different widths on the molecular scale.

3.3 Natural Gas Adsorption Capacity Test

The whole process was preceded by void volume calculation. The void volume at measuring cell is the free part located at the top of AC and between the sidelines space between activated carbon. Void volume testing used helium gas because it belongs to the noble gas group; has higher stability than other gases. The natural gas used in this study was obtained from SPBU which contained 87.76% of methane (CH_4) gas. The results for the void volume test can be seen in table 3.

Table 3. Void volume test

Activated Carbon Types	Void Volume (cm^3)
Unmodified	291
NiO-Modified	288
Commercial	299

From table 3 above, commercial AC had the highest void volume. This indicated that the number of filling of natural gas will be greater when measuring cell filled with commercial AC.

Figure 3 shows the amount of gas adsorbed at different pressures and temperatures carried out by modified AC-a.

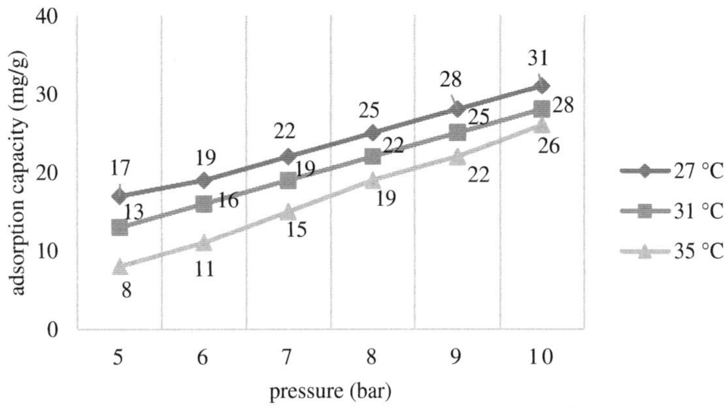

Figure 3. The adsorption capacity of modified AC

The variation of pressure in the isothermal system when natural gas adsorption on AC takes place shows the difference in the amount of natural gas that can be adsorbed. The phenomenon of increased adsorption capacity due to increased pressure, it can also be said that natural gas adsorption is physical adsorption [9] and there is a binding of gas molecules on the internal surface due to Van der Waals forces. Although NiO modified AC is neutral, nickel as a transition metal has electron acceptor properties and acts as Lewis acid so that methane which has Lewis alkaline properties will be attracted by acidic nickel [10].

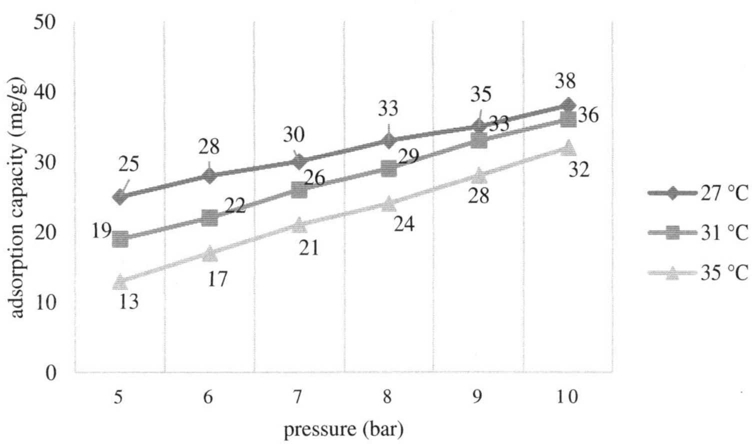

Figure 4. The adsorption capacity of commercial AC-a

The same results also occurred for modified AC. The same pattern as unmodified AC showed that the process runs in the same conditions.

The increase in temperature results in lower natural gas adsorption capacity, heat adsorption can affect the performance of natural gas storage in adsorbents [3] which in this case is caused by the exothermic nature of the adsorption so that an increase in temperature causes a decrease in the natural gas adsorbing capacity on activated carbon [11].

Pressure and temperature affect the adsorption capacity and release of activated carbon; the highest adsorption capacity is obtained at the highest pressure (10 bar) and the lowest temperature at variation (27 °C). The AC with the highest adsorption ability was the modified AC at an operating condition of 27 °C, 10 bars with an adsorption capacity of 31.9 mg/g.

3.4 Desorption Test

Table 5 provides an overview of the percentage efficiency of desorption/adsorption at 27 °C, 31 °C, 35 °C temperature and type of activated carbon, i.e. commercial (CAC), unmodified (UAC), and modified AC (MAC).

Table 4. Efficiency calculation results

T (°C)	Adsorption capacity (10 bar) [mg/g]			Desorption capacity (1 bar) [mg/g]			Efficiency (%)		
	CAC	UAC	MAC	CAC	UAC	MAC	CAC	UAC	MAC
27	38,4	31,9	30,9	18,5	12,6	10,4	48,24	39,62	33,23
31	35,8	29,1	27,9	17,9	11,9	9,6	49,87	40,66	34,10
35	31,8	27,0	26,0	17,0	11,4	9,4	51,77	41,87	36,97

Desorption is an endothermic process. Because of these characteristics, each type of AC at higher temperatures will release gases with a greater concentration than a lower temperature. The highest efficiency is achieved by CAC. CAC has a better surface structure with micropore which resulted in higher pore volume to store the methane gas.

4. Conclusion

Activated carbon made from straw waste has the potential to be further developed. The highest results obtained were non-modified AC with an S_{BET} of 741 m^2/g and with an iodine number of 578 mg/g and 2% NiO modified AC with an S_{BET} of 632 m^2/g with an iodine number of 535 mg/g. Pressure and temperature affect the adsorption capacity. The highest adsorption capacity was obtained at the highest pressure (10 bar) and the lowest temperature variation (27 °C). The highest number was detected in unmodified AC-a with 31.9 mg/g adsorption capacity. The highest efficiency was achieved by unmodified AC-a at 35 ° C with an efficiency of 41, 87%. NiO impregnation should produce a better AC quality with the modified process. Further study is needed to obtain more optimal results.

Acknowledgments

The authors acknowledge all the help and support from the colleagues and staff.

References

[1] Makal TA, Li J-R, Lu, W, and Zhou H-C. 2012 *Chem. Soc. Rev* **41**, 7761-7779.
[2] Eswara AK, Misra SC and Ramesh US 2013 *Annual Conference SNAME. Society of Naval Architects and Marine Engineers*, 201-217.
[3] Alhasan, S., R. Carriveau, and D. S-K. Ting 2016 *Int. J. of Env. Studies* 73(**1**), 1-14
[4] Nam H, Wang S and Jeong HR 2018 *Fuel,* **21** 186-194.
[5] Sadeek SA, Negm NA, Hefni HHH and Wahab MMA 2015 *Int. J. Bio. Macrom.,* **81** 400-409.

[6] Valente NJM, Gomes JA, Suhas, Carrott PJM, Laginhas C and Roman S 2009 *J. Hazard Mater.*,**167** 904–10.

[7] Miao Q, Tang Y, Xu J, Liu X, Xiao L, and Chen Q 2013 *J. Taiwan Ins. Chem. Eng.* **44** 458-466

[8] Yuliusman, Nasruddin, Afdhol MK, Amiliana RA, Hanafi A, and Rachmanda B 2017 *IOP Conf. Series: Mat. Sci. and Eng.,* **180**, 012282

[9] Yuliusman, Nasruddin, Sanal A, Bernama A, Haris F and Hardhi M 2017 Improved of Natural Gas Storage with Adsorbed Natural Gas (ANG) Technology Using Activated Carbon from Plastic Waste Polyethylene Terephthalate, *International Conference on Green and Renewable Energy Resources* **75**, 1-6.

[10] Djeridi W, Mansour NB, Ouerderni A, Llewellyn PL, and Mir LE 2015 *Int. J. Hydrogen Energy*, 1-12.

[11] Hamza UD, Nasri NS, Mohammed J and Majid ZA 2016 *International Conference on Chemical and Biological Sciences, MATEC Web of Conferences*, 1-5.

IOP Conf. Series: Materials Science and Engineering **742** (2020) 012036 doi:10.1088/1757-899X/742/1/012036

Synthesis and Characterization of Coffee Based-Activated Carbon with Different Activation Methods

Yuliusman[1*], A. Bernama[1] and A.R. Nafisah [1]

[1]Chemical Engineering Department, Faculty of Engineering, Universitas Indonesia, Kampus UI Depok, Indonesia, 16424.

E-mail: usman@che.ui.ac.id

Abstract. Activated carbon (AC) is used in many applications because it has many strengths such as high adsorption ability, inexpensive, and not hazardous. The use of agricultural waste in AC production has become a popular topic in recent years. One of the waste materials that can be utilized is coffee wastes. Indonesian have a high rate of coffee consumption which results in a high number of wastes. In general, the AC generation was done by three main steps; carbonization at 400 C, chemical activation using KOH and physical activation using CO_2 with a 100 ml/minute flow rate. The obtained results showed that AC has the highest iodine number of 443.58 mg/g and a surface area of 399.05 m^2/g. The SEM results indicated the pore had formed on the carbon surface. The carbon mass percentage was known at 67.5% from the EDX test. These results showed that coffee waste can be used as an alternative raw material and can be used for several needs.

1. Introduction

Adsorbents such as activated carbon is always an interesting topic to be developed. Activated carbon is an adsorbent with large internal pores and a wide area. But the main challenge is to find adsorbents with high storage performance at low pressures and ambient temperatures that allow for desorption under mild conditions [1]. AC already used in previous research, such as to remove metal from the aqueous environment [2], removal of organic pollutants [3], water purification [4], etc.

Nowadays, biomass waste became a popular raw material due to its renewable properties [5]. Coffee waste is one of the lignocellulosic residues produced in large numbers throughout Indonesia with a total plantation area of 1.3 million hectares (ha). Indonesia can produce at least 748 thousand tons per year or 6.6% of world coffee production in 2017. Therefore, research using coffee grounds to make activated carbon has good prospects because of its lignocellulosic content and large amount.

In this study, activated carbon producing was done with chemical, physical, and the combination of both activation process. Coffee waste as the material was carbonized in a furnace. KOH solution mixed with carbon serves as an activating agent that will oxidize carbon and damage the inner surface of carbon so that it will form pores and increase adsorption power [6]. The optimum condition such as temperature was controlled and monitored. The physical activation was done using CO_2 under high temperatures. This study aims to produce high quality activated carbon with good performance for adsorption or desorption that can be used for further research.

2. Materials and Methods

2.1 Activated Carbon Synthesis

Preparation was done with the dehydration process of coffee solution at 100 °C for 3 hours. Carbonation of coffee grounds was carried out at 400 °C for 2 hours using Furnace Vulcan A-550. There were two steps for activation, chemical and physical. Chemical activation used KOH as an activating agent. After activation using KOH, then the physical activation process was carried out by the thermal activation method, by heating activated carbon at high temperatures, above 600 °C with the help of gas carbon dioxide (CO_2). The activation process used a temperature of 600 °C with a CO_2 gas flow rate of 100 ml/minute with a heating rate of 5.75 °C /minute for 1 hour.

Then the samples were washed with HCl for 1 hour and rinsed with distilled water until the neutral pH. The AC next was dried in the oven for 1 hour with a temperature of 110 °C

2.2 Characterization Process

There were 3 types of characterization used in this study. Iodine number test to calculate the surface area of AC, *Field emission scanning electron microscopy* (FESEM) to analyze the surface structure which integrated with *Energy Dispersive X-Ray Spectroscopy* (EDX) to identify the component of AC, and BET to measure the actual surface area of the AC using Micromeritics ASAP 2020 V.4.02

Iodine adsorption was obtained by measuring the volume of a titration between $Na_2S_2O_3$ and I_2. From this volume, iodine (mg/g) was calculated using equation 1.

$$Adsorbed\ Iod\ \left(\frac{mg}{g}\right) = \frac{\left(V_{titrat} - \frac{v\,x\,n}{0.1}\right)x\ 12.69\ x\ 2}{W} \quad (1)$$

Where:

$V_{titrate}$: I_2 volume which mixed with the AC (5 ml)
v: natrium thiosulphate solution used (ml)
N: sodium thiosulphate solution concentration (N)
W: the weight of AC used (gr)

3. Results and Discussion

3.1 Carbonization of Material

The carbonization process would remove unwanted and volatile components [7]. The final yield from the carbonization process was listed in table 1.

Table 1. Carbonization process results

No	Coffee waste initial mass (gram)	Final mass (gram)	*Yield* (%)
1	100.36	28.54	28%
2	100.21	30.17	30%
3	100.25	30.38	30%
4	100.12	30.05	30%
5	100.45	32.14	32%
6	100.22	31.12	31%
7	100.05	28.18	28%
8	100.24	29.35	29%

9	100.16	30.21	30%
10	100.27	29.97	30%

Volatile contaminants can be defined by the mass loss after the carbonization process (burn-off). The functional groups of raw material were also removed in this process [8]. The value of mass lost can be seen in Table 2.

Table 2. Burn off calculation

No	Removed mass (gram)	*Burn Off* (%)
1	71.82	72%
2	70.04	70%
3	69.87	70%
4	70.07	70%
5	68.31	68%
6	69.1	69%
7	71.87	72%
8	70.89	71%
9	69.95	70%
10	70.3	70%

The average percentage of burn off was 70% which means over 70% of the total component removed from samples.

The carbon activation process was carried out with two steps, chemical followed by physical activation. For this reason, before the activated process, samples were crushed and sieved with 150 mesh size. The adsorption ability of AC was very affected by its size. The smaller size of samples will give a higher contact area in the adsorption process. After being crushed, carbon is activated chemically using a KOH solution. The results of activation using the KOH solution can be seen in table 3.

Table 3. Yield after chemical activation

No	Carbon mass (gr)	KOH concentration (M)	KOH volume (ml)	Carbon mass after activation (gr)	*Yield*
1	10.06	5	100	8.1279	81%
2	10.11	6	100	8.2041	81%
3	10.08	7	100	8.1133	80%
4	10.07	8	100	8.1937	81%

In chemical activation, carbon was mixed using a KOH solution at 65 °C and stirred using a magnetic stirrer for four hours. KOH solution mixed with carbon serves as an activating agent that will oxidize carbon and damage the inner surface of carbon so that it will form pores and increase adsorption capability. Also, activating agents will inhibit the formation of tar and reduce the formation of acetic acid, methanol and so on [8]. The reaction occurred in the activation process are:

$$6\,KOH + C \leftrightarrow 4K + CO_2 + H_2O \quad (2)$$

$$6KOH + C \leftrightarrow 2K + 3H_2 + 2K_2CO_3 \quad (3)$$

$$4KOH + 2CO_2 \leftrightarrow 2K_2CO_3 + 2H_2O \quad (4)$$

The chemical activation was followed by physical activation using CO_2. The main purpose of this process was to increase the pore volume of AC. The gases used for gasification were used to remove volatile substances which were impurities on the AC. The results were shown in table 4.

Table 4. Yield after physical activation

No	Initial mass (gr)	Final mass (gr)	*Yield %*
1	5.006	3.542	71%
2	5.011	3.76	75%
3	5.007	3.742	75%

The temperature chosen for this process was 600 °C as this was the optimal temperature for physical activation. If the temperature is too low then the activation process will not run optimally, while the temperature is too high will cause damage to the carbon structure (sintering), thereby reducing the surface area. The higher the CO_2 flow rate used; the more carbon atoms react to carbon monoxide (CO) gas. As a result of the reaction and carbon erosion, the yield of activated carbon will also decrease compared to the chemical activation process as in Table 3. The reaction between CO_2 gas and carbon (C) is stated in Equation 3.

$$C + CO_2 \rightarrow 2CO \tag{5}$$

Reactions that occur in the physical activation will remove the mass to form active carbon pores but if the reaction is excessive then the pore structure will be damaged [9]. Besides the nature of coffee grounds that are very sensitive to heat causes reduced surface area because it is easily oxidized by CO_2.

After that, AC was washed using a 1M HCl solution to remove K+ ions which were still attached to the carbon. The main goal was to clean impurities and reduce the ash content. Washing with HCl was done for 1 hour by immersing activated carbon in the HCl solution. After that, the AC was washed using distilled water to neutralize the remaining acid solution in the AC.

3.2 Characterization process

3.2.1 Iodine number calculation. The surface area of AC as an adsorbent is an important parameter. AC can be said as a good adsorbent if it has a high surface area. Iod number test is a method for determining surface area by performing sodium thiosulfate titration with an iodine solution that has been mixed with activated carbon [10]. The value of the iod number obtained is the absorption capacity of milligrams of iod per gram unit of activated carbon (mg/g). Iod number testing was carried out on AC which had done carbonization, chemical activation with variations in KOH molarity, and physical-chemical activation.

The results of the iodine number calculation could be seen in figure 1. It can be concluded that the surface area value is inversely proportional to the volume of the titration. The lower the titration volume, the greater the surface area.

The activation process of Physical Chemistry had failed because this process damages the pores of AC. It could be seen that this process damages the parts of the pores that have formed in carbonization and chemical activation. This caused the test result of the physical-chemical activation iod number was lower because of the reduced pores to absorb iodine.

IOP Conf. Series: Materials Science and Engineering **742** (2020) 012036 doi:10.1088/1757-899X/742/1/012036

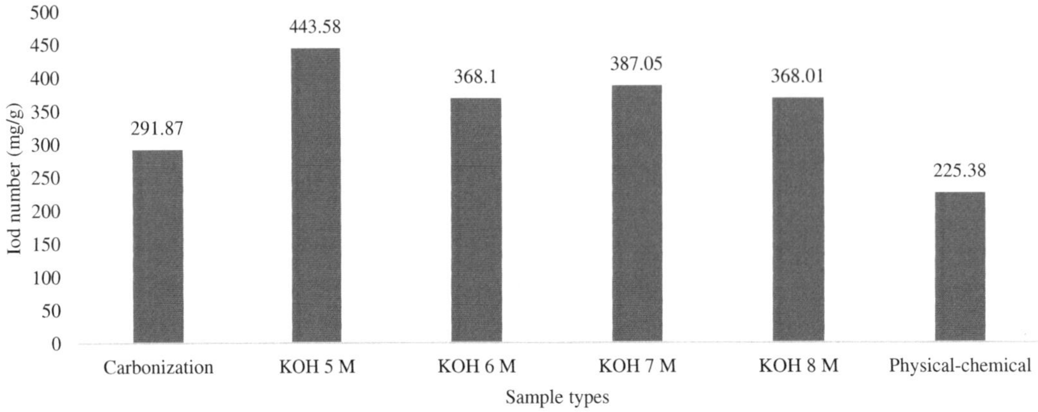

Figure 1. Iodine number calculation results

3.2.2 SEM, EDX and BET characterization. The characterization of AC using SEM aims to see the morphological structure of activated carbon that has been made [5]. Characterization is carried out on only carbonization samples and activated carbon which has undergone chemical activation and physical-chemical activation.

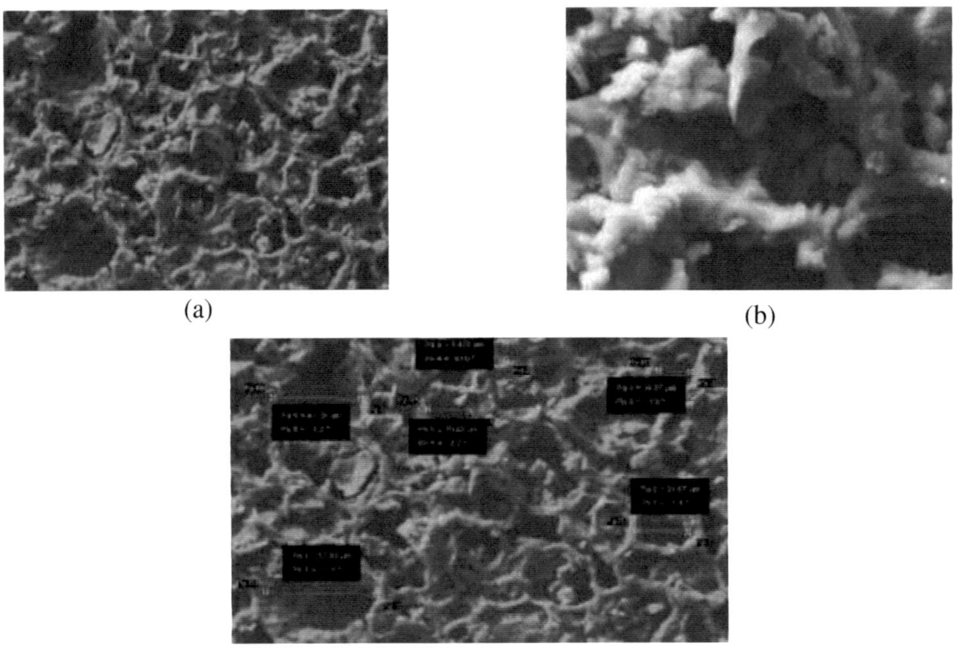

(c)

Figure 2. SEM Characterization of Carbonize samples; (a) 500x magnified, (b) 2000x magnified, (c) Pore size with 500x magnified.

The carbonization process forms the mesoporous part as shown in figure 1. The SEM test also conducted to the carbon which activated with KOH 5 M. From figure 2, it can be seen that the pores were much smaller than the first one. The structure of the pore is also more visible.

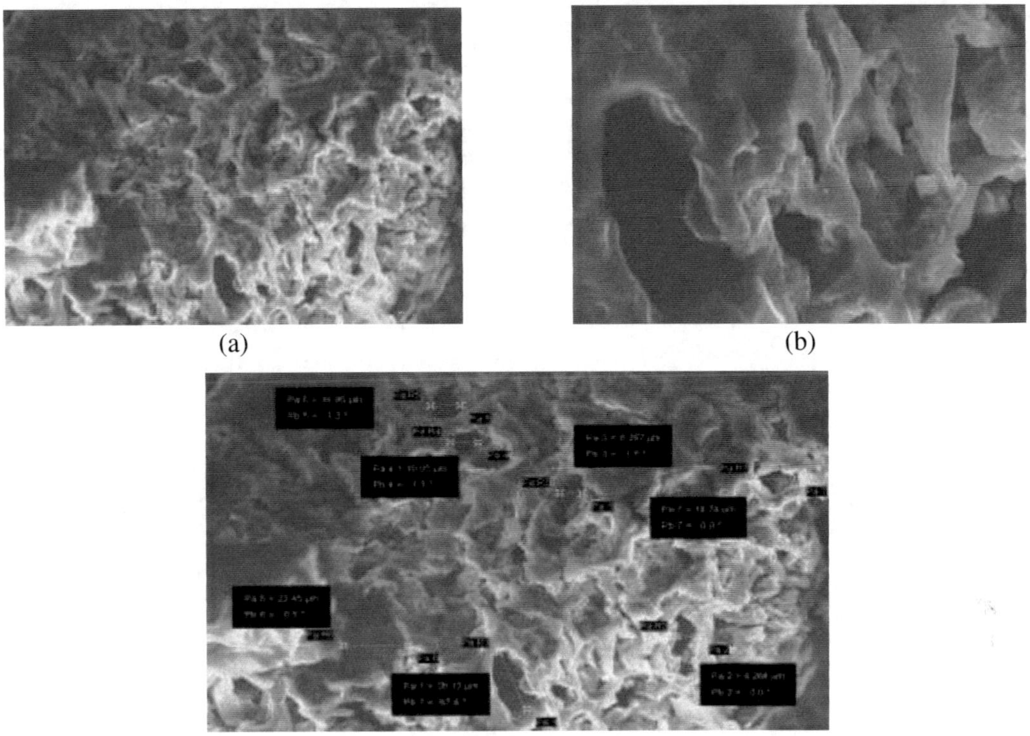

(a) (b) (c)

Figure 3. SEM Characterization of Chemically activated carbon; (a) 500x magnified, (b) 2000x magnified, (c) Pore size with 500x magnified.

From figure 3c, it can be seen that the pore size distribution was different from figure 2c. In figure 3c, the pore size distribution was smaller. This indicated that the chemical activation broke the carbon wall and made some new pores. Although from figure 2a, the surface area of the AC was still uneven.

Energy Dispersive X-ray (EDX) test was also done in this study. This test aims to determine the composition of a material by firing x-rays at the material. The (EDX) test results can be seen in table 5.

Table 5. EDX results

Elements	Mass percentage (%)	Atom percentage (%)
Carbon	67.5	72.88
Nitrogen	7.66	7.09
Oxygen	24.54	19.89
Aluminum	0.21	0.1
Sulfur	0.06	0.02
Chlorine	0.03	0.01

The carbon components in AC was relatively high. Nitrogen, Aluminum, and Sulfur are elements that are already contained in coffee beans. Chlorine was obtained from the washing process of activated carbon using 1M hydrochloric acid (HCl).

BET testing was done once using the sample with the best activation method previously tested with the iod number method, i.e. Chemical Activation of 5 M KOH. BET surface area characterization was carried out using Micromeritics ASAP 2020 V4.02. The surface area results obtained from chemical activation are 399.0507 m^2/g. This number is still low compared to other studies, so further research is needed to obtain optimal condition operation and activation process modification.

4. Conclusion

Coffee wastes had a potential future to replace the current AC raw materials. The AC production with chemical activation using KOH 5 M resulted in the highest surface of 399.05 m^2/g. But the physical activation ruined the pore structure in the carbon surface which decreases the iodine number of AC. The SEM Characterization showed that some mesopore had already created although the pore surface was still disorganized and the EDX test exhibited the high number of carbon present in the AC produced.

Acknowledgments
The authors thank the colleagues and staff for the help and support throughout the research processes.

5. References

[1] Erabee, Iqbal & Ahsan, Amimul & Wahid Zularisam, Abid & Idrus, Syazwani & Norsyahariati Nik Daud, Nik & Thirugnanasambantham, Arunkumar & Sathyamurthy, Ravishankar & Al-Rawajfeh, Aiman 2016 *Engineering Journal,* **21**, 1-14.
[2] Raval NP, Shah PU and Shah NK 2016 *J. Environ. Manag.* **179**, 1-20.
[3] Dai Y, Zhang N, Xing C, Cui Q and Sun Q 2019 *Chemospere* **223**, 12-27.
[4] Singh NB, Nagpal G, Agrawal S and Rachna 2018 *Environt. Tech. & Innov.* **11**, 187-240.
[5] Yuliusman, Nasruddin, M. K. Afdhol, R. A. Amiliana et al 2017 *IOP Conf Series: Mat. Sci. And Eng.* **180**, 012282.
[6] Hadi, R., Shohreh, F. and Javad T 2019 *J. Env. Chem. Eng.* 7(3).
[7] Moreno-Piraján JC, Bastidas-Barranco MJ and Giraldo L 2019 *J. Therm. Anal. Calorim.* 131-259.
[8] Nam H, Wang S and Jeong HR 2018 *Fuel*, **21** 186-194.
[9] Dorez G, Ferry L, Sonnier R, Taguet A and Lopez-Cuesta J-M 2014 *J. Analytic. & Appl. Pyro.* **107**, 323-331.
[10] Yuliusman, Sinto J, Nugroho Y W, and Naf'an H I 2018 *E3S Web of Conferences* 67, 03020.

IOP Conf. Series: Materials Science and Engineering **742** (2020) 012037 doi:10.1088/1757-899X/742/1/012037

Optimization of Ceric Ammonium Nitrate and Ferrous Ammonium Sulfate in the Synthesis of Chitosan-Graft-Maleic Anhydride

Milla Nadia[1], Daniel Timotius[1], Sang K. Wirawan[1], Nur Rofiqoh Eviana Putri[1], and Yuni Kusumastuti[1*]

[1]Department of Chemical Engineering, Faculty of Engineering, Universitas Gadjah Mada, Yogyakarta 55281, Indonesia

E-mail: yuni_kusumastuti@ugm.ac.id

Abstract. Amongst various methods for chitosan modification, grafting process is one of the promising techniques to alter chitosan surface properties. Maleic anhydride (MA) was grafted onto chitosan in the presence of an initiator to produce chitosan-graft-maleic anhydride (CTS-graft-MA). In this study, we compared and optimized two types of initiators: *Ceric Ammonium Nitrate* (CAN) and *Ferrous Ammonium Sulfate* (FAS) with various ratios in the synthesis of chitosan-graft-maleic anhydride. Chitosan was dissolved in dilute acetic acid solution. Subsequently, the initiator was added into solution to initiate the reaction and maleic anhydride was added immediately afterward. The reaction was carried out for 45 minutes at 70 °C with 430 rpm of agitation. The results of Fourier Transformed Infrared (FTIR) spectroscopy showed an appearance of a new peak at 1705.07 cm^{-1} which belongs to C=O bands for both initiators. To find out the optimum results from CAN and FAS, the back-titration method was performed. The optimum ratio for CAN is 2.51% with the acidic content 5.4421 mmol/gr, while FAS is 0.268% with the acidic content 4.4673 mmol/gr. In conclusion, CAN is more effective than FAS to initiate the grafting process on CTS-graft-MA.

1. Introduction

Chitosan is the most abundant natural polysaccharide after cellulose in terms of utilization and distribution [1]. Chitosan can be produced by deacetylation of chitin. It has various characteristics such as biocompatible, biodegradable, non-toxic degradable product, non-immunogenic, and non-carcinogenic [2]. Because of these characteristics, chitosan has been widely utilized in food, biomedical, water treatment, functional membrane, and flocculation process [3],[4]. Utilization of chitosan, however, is not easy since it cannot dissolve in water but is able to dissolve in diluted acid. There are several acids that usually used to dissolve chitosan such as acetic acid, formic acid, lactic acid, and propionic acid. However, in terms of formation of flexible and transparent film, only acetic acid and formic acid are favourable [5]. Therefore, to help modify the properties of chitosan, physical and chemical treatment can be applied. One of the most promising methods is by grafting process.

Chitosan can be grafted using several chemicals, which depends on the final expected properties. Maleic anhydride (MA) is widely used to be grafted into chitosan. Chitosan-graft-maleic anhydride (CTS-graft-MA) can be applied in several applications such as pervaporation [6], heavy metal adsorption [7], and biomaterials [8]. In pervaporation, CTS-graft-MA significantly increases the selectivity of water-ethanol pervaporation than unmodified chitosan. CTS-graft-MA can also be utilized to adsorb copper (II) and lead (II) in wastewater treatment. The hydrophilicity of CTS-graft-MA due to the introduction of an additional carboxyl group into the chitosan backbone is also favorable in biomaterial application

To produce CTS-graft-MA in a more efficient way, the use of initiator is needed. The reaction between the initiator and chitosan will produce free radicals, which makes the amine group of chitosan is more reactive to react with MA. Grafting is started by producing one or more free radicals in the chitosan chain and allows these radicals to react with polymerizable monomers [9]. Recently, a number of initiators have been developed for grafting such as Ferrous Ammonium Sulfate (FAS) [10], Ceric Ammonium Nitrate (CAN) [8], and Ammonium Persulfate (APS) [11].

Amino groups of chitosan can be covalently grafted with MA using CAN as an initiator. There are 3 steps in grafting process using CAN as an initiator: (1) the solvation of water into chitosan, (2) the formation of the complex between solvated chitosan and CAN, lastly (3) grafting initiation by radicals from the complex [12]. As an initiator, FAS can prevent homopolymerization for the preparation of chitosan-graft [10]. In particular, the use of FAS as a reducing agent has shown to be a convenient method of controlling the extent of grafting [13].

In this research, we compared CAN and FAS as the initiator for the grafting process of chitosan and maleic anhydride. In order to optimize the condition, the acidic content from several ratios of CAN and FAS was observed.

2. Materials and Methods

2.1. Materials
Chitosan low molecular weight (DD ≥ 75%; MW = 50.000-190.000 Da) was purchased from Sigma Aldrich, USA. Maleic Anhydride (MA) was purchased from Nacalai Tesque inc, Japan and used without purification. Acetic acid (CH_3COOH), ferrous ammonium sulfate (FAS), and ceric ammonium nitrate (CAN) were purchased from Merck, Germany.

2.2. Film preparation
The method used in this research was similar to other studies with a slight modification [7]. About 0.6 grams of chitosan were dissolved in 30 mL acetic acid solution (1% v/v) followed by stirring with a magnetic stirrer at 70 °C to obtain a homogeneous solution. After that, the initiator (CAN or FAS) was added into the solution to initiate the reaction. The variations of CAN in this study were 0%; 0.83%; 1.67%; 2.51%; and 3.33% (% w/w), while for FAS 0%; 0.089%; 0.178%; 0.268%; and 0.357% (% w/w), with all ratio to chitosan weight. Immediately after, 0.3 grams of MA were added to the solution and the reaction was carried out for 45 minutes at 70 °C and 430 rpm. The obtained homogeneous solution then casted on a petri dish and dried overnight in an oven at 50 °C. In order to remove the excess of maleic anhydride, the dried film was washed by acetone (25% v/v).

2.3. Fourier Transform Infrared (FTIR) Spectroscopy
The functional groups of samples were measured by FTIR spectroscopy. The FTIR study was performed using SHIMADZU IR Prestige-21 in the range of 400 – 4000 cm^{-1}. All the samples were scanned with a resolution of 4 cm^{-1} and 10 number of scans.

2.4. Acidic Content Measurement
In order to measure the amount of carboxyl group in CTS-graft-MA film, the back-titration method was performed [14]. The content of carboxyl groups (mmol/g) was determined by titration of 0.01 g sample dissolved in 10 mL NaOH solution (0.01 M) with HCl solution (0.005 M). The acidic content (AC) was calculated by equation (1) [15]:

$$AC = \frac{C_{NaOH} \times V_{NaOH} - C_{HCl} \times V_{HCl}}{m_{sample}} \qquad (1)$$

3. Result and discussion

3.1. FTIR Spectroscopy

FTIR spectroscopic measurements were carried out to confirm the formation of grafting, based on changes in the chemical structure of chitosan. Figure 1 shows the FTIR spectra for (a) chitosan, (b) CTS-graft-MA using CAN initiator, and (c) CTS-graft-MA using FAS initiator. The chitosan FTIR spectrum (a) shows that peaks at 1643.4 cm^{-1} and 1558.48 cm^{-1} belong to amide I and amide II functional groups, while peak around 1373.32 cm^{-1} belong to C-N bond. This result is similar to other studies [10]. The results from figure 1(b) and (c) show the appearance of a new peak at 1705.07 cm-1 which belongs to C = O bands from the carboxyl group grafted on the chitosan backbone. The presence of the C = O band in each sample is confirmed that this method is successful in producing chitosan-graft-maleic anhydride [7]. The interaction of chitosan, initiator and maleic anhydride is illustrated in figure 2. Chitosan was oxidized by Ce(IV) to form a radical on its amine and hydroxyl group. Radical chitosan then reacted with maleic anhydride and produced chitosan-graft-maleic.

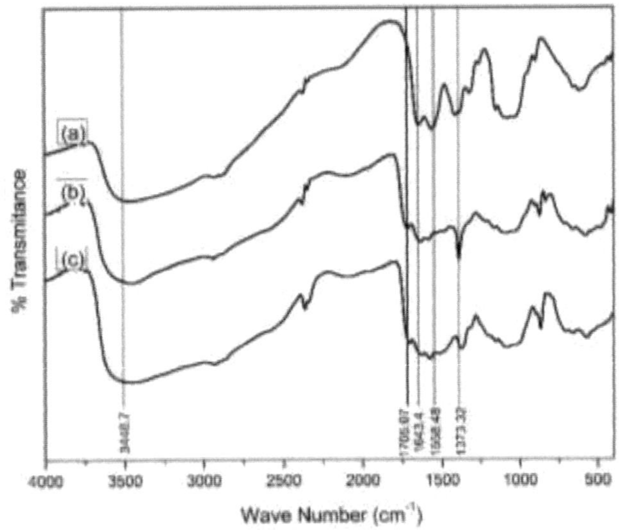

Figure 1. FTIR Spectrum for pure CTS (a), CTS-g-MA using CAN (b), CTS-g-MA using FAS (c).

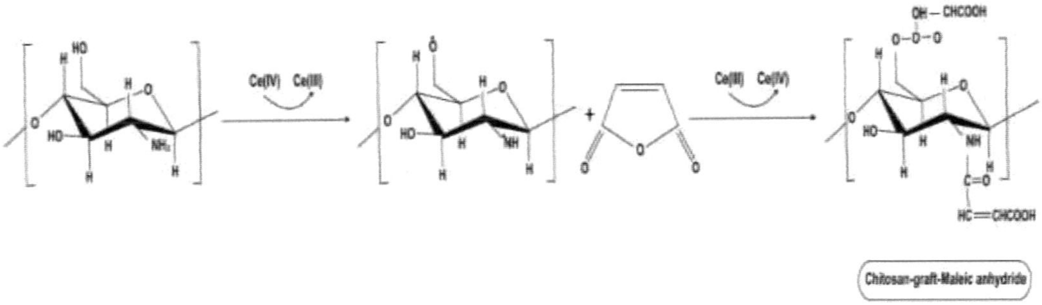

Figure 2. Schema of Interaction between Chitosan, Initiator, and Maleic Anhydride [7].

3.2. Acidic Content Measurement

To find out the optimum results from CAN and FAS initiator, the back-titration method was performed. Grafting was started by the formation of free radicals in the chitosan chain. These free radical electrons allow initiation of copolymerization of chitosan with MA through ring-opening mechanism. The total acidic content of each sample is determined by back-titration using Equation (1). Different values of acidic content (AC) occur because of the variations in the initiator ratio. The acidic content result of CAN is shown in Figure 3(a). Higher ratio of CAN leads to an increase of the AC in the membrane due to the new carboxylic groups on the chitosan backbone. However, there is a ratio of CAN which gives the optimum value of AC. The optimum value of AC obtained when the CAN ratio is 2.51%. This result is similar to other studies that used eugenol to modify chitosan in the form of hydrogels [9]. The graft yield increase along with the ratio of CAN and decreased after reaching the optimum value. The use of CAN initiator ratio greater than 3.33% leads to damaged film. AC results from FAS initiator have similar trend. However, the optimum AC value of CAN is higher than the AC value of FAS. Therefore it can be concluded that CAN is more effective than FAS to initiate the grafting process on CTS-graft-MA.

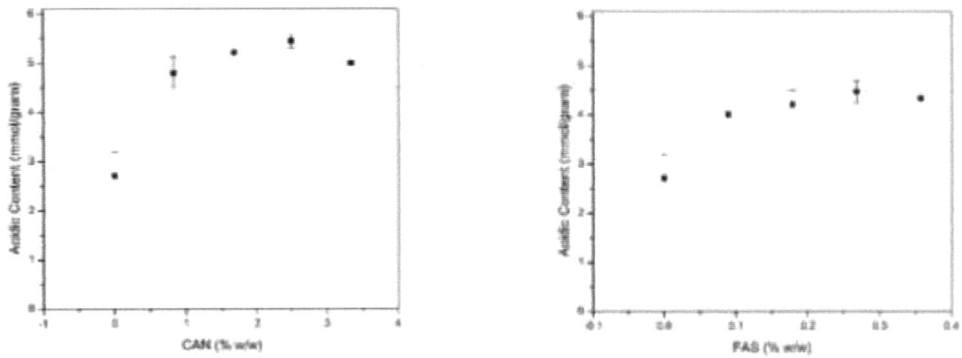

Figure 3. Acidic Content of CTS-graft-MA using CAN (a) and FAS (b).

4. Conclusions

The CTS-graft-MA using CAN and FAS as the initiator is successfully produced and had been confirmed using FTIR. The results show an appearance of the new peak that belongs to the carboxyl group of MA for both initiators. CAN is more effective than FAS with the optimum value of AC was obtained with CAN ratio 2.51%. The use of the CAN ratio greater than 3.33% leads to damaged film.

Acknowledgment

This research was supported by Ministry of Education and Culture of Indonesia through PDUPT scheme with contract number No. 2573/UN1.DITLIT/DIT-LIT/LT/2019.

References

[1] Elgadir M A, Uddin M S, Ferdous S, Adam A, Chowdhury A J K and Sarker M Z I 2014 *Impact of chitosan composites and chitosan nanoparticle composites on various drug delivery systems: A review J Food Drug Anal.* **23** 1–11

[2] Manna U, Bharani S and Patil S 2009 *Layer-by-Layer Self-Assembly of Modified Hyaluronic Acid/ Chitosan Based on Hydrogen Bonding Biomacromolecule.* **10** 2632–2639

[3] Jayakumar R, Prabaharan M, Reis R L, Mano J F 2005 *Graft copolymerized chitosan—present status and applications* s *Carbohydr. Polym.* **62** 142–158

[4] Kusumastuti Y, Petrus H T B M, Yohana F, Buwono A T, Zaqina R B 2017 *Synthesis and Characterization of Biocomposites Based on Chitosan and Geothermal Silica AIP Conf. Proc.***1823** 020127-1–020127-6

[5] Singla A K and Chawla M 2001 *Chitosan: some pharmaceutical and biological aspects – an update J. Pharm. Pharmacol.* **53** 1047–1067

IOP Conf. Series: Materials Science and Engineering **742** (2020) 012037 doi:10.1088/1757-899X/742/1/012037

[6] Zhang W, Li G, Fang Y and Wang X 2007 *Maleic anhydride surface-modification of cross linked chitosan membrane and its pervaporation performance J. Memb. Sci.* **295** 130–138

[7] Lavanya R, Gomathi T, Vijayalakshmi K, Saranya M, Sudha P N and Anil A 2017 *Adsorptive removal of copper (II) and lead (II) using chitosan-g-maleic anhydride-g-methacrylic acid copolymer Int. J. Biol. Macrom*ol. **104** 1495–1508

[8] Hasipoglu H N, Yilmaz E, Yilmaz O and Caner H 2005 *Preparation and characterization of maleic acid grafted chitosan Int. J. Polym. Anal. Charact.* **10** 313–327

[9] Hamedi H, Moradi S, Hudson S M and Tonelli A E 2018 Chitosan based hydrogels and their applications for drug delivery in wound dressings: A review Carbohydr. Polym. **199** 445–460

[10] Ibrahim A G, Saleh A Sh, Elsharma E M, Metwally E and Siyam T 2019 Chitosan-g-maleic acid for effective removal of copper and nickel ions from their solutions Int. J. Biol. Macromol. **121** 1287-1294

[11] Kyzas G Z, Siafaka P I, Lambropoulou D, Lazaridis N K and Bikiaris D N 2013 *Poly(itaconic acid) grafted chitosan adsorbents of different cross–linking for Pb(II) and Cd(II) uptake Langmuir.* **30** 120-130

[12] Jung B, Chung S and Lee S B 2006 *Preparation and Characterization of Eugenol-Grafted Chitosan Hydrogels and Their Antioxidant Activities J Appl Polym Sci.* **99** 3500 –3506

[13] Yazdani-Perdram M, Retuert J and Quijada R 2000 Hydrogels based on modified chitosan, 1Synthesis and swelling behavior of poly(acrylic acid) grafted chitosan Macromol. Chem. Phys. **201** 923–930

[14] Vasi A M, Popa M I, Butnaru M, Dodi G and Verestiuc L 2014 *Chemical functionalization of hyaluronic acid for drug delivery applications Mater. Sci. Eng. Mater. Biol Appl. C* **38** 177–185

[15] Timotius D, Rochmadi and Kusumastuti Y 2018 *Preparation and Characterization of Local Indonesian Chitosan-graft-Maleic Anhydride as Drug Carrier IOP Conf. Ser.: Mater. Sci. Eng.* **599** 012029

IOP Conf. Series: Materials Science and Engineering **742** (2020) 012038 doi:10.1088/1757-899X/742/1/012038

Effect of Reaction Time on Modification of Multi-walled Carbon Nanotubes with HNO_3

R Greta Putri [1], Y Syafera[2], F Larasati[3], N Rofiqoh Eviana Putri[4], M Handayani[5], Rochmadi[6] and Y Kusumastuti[7]

[1, 2,3, 4,6,7] Chemical Engineering Department, Faculty of Engineering, Universitas Gadjah Mada, INDONESIA
[5] Research Center for Metallurgy and Materials, Indonesian Institute of Sciences, INDONESIA

E-mail: yuni_kusumastuti@ugm.ac.id

Abstract. The functionalization of Multi-walled Carbon Nanotubes (MWCNTs) is addressed to improve its interaction capability with other materials for further applications. The oxidation of MWCNTs by nitric acid is one of the promising methods for its surface modifications. In this research, MWCNTs functionalization was carried out in several steps such as oxidation, filtration, washing, and drying. The optimization of reaction time for MWCNTs functionalization was conducted to achieve the optimum condition which efficiently functionalizes MWCNTs surface. The sample characterizations were investigated by Boehm titration, Fourier Transform Infrared (FTIR) Spectroscopy, Thermo Gravimetric Analysis (TGA), and dispersion test. As a result, carboxyl groups were successfully attached onto the MWCNTs surface and stable dispersion of MWCNTs was obtained after 12 hours. Oxidation with nitric acid for 12 hours at room temperature is the optimum treatment time for surface modification.

1. Introduction

Multi-walled Carbon Nanotubes (MWCNTs) have attracted great attention because of their unique characteristics including high aspect ratio, high surface area, and stability of nanoparticles. However, MWCNTs are hydrophobic and are poorly dispersed in either water or organic solvent. Because of its fine size, high surface energy of MWCNTs and with its intrinsic *Van der Waals* forces, raw MWCNTs are tend to aggregate and entangle [1]. These characteristics will limit of their application. Covalent surface modification displays functional groups such as –COOH, -COH and –OH groups that are attached to the sidewall and side end of MWCNTs by covalent bonds [2]. Chemical oxidation is carried out for functionalization of MWCNTs. MWCNTs can be treated with strong oxidant such as HNO_3, $KMNO_4$, or HNO_3/H_2SO_4 to form fractional concentration of carboxyl groups and this can increase the MWCNTs solubility characteristics.

Previous studies described numerous methods for chemical functionalization of carbon nanotubes (CNTs). Shen et al. [1] reported that amino-functionalized MWCNTs can improve their dispersion in H_2O. Olugbenga et al. [3] showed that using of organic acid-inorganic acid mixture is potential for the purification of CNTs. L. Thi Mai Hoa [4] oxidized MWCNTs in a mixture of HNO_3/H_2SO_4 to study the influence of different oxidizing conditions on the morphological and chemical-physical properties of MWCNTs.

In this paper, we are concerned with optimization of reaction time for MWCNTs functionalization by oxidizing the MWCNTs with nitric acid and investigate the influence of reaction time on the amount of carboxylated groups in modified MWCNTs.

2. Experimental

2.1. Material

Multi-walled carbon nanotubes (MWCNTs with an OD 6-9 μm, length 5 μm, and carbon basis >90%) was purchased from Cheap Tubes Inc, USA. Nitric acid (HNO_3 65% v/v), glacial acetic acid, and sodium hydroxide were purchased from Sigma Aldrich, USA. Hydrochloric acid (HCl 37% v/v) was purchased from Mallinckrodt Pharmaceuticals, UK. The solution of 0.05 N sodium hydroxide and 0,05 N hydrochloric acid for titration process are prepared.

2.2. Oxidation of the MWCNTs

Oxidation of MWCNTs was conducted according to the previous method reported by M. Habibizadeh et al. [5]. Raw MWCNTs were functionalized by adding 500 mg of MWCNTs in a round bottom flask with 100 mL of 14,4 M HNO_3. A bottom flask equipped with reflux was immersed in oil bath on the top of hot plate. This process was run by vigorously magnetic stirring. After the certain reaction time, MWCNTs were filtered and washed with aquadest to neutral pH. Functionalized MWCNTs were dried in oven at 80°C for 24 hours. For the study of effect of reaction time, the mixture was treated by magnetic stirring for 6, 9, and 12 hours at room temperature which represent as Sample A, B and C, respectively.

2.3. Characterization of MWCNTs

The functional groups of functionalized MWCNTs were analyzed using Fourier Transform Infrared Spectroscopy (FTIR) (Spectrophotometer Shimadzhu FTIR-8201 PC, Country) with KBr method. Boehm titration was used in order to calculate acidic groups in functionalized MWCNTs which represent the relative concentration of acidic groups on the surface. The oxidized MWCNTs were immersed in 0.05 N NaOH solution and stirred continuously for 24 h to reach the equilibrium state, then dropwise titrated with 0.05 N HCl solution until the neutral point in order to determine the unreacted NaOH solution. The concentration of acidic groups on the surface MWCNTs is equal to the amount of reacted NaOH. Goertzen et al. [6] has introduced the total amount of reacted surface functionality (n_{CSF}) with equation 1 as below:

$$n_{csf} = N_B \times V_B - N_T \times V_T \times \frac{V_B}{V_a} \quad (1)$$

where N_B and V_B are the concentration and volume of the reaction base mixed with the carbon. n_{csf} denotes the moles of carbon surface functionalities on the surface of the carbon that reacted with the base during the mixing step. V_a is the volume of the aliquot taken from V_B. N_T and V_T are the concentration and volume of the acid added to the aliquot.

Thermo Gravimetric Analysis (TGA) was used to estimate the degree of functionalization and investigate the thermal stability of samples. The weight of the samples was approximately 10 mg. The test was performed under N_2 (100 mL/min). The sample was heated at 10°C min^{-1} from 30°C to 800°C. The grafting percentage (GP) of the MWCNTs-COOH surface was determined using the following equation [4]:

$$GP = (\Delta W_{MWCNT\text{-}COOH} - \Delta W_{MWCNTs}) \times 100\% \quad (2)$$

$\Delta W_{MWCNT-COOH}$ and ΔW_{MWCNTs} are the percentage weight loss of MWCNTs-COOH and MWCNTs, respectively.

The dispersion test was carried out to determine the improvement of MWCNTs solubility after being modified. The oxidized MWCNTs (100 mg) were dispersed in 5 mL water by vigorously shaking for 10 minutes at room temperature. The mixture was left to stand for 48 hours to observe its stability.

3. Result and Discussion

Raw MWCNTs could be functionalized with oxygen-containing groups like -COOH and -OH at the MWCNTs ends and at the outer sidewall, where the defects are introduced through oxidation [5]. These could be further used for the reaction to conjugate more functional groups as shown in Figure 1.

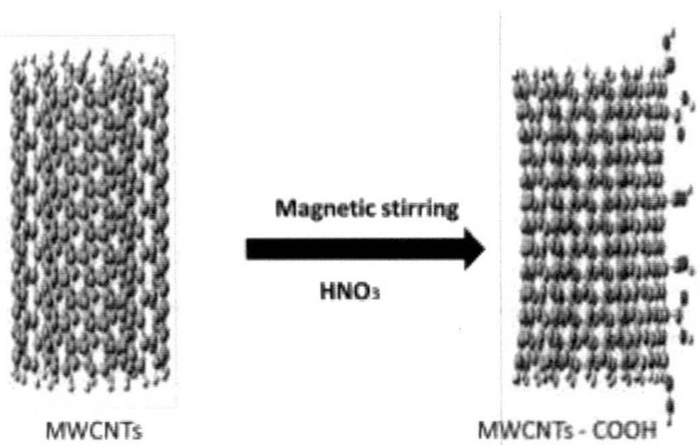

Figure 1. Reaction scheme for functionalization of MWCNTs.

To analyze total acidic groups in carboxylates MWCNTs, Boehm titration was performed. Figure 2 shows the total amount of -COOH groups in functionalized MWCNTs that increase along with the reaction time. The mmol of acidic functional groups (-COOH) per gram of functionalized MWCNTs materials is given in Figure 2.

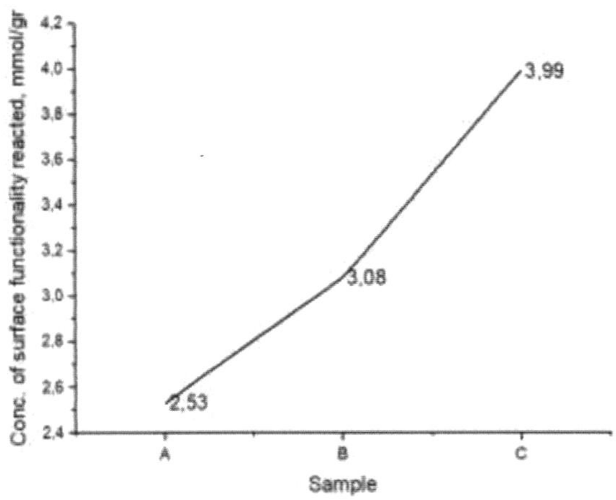

Figure 2. The concentration of -COOH groups estimated by Boehm acid-base titration method.

As seen from the data, the extent of –COOH functional group of every sample gradually increase along with the treatment duration from 2,53 mmol/g on sample A to 3,08 mmol/g on sample B to 3,99 mmol/g on sample C.

The functionalization of MWCNTs was confirmed by FTIR Spectroscopy. Table 1 shows the FTIR spectra characteristics of raw and functionalized MWCNTs. As can be observed in FTIR spectra, the characteristic peaks, named X and Y band, are identified. The first peak, X band, is corresponded to the C-O stretching vibration. It is located at 1126.43; and 1111 cm^{-1} for samples B, and C, respectively. The second peak is Y band, which is related to the stretching of C=C bonds [4]. The peak is located at 1527.62 cm^{-1} for all samples. No shifting occurs in modified MWCNTs samples, but the intensity changed along with the treatment duration. Furthermore, the peak at 3448.72 and 1635.64 cm^{-1} are corresponded to hydroxyl (O-H) and carbonyl groups (C=O), respectively [5]. It can be observed from Table 1 that the position of X and Y band are shifted for raw MWCNTs compared to functionalized MWCNTs.

Table 1. FTIR spectra and intensity ratios for functionalized MWCNT s.

Sample	Peak, cm^{-1}	Intensity	Intensity Ratio, I_X/I_Y
A	X: -	-	-
	Y: 1527.62	1.438	
B	X: 1126.43	9.385	1.222
	Y: 1527.62	7.682	
C	X: 1111.00	29.320	1.012
	Y: 1527.62	28.960	

The shifting can be caused by treatment effect upon strong acid (HNO_3). The functionalization process of MWCNTs, especially with strong acid, will breaks some of its bonds and attaches functional groups that can be interpreted as defects on the structure. When the functionalization process is carried out, these peaks can still be identified. This proves that the modification process does not damage the structure of MWCNTs [4]. Comparing the X and Y band for functionalized MWCNTs samples, it is clear that the intensity increased. It shows that the carboxyl group was successfully attached to MWCNTs. In addition, change in intensity ratio I_X/I_Y can also further support the formation of carboxylated MWCNT.

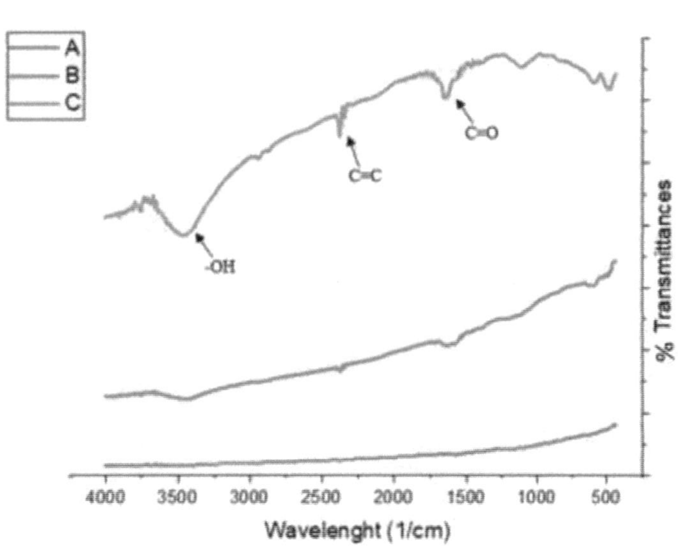

Figure 3. FTIR spectra for functionalized MWCNT

The influence of treatment duration is showed in FTIR spectrum as shows in Figure 3. From the IR absorption spectra, it was found that the increase of treatment duration will strengthen the peak intensity of carboxyl groups. The absorption spectrum of MWCNT-COOH mainly consisted of –OH

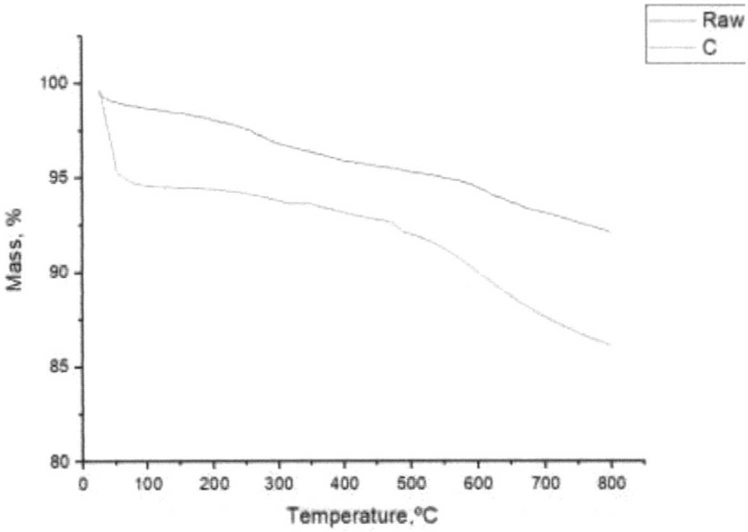

Figure 4. TGA curves of raw MWCNTs and functionalized MWCNTs.

stretch at 3448 cm^{-1} and carboxyl >C=O stretch at 1635 cm^{-1}. The result was in agreement with the explanation above.

In order to confirm the grafting rate of -COOH on the surfaces of MWCNTs, the thermal gravimetric analyses (TGA) are performed. We did comparison the TGA analysis of raw material of MWCNTs and the selected functionalized MWCNT based on the result from Boehm titration (sample C). Figure 4 shows the TGA analysis of raw MWCNT, and Sample C of MWCNT-COOH that presents the curves corresponding to the weight loss as a function of temperature. We observed a weight loss of 7.51% for raw MWCNT in TGA curves. For the MWCNTs-COOH obtained after modification, the final weight loss for Sample C is 13.45%. As shown in TGA curve, the oxidized MWCNTs start decomposing earlier than raw MWCNTs. It can be attributed to the presence of functional groups on the nanotubes [4]. As calculated by equation (2), the grafting percentage (GP) corresponding to C is 5.93%, which reveals that an increase of the thermal degradation is observed for functionalized MWCNTs.

Figure 5 shows the dispersion images of MWCNTs-COOH in comparison with the raw MWCNTs in water (100 mg/5 mL). Raw MWCNTs could not disperse because of the strong intrinsic Van der Waals forces between them. Raw MWCNTs was totally separated from water and form bundles. However, sample A, B, and C could only maintain its stable dispersion for 10 minutes and after 48 hours, the samples were completely precipitated. This means that the increase of reaction time in oxidation reaction has not yet effectively reduced the apparent activation energy of MWCNTs, which can prevent the aggregation phenomenon [7].

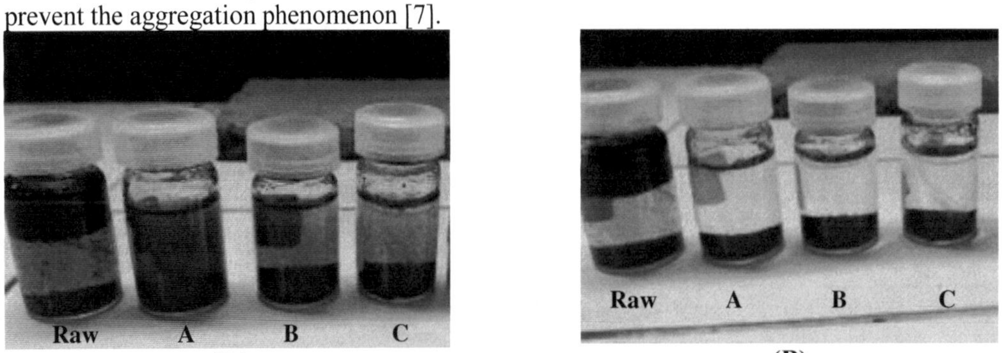

(A) (B)

Figure 5. The dispersion images of raw MWCNTs and MWCNTs-COOH **(A)**: after 10 minutes; **(B)**: after 48 hours.

4. Conclusion

The usefulness of carbon nanotubes can be limited by its hydrophobic nature. Most of the purification and functionalization methods for the improvement of the hydrophilicity as resulted to the destruction of the carbon nanotube side walls and shortening of the chain length. The purification and functionalization methods adopted in this research were able to achieve the grafting of -COOH groups into MWCNT's surface. The result shows that functional groups will increase with the rise of reaction time. The highest result of functionalization occurred under the reaction time 12 hours with the concentration of acidic groups reaches 3.99 mmol/gram and grafting percentage reaches 5.93%.

Acknowledgments

This research was supported by Ministry of Education and Culture of Indonesia through PDUPT scheme with contract number 2526/UNI.DITLIT/DIT-LIT/LT/2019.

Reference

[1] Shen J, Huang W, Wu L, Hu Y and Ye M 2007 Thermo-physical properties of epoxy nanocomposites reinforced with amino-functionalized multi-walled carbon nanotubes *Applied Science and Manufacturing,* Part A **38**, 1331-36.

[2] Zhao Z, Yang Z, Hu Y, Li J and Fan X 2013 Applied surface science multiple functionalization of multi-walled carbon nanotubes with carboxyl and amino groups *Applied Surface Science,* **276**, 476–481.

[3] Olugbenga O, Michael D and Suny I 2018 Investigation on purification potential of multiwalled carbon nanotubes using organic-mineral acid mixture *Journal of Nanoscience and Nanoengineering*, **4**, No. 1, pp. 1-8.

[4] Hoa, L 2018 Characterization of multi-walled carbon nanotubes functionalized by a mixture of HNO3/H2SO4 *Diamond and Related Materials*, **89**, 43-51.

[5] Habibizadeh M, Rostamizadeh K, Dalali N and Ramazani A 2017 Preparation and characterization of PEGylated multiwall carbon nanotubes as covalently conjugated and non-covalent drug carrier: A comparative study *Materials Science and Engineering*: C, **74**, 1-9.

[6] Goertzen S L, Thériault K D, Oickle A M, Tarasuk A C and Andreas H A 2010 Standardization of the Boehm titration. Part I. CO2 expulsion and endpoint determination *Carbon* **48**(4), 1252–61.

[7] Ma W, Zhao Y, Li Y, Zhang P, Cao Z, Yang H, Liu C, Tao G, Gong F and Matsuyama H 2018 Synthesis of hydrophilic carbon nanotubes by grafting poly(methyl methacrylate) via click reaction and its effect on poly(vinylidenefluoride)-carbon nanotube composite membrane properties *Applied Surface Science*, **453**, 79-90.

IOP Conf. Series: Materials Science and Engineering **742** (2020) 012039 doi:10.1088/1757-899X/742/1/012039

Synthesis of Pyrogallol Derivative as a Soluble Antioxidant for Biodiesel Additive

Regina Christianty Handayani[1], Akhmad Darmawan[2], Mohammad Nasikin[3] and Hery Sutanto[*1]

[1]Department of Chemical Engineering, Faculty of Life Sciences and Technology, Swiss German University, Tangerang Selatan, Banten, Indonesia.
[2]Research Center for Chemistry, Indonesian Institute of Sciences, Kawasan PUSPIPTEK Serpong, Tangerang Selatan, Banten, Indonesia.
[3]Chemical Engineering, Department of Chemical Engineering, Universitas Indonesia, Depok, Indonesia

[*]Corresponding Author : Hery.sutanto@sgu.ac.id

Abstract. Biodiesel as a promising alternative fuel has a major technical drawback which is its instability during storage. One of the solutions to this problem is the addition of antioxidants to improve storage stability. However, the commonly used antioxidant, pyrogallol, has poor solubility in biodiesel due to its polarity difference with biodiesel. Hence this research is focused on modification of pyrogallol with methyl linoleate through a free radical mechanism using benzoyl peroxide as the radical initiator to increase pyrogallol solubility. The presence of the proposed product (10E,12Z)-9-(2,3,4-trihydroxyphenyl) octadeca-10,12-dienoate and its isomer methyl (9Z,11E)-13-(2,3,4-trihydroxyphenyl) octadeca-9,11-dienoate was confirmed using TLC, HPLC and LC-MS/MS which yielded 0.2%. UV-Vis Spectrophotometer was used for the solubility test and the result showed that the solubility limit of the product 2500ppm which was higher than pyrogallol. For the antioxidant activity, the Rancimat test was performed to see the product's performance in accelerated oxidation as well as acid value determination during four weeks of storage. The Rancimat test result shows the product has higher induction period compared to pyrogallol and fulfilled the oxidation test standard. However, the product showed lower acid value stability compared to pyrogallol and biodiesel.

1. Introduction

Biodiesel, also called Fatty acid methyl ester (FAME) is a renewable fuel derived from biological sources through transesterification reaction that has lower smoke emission [1]. However, apart from the advantages biodiesel has, it has a major technical drawback that affects biodiesel's quality during storage [2]. The high content of unsaturated fatty acid chains and the presence of carbon double bonds in the biodiesel leads to high reactivity with O_2 [3]. Biodiesel oxidative degradation leads to acid sediments formation which causes blockage in engine fuel line and filters. Therefore, antioxidants are required to improve the oxidation stability of biodiesel. The most used Phenolic antioxidants are Pyrogallol, Propyl gallate, TBHQ, BHA and BHT[4]. Due to its long hydrocarbon chain, biodiesel is relatively non-polar. In contrast, phenolic antioxidants which are hydroxyl groups are polar compounds. The difference in polarity between biodiesel and phenolic compounds creates an electronegativity gap which results in low solubility of phenolic antioxidants in biodiesel [5]. In recent years, there has been a rise in research regarding the modification of phenolic compounds to increase the solubility of phenolic antioxidant. One of them is phenolic compound modification using a free radical mechanism [6]. One of the commonly used initiators for free radical mechanism is benzoyl peroxide. Upon heating, benzoyloxy radicals were formed from the dissociation of benzoyl peroxide which will initiate the reaction [7], [8]. In this research, pyrogallol will be modified using methyl linoleate using free radical mechanism with benzoyl peroxide as the initiator. The new product is theoretically expected to have a higher solubility than pyrogallol.

2. Research Methods

2.1. Preliminary Reaction
The optimization of the coupling reaction between pyrogallol and methyl linoleate was performed by variating the temperature and the sequential addition of methyl linoleate. The reaction was conducted using a 100 ml boiling flask connected to reflux and equipped with a magnetic stirrer. The result of the reaction was qualitatively observed with thin layer chromatography (TLC).

2.2. Coupling Reaction
The synthesis reaction was prepared by weighing 0.24 g of Benzoyl Peroxide into a boiling flask and diluting it in 10 ml of Ethyl Acetate. Afterwards, 0.33ml of methyl linoleate was added into the flask. The air inside was purged by injecting nitrogen into the flask to remove all the oxygen inside. The mixture was then heated until 90°C in a water bath and stirred continuously for 2 hours. Next, in a 50 ml beaker, 0.126 g of Pyrogallol was diluted in 1ml of ethyl acetate. The pyrogallol solution was then added to the reaction mixture and was stirred for another 15 minutes.

2.3. Solubility Test
The reaction mixture was added to biodiesel with various concentrations and diluted in *n*-hexane with 1:100 dilution factor. Each diluted sample solution were then scanned using UV/Vis spectrophotometer to find the wavelength of the maximum absorbance. Lastly, the absorbance of each concentration was measured in that wavelength where the absorbance will keep increasing until a certain concentration where it will remain constant or stopped increasing which indicates the antioxidant solubility limit in biodiesel. The solubility limit data will later be compared to the solubility limit data of pyrogallol

2.4. Antioxidant Activity Qualitative Analysis
The antioxidant qualitative analysis uses the method of dot blot assay. First of all, a solution of methanolic DPPH was prepared, 0.1 g of DPPH was diluted in 50 ml methanol in a volumetric flask to make a 0.2% methanolic DPPH solution. Afterward, pyrogallol and product prepared and diluted to ethyl acetate to make a solution with a concentration of 1500 ppm. The pyrogallol and the product were then spotted to TLC paper then sprayed with the methanolic DPPH solution prepared earlier. The color change was then further analyzed.

2.5. Rancimat Test
The Rancimat method is also called the automated Swift test or accelerated oxidation test was conducted to determine the effectiveness of antioxidants to restrict biodiesel to oxidation. The test was performed using Metrohm 873 based on standard test method EN 14112 [2], [9]. The method involves passing airflow at a steady rate of 10l/h and heating with a constant temperature of 110 °C which leads to the oxidation of biodiesel, with volatile, water-soluble short chain carboxylic acids formed as the oxidation products. The oxidation products were transported in the stream of air to a vessel containing distilled water and the conductivity of this vessel was continuously monitored. The time duration of

steep increase in the conductivity was measured in rancimat method, it is known as the induction period (IP) and it has been an element of standard for biodiesel oxidative stability; in EN 14214 biodiesel fuels must meet an induction period requirement of 6 h and in ASTM D6751 must meet an induction period requirement of 3 h.

2.6. *Acid Value Analysis*

Both pyrogallol and the product are subjected to acid value analysis, a sample with 1500 ppm concentration mixed with biodiesel were prepared for each antioxidant. A sample of pure biodiesel was also prepared as a comparison with the antioxidant. The samples were labeled: Biodiesel (B), Biodiesel-Pyrogallol (B-PY) and Biodiesel-Product (B-PR). The samples were analyzed weekly during four weeks of storage time. In an Erlenmeyer flask, 10ml of 2-Propanol was prepared. Afterward, 1 ml of sample was added and mixed until homogeneous. A calibrated pH meter was prepared. Then, the solution was titrated with a standard 0.1 N NaOH with increasing volume to obtain the diagram of NaOH volume used to make the pH 7 (V1). After plotting the data, the required NaOH volume to change the pH of the solution into 7 was calculated using the equation of the curve. The acid values method was a modification of Indonesia National Standard for oil and lipid SN1 01-30199. Titration for 2-propanol was done to neutralize the 2-propanol slight acidity. The volume of NaOH used to neutralize 2-propanol was noted as (Vo). At last, the acid values of the samples were calculated using the equation below:

$$Acid\ Value = \frac{V\ x\ N\ x\ 56.1}{m} \quad (1)$$

3. Results and Discussion

3.1. *Preliminary Reaction*

Benzoyl Peroxide as a thermal initiator required heat to decompose, therefore to obtain the optimum condition for the reaction, variating the temperature reaction was performed. According to Odian (2004) benzoyl peroxide half-life ($t_{1/2}$) at 70°C was 7.3 h, which is why the TLC spots did not show a positive result because the benzoyl peroxides did not encounter enough heat to decompose. The sequence addition of methyl linoleate was also varied because methyl linoleate as a long chain methyl ester took longer time to become radical therefore reaction with addition of methyl linoleate before and after heating was observed. From the variations, the optimum condition of the reaction was observed to be at temperature 90°C with the addition of methyl linoleate before heating.

Table 1. Results of Preliminary Reaction

Temperature (°C)	Methyl linoleate addition	New Spots and Intensity
90	After heating	+
	Before heating	++
70	After heating	-
	Before heating	-

3.2. *Coupling Reaction*

Synthesis of pyrogallol with methyl linoleate was executed via coupling reaction by using benzoyl peroxide as the initiator of the reaction. In the synthesis reaction, ethyl acetate was used as a solvent because it is a semi-polar solvent that can dilute all the components for this reaction. Benzoyl Peroxide was heated for 2 hours in a water bath with a temperature of 95°C to make sure that the benzoyl peroxides decomposed into benzoyloxy radicals. During the whole reaction, reflux was used to condense the ethyl acetate because of its boiling point at 78°C. Nitrogen was used because during heating, methyl linoleate radical tends to bond with oxygen. Therefore, oxygen inside the flask was removed to create an optimum condition for the reaction. The reaction mixture between methyl linoleate and benzoyl peroxide produced a transparent non-viscous solution. Pyrogallol was not heated together with benzoyl peroxide and methyl linoleate because pyrogallol as secondary alcohol would decompose into ketone upon heating which would cause pyrogallol to lose its antioxidant ability. When pyrogallol-ethyl acetate solution was added into the benzoyl peroxide and methyl linoleate solution, the color of the product mixture changed into light brown transparent solution. The color change occurs because some of the pyrogallol might have absorbed some oxygen (Graf, 2006). After a few hours, the color of the solution changed to dark red-black color which indicates that some of the pyrogallol might have been oxidized.

3.3. Thin Layer Chromatography (TLC) Identification
As a pre-identification, the reaction product was subjected to a Thin Layer Chromatography (TLC) analysis to observe the retention factors difference. The solvent used was a combination of n-hexane and ethyl acetate with a ratio of 1:5. The combination was chosen because ethyl acetate is categorized as a semi-polar solvent and n-hexane is a polar solvent which results in a fairly non-polar mobile phase. The stationary phase of the TLC paper was the polar silica gel.

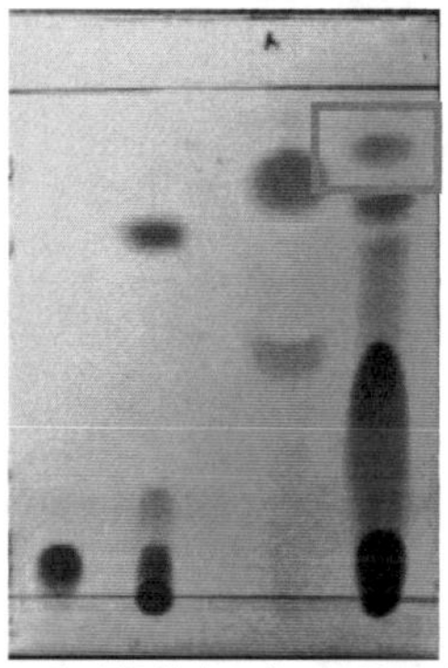

Figure 1. TLC Analysis Result

In the TLC paper, spot (a) was pyrogallol, (b) was benzoyl peroxide, (c) was methyl linoleate, and (d) was the product. The calculated Rf shown on the reaction mixture spot were (a) was 0.1, benzoyl peroxide (b) 0.73, methyl linoleate (c) showed 2 spots one with an Rf value of 0.55 and the other one with an Rf value of 0.83. The product (d) showed 4 spots, each has Rf value from the bottom to the top were 0.13, 0.5, 0.78 and 0.9. Since the predicted product was a combination of pyrogallol and methyl linoleate, the Rf value of the product should be greater than pyrogallol as the product will be more non-polar compared to pyrogallol. Hence, the Rf value of 0.5 or 0.9 is still possible for the predicted product.

To determine which of these Rf values belong to the product, partition coefficient (log P) can be used as an indicator to determine how hydrophilic or hydrophobic a molecule is. A polar compound will have a lower log P and non-polar compounds will have higher log P. Log P of the predicted product was calculated in ChemDraw which was (a) 6.58. Pyrogallol as a polar molecule has a log P value of (b) 0.87 and methyl linoleate has a log P value of (c) 6.23 this confirms that the predicted product was more non-polar than both pyrogallol and methyl linoleate. Therefore, the Rf value of the product taken was 0.9 because the spot was higher or in other words more non-polar than methyl linoleate.

IOP Conf. Series: Materials Science and Engineering **742** (2020) 012039 doi:10.1088/1757-899X/742/1/012039

3.4. HPLC Identification
In order to further identify the product, HPLC analysis was performed at a wavelength of 280 nm and 0.5ml/min flow rate. The qualitative analysis was performed by comparing the retention time between the product, pyrogallol, methyl linoleate, and benzoyl peroxide.

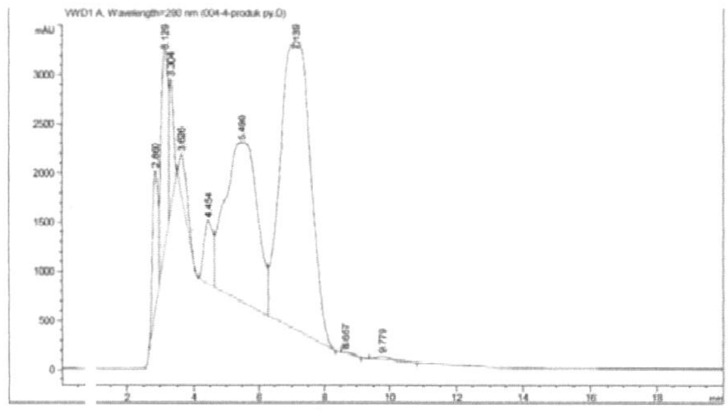

Figure 2. HPLC Spectrum

From the analysis, there were 9 appearing peaks as shown in figure 2. The retention time of pyrogallol was 2.867 min, which means that peak with retention time 2.860 can be eliminated as pyrogallol. Peak with retention time of 7.139 has a similar retention time to benzoyl peroxide retention time which is7.263 min, Methyl linoleate has a retention time of 8.399 min, and as the predicted product has higher Log P value than all methyl linoleate, pyrogallol and benzoyl peroxide, it is expected that the product will have longer retention time. Hence, a peak with a retention time of 9.779 min was taken out as the retention time of the product with a yield of 0.2%.

3.5. LCMS Identification
For further identification of the product, LC-MS/MS analysis was performed to determine the molecular mass of the components in the sample, the product was subjected to LC-MS/MS. The determination of the product was done using mass spectral fragmentations. The results of LC-MS/MS are shown in the figure below:

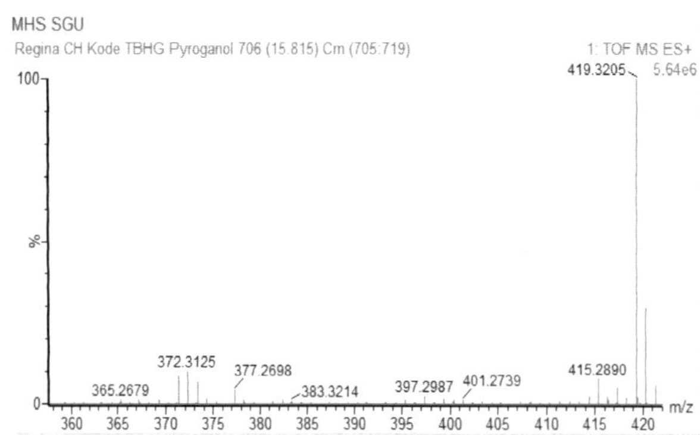

Figure 3. Mass Spectrum of LC-MS/MS

Theoretically, the predicted product ($C_{25}H_{38}O_5$) has a molecular weight of 418.589 g/mol. However, in order to be detected a molecule should be in its charged form. Therefore, metal cation or anions are commonly used to add a charge of the molecule. In this instrument (H^+) was used which results in addition of 1 m/z in order to obtain the real molecular weight of the product. From the result, it could be seen in Figure 3 the spectrum shows the candidate mass of the product ($C_{25}H_{38}O_5$) which was 419.3207 g/mol. The candidate mass number is close to the molecular weight prediction of the product, therefore the presence of the desired product can be confirmed.

3.6. Proposed Mechanism for the coupling reaction

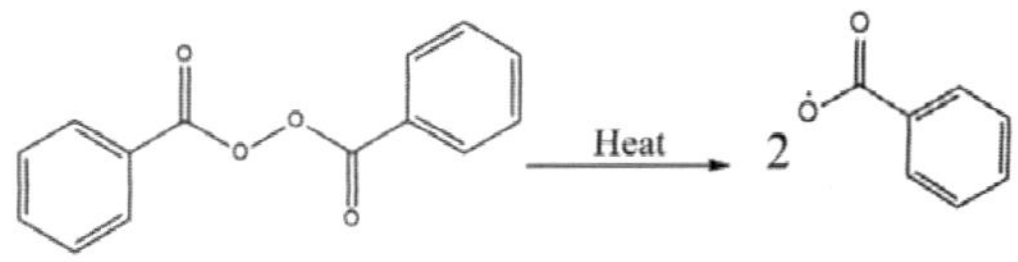

Figure 4. Benzoyloxy radical formation

The first step of radical chain reaction is the radical formation. Upon heating, benzoyl peroxide became unstable and decompose into benzoyloxy radicals. The unpaired benzoyloxy radicals became radical initiators as they are chemically reactive with the presence of other molecules. The next step is initiation, where pyrogallol and methyl linoleate as a non-radical compound became radical by the presence of benzoyloxy radicals. Free-radical molecules are not stable, therefore both pyrogallol and methyl linoleate will go through radical resonance or delocalization to stabilize the molecule.

Pyrogallol as the main as a phenolic compound has three -OH substituents. A substituent on a benzene ring had an effect on the additional substituent placement, which was the substitution of hydrogen with methyl linoleate radical. The substituent of pyrogallol affected the orientation of the new substituent. In general, when the substituent group attached to the benzene ring had at least one pair of the unshared electron, it is *ortho-para* directing. On the contrary, when the substituent group has either a partial positive charge or full positive charge, it is *meta* directing [10]. In this situation, the substituents attached to the ring was a hydroxyl group (-OH) with two pairs of unshared electrons, which was an *ortho-para* director. Since pyrogallol had three -OH substituents, there were several possibilities of the methyl linoleate radical orientation with each -OH substituent directing *ortho-para*.

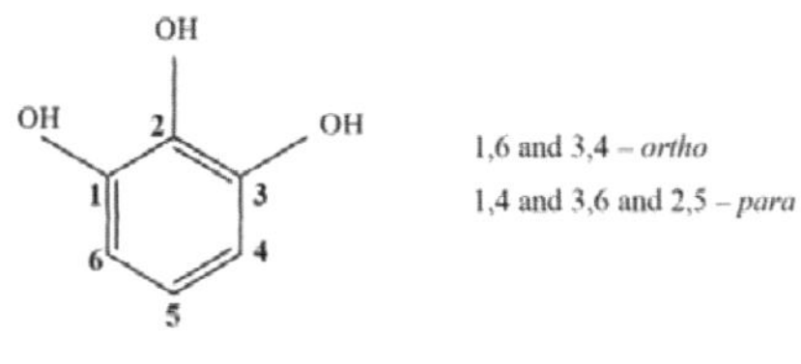

Figure 5. Orientation possibilities of methyl linoleate in pyrogallol

As seen in the figure above, atom C-4 and C-6 had more possibilities to be substituted compared to C-5. Hence, in Figure 6 the radical delocalization mechanism proceeded towards the substitution of atom C-4 (or C-6 which can be countered as C-4 because pyrogallol is a symmetric molecule), which was a *para* position from -OH number 1 and *ortho* position form -OH number 3.

The resonance delocalization of pyrogallol after radical attack by benzoyloxy radical can be seen in the figure below:

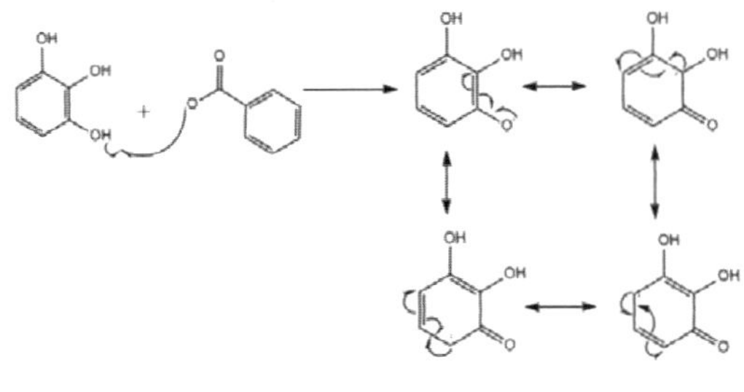

Figure 6. Pyrogallol resonance delocalization after the radical attack

From figure 6, it can be seen that there were two forms of product, in keto and in enol form. For aldehydes and ketones, keto form is considered because of it more stable compared to enol form. However, in phenolic compound, the keto-enol equilibrium lies entirely on the enol side due to thermodynamic stabilization by the aromaticity [11].

Methyl linoleate as a polyunsaturated fatty acid with two double bonds located in C-9 and C-13 atom has methylene hydrogen in C-11 atom, which is prone to radical attack. When C-11 became radical, resonance delocalization will occur to the C-9 and C-13 atom as illustrated in Figure 7.

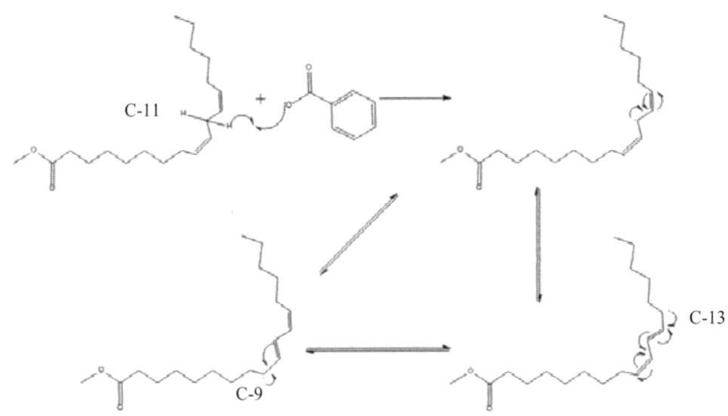

Figure 7. Methyl Linoleate resonance delocalization after the radical attack

Theoretically, with the optimum condition and proper mixing of the reaction, the radical pyrogallol and radical methyl linoleate were expected to encounter the last step of radical polymerization which is the coupling termination, where pyrogallol and methyl linoleate forms a bond as seen in Figure 8 and Figure 9. The product from the reaction will be a pyrogallol molecule attached by a methyl linoleate molecule.

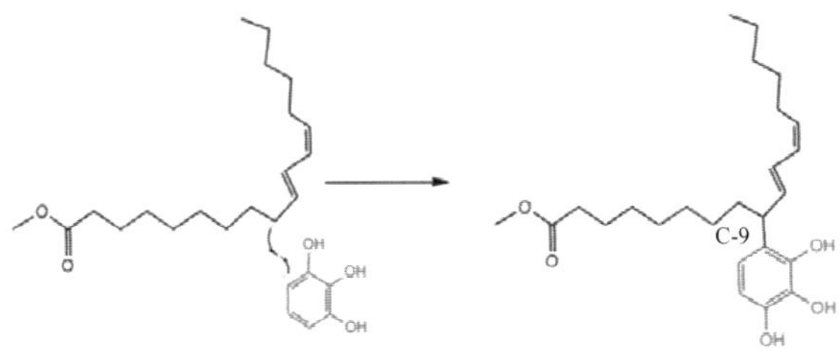

Figure 8. Termination of radical pyrogallol and radical methyl linoleate in C-9

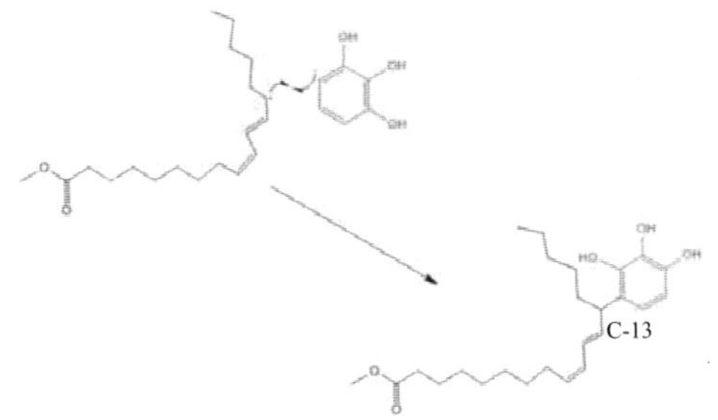

Figure 9. Termination of radical pyrogallol and radical methyl linoleate in C-13

3.7. Solubility Analysis

In this research, pyrogallol was synthesized with methyl linoleate with the purpose to increase its solubility in biodiesel. Therefore, the reaction product was subjected to a solubility test to compare the solubility limit of pyrogallol and the reaction product.

Before the solubility measurement, pyrogallol, methyl linoleate, and the new product were scanned in the UV-VIS spectrophotometer with a wavelength range of 190 nm to 400 nm to determine each maximum wavelength. Two concentrations (lowest and highest) were used to identify the peak. Based on the scanning result, the wavelength of maximum absorbance was 274 nm for the reaction product and 205 nm for methyl linoleate. After the maximum wavelengths were obtained, both pyrogallol and the reaction product was weighed in various concentrations ranging from 500ppm to 3000 ppm. Based on observation, all concentrations of pyrogallol and the reaction product were visually soluble in biodiesel. Hence, the solubility was measured with UV-Vis Spectrophotometer. A dilution of 1:100 was used to measure the absorbance value of each concentration.

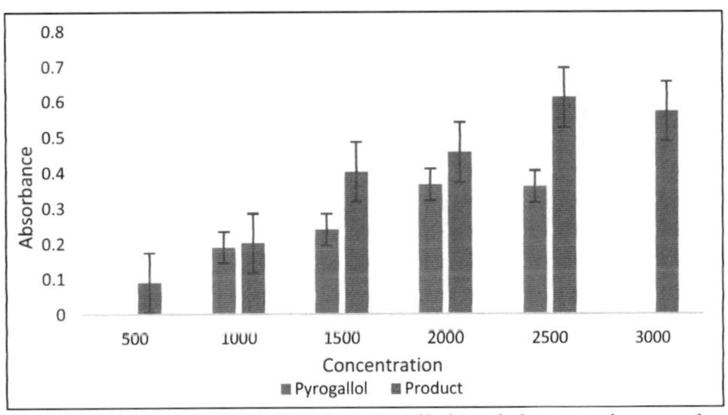

Figure 10. Solubility limit of pyrogallol and the reaction product

As seen in Figure 10 the absorbance value of pyrogallol stopped increasing at 2000 ppm, which implied that the concentration of the antioxidant did not add up in biodiesel after 2000 ppm. Therefore, it could be assumed that pyrogallol's solubility limit in biodiesel was 2000 ppm or 0.002 g/ml in biodiesel. The same analysis was performed with the reaction product's absorbance. The absorbance of the reaction product keeps increasing until 2500 ppm, while pyrogallol's absorbance stopped increasing at 2000 ppm. Hence, the reaction products solubility limit was 2500 ppm or 0.0025 g/ml in biodiesel. This concludes that the solubility of the reaction product in biodiesel was higher than pyrogallol.

3.8. Antioxidant Qualitative Analysis

Although the main objective of this research was increasing the solubility of pyrogallol in biodiesel, the antioxidant activity of the new product must also be evaluated to justify that the new product did not lose its antioxidant ability. Therefore, pyrogallol and the reaction product was analyzed qualitatively. The purpose of this qualitative analysis is to evaluate the DPPH change of color.

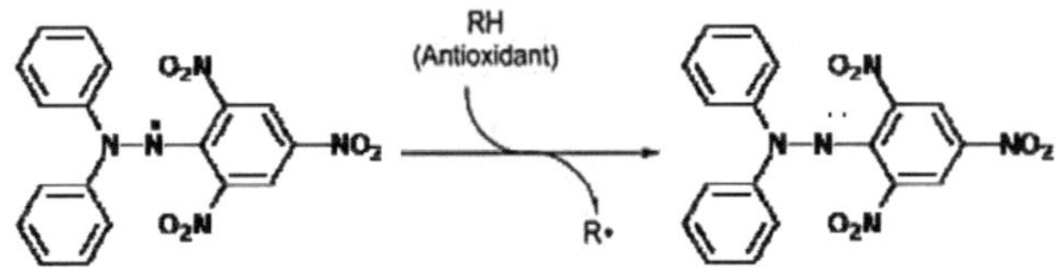

2,2-diphenyl-1-picrylhydrazyl
Free Radical (DPPH•)

2,2-diphenyl-1-picrylhydrazine
(DPPH)

Figure 11. DPPH Antioxidant Mechanism

DPPH as a stable free radical has a violet color in methanolic solution. In the presence of an antioxidant, the color violet (2,2-diphenyl-1-picrylhydrazyl) will change into yellow (2,2-diphenyl-1-picrylhydrazine) because DPPH as a free radical will react with hydrogen which leads to DPPH losing its free radical property. The results showed, pyrogallol as one of the best antioxidants indicated the presence of antioxidant activity with the presence of color change as expect. As for the reaction product, the color of the methanolic DPPH also changed from violet to yellow, which can be qualitatively concluded that the new product had antioxidant activity.

3.9. Rancimat Test

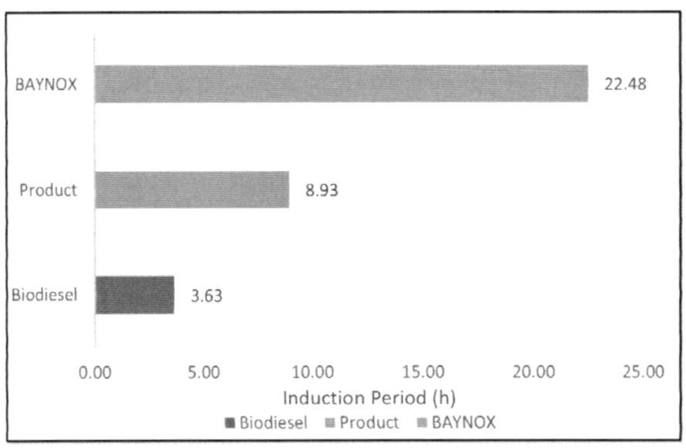

Figure 12. Induction period results from rancimat test

The Rancimat test was conducted based on EN 14112 method, the samples prepared was blank biodiesel, the reaction product, and BAYNOX as the commercial antioxidant for comparison. The concentration for BAYNOX and the reaction product was 2000 ppm in biodiesel. The result confirmed that the reaction product can slow down the oxidation of biodiesel as the reaction product has a higher induction period compared to blank biodiesel. Based on the international standard for biodiesel, the induction period standard successfully fulfills the ASTM D6751 (United States) standard which is induction period higher than 3 h and successfully fulfilled the national standard of biodiesel SNI 7182 which was induction period higher than 8 h.

3.10. Acid Value

Acid Value is used as an indicator to determine the amount of free fatty acid (FFA) in biodiesel sample. Acid value monitors the degradation of biodiesel during storage since acid value increases with degradation. The Acid value was calculated using Eq. 1.

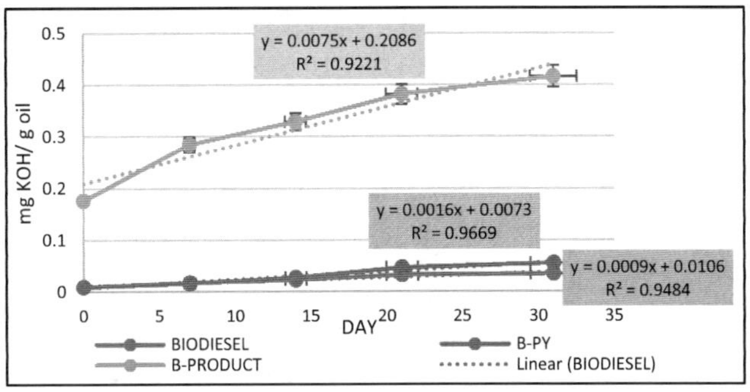

Figure 13. Acid value vs Biodiesel additive

As seen on Figure 15, biodiesel with no additives and biodiesel-pyrogallol has similar acid values (around 0.0089) at day 0, while the acid value of the biodiesel- reaction product was much higher which is 0.17. The reaction product had much higher acid value compared to biodiesel and biodiesel-pyrogallol, possibly because of the formation of formic acid which was a product from biodiesel secondary oxidation. The oxidation of biodiesel might have been caused by the remaining unreacted benzoyl peroxide in the mixture since benzoyl peroxide is a strong oxidizing agent.

All samples have an increase in acid value during the four weeks of analysis. On the 31st day, biodiesel-product has a higher acid value compared to biodiesel-pyrogallol and biodiesel with no additives which doesn't show a positive result of the product in slowing down the increase of acid number caused oxidation. However, the biodiesel-product at the end of analysis has an acid value of 0.41 which was not more than 0.5. Based on EN 14214 and ASTM D6751 acid value of biodiesel must not exceed 0.5 to fulfill the antioxidant requirement, thus the new product can still be considered as an antioxidant for biodiesel.

4. Conclusion

The synthesis between pyrogallol and methyl linoleate to produce a molecule with higher solubility in biodiesel produced a molecule (10*E*,12*Z*)-9-(2,3,4-trihydroxyphenyl) octadeca-10,12-dienoate and its isomer methyl (9*Z*,11*E*)-13-(2,3,4-trihydroxyphenyl) octadecyl-9,11-dienoate. However, from the HPLC result, it was found that the yield of the product was only 0.2%.

The TLC analysis indicates that the new product was more non-polar compared to pyrogallol and methyl linoleate because of its higher Log P value. The HPLC showed the product's peak at 9.779 min with a yield of 0.2%, confirming that the product was a non-polar compound. As for the LC-MS/MS, the result shows the desired product molecular mass which was 418 g/mol.

The solubility test cannot be performed to evaluate the solubility limit of the desired product, however, the test result revealed that the reaction product had a higher solubility limit which was 2500 ppm in biodiesel, compared to pyrogallol's solubility limit at 2000 ppm. The antioxidant activity evaluation concluded that the reaction product fulfilled the ASTM D6751 (United States) standard, European EN 14214, and Indonesia National Standard SNI 7182 induction period requirements of 3 h, 6 h and 8 h. However, the reaction product had a lower acid value stability compared to pyrogallol and blank biodiesel.

IOP Conf. Series: Materials Science and Engineering **742** (2020) 012039 doi:10.1088/1757-899X/742/1/012039

References

[1] N. Kapilan, T. P. Ashok Babu, and R. P. Reddy, "Technical aspects of biodiesel and its oxidation stability," Int. J. ChemTech Res., vol. 1, no. 2, pp. 278–282, 2009.

[2] J. Pullen and K. Saeed, "An overview of biodiesel oxidation stability," Renew. Sustain. Energy Rev., vol. 16, no. 8, pp. 5924–5950, 2012.

[3] N. Kumar, "Oxidative stability of biodiesel: Causes, effects and prevention," Fuel. 2017.

[4] Supriyono, H. Sulistyo, M. F. Almeida, and J. M. Dias, "Influence of synthetic antioxidants on the oxidation stability of biodiesel produced from acid raw Jatropha curcas oil," Fuel Process. Technol., vol. 132, pp. 133–138, 2015.

[5] W. H. Brown and T. Poon, Introduction to Organic Chemistry, 6th Edition. Wiley, 2016.

[6] H. Sutanto, B. H. Susanto, and M. Nasikin, "Solubility and antioxidant potential of a pyrogallol derivative for biodiesel additive," Molecules, vol. 24, no. 13, pp. 1–16, 2019.

[7] W.-F. Su, "Radical Chain Polymerization BT - Principles of Polymer Design and Synthesis," W.-F. Su, Ed. Berlin, Heidelberg: Springer Berlin Heidelberg, 2013, pp. 137–183.

[8] D. B. Denney and D. Z. Denney, "Studies of the Mechanisms of the Reactions of Benzoyl Peroxide with Secondary Amines and Phenols," J. Am. Chem. Soc., vol. 82, no. 6, pp. 1389–1393, 1960.

[9] T. C. P. M. Ramos, E. F. De Souza, C. C. Pina, A. A. Cavalheiro, A. R. Fiorucci, and M. S. Da Silva, "Evaluation of Natural Antioxidants Action in Oxidative Stability of Commercial Biodiesel," Orbital Electron. J. Chem., vol. 10, no. 1, 2018.

[10] S. A. Solomons, T. W. Graham; Fryhle, Craig B.; Snyder, "'Organic Chemistry,'" in J. Chem. SOC, 2013.

[11] K. Malinda, H. Sutanto, and A. Darmawan, "Characterization and antioxidant activity of gallic acid derivative," in AIP Conference Proceedings, 2017.

IOP Conf. Series: Materials Science and Engineering **742** (2020) 012040 doi:10.1088/1757-899X/742/1/012040

Effect of Temperature on Rare Earth Elements Recovery from Coal Fly Ash Using Citric Acid

Pramesti Prihutami[1], Wahyudi Budi Sediawan[1], Widi Astuti[2] and Agus Prasetya[1]

[1]Department of Chemical Engineering, Universitas Gadjah Mada, Yogyakarta, Indonesia.
[2]Research Unit for Mineral Technology, Indonesian Institute of Sciences, Lampung Selatan, Indonesia.

E-mail : aguspras@ugm.ac.id

Abstract. Production of fly ash as a solid waste of coal combustion rises as the need of electricity in Indonesia increases. In 2020, the amount of fly ash produced is estimated to reach 9.72 million tons. Utilization of coal fly ash is still limited while poor handling on abundant amount of this material can cause air, soil and water pollution. This research aims to recover rare earth elements contained in coal fly ash. In this study, the recovery of La, Ce, Nd, Dy and Y was done using citric acid. Magnetic phase of fly ash with particle size less than 38 μm from PLTU Tanjung Awar-awar Tuban was leached with 300 mL citric acid of 0.5 M using solid/liquid ratio of 1:10. The experiment was carried out in various temperature of 26°C and 45°C. Liquid samples were taken at 2, 5, 10, 15, 30, 60, 90, 120 and 240 minutes to be analyzed. The five elements recovered was found to rise until the end of the leaching period. Value of relative recovery also increases along with leaching temperature. The greatest recovery value is reached by yttrium that are 51.00% for leaching at 26°C and 83.35% at 45°C.

1. Introduction

Indonesia is a country who meets 58.1% its need of electricity using coal as its source of energy [1]. The utilization of coal is predicted to increase every year due to government's electricity supply business plan [2]. However, the combustion of coal produces fly ash which considered as solid waste with still limited utilization [3]. Total production of the waste is estimated to reach 9.72 million tons in 2020 [4], [5]. Poor handling on fly ash can cause air, soil and water pollution due to its ability to generate dust and heavy metal composition [6].

Aside from containing heavy metal, fly ash also contains rare earth elements [7], [8], [9]. Rare earth elements consist of 15 lanthanides, yttrium and scandium in which all have similar properties except for promethium which is a radioactive [10]. The demand for these elements rises 5.3% every year since its unique catalytic, magnetic and phosphorescent properties are critical for modern technology [11], [12], [13]. In general, world's coal fly ash is foreseen to have 445 ppm rare earth elements and is potential to be source of rare earth elements when the amount is comparable to conventional ore [7].

One of the method that can be used in order to recover rare earth elements from its source is leaching method using acid [14], [15], [16]. Organic acid can be used to replace inorganic acid which may produce dangerous gasses as well as harm ecosystem when penetrates into soil and water [17], [18]. In this study, the acid used is citric acid which is one of the common acid produced from metabolism activity in living being including microbes [19]. Leaching activity in different temperature was done to observe its effect on rare earth elements recovery from coal fly ash.

2. Materials and Methods

2.1. Materials

Coal fly ash was obtained from PLTU Awar-awar Tuban, Jawa Timur, Indonesia. Fly ash has been screened on sieve trays to get raw material with particle size of less than 38 μm. The raw material has been separated using magnetic separator via wet method to obtain magnetic phase of fly ash that later was used in the study. Citric acid monohydrate ($C_6H_8O_7.H_2O$) used as leachant was of analytical grade bought from Merck with CAS number of 5949-29-1 and was prepared using distilled water.

2.2. Coal fly ash preparation

Coal fly ash was dried in oven at 125°C for an hour.

2.3. Citric acid leaching

As much as 300 mL of citric acid 0.5 M was put into a three neck flask placed on waterbath. Fly ash of 30 gram was then added. The mixture was stirred using glass stirrer of 400 rpm for 4 hours. Samples of 3 mL were drawn using syringe at 2, 5, 10, 15, 30, 60, 90, 120 and 240 minutes. Solid and liquid phase were later separated using centrifuge at 4,000 rpm for one minute. The experiment was repeated for temperature of 45°C. Three neck flask filled with 300 mL of citric acid 0.5 M was heated to experiment temperature before fly ash was added and the temperature was kept constant using waterbath.

2.4. Liquid sample preparation

Liquid sample was filtered using microfilter. As much as 1 mL of sample was then drawn and diluted using distilled water to produce 10-fold dilution before analyzed using ICP-EOS PQ 9000 Elite.

3. Result and discussion

Relative recovery of rare earth elements is described as the amount of element leached into liquid body at every sampling time compared to maximum amount of element that can be leached by the system. The value of relative recovery is calculated using Equation (1).

$$\text{Recovery} = \frac{C_t}{C_{max}} \text{x } 100\% \qquad (1)$$

where Recovery is relative recovery in percent (%), C_t is concentration of rare earth element at every sampling time (mg/L), C_{max} is maximum concentration of rare earth element that can be leached into citric acid solution (mg/L).

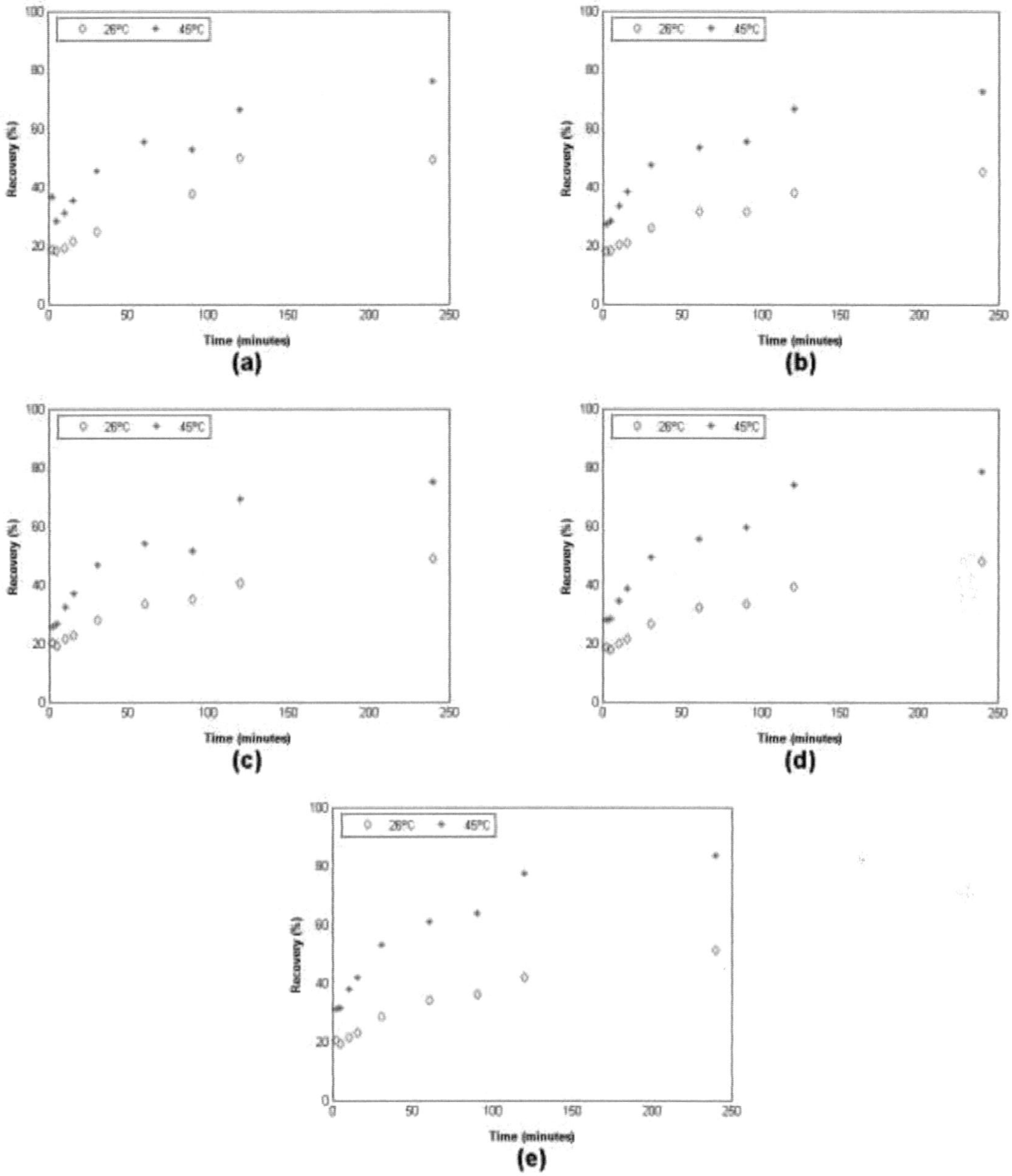

Figure 1. Recovery of (a) Lanthanum, (b) Cerium, (c) Dysprosium, (d) Neodymium and (e) Yttrium Using Citric Acid Solution

Figure 1 showed that recovery of five rare earth elements studied increase as leaching time increases. The longer the leaching process, the longer contact time between citric acid and fly ash that enables more rare earth elements to be leached out. The recovery of rare earth elements is shown to be rapid at earlier time and later becomes comparably slow. Citric acid having 3 carboxyl group and 1 hydroxyl group leaches rare earth elements out of fly ash by forming soluble metal-ligand complex. This chemical reaction can only happen when the reactants undergo effective collision. At first, the amount of citric acid is still abundant. The amount becomes less as it leaches out rare earth elements as well as other leachable elements. As consequence, the number of collisions between reactants decreases making the leaching rate to become slower with time.

Figure 1 also showed that the recovery of lanthanum, cerium, dysprosium, neodymium and yttrium are all greater at 45°C compared to room temperature leaching. Heating the system to 45°C increases kinetic energy of the reactants, allows them to collide and react faster with citric acid molecules. Therefore, this results in faster leaching rate.

Table 1. Recovery for Every Leaching Temperature at 240 Minutes

Elements	Recovery (%)	
	26°C	45°C
La	49.06	75.95
Ce	45.10	72.36
Dy	48.78	75.05
Nd	47.93	78.47
Y	51.00	83.35

Among five elements studied, yttrium gives greatest recovery values. At room temperature leaching, 51.00% of yttrium can be recovered. The value goes even higher for leaching at 45°C and reaches 83.35% after 240 minutes. This is due to the small ionic radii of yttrium compared to the rest of the elements. Ion of Y^{3+} has ionic radii of 0.9 Å while La^{3+}, Ce^{3+}, Dy^{3+} and Nd^{3+} have radii of 1.032 Å, 1.01 Å, 0.983 Å and 0.912 Å, respectively [20]. Small radii results in smaller size of yttrium-ligand complex formed after reaction thus makes it possible to diffuse more easily into liquid body. Bigger ionic radii will produce immense complex with citric acid resulting in slower diffusion and smaller amount of elements leached into liquid body.

Leaching kinetics of rare earth elements from the solid surface to liquid body is also studied. C_{max} is defined as concentration of rare earth element in liquid film that is always in equilibrium with its concentration in solid. Meanwhile, C_t is rare earth element concentration in liquid body at any given time. The difference between C_{max} and C_t is the driving force for rare earth element leaching.

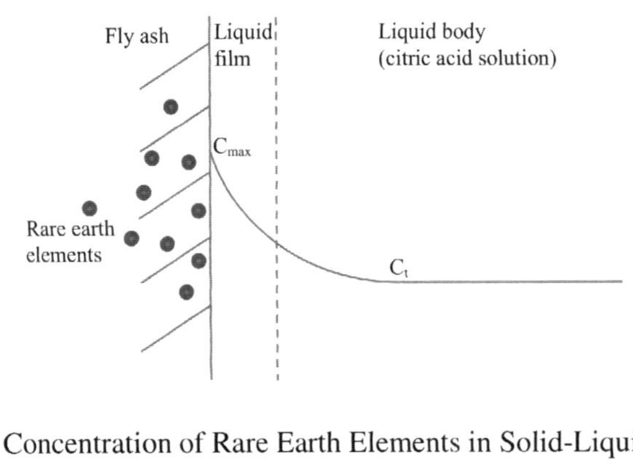

Figure 2. Concentration of Rare Earth Elements in Solid-Liquid System

Assuming that the chelate formation of rare earth element – citric acid in solution is faster than the mass transfer rate of rare earth element in liquid film, the rate of leaching of rare earth element to liquid body for batch system can be described by Equation (2), assuming that the leaching rate follows first order mass transfer kinetics.

$$V\frac{dC_t}{dt} = k_{app}(C_{max} - C_t)V \quad (2)$$

where V is liquid volume (L), C_t is rare earth element concentration in liquid body at any time t (mg/L), t is leaching time (min), k_{app} is the apparent leaching rate constant (min^{-1}) and C_{max} is maximum concentration of rare earth element in liquid (mg/L).

IOP Conf. Series: Materials Science and Engineering **742** (2020) 012040 doi:10.1088/1757-899X/742/1/012040

Equation (2) can be written in relative recovery form to obtain Equation (3).

$$\frac{dx}{dt} = k_{app}(1 - x) \quad (3)$$

where x is relative recovery as described by Equation (1).

The solution for Equation (3) results in linear equation.

$$-\ln(1 - x) = k_{app}t + c \quad (4)$$

where x is the value of relative recovery, k_{app} is leaching rate constant (min^{-1}), t is leaching time (min) and c is intercept.

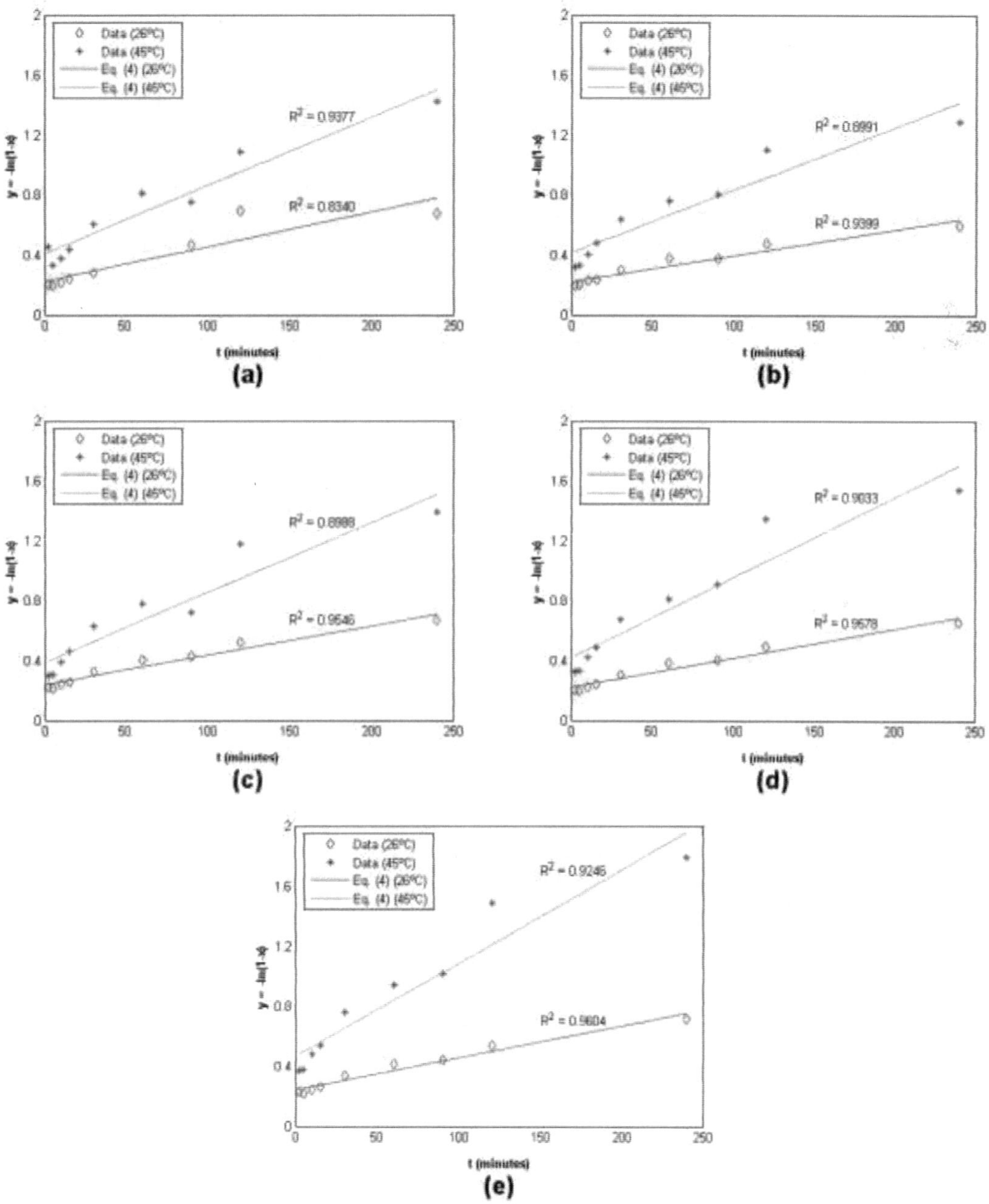

Figure 3. Leaching Kinetics of (a) Lanthanum, (b) Cerium, (c) Dysprosium, (d) Neodymium and (e) Yttrium into Citric Acid Solutions

As shown by Figure 3 (a) to (e), the leaching kinetics for all rare earth elements investigated fit well with the first order mass transfer kinetics model.

The apparent leaching rate constant (k_{app}) is assumed to follow the form of Arrhenius equation [21].

$$k_{app} = A' e^{-\frac{E}{RT}} \quad (5)$$

where A' describes the relative intensity of interactions between rare earth element and citric-acid molecules, E describes 'energy barrier' for leaching process, R is gas constant and T is absolute leaching temperature. The value of A' as well as E can be calculated using Equation (5).

Table 2. The Mass Transfer Kinetics Data of Rare Earth Elements into Citric Acid Solution

Element	k_{app} (min^{-1})		A	E/R (K)
	26°C ($x10^3$)	45°C ($x10^3$)		
La	2.31	4.55	196.78	3,396.31
Ce	1.73	4.16	4,238.58	4,401.61
Dy	1.94	4.67	4,798.55	4,403.81
Nd	1.94	5.34	45,397.37	5,076.45
Y	2.12	6.21	142,630.26	5,392.52

The value of k_{app} for different temperature show some differences. Leaching at 45°C causes the value of k_{app} to become 2 to 3 times higher than the ones at room temperature. The differences among k_{app} value of all rare earth elements at room temperature are practically not significant. Meanwhile, the distinction of the leaching rate constant for each element is noteworthy at elevated temperature. It differs from $4.16x10^{-3}$ to $6.21x10^{-3}$ min^{-1} with yttrium having the highest leaching rate constant. This corresponds with earlier data where recovery of yttrium is the highest at the same given time. The small size of yttrium ion makes it collide more frequently and it is enough to overcome its high activation energy.

4. Conclusions

Recovery of five rare earth elements from coal fly ash using citric acid was studied. The results of the present study showed that citric acid had a promising potential as leachant for rare earth elements. The amount of rare earth elements leached increases with time. Elevated temperature has positive effect on leaching La, Ce, Dy, Nd as well as Y shown by the escalation of recovery value. The leaching rate of each element is faster as the temperature becomes higher. At 26°C, leaching rates of the rare earth elements practically have no significant differences but at 45°C the leaching rates turn to be more varied. The highest value of recovery is achieved by yttrium which has the smallest ionic radii with a value of 83.35% at 45°C after 240 minutes of leaching.

Acknowledgment

This study was supported by Research Unit for Mineral Technology from Indonesian Institute of Sciences (BPTM LIPI) Lampung and Universitas Gadjah Mada, Yogyakarta, Indonesia.

References

[1] Arinaldo D and Adiatma J C 2019 Indonesia's Coal Dynamics: Toward A Just Energy Transition ed F Tumiwa (Jakarta: Institute for Essential Services Reform (IESR))

[2] Hasan M H, Mahlia T M I and Nur H 2012 A review on energy scenario and sustainable energy in Indonesia Renew. Sustain. Energy Rev. 16 2316–28

[3] Bhatia A and Gakkhar N 2017 Eco-friendly fly ash utilization: properties and its applications in various field to reduce the environmental pollution J Env. Life Sci. 2 66–76

[4] Kurniawan A R 2017 The Behavior of Heavy Metals Content in Coal Combustion Products (CCPs) and Its Leachate From Indonesia Coal Power Plants Indones. Min. J. 17 75–86

[5] Abdullah M H, Rashid A S A, Anuar U H M, Marto A and Abuelgasim R 2019 Bottom ash utilization: A review on engineering applications and environmental aspects IOP Conf. Ser. Mater. Sci. Eng. 527 1–9

[6] He Y, Luo Q and Hu H 2012 Situation Analysis and Countermeasures of China's Fly Ash Pollution Prevention and Control Procedia Environ. Sci. 16 690–6

[7] Franus W, Wiatros-motyka M M and Wdowin M 2015 Coal fly ash as a resource for rare earth elements Env. Sci Pollut Res 22 9464–74

[8] Dai S, Zhao L, Peng S, Chou C, Wang X, Zhang Y, Li D and Sun Y 2010 Abundances and distribution of minerals and elements in high-alumina coal fl y ash from the Jungar Power Plant, Inner Mongolia, China Int. J. Coal Geol. 81 320–32

[9] Hower J C, Groppo J G, Graham U M, Ward C R, Kostova I J, Maroto-valer M M and Dai S 2017 Coal-derived unburned carbons in fly ash: A review Int. J. Coal Geol. 179 11–27

[10] Taggart R K, Hower J C, Dwyer G S and Hsu-Kim H 2016 Trends in the Rare Earth Element Content of U.S.-Based Coal Combustion Fly Ashes Environ. Sci. Technol. 50 5919–26

[11] Alonso E, Sherman A M, Wallington T J, Everson M P, Field F R, Roth R and Kirchain R E 2012 Evaluating rare earth element availability: A case with revolutionary demand from clean technologies Environ. Sci. Technol. 46 3406–14

[12] Charalampides G, Vatalis K I, Apostoplos B and Ploutarch-Nikolas B 2015 Rare Earth Elements: Industrial Applications and Economic Dependency of Europe Procedia Econ. Financ. 24 126–35

[13] Balaram V 2019 Rare earth elements: A review of applications, occurrence, exploration, analysis, recycling, and environmental impact Geosci. Front. 10 1285–303

[14] Panda R, Kumari A, Jha M K, Hait J, Kumar V, Kumar J R and Lee J Y 2014 Leaching of rare earth metals (REMs) from Korean monazite concentrate J. Ind. Eng. Chem. 20 2035–42

[15] Walawalkar M, Nichol C K and Azimi G 2016 Process investigation of the acid leaching of rare earth elements from phosphogypsum using HCl, HNO3, and H2SO4 Hydrometallurgy 166 195–204

[16] Kashiwakura S, Kumagai Y, Kubo H and Wagatsuma K 2013 Dissolution of Rare Earth Elements from Coal Fly Ash Particles in a Dilute H2SO4 Solvent Open J. Phys. Chem. 03 69–75

[17] Semrau K T 1971 Control Of Sulfur Oxide Emissions From Primary Copper, Lead And Zinc Smelters — A Critical Review J. Air Pollut. Control Assoc. 21 185–94

[18] Hassan P, Jusop S, Ismail R, Aris A Z and Panhwar Q A 2016 Soil and water quality of an acid sulfate soil area in Kelantan Plains, Malaysia and its effect on the growth of rice Asian J. Agric. Food Sci. 96 124–38

[19] Show P L, Oladele K O, Siew Q Y, Zakry F A A, Lan J C-W and Ling T C 2015 Overview of citric acid production from Aspergillus niger Front. Life Sci. 8 271–83

[20] Shannon R D 1976 Revised Effective Ionic Radii and Systematic Studies of Interatomic Distances in Halides and Chaleogenides Acta Crystallogr. A32 751–67

[21] Laidler K J 1984 The development of the Arrhenius equation J. Chem. Educ. 61 494–8

IOP Conf. Series: Materials Science and Engineering **742** (2020) 012041 doi:10.1088/1757-899X/742/1/012041

Amorphous Silicate Decomposition from Non-Magnetic Coal Fly Ash using Sodium Hydroxide

H Manurung[1], W Rosita[1], H B T M Petrus[1,2] and I M Bendiyasa[1,2]

[1]Departemen Teknik Kimia (Sustainable Mineral Processing Research Group), Fakultas Teknik, Universitas Gadjah Mada, Jl. Grafika No. 2,Yogyakarta 55281 Indonesia
[2]Unconventional Geo-resources Research Center, Faculty of Engineering, Universitas Gadjah Mada, Jl. Grafika No. 2,Yogyakarta 55281, Indonesia

E-mail : imbendiyasa@ugm.ac.id

Abstract. In five years, from 2015 to 2020, the need of coal in Indonesia is doubled to 166.2 million tonnes and increasing up to 2050 as stated in the Indonesian Energy Mix policy. Generally, coal is used in the coal-fired power plan. After the combustion, there are some residual materials such as mineral matter, fly ash and bottom ash (FABA) listed as hazardous materials due to the toxicity and metal contents. Referring to the amount of coal consumption in 2020, the fly ash and bottom ash will be of about 19.92 million tonnes. The utilization of FABA is then become a must thing to do. One of the alternatives is rare earth element (REE) extraction. Indonesian coal contains sufficient amount of REE to extract that is concentrated in the FABA to the value of more than 400 ppm. Regarding to the REE mode of occurrence that is mostly concentrated in the glass form or amorphous silicate, this study was conducted in order to observe the effect of sodium hydroxide to breakage the amorphous silicate structure. In order to obtain the amorphous silicate, magnetic separator was applied. Based on the XRF analysis, the non-magnetic fly ash (amorphous silicate of glass form) contained major elements of SiO_2, Al_2O_3, Fe_2O_3, CaO, MgO and the rest were minor and trace elements such as Na_2O, K_2O, Cr_2O_3, MnO, SrO and P_2O_5. The particle size of the amorphous silicate is less than < 38 µm (- 400 mesh). To run the experiments, sodium hydroxide was used as the reagent with varied concentrations of 6 M, 8 M and 10 M. Other variables were temperature of 28 ^{0}C (room temperature), 60 ^{0}C and 90 ^{0}C and residence time of 2, 4, and 6 hour. Based on the results analysis, the amorphous silicate was decomposed to the most of about 22 % at 90 ^{0}C of temperature, 10 M of sodium hydroxide concentration, and 6 hour of residence time.

1. Introduction

Coal is one of the main energy resource in the world. Indonesia is listed as a top ten of coal producer in the world. In 2019, Indonesia has 50,763.52 million tonnes of resources and 39,890.95 million tonnes of coal reserves[1]. Indonesian coal mostly in medium rank until low rank due to high content of impurities such as mineral matter, moisture content, ash and sulfur content[2]. In five years, from 2015 to 2020, the need of coal in Indonesia is doubled to 166.2 million tonnes and increasing up to 2050 as stated in the Indonesian Energy Mix policy. Generally, coal is used in the coal-fired power plan. After the combustion, there are some residual materials such as mineral matter, fly ash and bottom ash (FABA) listed as hazardous materials due to the toxicity and metal contents. Referring to the amount of coal consumption in 2020, the fly ash and bottom ash will be of about 19.92 million tonnes[3]. The utilization of FABA is then become a must thing to do. Coal fly ash were listed as a one of alternative resource of rare earth element (REE)[4]–[6]. Ferian et al[7] recently reported that fly ash and bottom ash from coal power plant in Java are promising raw material for rare earth element extraction.

The existence of rare earth element in coal fly ash were influenced by the coal rank and mineral matter such as quartz, clay mineral, sulphide and phosphate[8]. Kashiwakura et al[9] reported that due to high temperature in combustion process, rare earth element were melting and distributed into the fly ash such as amorphous phase, quartz (SiO_2) and mullite ($3Al_2O_3$). Due to form amorphous silicious and quartz, rare earth element was found that to be a slight enrichment in the samaller non-magnetic component of coal fly ash[10].

Indonesian coal contains sufficient amount of REE to extract that is concentrated in the FABA to the value of more than 400 ppm[6], [7]. Regarding to the REE mode of occurrence that is mostly concentrated in the glass form or amorphous silicate, this study was conducted in order to observe the effect of sodium hydroxide to breakage the amorphous silicate structure. In order to obtain the amorphous silicate, magnetic separator was applied.

The aim of this study is to identify the optimum condition of sodium hydroxide to breakage the amorphous silicate. The variable studied were reagent concentration, residence time and temperature. In addition sodium hydroxide not only decomposed the amorphous silicate but also to change the REE bound to from REEOH.

2. Research methodology

2.1. Material

Coal fly ash were collected from PLTU Tuban East java. Each particle size were separated using sieve. The particle size less than 38 μm (-400 mesh) were used as a raw material. In order to obtain non-magnetic component (amorphous silicate), wet magnetic separator was applied in Centre for Research and Technology Development of Mineral and Coal (PUSLITBANG TekMIRA) Bandung, West Java. XRF analysis were applied for raw material characterization to identify the major and minor component. ICP-MS also were done to identify the rare earth element concentration in ALS Global – Geochemistry Analytical Laboratory in North Vancouver, BC, Canada.

Table 1 Mayor element and Minor element in non-manetic coal fly ash[11]

Mineral	Wt (%)
SiO_2	53.20
Al_2O_3	31.90
Fe_2O_3	5.58
CaO	3.08
MgO	2.11
Na_2O	0.90
K_2O	0.99

2.2. Silicate digestion procedure

Silicate digestion process was conducted in a erlenmeyer 100 mL which was placed on a water bath equipped with shaking equipment. 10 gr sample with 50 mL sodium hydroxide (solid-to-liquid 20 %) and heated until to desired temperature. The variable studied were reagent concentration (6M, 8M and 10 M), temperature (30^{0}C (room temperature), 60 ^{0}C and 90 ^{0}C), and residence times(2 hour, 4 hour and 6 hour). The solid digested were washed using destilated water (60x) based on dry weight sample in 30 minutes and stirred using magnetic stirrer.

2.3. *Data analysis and calculation*

The data analysis for this experiment were performed via MINITAB. The method used was DOE using Response Surface Method as the basis to find the optimum condition.

3. Result and discussion

3.1. Material characterization

The major elements concentration in the form of oxide as given in Table 1. Silicate oxide is the major oxide in the non-magnetic coal fly ash followed by alumina oxide, iron oxide, calcium oxide, magnesium oxide, sodium oxide and kalium oxide. The concentration of Rare Earth Element and Yittrium (REY) in raw CFA around 260.38 ppm and REY concentration in non-magnetic CFA around 314.15 ppm. It was enriched around 20.65% after magnetic separator. The amount of each rare earth element can be seen as in Figure 1.

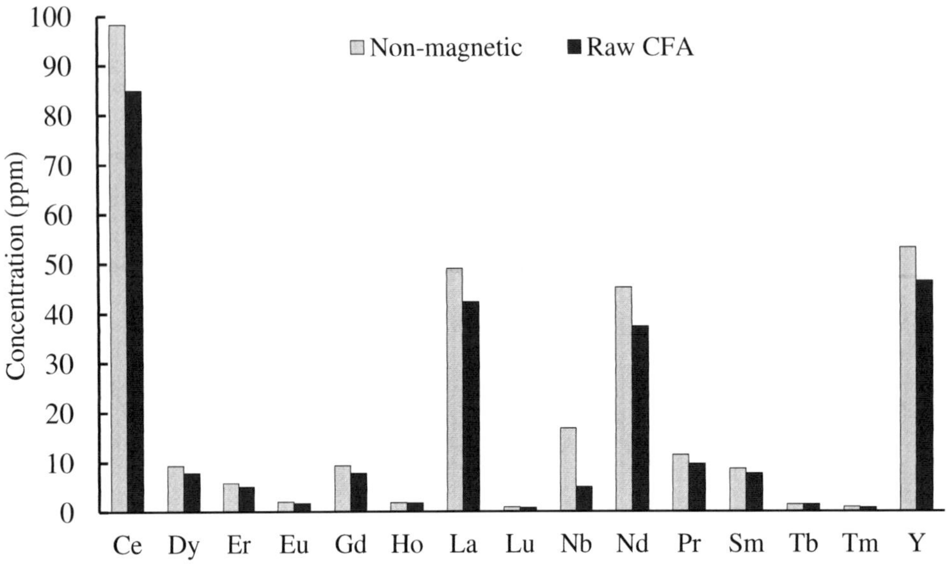

Figure 1. Rare earth and Yittrium concentration in raw CFA and non-magnetic CFA

Figure 1 shows that rare earth and yittrium concentrations were slightly increase from raw CFA after physical separation using magnetic separator. to non-magnetic CFA component. Blisset et al[10] also reported that, rare earth element was found to be enrichment in the non-magnatic coal fly ash. The enrichment processes consist due to mineralization of REY in the coal fly ash after combustion process and than distributed to amorphous mineral such as silicious, clay mineral, mullite[9], [12], [13].

3.2. Silicate digestion

The experiment result related to silicate digestion at various reagent concentration, temperature and residence time are shown in Figure 2 and Figure 3. Sodium hydroxide were used to decrease and breakage the amorphous silicate content. Generally, the higher concentration the higher SiO2 decomposed, thin consist in all residence time and all temperature. The maximum SiO2 decomposed around 21% at temperature 90 ^{0}C, sodium hydroxide concentration 10M and residence time 6 hours. Based on the coal fly ash characterization shows that REY were distributed in the amorphous mineral, than the chemical formulation can be wrote as SiO_2.REY. Na+ and OH- from sodium hydroxide will react with SiO_2.REY, the assumption reaction can be Describe as equation 1 below.

$$SiO_2.REY_{(s)} + NaOH_{(aq)} \rightarrow NaSiO_{3\,(aq)} + REY(OH)_{3(s)} + H_2O_{(l)} \quad (1)$$

Equation 1 shows that Na+ from sodium hydroxide react with SiO_2 and form $NaSiO_3$ in aqueous phase. In addition, OH- from sodium hydroxide were modify the REY bound to form $REY(OH)_3$ in the solid phase. Figure 2 shows the amount of SiO_2 were decomposed in the various temperature and concentration at 6 hour residence time.

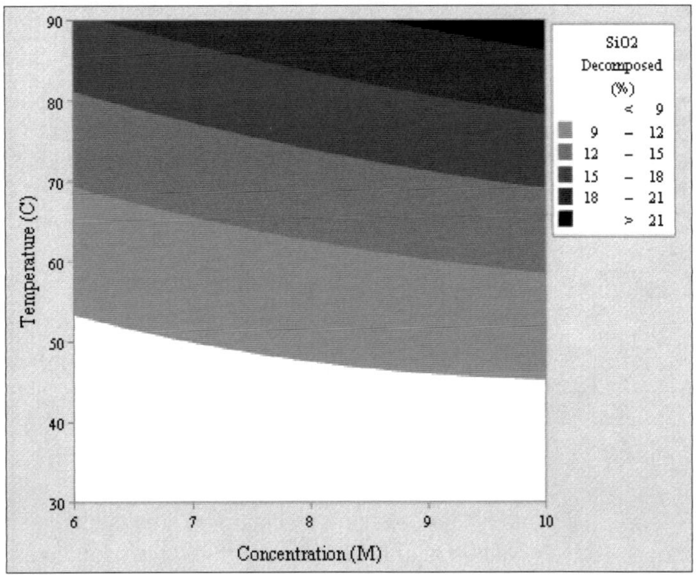

Figure 2 Contour plot of SiO_2 decomposed vs temperature (C); Concentration (M) at 6 hour residence time

Figure 2 shows that temperature and concentration influence the amount of SiO_2 decomposed. In this part, residence time at 2 hour and 4 hours are not shown due to the similar phenomenon with 6 hours residence time. For predicting the SiO_2 decomposed at 2, 4 and 6 hours residence time of silicate digestions have developed models and given as Equaiton 2, Equaition 3 and Equation 4 below..

For 2 hour residence time

$$SiO_2 \text{ Decomposed } (\%) = 27{,}29 - 2{,}90x - 0{,}4935y + 0{,}109x^2 + 0{,}004491\, y^2 + 0{,}02240\, xy \quad (2)$$

R-sq = 99,32 %

For 4 hour residence time

$$SiO_2 \text{ Decomposed } (\%) = 20{,}3 - 3{,}43x - 0{,}091y + 0{,}224x^2 + 0{,}00272y^2 + 0{,}0048xy \quad (3)$$

R-sq = 93,23 %

For 6 hour residence time

$$SiO_2 \text{ Decomposed } (\%) = 9{,}3 - 0{,}01\, x - 0{,}229y + 0{,}042x^2 + 0{,}002379y^2 + 0{,}02087xy \quad (4)$$

R-sq = 96,82 %
Where:
x = concentration (M)
y = temperatur (^{0}C).
Based on the Equation 2,3 and 4 shows that intercept for each residence times were different, they were 27,29 (2 hour), 20,3 (4 hour) and 9,3 (6 hour).

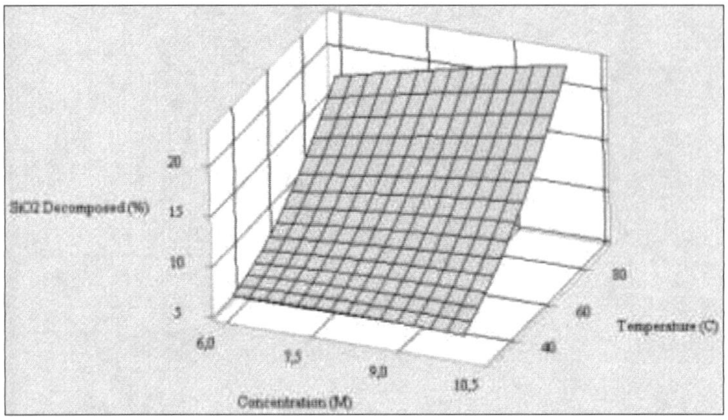

Figure 3 Surface plot of SiO_2 decomposed (%) vs Temperature (C); Concentration (M) at 6 hour residence time.

Figure 3 show the surface plot of SiO_2 decomposed at various temperature and concentration. The influence of temperature on the SiO_2 decomposed shows that increasing temperature from room temperature to 60 ^{0}C and 90 ^{0}C were increase the SiO_2 decomposed. This is consist with Arrhennius rate law, wich states that, reaction rate will increase twice when increasing temperature 10 ^{0}C[14], [15]. This data also shows that, the solubility of amorphous silicate were increase simultaneously during increasing temperature until 90 ^{0}C.

The influence of residences time on SiO2 decomposed shows that, residence times were influence the SiO_2 decomposed. Increasing residence time were increase the SiO_2 solube. The maximum SiO_2 decomposed 15,16%, 17,4% and 22,01 % were achieved at 2, 4 and 6 hour residence time respectively at temperature 90 ^{0}C and concentration 10 M..

The influence of concentration on SiO_2 decomposed shows that (Figure 2 and Figure 3), there were no significant effect of sodium hydroxide concentration in room temperature reaction. Increasing temperature from room temperature to 60 ^{0}C where increase the SiO_2 decomposed from 9% to 15%. Increasing temperature until 90 ^{0}C shows that, there were significant effect to SiO_2 decomposed, where SiO_2 decomposed were increase from 15 to 21%. This phenomenon consist with thermodynamic rule which states that, reaction rate (k) is a function of temperature[16]. However, in the highest temperature (90^{0}C), concentration were slightly influence the SiO_2 decomposed, where the higher concentration the more SiO_2 dissolve.

4. Conclusion

The experiment results in this work conclude that physical separation using magnetic separator was increase the REY concentration around 20.65 %. While the silicate digestion is strongly affected by temperature, concentration and residence time. The maximum decomposition of SiO_2 (21%) was achieved at temperature of 90 ^{0}C, sodium hydroxide concentration of 10 M and residence time of 6 hour. The increasing concentration of sodium hydroxide from 6 M up to 10 M does not provide any significant effect for decreasing SiO_2 content. However, only temperature that strongly affect the rate of silicate decomposition. Increasing residence time more than 2 hour slightly increases the rate of silicate decomposition.

Acknowledgments

This research was financially supported by Kementerian Riset. Teknologi. dan Pendidikan Tinggi Republik Indonesia on Pendanaan Tesis Magister Program tahun 2019 with contract number 2869/UN1.DITLIT/DIT-LIT/LT/2019.

References

[1] A. C. Adi, F. Lasnawatin, A. B. Pranton, and V. M. Suzanti, "Handbook of energy & economic statistics of indonesia," Jakarta: Ministry of Energy and Mineral Resources Republic of Indonesia, 2019.
[2] Hadiyanto, "Anatomi Sumber Daya Batubara Serta Asumsi Pemanfaatan Untuk PLTU di Indonesia," Jakarta, 2016.
[3] S. Edi, Handbook of Energy and Economic Statistic of Indonesia. Jakarta: Ministry of Energy and Mineral Resources Republic of Indonesia, 2017.
[4] S. Peramaki, "Method Development for Determination and Recovery of Rare Earth Elements from Industrial Fly Ash," 2014.
[5] F. Anggara, D. H. Amijaya, A. Harijoko, T. N. Tambaria, A. A. Sahri, and Z. A. N. Asa, "Rare earth element and yttrium content of coal in the Banko coalfield, South Sumatra Basin, Indonesia: Contributions from tonstein layers," Int. J. Coal Geol., vol. 196, no. July, pp. 159–172, 2018.
[6] A. Ferian and M. Petrus Himawan T., "Potential Source of Rare Earth elements (REE) from Sangatta Coal, Kuta Basin, Indonesia," Annu. Meet. Soc. Org. Pet., vol. 32, no. 170, pp. 9–10, 2016.
[7] A. Ferian, B. D. A. Ayu, R. Widya, and P. H. B. T. Murti., "The Composition and mode of occurence of rare earth elements and yittrium in fly ash and bottom ash from coal-fired plants in Java, Indoneisa," Soc. Oorganic Petrol. Annu. Meet., 2018.
[8] R. B. Finkelman, C. A. Palmer, and P. Wang, "International Journal of Coal Geology Quantification of the modes of occurrence of 42 elements in coal," Int. J. Coal Geol., vol. 185, no. September 2017, pp. 138–160, 2018.
[9] S. Kashiwakura, Y. Kumagai, H. Kubo, and K. Wagatsuma, "Dissolution of Rare Earth Elements from Coal Fly Ash Particles in a Dilute H2SO4 Solvent," vol. 2013, no. May, pp. 69–75, 2013.
[10] R. S. Blissett, N. Smalley, and N. A. Rowson, "An investigation into six coal fly ashes from the United Kingdom and Poland to evaluate rare earth element content," FUEL, vol. 119, pp. 236–239, 2014.
[11] H. Manurung, W. Rosita, I. M. Bendiyasa, A. Prasetya, H. Bayu, and T. Mutri, "Recovery REY from Non-magnetic Coal Fly Ash with Acetic Acid Solvent," no. April, pp. 1–5, 2019.
[12] K. C. Galbreath, D. L. Toman, C. J. Zygarlicke, and J. H. Pavlish, "Trace Element Partitioning and Transformations During Combustion of Bituminous and Subbituminous U . S . Coals In a 7-kW Combustion System," Energy & Fuels, vol. 1, no. 9, pp. 1265–1279, 2000.
[13] R. L. Davison, D. F. S. Natusch, and J. R. Wallace, "Trace Elements in Fly Ash Dependence of Concentration on Particle Size," vol. 8, no. 13, 1974.
[14] O. Levenspiel, Chemical Reaction Engineering. United States of America: John Wiley & Sons, 1999.
[15] H. S. Fogler, Elements of Chemical Reaction Engineering, Fourth Edi. Michigan United States America: Prentice Hall Proffessional TEchnical Reference, 2005.
[16] S. J., V. Ness, and A. M. M, Introduction to Chemical Engineering Thermodynamics, Seventh Ed. Mc Graw Hills Chemical Engineering Series, 2009.

IOP Conf. Series: Materials Science and Engineering **742** (2020) 012042 doi:10.1088/1757-899X/742/1/012042

Fly Ash and Bottom Ash Utilization as Geopolymer: Correlation on Compressive Strength and Degree of Polymerization Observed using FTIR

A S Adelizar[1], M Olvianas[1], D M Adythia[1], M F Syafiyurrahman[1], I G A A N Pratama[1], W Astuti[2], and H T B M Petrus[1,3]

[1]Departemen Teknik Kimia (Sustainable Mineral Processing Research Group), Fakultas Teknik, Universitas Gadjah Mada, Jl. Grafika No. 2,Yogyakarta 55281 Indonesia
[2]Research Unit for Mineral Technology, Indonesian Institute of Science, Jl. Ir. Sutami Km. 15,Tanjung Bintang 35361, Indonesia
[3]Unconventional Geo-resources Research Center, Faculty of Engineering, Universitas Gadjah Mada, Jl. Grafika No. 2,Yogyakarta 55281, Indonesia

E-mail : bayupetrus@ugm.ac.id

Abstract. Concerning on the dependancy of coal on electric generation wordwidely, study on fly ash, thesolid waste of fired coal electricity generation, has been comprehensively conducting, i.e. geopolymer, cenosphere, soil remediation, and adsorbent. However, only few studies dealing with the utilization of bottom ash. Having of about 53% of coal in the 2050 energy mix, Indonesia will be dealing with more than 200 million tonnes of fly ash and bottom ash (FABA). In this study, FABA was mixed in certain proportion of 50:50, 75:25, dan 100:0. The formation of geopolymer was conducted with sodium hydroxide and sodium silicate in certain concentration. Once the mixture had been homogenized, it was casted and cured in varied temperature of 30, 60, dan 90 oC with curing time of 1, 3, 5, 14, 21, dan 28 days. In each curing time, sample was then analyzed using compresive strength apparatus and FTIR in order to observed the geopolimerization of the FABA mixture. It was found that compressive strength representing the macro characteristic of the geopolymer is in line with the CORR analysis of the FTIR results showing that the formation of aluminosilicate is responsible for the mechanical strength of the geopolymer. The higher the bottom ash ratio in the mixture with decrease the strength of the geopolymer in the order of one per second. Higher temperature results higher compresive strength to the value of 42 Mpa (FABA ratio is 75:25 at temperature of 90 oC) above the Indonesian standard for concrete. From this study, kinetics of FABA geopolymerization can be generated in accordance todiffusion model with R2 of 0.9135. Thus the utilization of FABA for geopolymer is going to be a potential solution to overcome the increasing amount of FABA resulted from fired coal electrical generation in Indonesia.

1. Introduction

Energy is the most crusial need for mankind worldwidely. The need of energy is driven by the increase of population requiring high productivity in which energy is highly required. Based on the data collected by Energy Information Administration, there will be increase by 56% on energy demands in the year of 2040 as depicted in Figure 1.

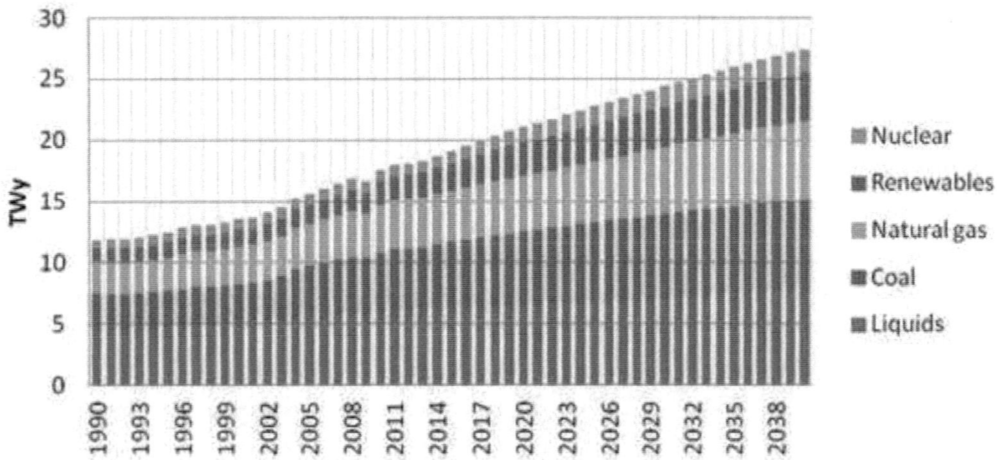

Figure 1. The prediction of world's energy demand for all energy resources.

From Figure 1, it can be seen that the dependancy on fossil fuel to fulfill the energy demand is still high. Among those fossil fuel sources is coal possesing the lowest production cost in electricity generation. Likewise, Indonesia has similar tendency in coal usage as energy source.

Based on the energy mix to the year of 2025, about 50% of the energy supply in Indonesia will be generated from the coal buring power generation in order to fulfill the 35 GW of electricity (Keputusan Menteri Energi dan Sumber Daya Mineral Republik Indonesia, 2016). Increasing needs of coal to fulfill energy sector will implicate to the increasing amount of coal ashes resulted. In general, the ash burning coal will be of about 10% of the total amount of coal used in the coal fired power plant. Amounting of about 10 million tonnes in 2019, this problematic coal fly ash will accumulate to its peak in the year of 2050.

In order to eliminate and or mitigate the potential danger of fly ash, some scientists have been developing utilization of coal fly ash. Not only mitigating the potential danger of fly ash, the studies are also dealing with the potential economic benefits resulted from coal fly ash, i.e., cenospheres recovery (Hirajima et al., 2008; Hirajima et al., 2010; Petrus et al., 2011), fly ash for heavy metals adsorbent (Dasmahapatra et al., 2001; Alinnor, 2007), and for building materials including pozzolanic matter (Singh and Garg, 1999), and geopolymer as for called green concrete (Wilson et al., 1994; Davidovits, 1999, Adelizar et al, 2018, Petrus et al, 2019). The newest trend of coal fly ash utilization is for rare earth element resources (Ferian et al, 2016, Ferian et al., 2018, Ferian et al., 2018, Seredin, 2010, Seredin et al., 2012). Aside from many studies conducted on coal fly ash utilization, it is the coal bottom ash that has not been well utilized due to its characteristic with higher unburned carbon content. Thus, in this study, geopolymer based on fy ash and bottom ash mixture has been developed. Different fromother studies on geopolymer, the characterization will include micro phenomenon in the form of Si- O-Al functional growth observation using FTIR (Olvianas et al, 2017, Petrus et al, 2019). The result will then compared with the compresive strength of geopolymer resulted. Thus it canbe described the effect of bottom ash in the geopolymerization of coal fly ash.

2. Research methodology

Equipment used for the synthesis process in this research include ball mill, Avery-Denison universal testing machine, mold cube with the size of 5x5x5 cm^3, Memmert UN oven, sieve 300 mesh, mortar, stirrer, and digital analytical balance, while for the equipment used for the characterization of the results are the Fourier Transform Infra Red (FTIR) 8201PC Shimadzu, X-Ray Diffraction (XRD), and X-Ray Fluoressence (XRF) spectrophotometers.

The materials used in this research include fly ash and bottom ash obtained from PLTU Tanjung Jati Jepara, NaOH pellet and sodium silicate (waterglass) solution purchased from PT. Brataco Chemical with technical grade chemicals.

2.1. Preparation of raw materials

Fly Ash and bottom ash are dried first with an oven to reduce moisture content, then ground with a ball mill and sifted with a 300 mesh sieve. Particles of raw materials that have not yet reached 300 mesh are re-milled with ball miles and sifted continuously.

2.2. Chemical and Mineralogical Analysis of Raw Materials

The raw materials to be used in this study are fly ash and bottom ash need to be analyzed in order to identify the character and composition of the material. The instruments used includes the Fourier Transform Infra Red (FTIR), the X-Ray Diffraction (XRD), and the X-Ray Fluoressence (XRF).

FTIR of raw materials are carried out in order to compare the spectrum before and after the geopolymerization process so that it can be modelled kinetics by calculating the value of the peak absorbance of corr at certain wavelengths indicating the function group of aluminosilicate. XRD analysis is conducted to know the mineral content of raw materials to know the reactivity of ingredients. XRF analysis is conducted to determine the content of oxide compounds in raw materials especially the content of SiO2 and Al2O3, so that each variation of fly ash and bottom ash composition is used always in the ratio of SiO2/Al2O3 which ranges from 3.0 to 3.8.

2.3. Polymerization experiments

Experiments were conducted by varying the comparison of fly ash and bottom ash composition (100:0, 75:25, 50:50). The concentration of NaOH used in the mixture is 12 M. Geopolymer paste is stirred until homogeneous and printed on the cube mold in the size of 5x5x5 cm^3. The Geopolymer paste mold is left 2 days, then dried with a variation of curing temperature 30 (room temperature), 60, and 90^oC. Samples of dried geopolymer paste, taken for analysis of pressure strength and FTIR tests at any time (1, 3, 5, 14, 21, 28 days). Each experiment with each variation of the fly ash and bottom ash composition is performed in 2 times repetition to validate the test data.

2.4. Compressive Strength Analysis

The geopolymer being resulted is tested by robust test press with the Universal Testing Machine Avery-Denison performed against the entire sample according to the long duration of drying of each sample.

2.5. Geopolymer Bond Formation Analysis

Geopolymer materials that have passed the drying process (curing) is further analyzed Geopolimernya bonds using FTIR instruments. This analysis will result in the spectrogram data of geopolymer bonds. Through the analysis of the FTIR will be obtained peaks on the number of waves that indicate a specific group of functions or bonds, so that this analysis can be used as a standard for viewing the fingerprint of a compound. Geopolymer samples shall be compared with zeolite to view fingerprint suitability and see the formation of aluminosilicate bonds. The number of waves that can indicate the formation of Si-O-T (T: Si or Al) bonds is 900-1300 cm^{-1}. Based on the peak on the number of such waves, the correlation value (CORR) is calculated with the equation of 2.1 and 2.2 against the zeolite which is then modeled by the isoconversional method. These results are then compared with the value of kinetics based on compressive strength. Geopolymer concrete produced from each variation is tested for concrete compressive strength and Fourier Transform Infrared (FTIR). Geopolymerization reaction is approached using iso-conversion method, where this method is used to facilitate kinetic modeling

IOP Conf. Series: Materials Science and Engineering **742** (2020) 012042 doi:10.1088/1757-899X/742/1/012042

for reactions that take place in a solid-state. The model of FTIR characterization is obtained based on the absorbance peak at each wave number that shows the characteristics of alumino-silicate.

Table 1. FTIR Interpretation for aluminosilicate content characterization (Krol, dkk., 2016)

Wave Number	Interpretation
900-1300 cm^{-1}	Si-O-T (T: Si, Al Tetrahedral) *assymetric stretching*
1650 cm^{-1}	H-OH *bending*
3400-3650 cm^{-1}	-OH *symmetric and asymmetric stretching*

The results of FTIR characterization were processed using a correlation equation (corr) to compare geopolymer samples with standard zeolites (Varmuza, dkk., 2002).

$$r = \frac{(a_i-\bar{a})(b_i-\bar{b})}{\sqrt{(a_i-)^2(b_i-\bar{b})^2}} \quad (1)$$

$$Corr = \frac{999(r+1)}{2} \quad (2)$$

3. Results and Discusssion

Raw material characterization is an important stage that has to be accomplished in order to predict the potential utilization of the fly ash-bottom ash as geopolymer raw material. Furthermore, FTIR analysis shows both fly ash and bottom ash has relatively low Si-O-Al component in comparison to the standard zeolite used in the experiment, as seen in figure 2.a.

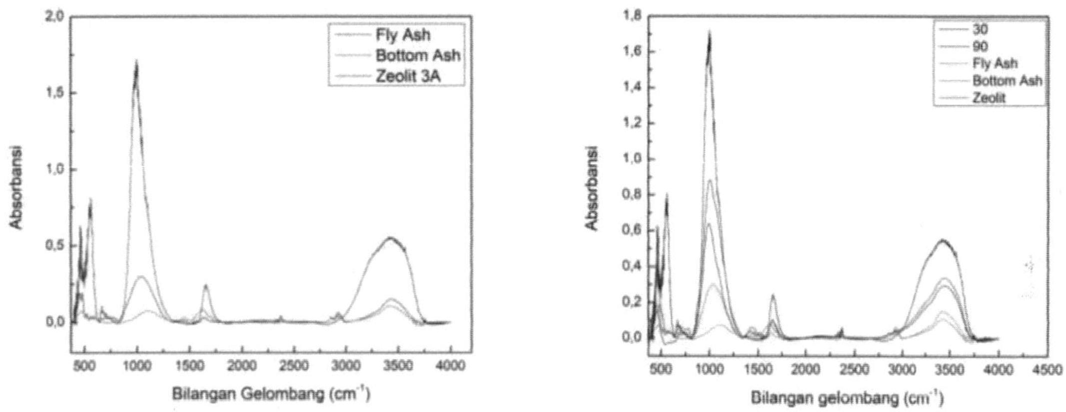

(a) (b)

Figure 2. FTIR characterization of fly ash and geopolymer samples (curing temperature of 30° and $90^{O}C$) in comparison with zeolites

In addition, silica and alumina are found abundantly in the fly ash and bottom ash with the chemical composition as shown in table 2.

Table 2. Chemical composition of fly ash and bottom ash

Chemical composition	SiO2	Al2O3	P2O5	SO3	K2O	CaO	Fe2O3	LOI	Others
Fly ash (%)	37.5	12.5	1.89	2.07	1.97	19.7	19.9	1.87	2.5
Bottom ash (%)	46.7	14.7	1.97	0.45	2.64	2.37	15.8	12.9	2.42

As shown in table 2, it can be counted taht the SiO2/Al2O3 in fly ash is 3,01 while it is found of about 3,31 in bottom ash. According to De Silva et al (2007), in order to have a good geopolymerization phenomenon, the SiO2/Al2O3 of the raw material should be in the range of 3.0-3.8. Thus, the fly ash and bottom ash from PLTU Tanjung Jati Jepara possess sufficient amount of SiO2/Al2O3 ratio to form high performance geopolymer. In term of percentage of loss on ignition (LOI) at which it was found that the LOI in bottom ash is higher than that of fly ash in the ratio of twelve time fold showing the abundance existence of unburned carbon in bottom ash. The existence of unburned carbon in the raw material of geopolymer will decrease its compressive strength due to low affinity with silica aluminate structure as the backbobe of geopolymer structure (Ha et al., 2005). As for the geopolymer product, longer time sequence creates greater absorbance which indicate larger geopolymerization of aluminosilicate. The compressive strength of geopolymer sample represents the degree of aluminosilicate formation. As it can be seen from Figure 3 that the compressive strength increases as the temperature increase for all fly ash and bottom ash composition. Howeber it can be seen that the higher the bottom ash addition into the geopolymer product, the lower the compressive strength as shown in figure 3.c. Similar tendency was also observed in the CORR FTIR product.

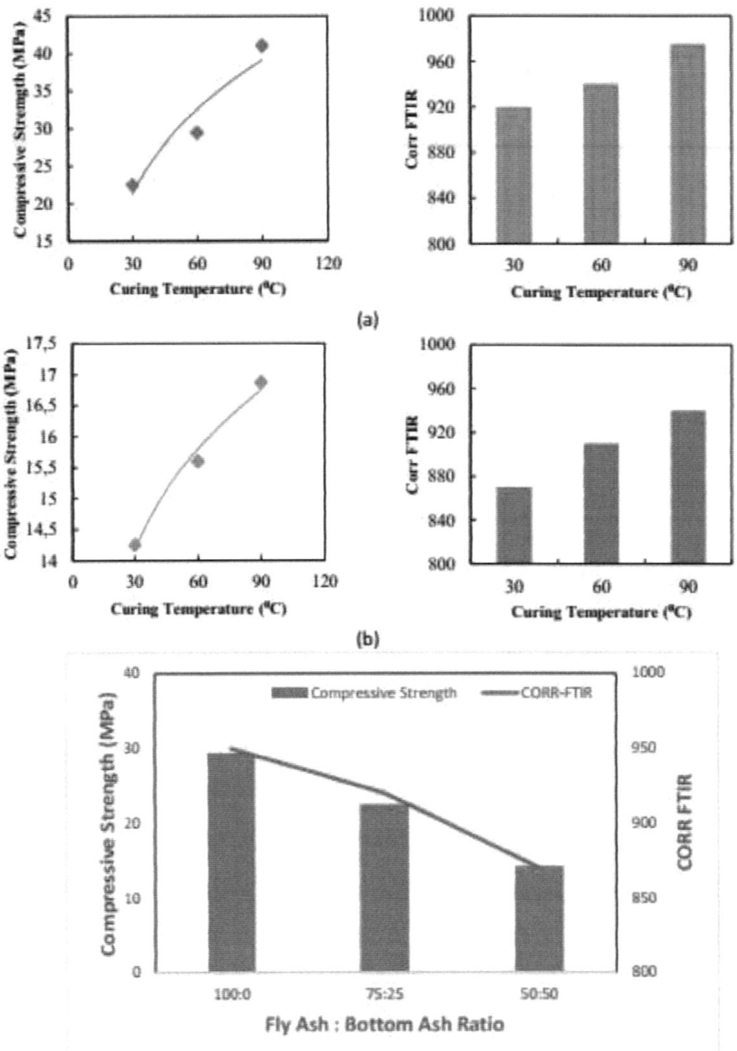

(c)

Figure 3. The comparison between compresive strength and Corr FTIR on fly ash to bottom ash ratio of (a) 75:25, (b) 50:50, and (c) Compresive Strength and CORR FTIR comparison at room temperature of curing.

Conversion factor (α) inserted into each of kinetic model for every composition to determine the fittest kinetic model for fly ash – bottom ash geopolymerization model, which is diffusion model with the highest R^2 value of 0.9888. This statement supported by previous research that the diffusion model mostly used for pozzolanic reaction (De Silva et al, 2007; Khawam and Flanagan, 2006).

Table 3. Arrhenius equation value for each FA:BA composition

FA:BA Composition	ln A	A (day^{-1})	-E/R	E (J/mol)
50:50	-0,6651	0,514222	-2750,7	22869,32
75:25	1,9725	7,188626	-3061,5	25453,31
100:0	5,1664	175,2827	-3777,2	31403,64

Based on Table 3, the increase of bottom ash weight ratio causes lower Arrhenius reaction rate constant. The unburned carbon in bottom ash causes low formation of aluminosilicate, thus decreasing the compressive strength.

4. Conclusions

Geopolymer successfully formed using fly ash from PLTU Tanjung Jati B Jepara with the highest compresive strength of 58 MPa resulted by 100:0 fly ash – bottom ash ratio with curing temperature of $90^{o}C$ for 28 days. The increase of compresive strength as a function of temperature and series of time sequence is in accordance with the semi quantitative FTIR analysis represented by corr value. The addition of bottom ash in the geopolymer formation decreases the compresive strength due to its unburned carbon content. The kinetics of fly ash geopolymerization has fitted with the diffusion model, in which diffusion control in the process including dissolution, gel formation, and solidification. The diffusion model fits the phenomenon with highest R^2 value of 0.9888. Greater bottom ash concentration causes lower compressive strength of geopolymer sample with the same curing temperature and time sequence.

Acknowledgement

Highly appreciation to the Minister of Higher Education of Indonesia for the financial support to conduct this project under PTUPT scheme with contract number 2702/UN1/DITLIT/DIT-LIT/LT/2019.

References

[1] Adelizar, A.S., Petrus, H.T.B.M., Olvianas, M., Widiyatmoko, A., 2018, Geopolymerization Kinetics of Fly Ash using Isoconversional Kinetics, Proceeding Book of the 8th annual basic science international conference, pp. 196-200.

[2] Alinnor, I., J., 2007, Adsorption of heavy metal ions from aqueous solution by fly ash, Fuel, vol. 86, 2007, pp. 853–857.

[3] Dasmahapatra, G., P., Pal, T., K., Bhadra, A., K., and Bhattacharya, B., 1996, Studies on separation characteristics of hexavalent chromium from aqueous solution by fly ash, Sep.Sci.Technol., vol. 31, 1996, pp. 2001–2009.

[4] Davidovits, J., Chemistry of Geopolymeric Systems, Geopolymer 99 International Conference, 1999.

[5] De Silva, P., Sagoe-Crenstil, K., dan Sirivivatnanon, V. (2007). Kinetics of Geopolymerization: Role of Al2O3 and SiO2. Cement and Concrete Research, 37, 512- 518.

[6] Direktorat Jenderal Ketenagalistrikan. (2016, February 16). Kementerian Energi dan Sumber Daya Mineral, Direktorat Jenderal Ketenagalistrikan. Dipetik April 12, 2017, dari Pemerintah Dorong Pemanfaatan Limbah PLTU: http://www.djk.esdm.go.id/index.php/detail-berita?ide=4060

[7] Ferian, A., Amijaya, D.H., Harijoko, A., Tambaria, T.N., Sahri, A.A., and Nur Asa, Z.A., 2018.

IOP Conf. Series: Materials Science and Engineering **742** (2020) 012042 doi:10.1088/1757-899X/742/1/012042

"Rare Earth Element and Yttrium Content of Coal in the Banko Coalfield, South Sumatra Basin, Indonesia: Contributions from Tonstein Layers." International Journal of Coal Geology 196 (July): 159–72.

[8] Ferian, A., Bestari Dea, A.A., Rosita, W., and Petrus, H.T.B.M., 2018. "The Composition and Mode of Occurence of Rare Earth Elements and Yttrium in Fly Ash and Bottom Ash from Coal-Fired Plants in Java, Indoneisa." The Society for Oorganic Petrology Annual Meeting.

[9] Ferian, A., and Petrus, H.T.B.M., 2016. "Potential Source of Rare Earth Elements (REE) from Sangatta Coal, Kuta Basin, Indonesia." Annual Meeting of The Society for Organic Petroligy 32 (170): 9–10.

[10] Ha, T.-H., Muralidharan, S., Bae, J.-H., Ha, Y.-C., Lee, H.-G., Park, K., dan Kim, D.-K. (2005). Effect of Unburnt Carbon on the Corrosion Performance of Fly Ash Cement Mortar. Construction and Building Materials, 509-515.

[11] Hirajima, T., Oosako, Y., Nonaka, M., Petrus, H., Sasaki, K., and Ando, T., Recovery of hollow and spherical particles from coal fly ash by wet separation process, Journal of MMIJ 124 (12), 2008, pp. 878-884

[12] Hirajima, T., Petrus, H., Oosako, Y., Nonaka, M., Sasaki, K., and Ando, T., Recovery of cenospheres from coal fly ash using a dry separation process: Separation estimation and potential application, International Journal of Mineral Processing 95 (1-4), 2010, pp. 18-24.

[13] Khawam, A., dan Flanagan, D., Solid-State Kinetic Models: Basics and Mathematical Fundamentals, J. Phys. Chem. B, 110(35), 2006, pp. 17315-17328.

[14] Krol, M., Minkiewicz, J., dan Mozgawa, W., IR Spectroscopy Studies of Zeolites in Geopolymeric Materials derived from Kaolinite, Journal of Molecular Structure, 2016, pp. 200-206.

[15] Olvianas, M., Widiyatmoko, A., and Petrus, H.T.B.M., IR spectral similarity studies of geothermal silica-bentonite based geopolymer, AIP Conference Proceedings, 2017, 1887 (1).

[16] Petrus, H., Hirajima, T., Oosako, Y., Nonaka, M., Sasaki, K., and Ando, T., Performance of dry-separation processes in the recovery of cenospheres from fly ash and their implementation in a recovery unit, International Journal of Mineral Processing 98 (1-2), 2011, pp. 15-23.

[17] Petrus, H.T.B.M., Adelizar, A.S., Widiyatmoko, A., Olvianas, M., Suprapta, W., Perdana, I., Prasetya, A., and Astuti, W., 2019, Kinetics of Fly Ash Geopolymerization using Semi Quantitative Fourier-Transform Infrared Spectroscopy (FTIR); Corr Data, IOP Conference Series: Materials Science and Engineering, 532, 1, p. 012001

[18] Seredin, V V. 2010. "SHORT COMMUNICATIONS A New Method for Primary Evaluation of the Outlook for Rare Earth Element Ores" 52 (5): 475–80.

[19] Seredin, Vladimir V, and Shifeng Dai. 2012. "International Journal of Coal Geology Coal Deposits as Potential Alternative Sources for Lanthanides and Yttrium." International Journal of Coal Geology 94: 67–93.

[20] Singh, M., and Garg, M., 1999, Cementitious binder from fly ash and other industrial wastes, Cem.Concr.Res., vol. 29, 1999, pp. 309–314.

[21] Varmuza, K., Demuth, W., dan Karlovits, M., Structural and Spectral Similarity, CHEM 02 - Biannual Conference on Chemistry. Cairo: Chemistry Department Cairo University, 2002.

[22] Wilson, M., Cabrera, J., G., and Zou, Y., 1994, The process and mechanism of alkali-silica reaction using fused silica as the reactive aggregate, Adv.Cem.Res., vol. 6, 1994, pp.117– 25

IOP Conf. Series: Materials Science and Engineering **742** (2020) 012043 doi:10.1088/1757-899X/742/1/012043

Valuable Metals Precipitation of Low Grade Nickel Ore Leachate Using Sodium Hydroxide

Surianti[1], A N Utami[1], M A Permatasari[1], I M Bendiyasa[1], W Astuti[2], and H T B M Petrus[1] *

[1]Department of Chemical Engineering (Sustainable Mineral Processing Research Group), Universitas Gadjah Mada 55281 Yogyakarta, Indonesia
[2]Research Unit for Mineral Technology, Indonesian Institute of Sciences (LIPI), Jl. Ir. Sutami Km. 15, Tanjung Bintang, Lampung Selatan, Indonesia

E-mail : bayupetrus@ugm.ac.id

Abstract. Nickel metal is widely used in various industries. Nickel sources come from nickel sulfide ores and nickel oxide ores (laterite). Nickel content in laterite ore is 1.5-2.5 % (weight basis), lower than the nickel content in sulfide ores. Indonesia is one of the major nickel producers with laterite nickel ore reserves of 1576 million tons or 15 % of total world reserves. The distribution of laterite nickel in Indonesia is found in several regions such as Maluku, Sulawesi and Papua. However, the low grade nickel processing in Indonesia has not been well established. Thus research on valuable metals extraction from ow grade nickel ore has put some interest to some researchers. In this study, precipitation of valuable metals from low grade nickel leachate was conducted using sodium hydroxide as the precipitation reagent. Nickel ore used in this experiment was obtained from Pomalaa, Southeast Sulawesi. After leaching at optimum condition resulting of about 75 % recovery of nickel using HCl (1 M at 90 °C and stirring for 1 hour), the leachate was then added sodium hydroxide to varied pH of 8, 10, and 12. Along with the pH conditioning, temperature was kept to the value of 55 °C, 70 °C, and 85 °C. The optimum nickel recovery of 83.97 % is obtained at pH 12 and temperature of 85 °C. The residual metals concentration in the solution after precipitation was analyzed using Inductively Coupled Plasma Atomic Emission Spectroscopy (ICP-AES).

1. Introduction

Nickel is a transition metal that is needed by various industrial fields. Nickel is often used as steel coatings, copper coatings, battery sources, and the automotive industry [1]. Based on research, as much as 66.2 % of Nickel metal consumption is dominated by the stainless steel industry [2]. Nickel on the earth is in the form of oxides and sulfides. Sulfide nickel has a higher Nickel content than Nickel Oxide (laterite Nickel), so the processing process is easier. Though nickel deposits in the soil are dominated by laterite nickel. Ore in the form of sulfide $(Ni, Fe)_9S_8$ there is only about 27 % of the total nickel ore in the world.

Indonesia is one of the countries with significant nickel resources, the resources of laterite ore in Indonesia reached 3.5 billion tons, equivalent to 52.2 million tons of Ni metal spread in Maluku,

Sulawesi and Papua [3]. Most nickel ore reserves in Indonesia are from laterite nickel. The processed nickel laterite profile comes from zone limonite (Ni 0.8 % -1.5 %) and low grade saprolite zone (1.5 % - 1.8 %). In general, nickel ore processing is carried out with two processes, namely hydrometallurgy and pyro metallurgy (smelting). The pyro metallurgical process of nickel processing in the limonite and saprolite zones requires high energy, so the costs incurred are also high. The hydrometallurgical process can be chosen as an alternative to processing low grade nickel ore <1.5 %. So far, many hydrometallurgical processes with a leaching process use high pressure HPAL / PAL with the main NiS products which are also costly and high energy. The Caron process is also one of the hydrometallurgical methods where the main product is NiO. The process of leaching in current atmospheric conditions has attracted many researchers. The leaching process in atmospheric conditions are a direct leaching of laterite nickel ore using organic and inorganic acids. Leaching using inorganic acids in atmospheric conditions has been carried out [4] and Leaching with organic acids [5], [6], [7], [8], [9], [10].

For nickel-rich leaching solutions, they are selectively deposited using the hydroxide precipitation, sulfuric precipitation, and hybrid precipitation methods [11]. The advantages of the Hydrochloric Acid leaching method include the reactant materials used are easily available, relatively inexpensive, and easy to do [6]. The optimum conditions for nickel laterite extraction were obtained at a temperature of 100 °C and a concentration of 5.5 M [12].

The Leaching process is carried out to extract nickel components in nickel ore. But not only nickel components are taken when doing leaching. Leaching solutions contain several metal ions such as Fe^{3+}, Fe^{2+}, Al^{3+}, Cr^{3+}, Ni^{2+}, Co^{2+}, Mn^{2+}, and Mg^{2+}. The solution obtained from leaching chloride was precipitated using NaOH. Precipitation can be carried out in the pH range 7.5 - 9.0 and at temperatures between 25 - 75 °C [13].

In this study the influence of temperature and pH on the deposition process of low grade nickel ore leaching solutions using NaOH will be observed. The choice of NaOH settling because it is a hygroscopic compound in which NaOH has the ability to absorb moisture from the environment.

2. Materials and Methods

2.1. Materials

The raw materials used are nickel ore and hydrochloric acid solutions. A 750 ml solution of 0.1 M hydrochloric acid was prepared and then put into a three neck flask equipped with a stirrer, thermometer and reverse cooler. Uniform nickel ore (-200 mesh particle diameter) weighs 150 grams. Furthermore, nickel ore is put into a three-neck flask. An acid solution containing nickel ore is heated to 90 °C. The leaching process is carried out for 4 hours. The results of leaching are then filtered using filter paper to separate solids and liquids. The filtered sample will then be analyzed using the AAS Inductively Coupled Plasma (ICP) [6]. Table 1 is the result of analysis using ICP-AAS low grade nickel ore leaching solution using hydrochloric acid.

The other raw materials used were NaOH obtained from PT Merck and aquades obtained from the Laboratory of Energy Conservation and Pollution Prevention (KEPP), Faculty of Engineering UGM Yogyakarta.

Table 1. ICP-AAS Analysis Results for Laterite Ore Nickel Leaching Solution

Component	Concentration (ppm)
Fe	4588.69
Al	587.09
Si	546.02
Ni	160.77

The results of the analysis of the initial leaching solution were used as a comparison to the precipitation results. The metals were analyzed by ICP AAS as many as 4 metals, namely Fe, Al, Si, and Ni. The metal was chosen because the percentage of concentration was the highest. Table 1 shows that Fe is the main metal composing leaching solution which is equal to 4588.69 ppm and Ni metal obtained is 160.77 ppm.

2.2. Methods

Leaching results that have been analyzed, then placed in a beaker. 15 % NaOH solution is prepared. Precipitation is done using a shaker. The extract solution (leaching) of 100 ml was put into Erlenmeyer then 15 % NaOH was added to pH 8. Erlenmeyer was then put into a shaker. Shakers are set at 55 °C and turned on for 1 hour. The results of the sediment are then filtered using filter paper and oven at 90 °C.

The filtered liquid is then analyzed using an Inductively Coupled Plasma Atomic Emission Spectroscopy (ICP-AES). The experiment was repeated under depositional conditions pH 10 and 12 and temperatures of 70 °C and 85 °C.

3. Results and Discussions

The deposition process is carried out by adding 15 % NaOH to the desired pH. The variations in pH used are 8, 10, and 12. The temperature used is 55 °C, 70 °C, and 85 °C. Deposition results show that the higher the pH, the more sediment formed darkens. At pH 12, the color of the precipitate formed tends to be close to black because more oxidized metal [14].

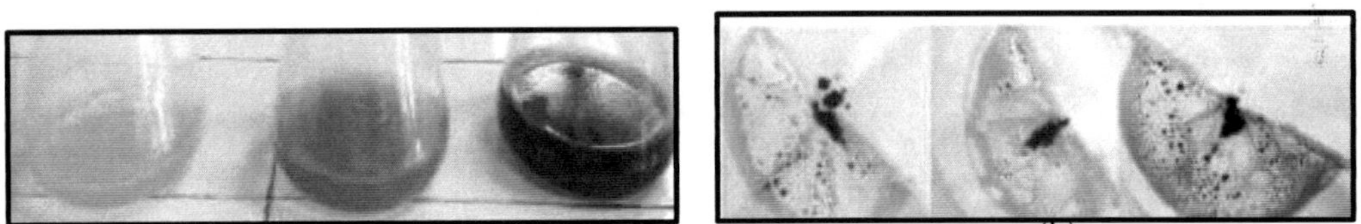

(a) (b)

Figure 1 Precipitate at pH 8, pH 10, pH 12, of (a) before Filtering and (b) after Filtering

The solution resulting from leaching chloride is precipitated using NaOH, for the metal deposits formed are hydroxides with solubility values as shown in Table 2. The following are the reactions that occur in the deposition of hydrochloric acid leaching solutions with NaOH [14]:

$$NiCl_2 + 2\ NaOH \rightarrow Ni(OH)2 + 2\ NaCl \quad (1)$$

$$FeCl_3 + 3\ NaOH \rightarrow Fe(OH)_3 + 3\ NaCl \quad (2)$$

$$AlCl_3 + 3\ NaOH \rightarrow Al(OH)_3 + 3\ NaCl \quad (3)$$

Table 2. Solubility of Hydroxide compounds [15]

Compound	KsP
$Ni(OH)_2$	1.6×10^{-16}
$Fe(OH)_3$	6.0×10^{-38}
$Al(OH)_3$	5.0×10^{-33}

The greater the Ksp value, the more soluble the compound will be, making it more difficult to settle. Based on table 2, the compounds $Fe(OH)_3$, $Al(OH)_3$, will precipitate because they have relatively small Ksp values compared to other compounds with relatively close values. Precipitation is carried out in a range of pH 7.5-9.0 and at temperatures between 25 - 75 °C [13].

The precipitated sample is then filtered by the sediment. The precipitate formed is oven at 90 °C then weighed to determine the mass formed. The filtrate obtained was analyzed using ICP-AES to find out the remaining metal which was not taken into sediment. In figure 2, 3, 4 and 5, it can be seen that the ppm concentrations of Fe, Al, Si, and Ni are left in the leaching solution. ICP AAS analysis is easy to do, but this analysis is very sensitive to impurities

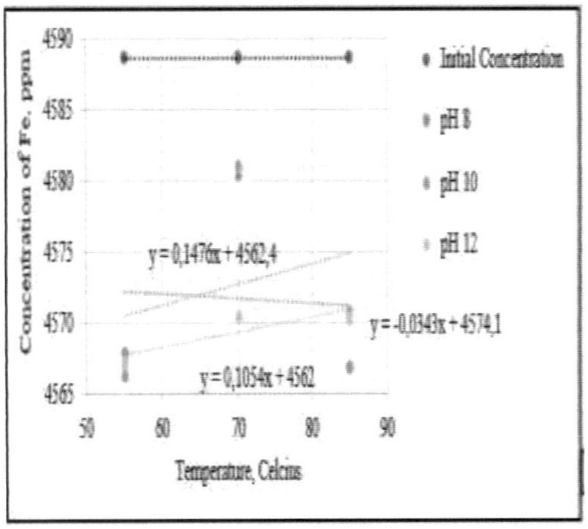

(a)

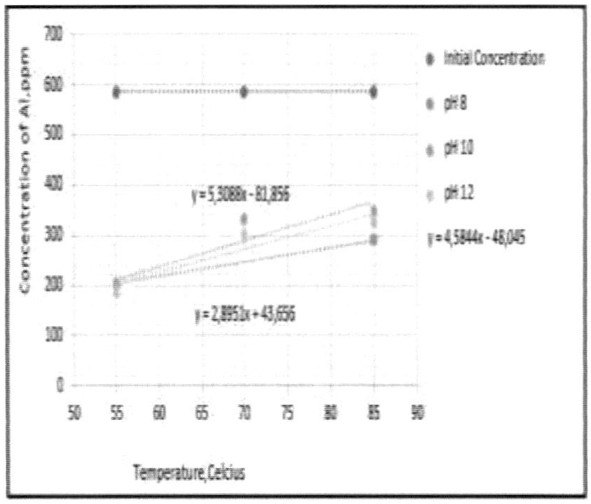

(b)

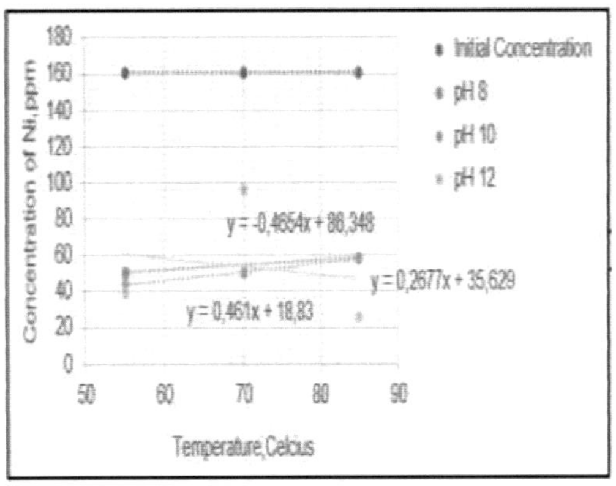

(c)

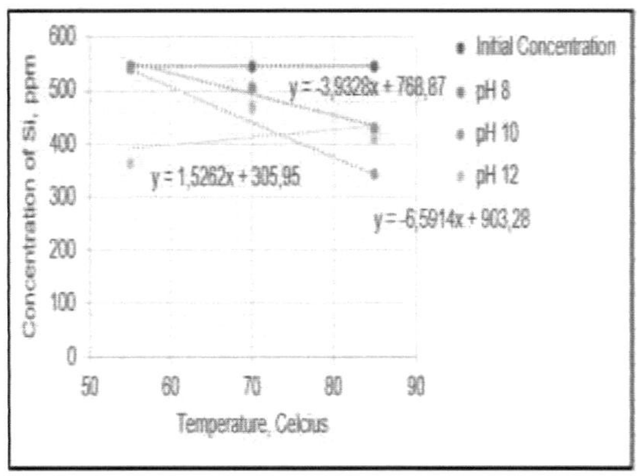

(d)

Figure 2. Graphic of the pH and Temperature Functions on Filtrate for (a) Fe Concentrations, (b) Al Concentration, (c) Ni Concentrations and (d) Si Concentrations.

Figure 2 (a) shows the relationship between the concentration of Fe in the filtrate solution and the temperature at various pH. For all pH, the amount of Fe metal at 70 °C is higher than the amount of Fe metal at 55 °C. At 85 °C the amount of Fe in the filtrate decreases. That is, the amount of Fe metal that settles at 70b °C decreases then rise again at 85 °C. This is because at a temperature of 70-75 °C, the precipitate formed switches from the form of hydroxide to oxide [13]. At pH 8 and 10, the amount of Fe metal remaining is almost the same. Whereas at pH 12, the remaining Fe metal is smaller, the higher the pH Fe the more it settles. Figure 2 (b) that the concentration of Al metal is left in the filtrate solution. The graph shows the higher the temperature, the less metal is drawn. Al solubility increases with increasing temperature.

Ni metal in filtrate solution can be seen from figure 2 (c). The graph shows that at pH 8 and pH 10 the concentration is almost the same at all temperatures. The Ni metal concentration of 70 °C at pH 12 increased from 55 °C, then returned to a temperature of 85 °C. This happens because at a temperature of 70 °C it is the transition temperature where $Ni(OH)_2$ changes to NiO. The higher the temperature, the higher the solubility of $Ni(OH)_2$, so that the less deposited Ni. Conversely, the higher the temperature, the smaller the solubility of NiO. So that NiO deposits are more stable [16]. Figure (d) shows that the higher the temperature of Si concentration in the filtrate solution decreases because the solubility decreases, so many Si metals are deposited.

The percentage of the results Fe, Al, Ni, and Si metals is calculated by comparing the concentration of the extract solution before and after deposited with 15 % NaOH. The results of Fe metal at pH 8, 10, and 12 are fairly low in the range of 0.10 - 0.50 %. Fe metal tends to settle at low pH, which is a pH range of 2.5-4.5. As for Al metal, the higher the temperature, the smaller the collection. The highest percentage of Al collection was obtained at pH 12 at 55 °C which was equal to 67.91 %. The results of Al metal collection are quite high because $Al(OH)_3$ is easier to settle at high pH.

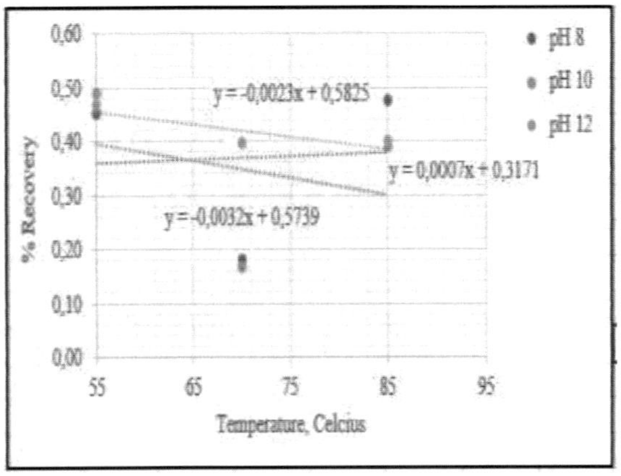

(a)

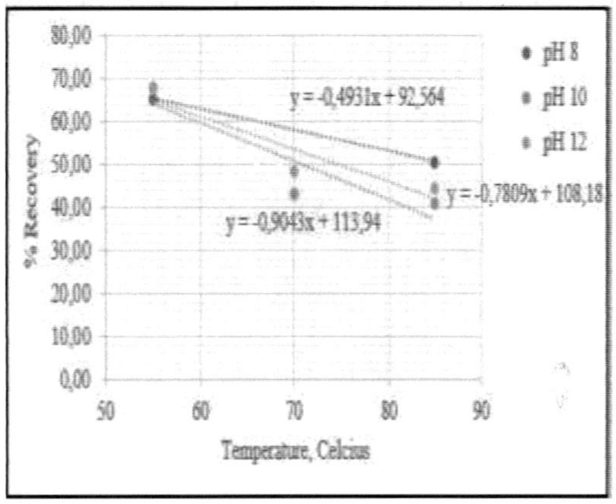

(b)

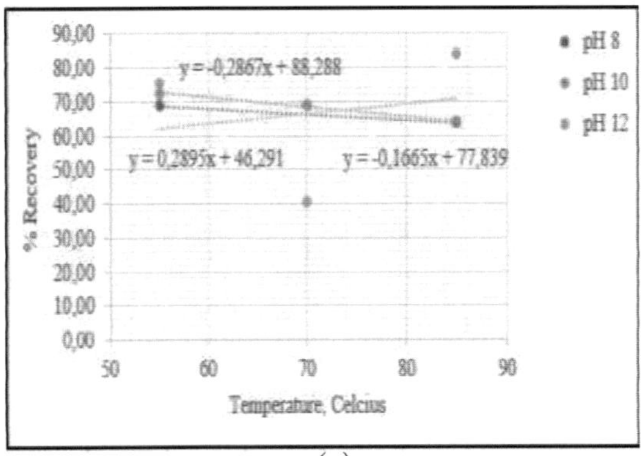

(c)

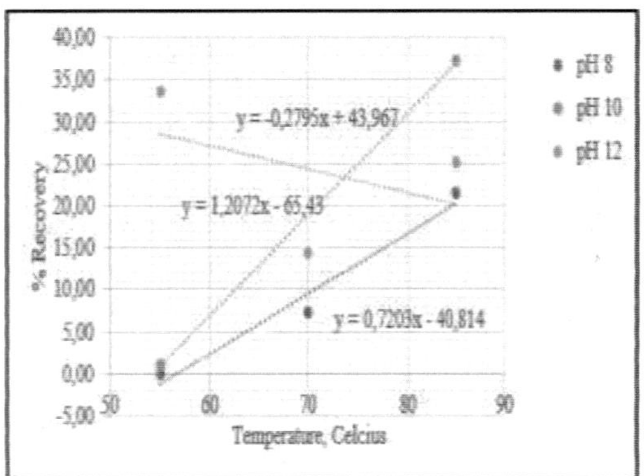

(d)

Figure 3. Graphic Percentage Metal Recovery of (a) Fe, (b) Al, (c) Ni and (d) Si

Nickel is the main product you want to get. Nickel in the pH range 8-12 with alkaline atmosphere can be precipitated in the form of $Ni(OH)_2$ and NiO. The higher the pH, the more precipitate formed. The effect of temperature can be seen that the higher the temperature, the deposition of Ni $(OH)_2$ decreases because Ksp increases (solubility increases), to a point $Ni(OH)_2$ changes to NiO. At pH 12 and a temperature of 70 °C it is the transitional point of Ni deposit formation. The maximum Ni collection results obtained at pH 12 temperature 85 °C which is equal to 83.97 %. The result of Si metal collection decreases with decreasing temperature.

Table 3. Analysis of Deposition Mass

pH	Deposition Mass, gram		
	50 °C	**70 °C**	**85 °C**
8	0.1435	0.1983	0.3417
10	0.1931	0.4622	0.4341
12	0.3445	0.4979	0.4216

The sediment mass is obtained after drying for 2 hours. The sediment mass at 70 °C increases from the mass of the sediment at 55 °C. However, at a temperature of 85 °C the sediment mass formed decreases. Whereas based on ICP AAS analysis the results of leaching filtrate showed that Fe, Al, Ni metal deposits decreased at 70 °C. This is because the metals analyzed are only Fe, Al, Ni, and Si. It is possible for other metals such as Mn, Mg, Cu, Co, etc. to settle more easily at 70 °C, so that the analysis of the total mass of the precipitate formed increases. Table 3 shows that the higher the pH, the more precipitate formed.

4. Conclusion

From the results of the experiment it can be concluded that NaOH is proven to be used to deposit Ni metal with optimum conditions obtained at pH 12 and 85 °C with recovery as much as 83.97 % and Temperature 70 °C as the transition temperature of $Ni(OH)_2$ to NiO so that at this point Ni recovery is the smallest.

References

[1] M. Vlachou, J. Hahladakis, and E. Gidarakos, "Effect of Various Parameters in Removing Cr and Ni from Model Wastewater by using Electrocoagulation," vol. 15, no. 4, pp. 494–503,2013.
[2] Z. H. U. De-qing, C. U. I. Yu, S. Hapugoda, K. Vining, and P. A. N. Jian, "Mineralogy and crystal chemistry of a low grade nickel laterite ore," vol. 22, 2012.
[3] P. Prasetyo, "Sumber Daya Mineral di Indonesia Khususnya Biji Nikel Laterit dan Masalah Pengolahannya Sehubungan dengan UU MINERBA 2009," no. November, pp. 1–10, 2016.
[4] D. B. and H. T. B. M. P. W Astuti, N M Prilitasari, Y Iskandar, "Leaching behavior of lanthanum, nickel and iron from spent catalyst using inorganic acids Leaching behavior of lanthanum, nickel and iron from spent catalyst using inorganic acids," pp. 0–7, 2018.
[5] W. Astuti, T. Hirajima, K. Sasaki, and N. Okibe, "Comparison of effectiveness of citric acid and other acids in leaching of low- grade Indonesian saprolitic ores Comparison of effectiveness of citric acid and other acids in leaching of low-grade Indonesian saprolitic ores," Miner. Eng., vol. 85, no. January, pp. 1–16, 2016.
[6] H. Sekar et al., "Pelindian Nikel dari Bijih Limonit Low-Grade Pomalaa Menggunakan Pelarut Asam Asetat," no. April, pp. 1–7, 2018.
[7] B. Journal, S. Sahu, N. C. Kavuri, and M. Kundu, "Dissolution Kinetics of Nickel Laterite Ore using Different Secondary Metabolic Acids," vol. 28, no. 02, pp. 251–258, 2011.
[8] K. C. Wanta, H. T. B. M. Petrus, and I. Perdana, "Uji Validitas Model Shrinking Core terhadap Pengaruh Konsentrasi Asam Sitrat dalam Proses Leaching Nikel Laterit," vol. 11, no. 1, pp. 30–35, 2017.
[9] L. Li et al., "Recovery of metals from spent lithium-ion batteries with organic acids as leaching reagents and environmental assessment," vol. 233, pp. 180–189, 2013.
[10] R. G. Mcdonald and B. I. Whittington, "Atmospheric acid leaching of nickel laterites review. Part II. Chloride and bio-technologies," vol. 91, pp. 56–69, 2008.
[11] F. Mohammadreza, N. Mohammad, and S. S. Ziaeddin, "Nickel extraction from low grade laterite by agitation leaching at atmospheric pressure International Journal of Mining Science and Technology Nickel extraction from low grade laterite by agitation leaching at atmospheric pressure," Int. J. Min. Sci. Technol., vol. 24, no. 4, pp. 543–548, 2019.
[12] T. S. Magnetik, "Pengaruh Konsentrasi Nikel pada Proses Leaching Mineral Geothite terhadap

Susceptibilitas Magnetik," pp. 1–5, 2011.

[13] Mubarok & Lieberto, "Precipitation of Nickel Hydroxide from Simulated and Atmospheric- leach Precipitation of Nickel Hydroxide from Simulated and Atmospheric-Leach Solution of Nickel Laterite Ore," Procedia Earth Planet. Sci., vol. 6, no. December 2013, pp. 457–464, 2015.

[14] Y. Chang, X. Zhai, B. Li, and Y. Fu, "Hydrometallurgy Removal of iron from acidic leach liquor of lateritic nickel ore by goethite precipitate ☆," Hydrometallurgy, vol. 101, no. 1–2, pp. 84–87, 2010.

[15] E. Care, "CRC Handbook of Chemistry and Physics."

[16] D. A. Palmer, P. Bénézeth, and D. J. Wesolowski, "Solubility of Nickel Oxide and Hydroxide in Water," pp. 264–269, 1980.

IOP Conf. Series: Materials Science and Engineering **742** (2020) 012044 doi:10.1088/1757-899X/742/1/012044

Synthesis and Characterization of Activated Carbon Derived from Salacca Peel Using $ZnCl_2$ Hydrothermal Carbonization and Chemical Activation with Microwave Heating

Hans Kristianto[1], Yoel Lavenki[1], Ratna Frida Susanti[1]*
[1]Department of Chemical Engineering, Faculty of Industrial Technology, Parahyangan Catholic University, Ciumbuleuit 94, Bandung 40141, Indonesia

E-mail : santi@unpar.ac.id

Abstract. In this paper we describe synthesis and characterization of salacca peel derived activated carbon. The salacca peel was carbonized using hydrothermal carbonization (HTC) at 225°C and 50 bar for 5 h with and without $ZnCl_2$ as HTC catalyst. The obtained hydrochar was furthermore activated using microwave assisted chemical activation at 700 W for 8 min under inert atmosphere. The ratio between hydrochar and impregnant ($ZnCl_2$) was varied between 1:2, 1:3, and 1:4. The samples were characterized using methylene blue equilibrium adsorption, Scanning Electron Microscopy (SEM), and Fourier Transform Infra-Red (FTIR) spectroscopy. It was obtained that the presence of catalyst during HTC could increase the methylene blue adsorption capacity (q_e) and visible pores on the surface. The increase of impregnation ratio gave increase the surface pores and q_e, with highest qe was obtained at impregnation ratio of 1:4. Furthermore, the obtained activated carbon had higher q_e and oxygen functional groups compared to that of commercial activated carbon, making it suitable for various applications.

1. Introduction

Hydrothermal process is irreversible decomposition of biomass using hot pressurized water. Recently, this process gains its fame due to various advantages, such as green solvent (water) and easily adjusted to obtain valuable liquid, gas, or solid products [1]. Hydrothermal carbonization (HTC) is specifically done to obtain solid, called hydrochar, as the main product. There are some advantages of HTC compared to conventional carbonization, such as lower operating temperature (below 300 °C) and ability to process biomass with high water content [1]. In the context of activated carbon synthesis, according to [2], HTC is an important step that helps pore development in activation process.

Usually HTC needs no catalyst in its process, as the water provide H^+ ion to hydrolyze the precursor. However, it is known that catalyst could be added to the HTC process to assist some reaction, such as hydrolysis, oxidation, dehydration, acylation, alkylation, etc. during process to obtain specific product (i.e. oil, gas, or char) and properties [3]. Some salts have been used in HTC, namely Na_2SO_4 [4], LiCl [5], $CaCl_2$, Ca-propionate, Ca-acetate, Ca-formate, Mg-acetate [6], and $ZnCl_2$ [7]. Combination of $ZnCl_2$-NaCl [8] and $ZnCl_2$-urea [9] also have been utilized to develop porosity on carbon during HTC. There are some evidence that $ZnCl_2$ could act as catalyst in cellulose hydrolysis and glucose decomposition [10], thus resulting in an increase of porosity [8].

Chemical activation with microwave heating was used in this study. Conventional heating depends on temperature gradient that makes long heating period thus resulting in high energy consumption. Microwave heating provides rapid, effective heat transfer, shorter heating period, lower activation temperature, and simple operation [11, 12]. Moreover, $ZnCl_2$ as impregnant acts as template for microporosity [13], as well as dehydrating agent, inhibiting tar and promoting the reaction to char formation [14].

In this study we utilize salacca peel as precursor to synthesize hydrochar and activated carbon. Salacca is an indigenous plant of Indonesia with large quantity of fruit production annually, namely 900,000 tons in 2018; making huge amount of peel waste is generated [15]. Previously, activated carbon from salacca peel has been utilized as adsorbent [16], Li-ion capacitor's [17] and Li-S battery's cathode [18], and Na-ion battery's anode [19], most of which are using conventional carbonization and activation. A combination of HTC and chemical activation with microwave heating was used to obtain activated carbon. The effect of the presence $ZnCl_2$ catalyst in HTC, and impregnation ratio in chemical activation to the morphology, functional groups, yield, and methylene blue adsorption capacity was investigated and presented in this paper.

2. Methodology

2.1. HTC process

Salacca fruits (*salak pondoh super* or *Salacca zalacca var Gaertner Voss*) were purchased from local markets in Bandung, Indonesia. The peel was separated from its edible parts, and washed repeatedly to remove impurities, and then oven dried for overnight at 110 °C. The dried peel then was crushed and sieved to obtained peel between size of -20+30 standard mesh, which was used as precursor to synthesize hydrochar and activated carbon in this study.

HTC process was done by following the procedure described in [17, 20]. A 5 g of dried salacca peel powder and 1.5 g of $ZnCl_2$ (Merck-p.a) were mixed in 50 mL of distilled water. The mixture was then loaded into a SUS316 subcritical water reactor with volume of 150 mL. The reactor was tightly closed and purged using N_2 gas to create inert condition. The pressure in the reactor was adjusted by feeding water into the reactor, using hydraulic hand pump (Sun run, Singapore), under continuous N_2 atmosphere for about 15 min. The heater was then turned on to achieve carbonization temperature of 225 °C and pressure of 50 bar. After 5 h, the reactor was cooled down until room temperature, and the obtained hydrochar (Z0) was separated, and oven dried for overnight at 110°C. To study the effect of $ZnCl_2$ during HTC process, the above procedure was repeated without $ZnCl_2$ (C0).

2.2. Chemical activation

The dried hydrochar (C0 or Z0) was impregnated for 24 h using $ZnCl_2$ at various impregnation ratio of hydrochar to $ZnCl_2$ 1:2, 1:3, and 1:4, denoted as Cx or Zx with x is the impregnation ratio and C or Z is the hydrochar without and with catalyst, respectively. The mixture was oven dried at 110 °C for 24 h. The mixture was then put into a quartz reactor and carbonized using modified food microwave oven (Panasonic) at irradiation power of 700 W for 8 min under inert atmosphere. The obtained activated carbon sample was washed using 0.1 M HCl, and repeatedly washed using distilled water until pH neutral was achieved. The sample was oven dried and stored in a desiccator.

2.3. Characterization

The hydrochar and activated carbon samples were characterized using scanning electron microscopy (SEM; JEOL JSM-IT-300) to observe the surface morphology, Fourier Transform Infrared Spectroscopy (FTIR; Prestige 21 Shimadzu Instruments) to analyse the functional groups. Methylene blue adsorption study was done to gain insight on hydrochar and activated carbon samples surface area. As much as 10 mg of sample was shaken overnight in a 250 mL methylene blue solution with initial concentration of 10 mg/L. The equilibrium adsorption capacity (qe; mg methylene blue/ g carbon) was calculated by using Eq 1, where Ci and Ce (mg/L) are initial and final concentration of methylene blue, V (L) is the solution's volume, and m (g) is the mass of carbon sample. The yield of activated carbon was calculated by following Eq 2.

$$q_e = (C_i - C_e) \times \frac{V}{m} \quad (1)$$

$$\% \, \boldsymbol{yield} = \frac{\boldsymbol{mass\ of\ activated\ carbon\ (g)}}{\boldsymbol{mass\ of\ hydrochar\ (g)}} \times \mathbf{100\%} \qquad (2)$$

3. Results and Discussions

3.1. Characterization of hydrochar

The morphology of C0 and Z0 is presented in Fig 1. a-b. It could be observed that more pores were formed on the surface of Z0, compared to C0. This observation is consistent with the qe value (Fig 1.c), where Z0 exhibited higher qe indicating more porous structure. Similar result was also observed by [9], where the presence of $ZnCl_2$ during HTC process help hydrolysis and decomposition of cellulose, promote development of porosity, resulting on increase of surface area. This phenomenon also explained lower Z0 yield than C0.

Based the FTIR spectra presented in Fig 1.d, it could be observed that there were some functional groups present in the hydrochar. Both C0 and Z0 samples possessed similar peak in the functional and fingerprint region, indicating similar structure albeit the catalyst in the HTC process. Noticeable peaks presented in the spectra are: broad peak of –OH stretching around 3400-3300 cm^{-1}, C-H vibrational stretching at 2951.8 cm^{-1}, 1693.5 cm^{-1} of C=O stretch, and 1602.85 cm^{-1} of C=C [21]. Comparing the spectra of C0 and Z0, we could observe that Z0 spectra had lower intensity that C0. It was possible due to the presence of $ZnCl_2$, which is an acidic catalyst, assists hydrolysis of C-O bond, thus lowering the peak at 1600 cm^{-1} and 1200-1000 cm^{-1} [8].

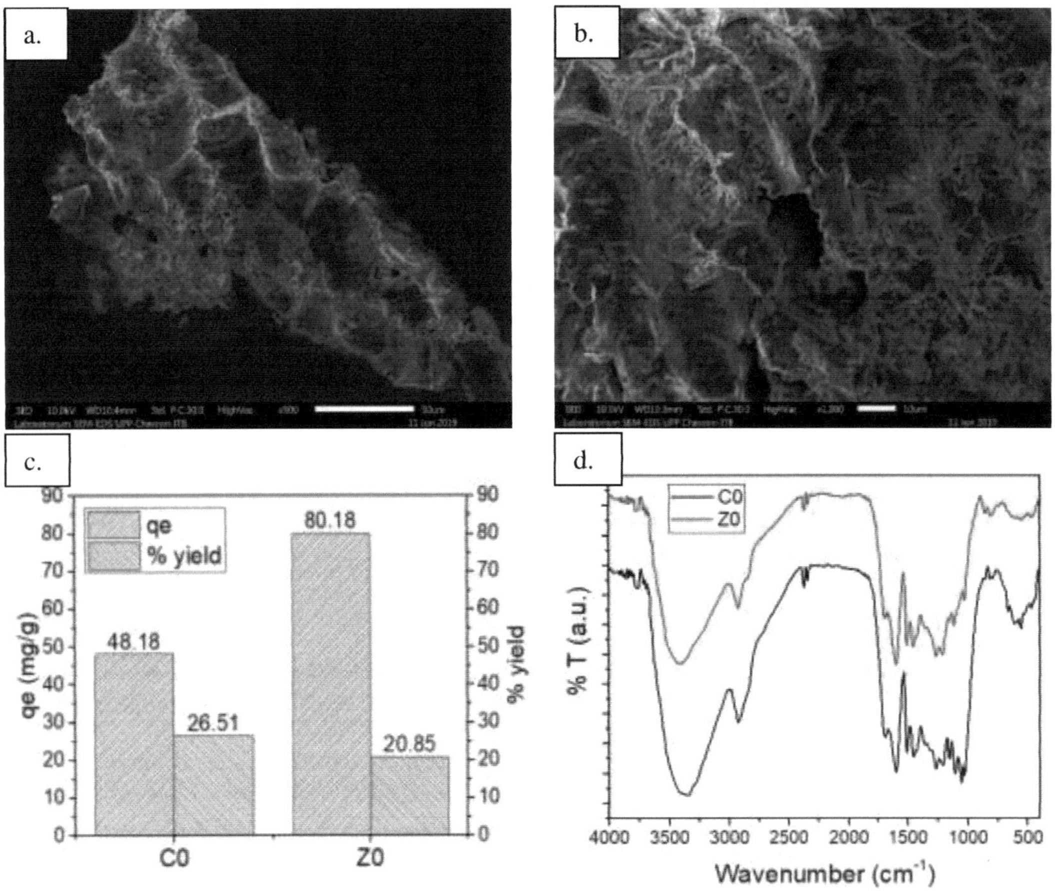

Figure 1. Morphology of C0 (a), and Z0 (b), yield and q_e (c), and FTIR spectra of C0 and Z0 (d)

3.2. Characterization of activated carbon

As shown in Fig 2.a, an increase of impregnation ratio has increased the qe. It is known that $ZnCl_2$ helps the development of porosity during carbonization [22]. It is known that at impregnation ratio below 2, $ZnCl_2$ gave strong inhibition effect to tar formation and promoted release of volatile gases, resulting in formation of micropores and mesopores [23], thus lowering the activated carbon yield (Fig 2.b). At higher of impregnation ratio, pore widening was more dominant than pore formation, thus creating more visible pores on the surface on activated carbon [24]. This result was in a good agreement with surface morphology, presented in Fig 3. It could be observed for both samples, that more pores were observed on the surface of activated carbon with increase of impregnation ratio. Comparing the effect of catalyst during HTC process, the Z samples generally had higher qe compared to C samples. We speculate that it was possible due to the initial hydrochar already had more pores in its structure, thus ease the diffusion of $ZnCl_2$ during impregnation process. Both C4 and Z4 samples had higher qe value compared to that of commercial activated carbon.

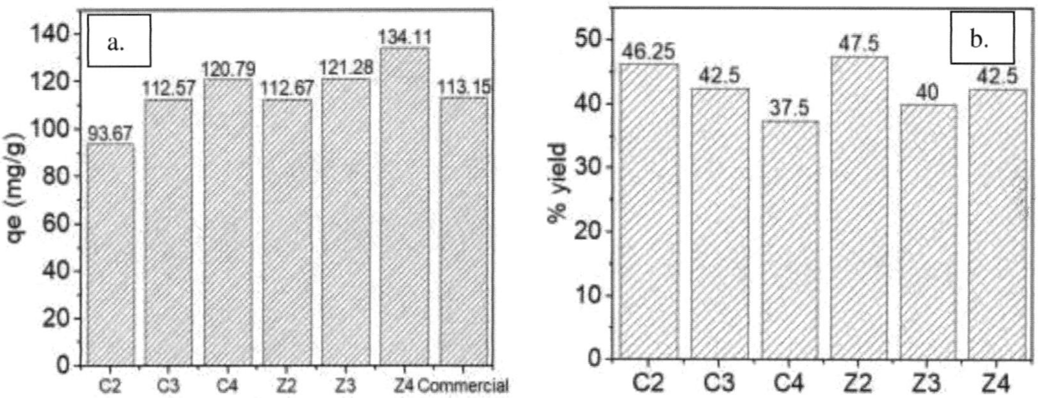

Figure 2. q_e of salacca peel AC compared to commercial AC (a), and salacca peel AC's yield (b)

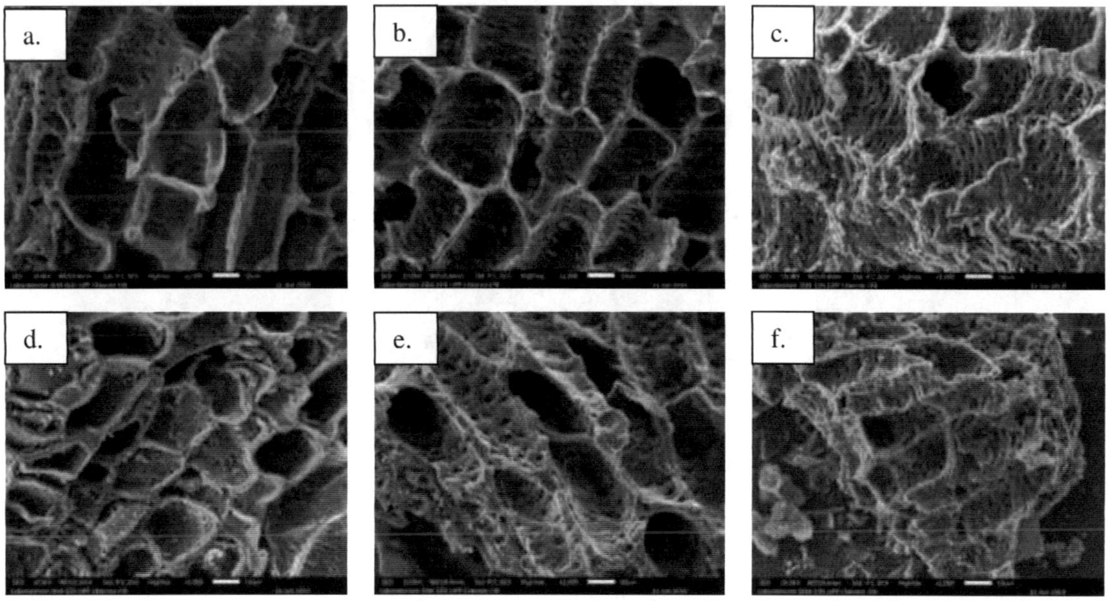

Figure 3. Morphology of C2 (a), C3 (b), C4 (c), Z2 (d), Z3 (e), and Z4 (f) at magnification of 1,000×

The FTIR spectra of C and Z samples are presented in Fig 4.a and b. It could be observed that both activated carbon samples exhibited different peaks in their respective hydrochar in the fingerprint region, indicating transformation of hydrochar to activated carbon during microwave heating activation process [25]. There are some peaks that persisted in the activated carbon, namely –OH (3400-3300 cm^{-1}) and C=O (1602.8 cm^{-1}), however at lower intensity compared to the respective hydrochar. It was possible due to decomposition of hydrochar structure to volatile gas (CH_4, CO, CO_2, H_2O) during activation process [26]. Further comparison of C4, Z4, and commercial activated carbon's FTIR spectra is presented in Fig 4.c. The activated carbons that were synthesized from hydrochar in this study had higher –OH peak than commercial activated carbon. This results are in a good agreement with previous studies [17, 20]. It was also observed that Z4 had higher intensity of –OH peak than C0. We speculate that the presence of $ZnCl_2$ during HTC lead more complete conversion of salacca peel to carbon structure, thus most of volatiles are already decomposed during this process, compared to C0. This made Z samples could preserved more functional groups during activation process. Further analysis is needed to confirm this statement.

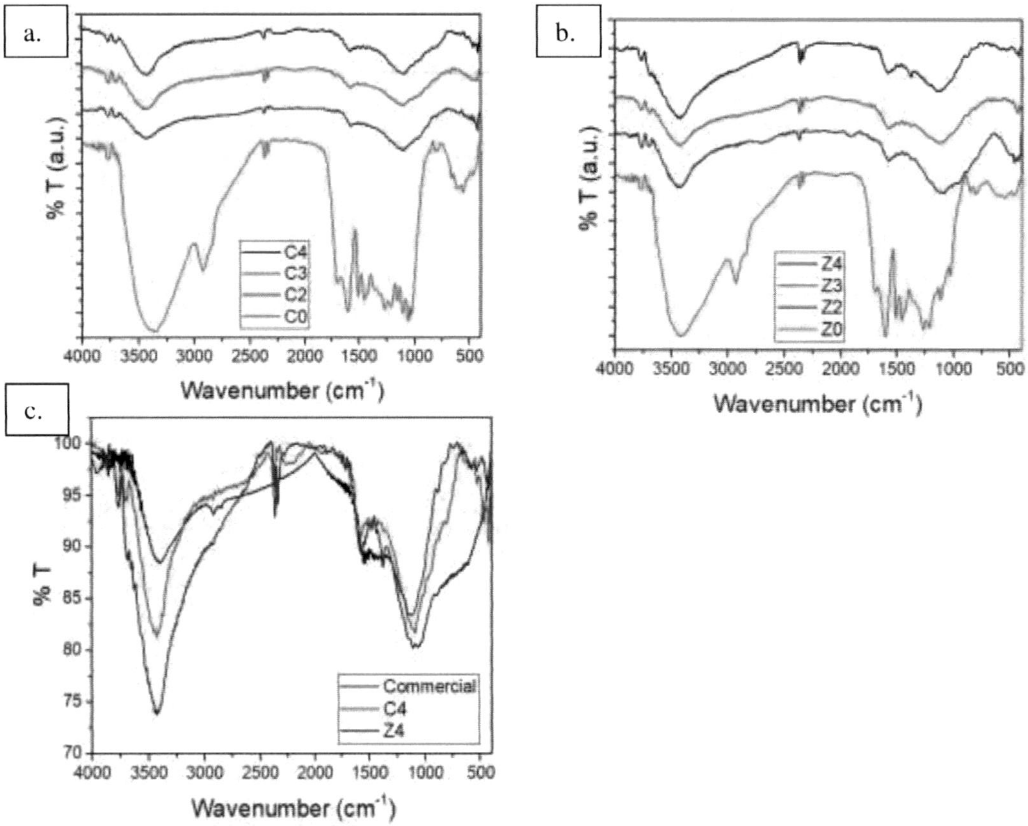

Figure 4. FTIR spectra of salacca peel AC carbonized without (a) and with catalyst (b), and comparison of C4 and Z4 with commercial AC

4. Conclusion

We have successfully synthesized activated carbon from salacca peel by using $ZnCl_2$ catalysed HTC and $ZnCl_2$ chemical activation with microwave heating. The presence of $ZnCl_2$ as catalyst during HTC promoted pore development, as observed in the morphology and higher qe, compared to sample without catalyst. Lower intensity of oxygen functional groups' peak was observed with Z0 sample due to the presence of $ZnCl_2$ helped the hydrolysis of biomass to gases. The presence of more pores in the Z0 sample helped further pore development in the $ZnCl_2$ chemical activation with microwave heating.

More pores were observed on the surface of the Z2-4 carbons compared to C2-4 samples as the initially well-developed pores in Z0 might help the impregnant diffusion into the carbon structure. The increase of impregnant in chemical activation resulted on increase of pores observed, and higher qe value indicating higher surface area was obtained. Compared to commercial activated carbon, the activated carbon derived from salacca peel in this study possessed higher qe value and oxygen functional groups, making it more suitable for various applications.

Acknowledgement
This research was funded by the Directorate of Research and Community Service, Directorate General of Research and Development Strengthening, Ministry of Research, Technology, and Higher Education of Indonesia under Research Scheme of *Penelitian Dasar Unggulan Perguruan Tinggi* (PDUPT) 2017-2019. The authors are thankful for the financial supports.

References

[1] Wang T, Zhai Y, Zhu Y, Li C, and Zeng G 2018 *Renew Sus Energ Rev* **90** 223–247
[2] Mbarki F, Selmi T, Kesraoui A, Seffen M, Gadonneix P, Celzard A, and Fierro V 2019 *Ind. Crops Prod.* **140** 111717
[3] Nizamuddin S, Baloch H A, Griffin G J, Mubarak N M, Bhutto A W, Abro R, Mazari S A, and Ali B S 2017 *Renew Sust Energ Rev* **73** 1289–1299
[4] Ming J, Wu Y, Liang G, Park J-B, Zhao F, and Sun Y-K 2013 *Green Chem* **15** 2722–2726
[5] Lynam J G, Coronella C J, Yan W, Reza M T, and Vasquez V R 2011 *Bioresour. Technol.* **102** 6192–6199
[6] Lynam J G, Reza M T, Vasquez V R, and Coronella C J 2012 *Fuel* **99** 271-273
[7] Yan J, Zhang H, Xie Z, and Liu J 2017 *AIP Conf. Proc.* **1864** 020218
[8] Teh S J, Hamid S B A, Lai C W, and Lim Y S 2015 *Nanosci. Nanotechnol. Lett.* **7** 611–615
[9] Guo S, Gao Y, Wang Y, Liu Z, Wei X, Peng P, Xiao B, and Yang Y 2019 *Environ Sci Pollut R* **In press** 1-9
[10] Zhu G, Zhu X, Fan Q, and Wan X 2011 *Resour Conserv Recy* **55** 409-416
[11] Hesas R H, Daud W M A W, Sahu J N, and Arami-Niya A 2013 *J Anal App Pyrol* **100** 1-11
[12] Ao W, Fua J, Mao X, Kang Q, Ran C, Liu Y, Zhang H, Gao Z, Li J, Liu G, and Dai J 2018 *Renew Sust Energ Rev* **92** 958–979
[13] González-García P 2018 *Renew Sust Energ Rev* **83** 1393–1414
[14] Liu B, Gu J, and Zhou J 2016 *Environ Prog Sustain Energy* **35** 133-140
[15] BPS. Dynamic Table in Horticulture (in Indonesian) 2019. 14th April 2019; https://www.bps.go.id/subject/55/hortikultura.html#subjekViewTab6.
[16] Arie A A, Vincent, and Putranto A 2016 *Adv. Mater. Lett.* **7** 226-229
[17] Susanti R F, Arie A A, Kristianto H, Erico M, Kevin G, and Devianto H 2019 *Ionics* **25** 3915-3925
[18] Arie A A, Kristianto H, Cengiz E C, and Demir-Cakan R 2019 *Ionics* **25** 2121-2129
[19] Arie A A, Kristianto H, Demir E, and Cakan R D 2018 *Mater. Chem. Phys.* **217** 254-261
[20] Susanti R F, Kevin G, Erico M, Kevien, Arie A A, Kristianto H, and Handoko T 2018 *J. Nanosci. Nanotechnol.* **18** 7263–7268
[21] Shen W, Li Z, and Liu Y 2008 *Recent Pat. Chem. Eng.* **1** 27-40
[22] Bouguettoucha A, Reffas A, Chebli D, Mekhalif T, and Amrane A 2016 *Desalin Water Treat* **57** 24056–24069
[23] Lua A C and Yang T 2005 *J Colloid Interface Sci* **290** 505-513
[24] Kamandari H, Rafsanjani H H, Najjarzadeh H, and Eksiri Z 2015 *Res Chem Intermed* **41** 71-81
[25] Chen J, Zhang L, Yang G, Wang Q, Li R, and Lucia L A 2017 *Bioresources* **12** 5928-5941
[26] Demiral I, Samdan C A, and Demiral H 2016 *Desalin Water Treat* **57** 2446–2454

IOP Conf. Series: Materials Science and Engineering **742** (2020) 012045 doi:10.1088/1757-899X/742/1/012045

Preparation of Salacca Peel Based Activated Carbon using $CeCl_3$ Catalyzed Hydrothermal Carbonization and Microwave Induced KOH Chemical Activation as Ni-Ion Capacitor Electrode

F C Ondy[1], C Chrismanto[1], R F Susanti[1]*, H Kristianto[1], H Devianto[2]

[1]Department of Chemical Engineering, Faculty of Industrial Technology, Parahyangan Catholic University, Ciumbuleuit 94, Bandung 40141, West Java, Indonesia
[2]Department of Chemical Engineering, Institut Teknologi Bandung, Ganesha 10, Bandung 40132, West Java, Indonesia

E-mail : santi@unpar.ac.id

Abstract. Activated carbon is a highly porous non-graphite carbon material. Activated carbon has been used for various applications, namely adsorbent, catalyst support and electrode. As supercapacitor's electrode, activated carbon has a high energy density and fast charge-discharge. In this study, salacca peel used as precursor to synthesized activated carbon through hydrothermal carbonization (HTC) and microwave induced chemical activation. The hydrochar (HC) was synthesized at 225°C, 5 bar for 5 h using cerium (III) chloride as catalyst. Furthermore, KOH was used as activating agent at various HC to KOH mass ratio at 1:2 (CE2), 1:3 (CE3), and 1:4 (CE4), under microwave irradiation at 700 W for 8 min. Activated carbon samples were analyzed using methylene blue adsorption capacity method, scanning electron microscopy (SEM), and Fourier-transform infrared spectroscopy (FTIR). The CE4 sample gave highest methylene blue adsorption capacity, indicating high surface area. This sample was used as nickel ion's capacitor electrode with nickel oxide as the reference electrode and 6 M KOH as the electrolyte. The electrochemical characteristic was studied using cyclic voltammetry (CV) with various scan rates (2 mV/s, 5 mV/s, 10 mV/s) for the nickel ion's capacitor electrode. It was obtained that CE4 provided higher capacitance compared to commercial activated carbon.

1. Introduction

Salacca fruit (*Salacca zalacca*) is one of the fruits that has a large commodity in Indonesia. High production and consumption of salacca fruits generate a huge amount of fruit waste, namely salacca peels. This part of salacca fruit is a potential source of biomass to synthesize activated carbon. Salacca peel has been previously used as precursor to synthesize activated carbon with some applications, such as adsorbent [1], Li-ion capacitor [2] and Li-S battery's electrode [3].

Hydrothermal carbonization (HTC) is a carbonization process that utilize water at subcritical condition[4]. Catalyst is sometimes added in HTC to accelerate hydrolysis and dehydration process of a biomass. In this study, HTC process was done by adding cerium (III) chloride as catalyst. Cerium (III) chloride could assist the HTC process in hydrolysis and dehydration process, as well as improved activated carbon surface area and adsorption capacity compared to other catalyst [5].

IOP Conf. Series: Materials Science and Engineering **742** (2020) 012045 doi:10.1088/1757-899X/742/1/012045

Chemical activation of HC from HTC process was done by microwave heat radiation. Microwave induced chemical activation differ from conventional pyrolysis method in the way of heating process. In conventional method, the heat source was placed outside the sample vessel. This makes heat travelled through the vessel, outer then finally inner side of the sample through thermal conduction[6]. Microwave heat radiation could instantly enter the sample to made heat entirely and fast provided inside the sample, resulted in lower energy usage compared to conventional pyrolysis[7]. This unique property of microwave radiation could solve the issues in using conventional pyrolysis method and furthermore could lower the activated carbon production cost. KOH was used in activation process as chemical activating agent to develop the microporosity in the carbon structure[8], making the activated carbon sample suitable for capacitor application. The obtained AC was characterized through methylene blue adsorption capacity method, scanning electron microscopy (SEM), and Fourier-transform infrared spectroscopy (FTIR). Furthermore, the best activated carbon was used as anode for nickel ion capacitor. The electrochemical characteristic of the supercapacitor was analyzed using cyclic voltammetry at various scan rates.

2. Methodology

2.1. Synthesis of activated carbon and characterization

The salacca fruits were acquired from local market in Bandung, Indonesia. The peel was separated from the fruit flesh, and washed thoroughly with running water. The peel that had been cleaned, was then dried in oven at 110°C for 24 h. The dried peel was crushed with food processor into powder. The powder was then sieved with -20+30 mesh until the desired size of salacca peel powder was obtained. The hydrothermal carbonization detailed procedure could be seen in another studies[2, 9]. To simplify, 5 g of salacca peel powder mixed with catalyst solutions consist of 10 mg cerium (III) chloride with 15 mL distilled water. This mixture was loaded into a 150mL subcritical reactor (SUS 316) and the reactor was closed and purged with nitrogen for 10-15 min to remove oxygen inside the reactor. The pressure in subcritical reactor was adjusted by pumping distilled water using hydraulic hand pump. After adjusting the pressure, the reactor was heated until reached 225°C. The desired pressure was achieved by pumping water into the system until 50 bar pressure was achieved. The HTC process was done for 5 h. After the process, the reactor was cooled down until room temperature was achieved. The obtained HC was separated and dried using oven at 110°C for 24 h.

The obtained HC was impregnated with KOH as activating agent. The impregnation ratio was done at various HC to KOH mass ratio of 1:2, 1:3, 1:4. This mixture was left for 24 h before oven dried at 120°C for 12 h. The dried mixture was placed in a quartz tube and carbonized using modified commercial microwave oven (Panasonic) under continuous stream of nitrogen at 700 W for 8 min. The obtained AC was washed using 0.1 M HCl and distilled water until neutral. The sample was dried in oven at 120°C for 12 h. In this study, the activated carbon samples are denoted as CEx; with x is the KOH mass ratio. The yield of activated carbon was calculated using equation (1).

$$\%\ \text{yield} = \frac{\text{mass of activated carbon}}{\text{mass of salacca peel}} \times 100\% \qquad (1)$$

The methylene blue adsorption capacity was measured by mixing 200 mL of 10 ppm methylene blue solution with 10 mg of obtained activated carbon samples in shaking water bath for overnight. The adsorption capacity was calculated using equation (2), where q is the adsorption capacity (mg/g activated carbon), Ci and Co are initial and equilibrium methylene blue concentration (mg/L), V is the solution volume (L), and m is the activated carbon's mass (g).

$$q = (Ci - Co) \times \frac{V}{m} \qquad (2)$$

The morphology and the functional group of obtained HC and AC was observed by using Scanning Electron Microscopy (SEM JEOL JSM-IT-300) and Fourier Transform Infrared Spectroscopy (FTIR - Prestige 21 Shimadzu Instruments), respectively.

IOP Conf. Series: Materials Science and Engineering **742** (2020) 012045 doi:10.1088/1757-899X/742/1/012045

2.2. *Preparation of electrode*

The electrode was prepared by mixing 0.05 g AC with 0.005 g polyvinylidene fluoride (PVDF) as binder, and ethanol as solvent to make the mixture become slurry. The slurry was applied evenly on stainless steel (SS) mesh. The counter electrode was prepared by mixing 0.05 g nickel oxide with 0.005 g PVDF, and ethanol as solvent. The mixture was applied evenly on carbon cloth. Both of the electrodes were oven dried at 80°C for overnight. The dried electrodes were assembled by using 6 M KOH as electrolyte and Nafion 212 as separator.

The electrochemical characterization was done by using cyclic voltammetry (CV) method which was measured using Gamry V3000 potensiostat. The CV measurement was done at scan rate of 2, 5, and 10 mV/s and voltage window 0-1V. The capacitance was calculated using equation (3), where I is the current (A), Δt is time (s), m is the mass of material in the electrode (g), and ΔV is the voltage window (V)

$$C = \frac{\Sigma|I|.\Delta t}{2.m.\Delta V} \quad (3)$$

3. Result and Discussion

3.1. *Morphology*

HC and CE samples were observed using SEM method to observe the morphology, presented in figure 1. It could be seen that there are formed pores which spread equally on the surface of HC compared to surface morphology of salacca peel at same magnification[2]. After chemical activation, more pores were developed in CE samples, compared to HC. Based on figure 1.b-d, the higher impregnation ratio, the more pores are observed on the surface of AC. At higher impregnation ratio, the carbon reaction with KOH became more intense. The result of this reaction is development of new pores, opening of tars - blocked pores, and expansion and deepening of the existed pores[10].

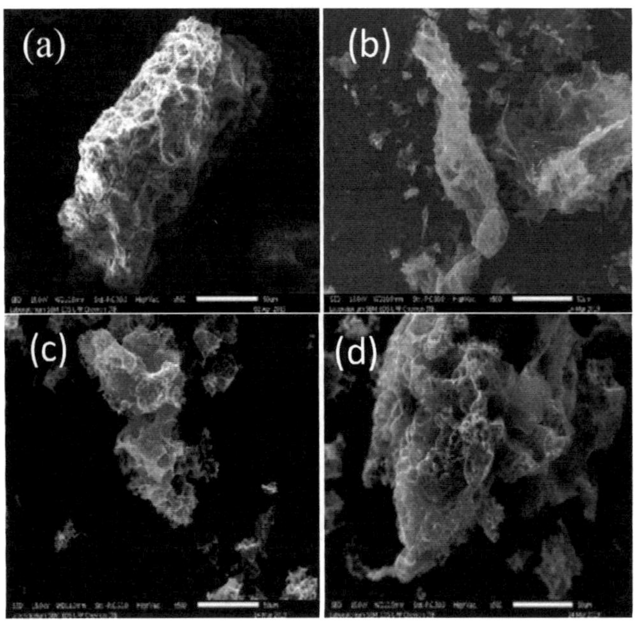

Figure 1. SEM result of HC (a), CE2 (b), CE3 (c), and CE4 (d) at magnification of 500×

IOP Conf. Series: Materials Science and Engineering **742** (2020) 012045 doi:10.1088/1757-899X/742/1/012045

3.2. FTIR spectra

FTIR spectroscopy was used to observe the functional groups contained on the HC and CE samples. In figure 2, it could be observed that there is a -OH carboxyl and carbonyl groups stretching vibration peak around 3000-3600 cm^{-1}, -CH aliphatic bonds peak around 2800-3000 cm^{-1}, highly C-H and C-O bonds around 1163-1034 cm^{-1}, and low intensity of aromatic C-H bonds at 600-800 cm^{-1}[11]. When HC compared to CE, it could be observed there is a decrease of O-H, C-H, C-O, and C-H aromatic bonds from HC to CE4, as well as disappearance of aliphatic bond as the result of chemical activation process. The decreased chemical bonds in CE occured due to thermal decomposition during the chemical activation into volatile gases[12].

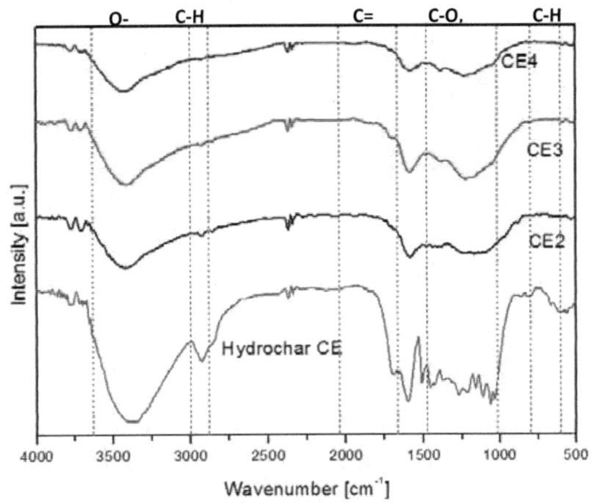

Figure 2. FTIR spectra of HC and CE samples

3.3. AC yield and adsorption capacity

The effect of impregnation ratio to the CE yield is presented in figure 3.a. It could be observed that with the increase of KOH for HC impregnation, CE's yield was decreasing. The higher amount of KOH used, the more carbon atoms in chemical activation process would be converted to CO and CO_2 following the KOH and HC reaction. Therefore, AC yield would decrease along with the higher KOH used in activation process[10]. The reaction between KOH and carbon is presented as followed[10]:

$$6\ KOH + 2C \rightarrow 2K + 3H_2 + 2K_2CO_3 \quad (4)$$

$$K_2CO_3 + 2C \rightarrow 2K + 3CO \quad (5)$$

$$K_2CO_3 \rightarrow K2O + CO_2 \quad (6)$$

$$K_2O + 2C \rightarrow 2K + CO \quad (7)$$

Methylene blue adsorption was used to observe HC and CE adsorption capacity that proportionally related to the surface area of the samples[13]. Based on figure 3.b, the CE adsorption capacity is higher than that of HC and commercial AC (Merck). As mentioned before, chemical activation process could help pores development, open blocked pores, and widen also deepen existed pores in activated carbon[10]. This result was also in a good agreement of SEM observation in figure 1 that showed CE samples' pores were more developed compared to HC. Furthermore, similar trend of the effect of chemical activating agent ratio to the pores development was also observed in this analysis.

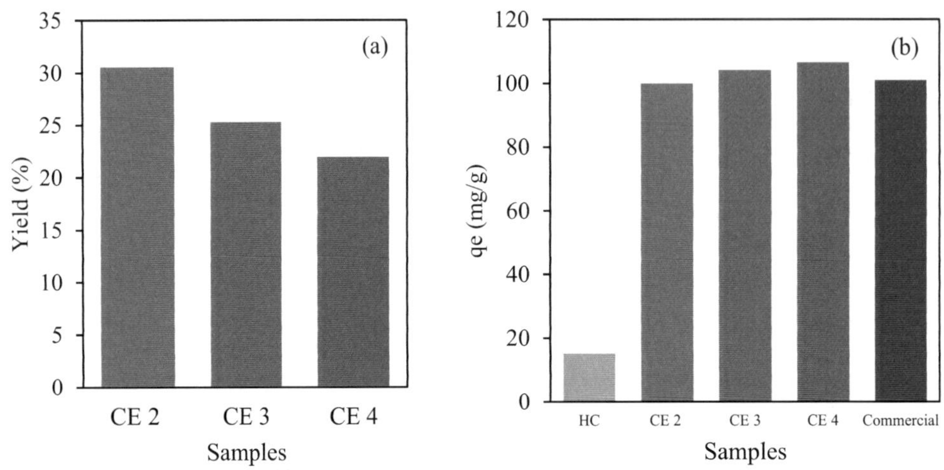

Figure 3. AC yield (a) and methylene blue adsorption capacity of HC and CE (b)

3.4. Cyclic Voltammetry Analysis

According to previous analysis result, CE4 gave the best result in terms of activated carbon's morphology, functional groups, and adsorption capacity. Therefore, CE4 was further used as supercapacitor's electrode that would be analyzed with cyclic voltammetry method. The voltammogram of CE4 is presented in figure 4. It could be observed that the voltammogram represented a leaf-like shape with a slope. There was no indication of anodic or cathodic peak observed in charge-discharge process, indicating ions adsorption-desorption process during the charge-discharge[1]. The shape of the voltammogram was a deviation to the rectangle shaped[14] ideal electric double layer capacitor voltammogram. The difference in voltammogram shape was possible due to resistance that occurred during adsorption-desorption process. Greater resistance made the voltammogram shape narrowed [15] and with bigger slope [16]. The CE4 capacitance was decreasing along with the increasing of scan rate used, namely 7.63 (2 mV/s), 3.6 (5 mV/s), and 2.2 F/g (10 mV/s). Compared to commercial activated carbon the capacitance of CE4 was higher (table 1). It is known that the capacitance value of a supercapacitor is depending on the surface area. Higher surface area could make more electrolytes were adsorbed during charging process, thus increasing the capacitance.

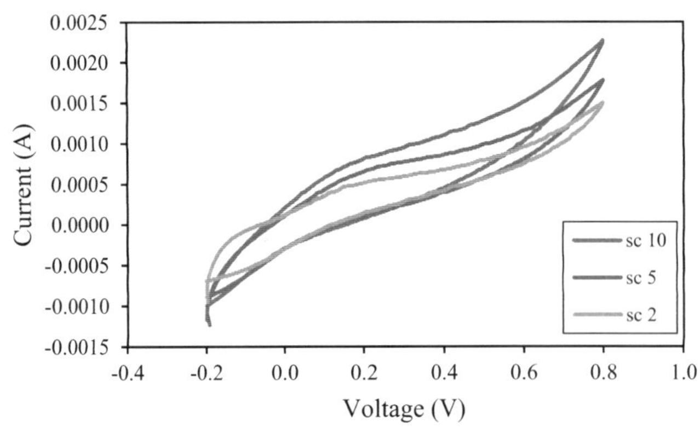

Figure 4. Voltammogram of CE4 at various scan rate

Table 1. Comparison of various electrode capacitance

Cathode	Capacitance (F/g)	Reference
AC CE4	7.63	This study
Commercial AC	1.49	[2]

4. Conclusions

This study highlighted the potential of salacca peel as a promising precursor for the preparation of activated carbon using hydrothermal carbonization and microwave induced chemical activation. The activated carbon structure became more porous with the increase of KOH usage as activating agent, as shown in SEM morphology and methylene blue adsorption capacity. The highest adsorption capacity was found in CE4 sample, which is also higher than commercial activated carbon. Meanwhile, the functional groups in CE samples decreased during the chemical activation. As Ni-ion electrode, CE4 sample exhibited a leaf-like shape with slope, and no anodic or cathodic peaks during charge-discharge, indicating adsorption-desorption process in the process. It was found that CE4 gave higher capacitance than commercial activated carbon, highlighting its promising application as supercapacitor electrode.

Acknowledgement

This research was funded by the Directorate of Research and Community Service, Directorate General of Research and Development Strengthening, Ministry of Research, Technology, and Higher Education of Indonesia under Research Scheme of *Penelitian Dasar Unggulan Perguruan Tinggi* (PDUPT) 2017-2019. The authors are thankful for the financial supports.

References

[1] Arie A and Vincent P A 2016 *Adv Mater Lett* **7** 226-9
[2] Susanti R F, Arie A A, Kristianto H, Erico M, Kevin G and Devianto H 2019 *Ionics* **25** 3915-3925
[3] Arie A A, Kristianto H, Cengiz E C and Demir-Cakan R 2019 *Ionics* **25** 2121-9
[4] Kambo H S and Dutta A 2015 *Renew Sust Energ Rev* **45** 359-78
[5] Shen W, Zheng J, Qin Z, Wang J and Liu Y 2004 *J Mater Sci* **39** 4693-6
[6] Thostenson E and Chou T-W 1999 *Compos Pt A: Appl Sci Manuf***30** 1055-71
[7] Xie Z, Yang J, Huang X and Huang Y 1999 *Journal of the European Ceramic Society* **19** 381-7
[8] Hesas R H, Daud W M A W, Sahu J and Arami-Niya A 2013 *J AnalAppl Pyrol* **100** 1-11
[9] Susanti R F, Kevin G, Erico M, Andreas A, Kristianto H and Handoko T 2018 *J Nanosci Nanotechnol* **18** 7263-8
[10] Basta A, Fierro V, El-Saied H and Celzard A 2009 2 *Biores Technol* **100** 3941-7
[11] Xu Q, Qian Q, Quek A, Ai N, Zeng G and Wang J 2013 *ACS Sust Chem Eng* **1** 1092-101
[12] Yagmur E, Ozmak M and Aktas Z 2008 *Fuel* **87** 3278-85
[13] Mezohegyi G, van der Zee F P, Font J, Fortuny A and Fabregat A 2012 *J Environ Manage* **102** 148-64
[14] Wang G, Zhang L and Zhang J 2012 *Chem Soc Rev* **41** 797-828
[15] Wendy G. Pell B E C 2001 *J Power Sources* **96** 57-67
[16] Gu W, Peters N and Yushin G 2013 *Carbon* **53** 292-301

Microwave Assisted KOH Activation of Salacca Peel Dericed Activated Carbons as Adsorbents for Methylene Blue Removal from Aqueous Phase

Johan Klemantan[1], Hans Kristianto[1] and Arenst Andreas Arie[1]*

[1]Department of Chemical Engineering, Faculty of Industrial Technology, Parahyangan Catholic University, Ciumbuleuit 94, Bandung, 40141, Indonesia.

E-mail : arenst@unpar.ac.id

Abstract. In this study, salacca peel derived activated carbons (SPAC) were prepared by microwave assisted- KOH activation process. In the preparation process, the fixed mass ratio between salacca peel and KOH of 1:2 was used. Next, the effects of activation time (5, 15, 20 and 25 minutes) and microwave power (30, 380 and 700 W) on the characteristics of SPAC were then investigated. The BET surface area of SPAC was about 1491 m^2/g for activation time of 25 minutes and microwave power of 700 W. The as-prepared SPAC was then then as adsorbents for methylene blue (MB) removal from aqueous phase. The adsorption data of SPAC were analyzed by Langmuir and Freundlich adsorption isotherms and the best correlation was shown by the Langmuir isotherm with maximum capacity of 327.9 mg/g. The analyses of kinetic data showed that the adsorption of MB on SPAC follow the pseudo-second order kinetic model.

1. Introduction

Activated carbon is one of the most well known carbon materials since it has been widely used in many industrial applications such as purification of liquid and gas, catalyst or (photo) catalyst support, electrodes in supercapacitors or redox batteries andhydrogen storage [1]. Activated carbons have some beneficial characteristics such as high surface area, easy availability and high thermal stability[2].

One of the main problems in commercial production of activated carbonsis seeking a low cost, renewable and abundant precursor.There have been many efforts to produce activated carbons from various kinds of biomass [3–5]. Even though biomass has great potencies for activated carbons production, the processing steps are critical [6]. The activation process is the most important part in the production of activated carbon. This process opens or increases porosity on the surface of carbons. In general, activated carbon can be prepared by one of the following two methods of activation which are by physical or chemical activation [7, 8].Both of them however, consumed much energy and took more than 6 hours to complete [9].

Another new promising technique that can produce equally high quality activated carbon that utilizes relatively less energy in shorter time is by microwave (MW) heating process [10 - 11]. In the MW device, the microwaves supply energy directly to the heated samples. Energy transfer is not by conduction or convection as in conventional heating, but microwave energy is readily transformed into heat inside the particles by dipole rotation and ionic conduction [12]. Microwave heating has been used to prepare activated carbon originated from many types of biomass such as, pineapple peel[13], jackfruit peel [14], mangosteen peel tobacco stem, rice husk,mangosteen peel[15],durian shell [16], orange peel [17], guava peel [18], sugarcane bagasse [19], siris seeds pod [20],rambutan peel [21], banana peel [22], Albizia lebbeck seed pods,wheat straw [23], Parkinsonea aculeata wood[24], wood chips [25], almond shell [26] and pine wood [27] precursors.

Although many kinds of biomass precursors have been used to produce activated carbon, the work related to the utilization of salacca peel as precursor using microwave heating process has not been reported elsewhere. Salacca fruit or also known as snake fruit reddish-brown scaly skin is a typical tropical fruit cultivated in Indonesia especially in Bali, Lombok, Maluku and Sulawesi [28-29]. After consumption of the fruit, the peels are usually discarded as waste and until now, there has not been utilized as raw materials for useful products [30]. Our previous works focused on the preparation of SPAC as adsorbent material for removal of methylene blue using conventional heating method [31].

The main objective of this study is to investigate the use of microwave assisted KOH activation process to prepare SPAC. The influences of activation time and microwave power in the preparation process were further investigated.The SPAC was then used as adsorbent for methylene blue (MB) removal from aqueous phase. The adsorption isotherms were used to determine the maximum capacity of MB on SPAC. The adsorption kinetics were studied by the pseudo first and second order model.

2. Experimental

The salacca peel was obtained from the local market in Bandung, Indonesia. The peels were then washed with distilled water, and dried in an oven at 110°C for 24 h to remove the moisture content. The dried salacca peel were crushed and sieved to a uniform particle size (100-200 meshes). The microwave-assisted process was conducted using a modified microwave oven (Electrolux). Mixtures of KOH and salacca peel powder with mass ratio of 2:1 were put in a quartz tube reactor placed in a microwave oven. The activation processes were then subjected to various microwave heating times (5, 15, 20,25 and 30 minutes) and microwave radiation powers (30,380 and 700 W) under nitrogen flow. The resulting material was washedrepeatedly with hot distilled water until solution pH reached 7.0. The obtained SPAC were dried at 80°C for 24 h and then stored for further analyses.

Characterization in terms of specific surface area, pore volume, and pore diameter of the obtained activated carbons was determined by N_2 adsorption at with surface area and pore size analyzer using the Brunnaeur-Emmet-Teller (BET) method. The micropore volume was calculated by using t-Plot micropore volume. The pore size distribution was determined by using Barrett–Joyner–Halenda (BJH) model. To study the effect of the initial metal ion concentration, 500 mL of Methylene Blue (MB) at various concentrations (10, 20, 30, 40 and 50 ppm) were mixed with 20 mg of SPAC in the Erlenmeyer flasks at room temperature. These mixtures were kept at room temperature with continuous shaking at 100 rpm. All samples were filtered and analyzed by UV-Vis spectrophotometer.

To investigate the effects of initial pH solution, the initial pH was adjusted by 1 mol/L HCl and NH_4OH with 10 mg of SPAC and 500 mL of MB solutions with initial concentrations of 50 ppm at room temperature. For all the adsorption experiments, they are conducted triplicate and the average values were used for the analysis. The adsorption performances were measured through the adsorption capacity which can be calculated using the following equation:

$$q_t = \frac{C_0 - C_t}{m} V \tag{1}$$

where, C_0 and C_t (mg/L) are the initial concentration and concentration at time t of MB, respectively. V(L) is the total solution volume, and m(g) is the mass SPAC-adsorbent. The study of adsorption kinetics was conducted at various initial MB concentrations of 10-50 ppm and 5 at room temperature. About 10 mg ASAC was put into Erlenmeyer flasks with 250 ml MB solution; the samples were separated by filtering at different contacting time. The residual concentrations of MB were measured, and adsorption amounts of ASAC were calculated by equation (1).

The experimental data were then analyzed using the pseudo-first-order [32] and second-order kinetic models [33]. The equation of the pseudo first order model is given by:

$$q_t = q_e(1 - e^{-k1.t}) \tag{2}$$

where q_e (mg/g) is the amount of MB adsorbed at equilibrium, k_1 (min^{-1}) is the pseudo-first-order rate constant, and t (min) is the contact time.

The form of the pseudo-second-order kinetic model is given by:

$$q_t = \frac{q_e^2 k_2 t}{1+q_e k_2 t} \tag{3}$$

where k_2 is the pseudo-second-order rate constant (g/mg min).

The capacity of an adsorbent can be represented by its equilibrium adsorption isotherm. The Langmuir and Freundlich adsorption models are commonly used to study the adsorption behavior of materials and the correlation among adsorption parameters. The Langmuir isotherm equation is represented by the following equation:

$$q_e = \frac{q_m K_L C_e}{1+K_L C_e} \tag{4}$$

where, C_e is the equilibrium concentration of (ppm), q_e is the amount of MB adsorbed (mg/g), q_m is the maximum adsorption capacity of MB (mg/g), and K_L is the Langmuir adsorption equilibrium constant (L/mg) related to the affinity of the binding sites.

The Freundlich isotherm equation is described by the following equation:

$$q_e = K_F C_e^{1/n} \tag{5}$$

where K_F and n are the Freundlich adsorption isotherm constants, which are indicators of adsorption capacity and adsorption intensity, respectively.

3. Results and discussion

Figure 1 shows the N_2 adsorption desorption profiles of SPAC prepared by Microwave assisted KOH activation method. It can be concluded the profiles follows the Type IV according to the International Union of Pure and Applied Chemistry (IUPAC) classification of adsorption isotherms [34]. This is a typical structure of mesoporous material and it is beneficial for adsorption purposes [35].

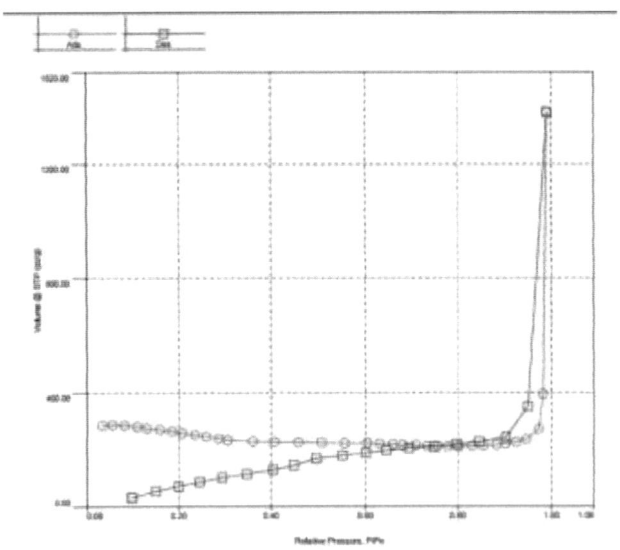

Figure 1 Isothermal N_2 Adsorption- Desorption of SPAC prepared by microve heating and KOH Activation.

Table 1 presented the BET surface area of SPAC prepared by microwave heating method at various activation time and microwave power. It can be seen that at short activation time of 5 minutes and all microwave power, the BET surface area of SPAC was lower than 200 m2/g. However, by setting icrowave power at 700 W and activation time to 25 minutes, we obtained SPAC with high surface area of 1491.04 m^2/g. For the adsorption of methylene blue from aqueous phase, this carbon sample as then used as adsorbents.

Table 1. BET Surface Area of prepared SPAC as function of activation time and microwave power

Run	**Activation Time (Minutes)**	**Microwave Power (W)**	**BET Surface Area (m^2/g)**
1	0	0	26.42
2	5	130	54.51
3	5	230	87.54
4	5	380	198.43
5	5	700	114.54
6	15	700	1302.87
7	20	700	1056.03
8	25	700	1491.04
9	30	700	1325.01

Figure 2 shows the SEM- morphology observation of SPAC prepared by the microwave assisted KOH activation method, it can be seen that the porous structure was found on the surface carbon samples, providing an adequate morphology for MB adsorption.

Figure 2 SEM image of SPAC prepared by the microwave assisted KOH activation method.

The adsorption behavior of MB onto SPAC at different pH values is shown in Figure 3. The effect of pH was done by varying the initial pH of MB solutions from 2 to 12. It can be seen that adsorption capacity increased with increasing pH.

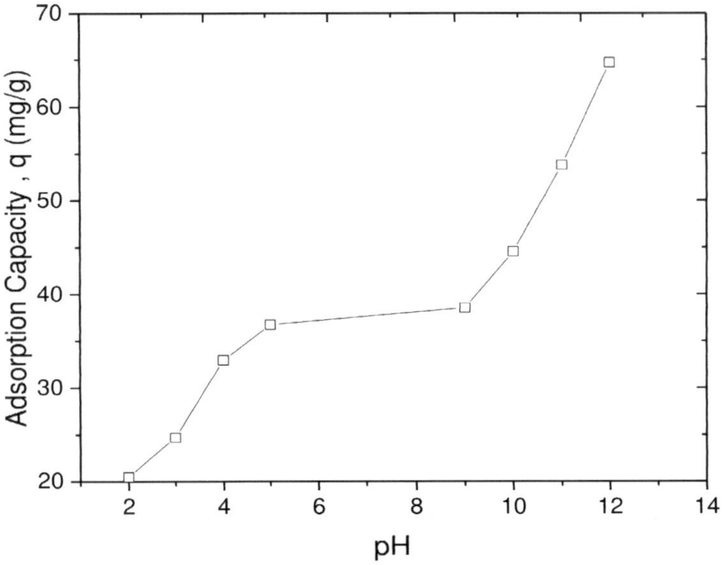

Figure 3 Profile of Adsorption Capacity versus initial pH solution of MB (Room Temperature, 10 mg of SPAC, 50 ppm of MB).

Adsorption capacity showed a very significant enhancement at pH of 6-9. This may be related to the protonation of MB in acidic medium, which creates a competition effect between excessive amounts of $H^+$and cationic dye on adsorption sites. Therefore, the surface of SPAC might become negatively charged as the solution pH increases. It can cause the formation of electric double layers which induces a surface polarity chang and finally it can enhance adsorption capacity.

Figure 4 shows the profile of adsorption capacity versus contact time as function of initial MB concentration (10-50 ppm). The data were then fitted using the pseudo-first-order and second-order models. Table 2 presents the fitting results for the adsorption kinetics models.

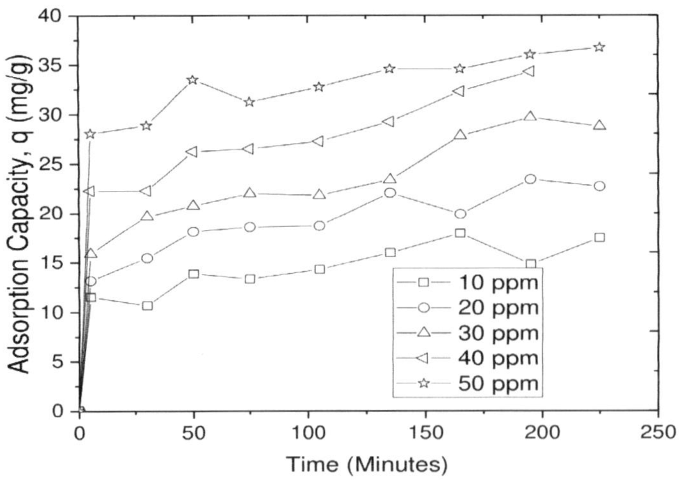

Figure 4 Profile of adsorption capacity versus contact time as function of initial concentration of MB(Room Temperature, 10 mg of SPAC, pH 6).

Table 2. Adsorption kinetic parameters

C_0 / ppm	Pseudo First Order Model			Pseudo Second Order Model		
	k_1 (minute^{-1})	qe (mg MB/g SPAC)	R^2	k_2 (g/mg min^{-1})	qe (mg MB/g SPAC)	R^2
10	0.0578	0.0186	0.533	0.2189	0.4588	0.9894
20	0.0602	0.0168	0.81	0.2258	0.4655	0.9797
30	0.0726	0.0133	0.719	0.2604	0.5189	0.9741
40	0.0888	0.0124	0.6298	0.3154	0.5745	0.9891
50	0.1049	0.0136	0.6551	0.3576	0.6105	0.9964

The data show that the adsorption of MB onto SPAC was best represented by the pseudo-second-order model. These fitting results implies that the MB-SPAC system followed the pseudo-second-order model and that the overall adsorption process rate was controlled by physisorption at the beginning of adsorption and next is the chemisorption.

The plots of the Langmuir and Freundlich adsorption isotherms models for the adsorption of MB onto SPAC at room temperature according to Eqs. (4) and (5) are shown in Fig 5. The model parameters were calculated from the linearized form of both models, as presented in Table 3. The equilibrium data were best fitted by the Langmuir isotherm model. It showed thatthe adsorption of MB onto SPAC from aqueous phaseoccurs via monolayer formation. Furthermore, from Langmuir model,the maximum capacity of MB onto SPAC can be estimated as 375.9 g MB/g SPAC.

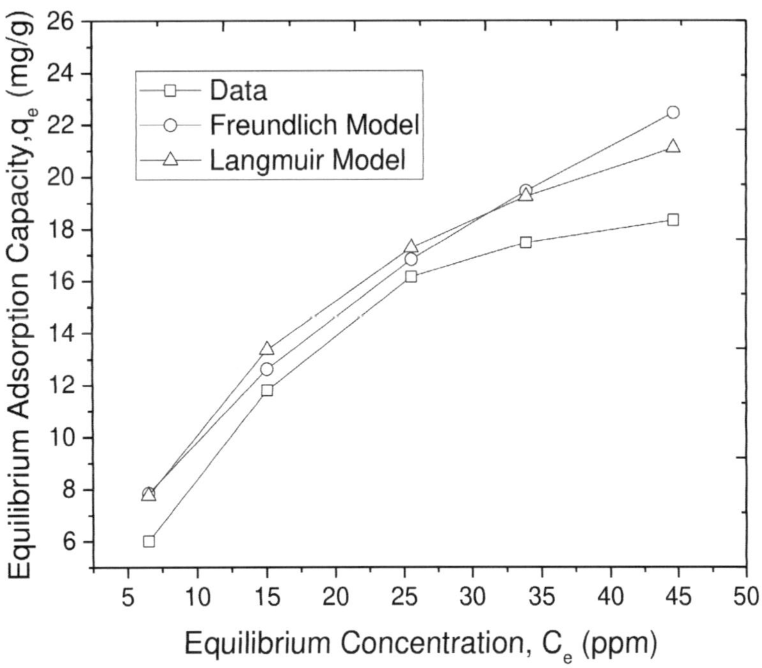

Figure 5 Plots of Langmuir and Freundlich models for adsorption of MB onto SPAC (pH 6, 10 mg of SPAC, varying initial concentration of MB from 10 to 50 ppm).

Table 3 Isothermal Adsorption Parameter Models

Langmuir model q_m (mg/g)	**K_L (L/mg)**	**R^2**	**Freundlich model K_F (mg/g)**	**n**	**R^2**
375.9	0.1709	0.9816	0.0951	2.7762	0.9011

4. Conclusions

This work studied the low cost and renewable adsorbent SPAC that was synthesized by microwave assisted KOH activation method from the salacca peel waste. The activation conditions which gives the highest BET surface area (1491.04 m^2/g) are as follows: a mass ratio (KOH:salacca peel) of 2:1 (w/w), a microwave heating time of 15 min, and a microwave power of 700 W. SPAC was then used as adsorbents for methylene blue (MB) removal from aqueous phase by varying initial pH solution and initial MB concentration. It was found that the adsorption process can be represented by Langmuir adsorption isotherms and the pseudo second order models.The maximum capacity was determined as 375.9 mg/g. Based on the experimental results, it can be said that the microwave assisted activation is an efficient and effective methodfor preparing SPAC as an excellent adsorbent which can be used for the separation and purification process in the industry.

Acknowledgements

This study is supported by funding from The Indonesia Toray Scientific Foundation through the Science and Technology Research Grant in 2015.

References

[1] M. Sevilla and R. Mokaya, "Energy storage applications of activated carbons: supercapacitors and hydrogen storage," *Energy Environ. Sci.*, vol. 7, no. 4, pp. 1250–1280, 2014.

[2] H. Marsh and F. Rodríguez-Reinoso, *Characterization of Activated Carbon*, no. 1. 2006.

[3] M. J. Ahmed, "Application of agricultural based activated carbons by microwave and conventional activations for basic dye adsorption: Review," *J. Environ. Chem. Eng.*, vol. 4, no. 1, pp. 89–99, 2016.

[4] M. A. Yahya, Z. Al-Qodah, and C. W. Z. Ngah, "Agricultural bio-waste materials as potential sustainable precursors used for activated carbon production: A review," *Renew. Sustain. Energy Rev.*, vol. 46, pp. 218–235, 2015.

[5] A. Jain, R. Balasubramanian, and M. P. Srinivasan, "Hydrothermal conversion of biomass waste to activated carbon with high porosity: A review," *Chem. Eng. J.*, vol. 283, pp. 789–805, 2016.

[6] J. S. Cha *et al.*, "Production and utilization of biochar: A review," *J. Ind. Eng. Chem.*, vol. 40, pp. 1–15, 2016.

[7] R. A. Canales-Flores and F. Prieto-García, "Activation Methods of Carbonaceous Materials Obtained from Agricultural Waste," *Chem. Biodivers.*, vol. 13, no. 3, pp. 261–268, 2016.

[8] A. A. Ahmad and A. Idris, "Preparation and characterization of activated carbons derived from bio-solid: a review," *Desalin. Water Treat.*, no. August, pp. 1–15, 2013.

[9] T. M. Alslaibi, I. Abustan, M. A. Ahmad, and A. A. Foul, "A review: Production of activated carbon from agricultural byproducts via conventional and microwave heating," *J. Chem. Technol. Biotechnol.*, vol. 88, no. 7, pp. 1183–1190, 2013.

[10] R. Hoseinzadeh Hesas, W. M. A. Wan Daud, J. N. Sahu, and A. Arami-Niya, "The effects of a microwave heating method on the production of activated carbon from agricultural waste: A review," *J. Anal. Appl. Pyrolysis*, vol. 100, pp. 1–11, 2013.

[11] J. Li *et al.*, "Biochar from microwave pyrolysis of biomass: A review," *Biomass and*

Bioenergy, vol. 94, pp. 228–244, 2016.

[12] J. Sun, W. Wang, and Q. Yue, "Review on microwave-matter interaction fundamentals and efficient microwave-associated heating strategies," *Materials (Basel).*, vol. 9, no. 4, 2016.

[13] K. Y. Foo and B. H. Hameed, "Porous structure and adsorptive properties of pineapple peel based activated carbons prepared via microwave assisted KOH and K2CO 3 activation," *Microporous Mesoporous Mater.*, vol. 148, no. 1, pp. 191–195, 2012.

[14] K. Y. Foo and B. H. Hameed, "Potential of jackfruit peel as precursor for activated carbon prepared by microwave induced NaOH activation," *Bioresour. Technol.*, vol. 112, pp. 143–150, 2012.

[15] K. Y. Foo and B. H. Hameed, "Factors affecting the carbon yield and adsorption capability of the mangosteen peel activated carbon prepared by microwave assisted K 2CO 3 activation," *Chem. Eng. J.*, vol. 180, pp. 66–74, 2012.

[16] K. Y. Foo and B. H. Hameed, "A cost effective method for regeneration of durian shell and jackfruit peel activated carbons by microwave irradiation," *Chem. Eng. J.*, vol. 193–194, pp. 404–409, 2012.

[17] K. Y. Foo and B. H. Hameed, "Preparation, characterization and evaluation of adsorptive properties of orange peel based activated carbon via microwave induced K2CO3 activation," *Bioresour. Technol.*, vol. 104, pp. 679–686, 2012.

[18] P. Singh, P. Raizada, D. Pathania, G. Sharma, and P. Sharma, "Microwave induced KOH activation of guava peel carbon as an adsorbent for congo red dye removal from aqueous phase," *Indian J. Chem. Technol.*, vol. 20, no. 5, pp. 305–311, 2013.

[19] K. Y. Foo, L. K. Lee, and B. H. Hameed, "Preparation of activated carbon from sugarcane bagasse by microwave assisted activation for the remediation of semi-aerobic landfill leachate," *Bioresour. Technol.*, vol. 134, pp. 166–172, 2013.

[20] M. J. Ahmed and S. K. Theydan, "Microwave assisted preparation of microporous activated carbon from Siris seed pods for adsorption of metronidazole antibiotic," *Chem. Eng. J.*, vol. 214, pp. 310–318, 2013.

[21] V. O. Njoku, K. Y. Foo, M. Asif, and B. H. Hameed, "Preparation of activated carbons from rambutan (Nephelium lappaceum) peel by microwave-induced KOH activation for acid yellow 17 dye adsorption," *Chem. Eng. J.*, vol. 250, pp. 198–204, 2014.

[22] P. Taylor, Z. Liu, Y. Liu, L. Chen, and H. Zhang, "Performance study of heavy metal ion adsorption onto microwave-activated banana peel," *Desalin. Water Treat.*, vol. 52, no. February 2014, pp. 7117–7124, 2014.

[23] H. Mao, D. Zhou, Z. Hashisho, S. Wang, H. Chen, and H. Wang, "Preparation of pinewood- and wheat straw-based activated carbon via a microwave-assisted potassium hydroxide treatment and an analysis of the effects of the microwave activation conditions," *BioResources*, vol. 10, no. 1, pp. 809–821, 2015.

[24] G. V. Nunell, M. E. Fernandez, P. R. Bonelli, and A. L. Cukierman, "Development and characterization of microwave-assisted activated carbons from Parkinsonea aculeata wood," *Adsorption*, vol. 22, no. 3, pp. 347–356, 2016.

[25] P. S. Thue *et al.*, "Preparation, characterization and application of microwave-assisted activated carbons from wood chips for removal of phenol from aqueous solution," *J. Mol. Liq.*, vol. 223, pp. 1067–1080, 2016.

[26] C. Du *et al.*, "Microwave-assisted preparation of almond shell-based activated carbon for methylene blue adsorption," *Green Process. Synth.*, vol. 5, no. 4, 2016.

[27] H. Mao *et al.*, "Adsorption of toluene and acetone vapors on microwave-prepared activated carbon from agricultural residues: Isotherms, kinetics, and thermodynamics studies," *Res. Chem. Intermed.*, vol. 42, no. 4, pp. 3359–3371, 2016.

[28] R. Haruenkit *et al.*, "Comparative study of health properties and nutritional value of durian, mangosteen, and snake fruit: Experiments in vitro and in vivo," *J. Agric. Food Chem.*, vol. 55, no. 14, pp. 5842–5849, 2007.

[29] S. Wijanarti, A. B. N. Putra, K. Nishi, E. Harmayani, and T. Sugahara, "Immunostimulatory activity of snake fruit peel extract on murine macrophage-like J774.1 cells," *Cytotechnology*,

IOP Conf. Series: Materials Science and Engineering **742** (2020) 012046 doi:10.1088/1757-899X/742/1/012046

vol. 68, no. 5, pp. 1737–1745, 2016.

[30] S. Gorinstein *et al.*, "The comparative characteristics of snake and kiwi fruits," *Food Chem. Toxicol.*, vol. 47, no. 8, pp. 1884–1891, 2009.

[31] A. A. Arie, Vincent, and A. Putranto, "Activated carbons from KOH-activation of salacca peels as low cost potential adsorbents for dye removal," *Adv. Mater. Lett.*, vol. 7, no. 3, 2016.

[32] R. L. Tseng, F. C. Wu, and R. S. Juang, "Characteristics and applications of the Lagergren's first-order equation for adsorption kinetics," *J. Taiwan Inst. Chem. Eng.*, vol. 41, no. 6, pp. 661–669, 2010.

[33] Y. S. Ho and G. McKay, "Pseudo-second order model for sorption processes," *Process Biochem.*, vol. 34, no. 5, pp. 451–465, 1999.

[34] M. Materials, "A Review : Fundamental Aspects of Silicate," no. January, 2016.

[35] P. I. Ravikovitch and A. V Neimark, "Characterization of nanoporous materials from adsorption," vol. 188, pp. 11–21, 2001.

IOP Conf. Series: Materials Science and Engineering **742** (2020) 012047 doi:10.1088/1757-899X/742/1/012047

Preparation of Activated Carbons Originated from Orange Peel Waste by Subcritical H_2O Activation Method

Dandy Hady[1], Ratna Frida Susanti[1], and Arenst Andreas Arie[1]*

[1]Department of Chemical Engineering, Faculty of Industrial Technology, Parahyangan Catholic University, Ciumbuleuit 94, Bandung, 40141, Indonesia.

E-mail : arenst@unpar.ac.id

Abstract. Low-cost and environmental friendly activated carbons were synthesized from orange peel waste by carbonization followed by activation process using subcritical water. The carbonization process of orange peel waste was conducted in the electrical furnace at temperature of 800°C for 2 h. Activation process of the impregnated orange peel was carried out in the tubular furnace for 1 h at activation temperature of 200°C and pressure variation of 120, 170 and 220 bar. Activated carbon with highest surface area of 491 m^2/g was obtained by water pressure of 170 bar. The activated carbons were then utilized as adsorbents for removal of methylene blue (MB) from aqueous solution. The batch adsorption study was carried out by varying the initial concentration of MB solution (2, 4, 6, 8 and 10 ppm). Experimental results showed that the adsorption kinetic of MB fitted the pseudo-second-order rate equation, whereas for the adsorption isotherm model followed two models i.e. the Dubinin-Radushkevich and Freundlich model. The adsorption mechanism was found to be governed by the intraparticle and surface diffusion mechanism.

1. Introduction

There are many types of dye used in large quantity in the textile, paper, beverages and food industries [1, 2]. Hence, those kinds of industries usually produce a quite large quantity of wastewaters containing toxic dyes. If the wastewaters are then discharged into water bodies without treatment, it will cause very serious environmental problems [3]. Generally, industrial dyes consist of complex aromatic structures, therefore, most of them are chemically stable and hard to be decomposed naturally [4, 5]. Methylene blue (MB) can be categorized as basic cationic dye because it releases a positively charged colored ion if it is dissolved in the solution. This type of dye may have serious effects on living aquatic creatures on short contact time.

Some conventional techniques can be used for the wastewater treatments such as coagulation, precipitation, biological based technique and advanced oxidation process. In comparisons with other techniques, wastewater treatment using adsorption has emerged as a quite popular method because it is very simple and effective to remove pollutants in small amount. There are many type of adsorbents for dye removal, one of them is activated carbon.

Activated carbon is the most widely used adsorbents due to the high surface area and easy availability. However, the high operating cost and difficulties in regeneration of activated carbon leads to many researchers to search for more economic adsorbents. Utilization of biomass resources such as agricultural and forestry waste, can be assumed as a renewable and abundant source for supplying the demands for production of functional materials. Recently, biomass has emerged as potential raw materials for the production of carbon adsorbents.

Commonly, the preparation of activated carbon can be done using physical or chemical activation. In the physical activation process, carbonization of a carbonaceous material is followed by the activation process at specific temperature using some typical oxidizing gases such as carbon dioxide or steam or their mixtures [6]. Hence physical activation process is quite slow since it consists of two steps of carbon preparations. In the chemical activation approach, carbonization is conducted after the carbon precursors are mixed with the chemical impregnates such as KOH, $ZnCl_2$ or H_3PO_4 [7]. This process is usually preferable than the physical method since it is a single operation and can be operated at lower operating temperature. However, this method is not environmental friendly due to the extensive use of chemical impregnates and it is also require high energy consumption [8].

The synthesis of activated carbon by sub critical water activation method can be an alternative method next to the physical and chemical activation method because of it is a green method. Specifically, it can produced activated carbons with a strong mechanical characteristics and more developed meso-porous structure [9]. From the literature, it can be said that sub/supercritical water can be used as a good and effective activating agent of carbon materials since it has a mixture properties between gas and liquid such as a low viscosity, a high diffusive properties, a high specific density, and a high dissolving property [10]. Therefore, it is expected that the sub/supercritical water activation method can be used for the preparation of activated carbons with a good mechanical strength [11].

In this work, low-cost and environmental friendly activated carbons were synthesized from orange peel waste by carbonization process followed by activation process using subcritical water. The orange peel derived activated carbons were then used in adsorbents for the removal of the methylene blue (MB) dyes from solutions.

2. Experimental

2.1 Synthesis of orange peel based activated carbons

Orange peel waste was firstly collected locally from the beverages stores in Bandung. It was then washed by distilled water, dried at the oven and crushed to the mesh size of +100/-200. The orange peel waste was then carbonized at 450°C using nitrogen gas for 1 h . After the final temperature was achieved, the activation was done for 2 h by changing the activation temperature (140 °C, 200 °C dan 250 °C) and maintaining the pressure at 80 bar. The carbon product was then cooled down to room temperature by the flow of nitrogen gas. It was then with distilled water to remove remaining chemical impregnates. The surface characterization of the activated carbon (AC) was examined by N_2 adsorption at 77K . The BET procedure was used to determine the surface area of activated carbons.

2.2 Analysis of methylene blue

For the synthetic dye solution, methylene blue (MB) was used as a dye adsorbate. Double distilled water was used to prepare the solutions and reagents. The UV spectrophotometer was the main instrument to measure the concentration of MB in the synthetic solution. The calibration curves were then constructed based on the Lambert-Beer Law which was reflected by the linear plot for the varying MB concentration.

2.3 Batch equilibrium studies

For the adsorption experiments, batch adsorption studies of MB onto the activated carbons were conducted by varying the initial concentration of MB. First of all, the equilibrium conditions was firstly determined by observing the concentration of MB until it reached a constant value. This experiment was done in the Erlenmeyer flasks in which MB solutions (200 ml) with different initial concentrations were placed in the flasks. Activated carbons with a mass of 0.5 g was added to the MB solutions under isothermal conditions (30 °C) for about 24- 48 h until it reached equilibrium conditions. The initial pH was adjusted to 7 by adding hydrochloric acid or sodium hydroxide. The flasks were then removed from the shaker. The final concentration of MB dyes solution was then measured using calibration curves. Before conducting the concentration measurements, the solution was filtered. Every experiment was repeated twice with the same conditions. The concentration of MB dyes at the equilibrium conditions ,q_e (mg g^{-1}), was determined by the following equations

IOP Conf. Series: Materials Science and Engineering **742** (2020) 012047 doi:10.1088/1757-899X/742/1/012047

$$q = \frac{(C_i - C_e)\,V}{m} \tag{1}$$

where C_i and C_e (mg/L) are the concentrations of MB dyes at initial and equilibrium conditions, respectively. V is the volume of the solution (L), and m is the mass of activated carbon used for adsorption (g).

2.4 Batch kinetic studies

The procedures of kinetic experiments were the same as the equilibrium experiments. The concentration of MB at time t, q_t (mg g^{-1}), was calculated by:

$$q_t = \frac{(C_i - C_t)V}{m} \tag{2}$$

where C_0 and C_t(mg/L) are the MB concentrations at initial and any time t, respectively. V is the volume of the solution (L), and m is the mass of activated carbon used for adsorption (g).

3. Results and discussion

Figure 1 showed the N_2 adsorption–desorption isotherms of the orange peel based activated carbons prepared by sub-critical H_2O activation method. The isothermal plots of activated carbons demonstrated a type III characteristics. This type is characterized by a well-defined plateau, which confirms its mesoporosity. It can be seen that hysteresis loop for the samples is found between the relative pressures (P/P_0) of 0.45 to 1.

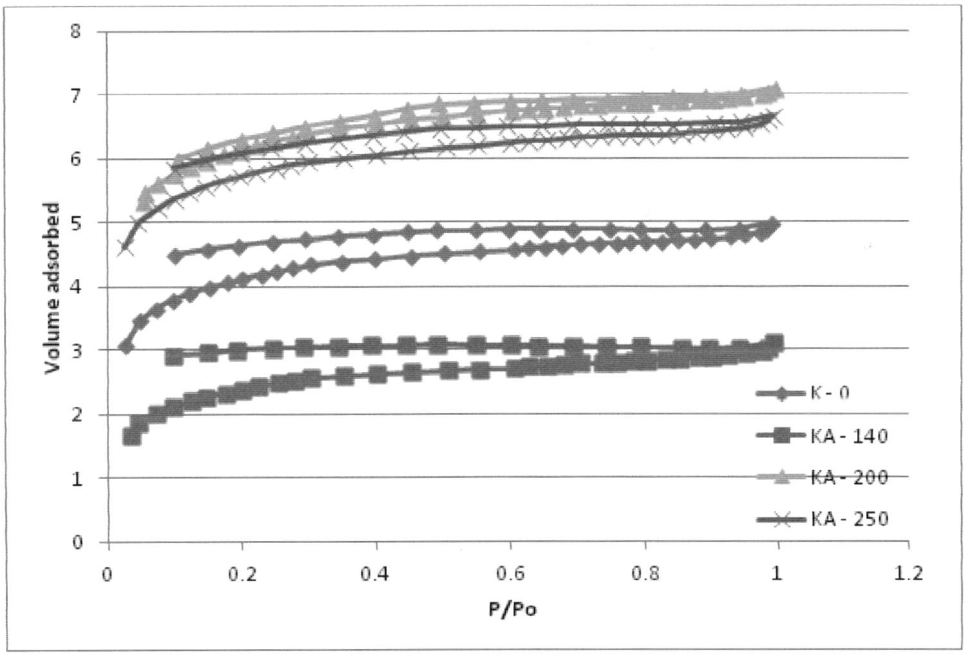

Figure 1 Adsorption/Desorption Isotherm of N_2 gas on the orange peel based activated carbons prepared at different activating temperature. (K-0 : Carbon samples before subcritical H_2O activation)

Table 1 shows the surface properties of the carbon products obtained by varying the activation temperature. It can be seen that the highest surface is shown by the sample activated at temperature of 200 °C.

Table 1 Surface characteristics of orange peel based activated carbons

Samples	Activation Temperature (°C)	BET Surface Area (m^2/g)
KA- 140	140	169.507
KA - 200	200	415.938
KA - 250	250	287.099

Figure 2 shows the plots of pore size distribution (PSD) of carbon samples synthesized by sub-critical H_2O activation method with varying activation temperature. It can be seen that all carbon samples can be classified as mesoporous carbons.

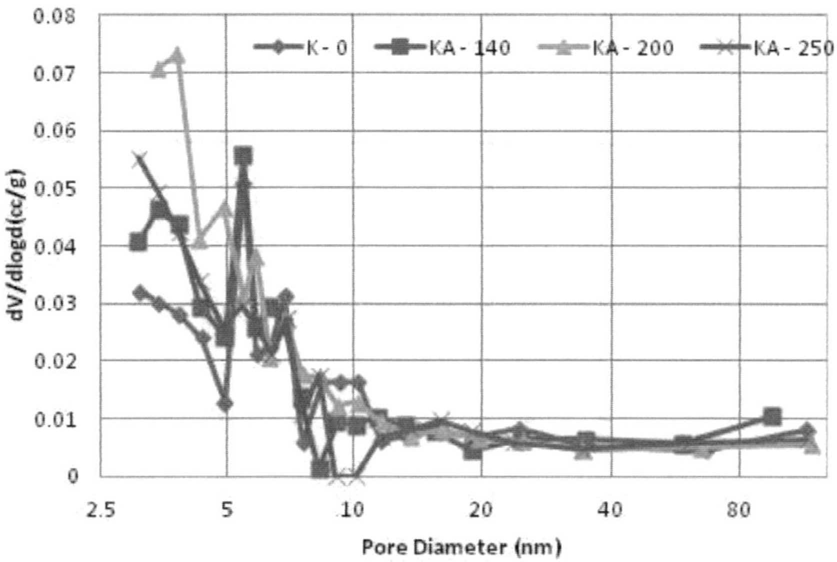

Figure 2 The plots of pore size distribution of the orange peel based activated carbons prepared at different activating temperature. (K-0 : Carbon samples before subcritical H_2O activation)

The adsorption isotherm models are usually used to correlate the concentration of MB dye in the the liquid and the solid phase at the equilibrium conditions. The data of equilibrium conditions were then fitted by different isotherm models. From these models, we can determine the maximum capacity of adsorption process [12]. Fig. 3 shows the adsorption isotherms of MB dye on the activated carbon. From the analysis of adsorption isotherm, we can use it further for investigating the interaction between solutes and adsorbents. It can be used further for optimization and design purpose.

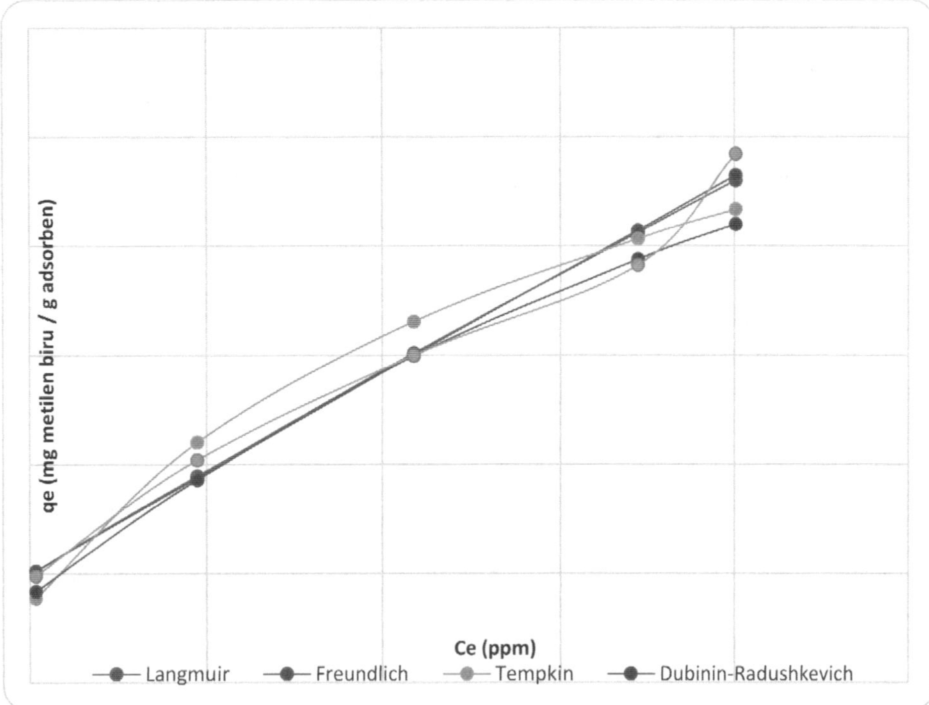

Figure 3. The plots of adsorption isotherm of methylene blue onto activated carbon

Adsorption isotherm study is carried out on the four type of isotherms, Langmuir, Freundlich, Tempkin and Dubinin Radushkevic model.The applicability of the isotherm equation is compared by observing the correlation coefficients. Based on the correlation coefficients, the most suitable model is the Dubinin Radushkevich isotherm model, as shown in Table 2.

Table 2 Dubinin Radushkevich Isotherms constant for Adsorption of MB onto activated carbons

Sample	Activation Temperature	q_m (mg/g)	K (L/mg)	E (KJ/mol)	R^2
K – 0	-	18,2	4×10^{-8}	3,54	0,961
KA – 140	140°C	12,0	7×10^{-8}	2,67	0,937
KA – 200	200°C	15,9	4×10^{-8}	3,54	0,988
KA – 250	250°C	13,3	5×10^{-8}	3,16	0,857

The Dubinin Radushkevich (D-R) model, which assume a heterogeneous surface or a non-constant adsorption potential as the Langmuir model [13], was used to test the experimental data. It could be used to differentiate between physical and chemical adsorption of MB. The following form of D-R models model is used :

$$\ln(q_e) = \ln(q_m) - K\varepsilon^2 \quad (3)$$

IOP Conf. Series: Materials Science and Engineering **742** (2020) 012047 doi:10.1088/1757-899X/742/1/012047

Where K is the coefficient correlated with the averaged free energy of adsorption (mol^2/J^2) and ε = is the Polanyi potential [J /mol] , which is defined as RT Ln (1+1/Ce) where R = universal gas constant (8.314 $J.mol^{-1}.K^{-1}$); and T = temperature (K)]. The constants of q_m and β can be calculated using linear regression method. Hence, it can be seen that the D-R equation represented the most representative to the experimental data than the other models. The average sorption energy, E, can be determined using the following equation :

$$E = \frac{1}{\sqrt{2K}} \tag{4}$$

It was found that the magnitude of E was less than 8 $kJ.mol^{-1}$. It indicated that the adsorption process was the type of physical adsorption.

Two kinetic models such as pseudo-first-order and pseudo-second-order kinetic were applied for the experimental results. The pseudo-first-order kinetic model proposed [14] was usually used to predict sorption kinetic and was defined as:

$$Ln(q_e - q_t) = Ln\ (q_e) - k_1 t \tag{5}$$

where q_e and q_t(mg/g) are the amounts of adsorbate adsorbed at equilibrium and at any time, t (h), respectively and k_1 (1/h) is the adsorption rate constant.
The pseudo-second-order equation [15] is expressed by :

$$\frac{t}{q_t} = \frac{1}{k_2\, q_e^2} + \frac{t}{q_e} \tag{6}$$

where k_2 (g/mg h) is the rate constant of second-order adsorption.

The values of all constants R^2obtained from the plots for adsorption of MB dye on the adsorbent at 25 °C are reported in Table 3. It was observed that the R^2 values obtained for the pseudo-second-order model was higher than that of first order model. This shows that the adsorption of MB dye on the adsorbent follows apseudo second-order kinetic model.

Table 3. Adsorption Kinetic Model Parameters

Pseudo first order model					
No.	Sample	Activation Temperature	qe (mg/g)	K_1 (min^{-1})	R^2
1	K – 0	-	2.03	7.68	0.993
2	KA – 140	140°C	1.89	43.4	0.940
3	KA – 200	200°C	2.01	25.3	0.955
4	KA – 250	250°C	1.97	34.6	0.963
Pseudo second order model					
No.	Sample	Activation Temperature	qe (mg/g)	K2 (min^{-1})	R^2
1	K – 0	-	2,03	0,0641	0,999
2	KA – 140	140°C	2,17	0,00748	0,992
3	KA – 200	200°C	2,15	0,0142	0,998
4	KA – 250	250°C	2,19	0,00964	0,996

4. Conclusions

The present investigation showed that orange peel can be effectively used as a raw material for the preparation of activated carbon using subcritical CO_2 activation process. Activated carbons was then used as adsorbents for the removal of methylene blue dye from aqueous solution. Methylene blue is found to adsorb strongly on the surface of activated carbon. Adsorption behaviour is described by a Dubinin Radushkevich type isotherm. Kinetic data follows pseudo second-order kinetic model.

Acknowledgements
This study is supported by funding from The Indonesia Toray Scientific Foundation through the Science and Technology Research Grant in 2013.

References

[1] K.R. Ramakrishna, T. Viraraghavan, "*Dye removal using low cost adsorbents*", *Water Sci. Technol.* 36 (2–3) , 1997, 189–196.
[2] P.N. Cheremisinoff, F. Ellerbusch (Eds.), Carbon Adsorption Handbook, Ann Arbor Science, Ann Arbor, Michigan, 1980.
[3] R.T. Yang, Adsorption, John Wiley & Sons Inc., Hoboken, New Jersey, 2003.
[4] D.M. Ruthven, Principles of Adsorption and Adsorption Process, Wiley,New York, 1984.
[5] P. Chingombe, B. Saha, R.J.Wakeman, Surface modification and characterisation of a coal-based activated carbon, Carbon 43 (2005) 3132–3143.
[6] P. Nigam, G. Armour, I. Banat, D. Singh, R. Marchant (2000). Physical removal of textile dyes and solid state fermentation of dye adsorbed agricultural residues, Bioresour. Technol. 272 p. 219.
[7] C. Lee, K. Low, P. Gan. Removal of some organic dyes by acid treated spent bleaching earth, Environ. Technol. 1999, 20 p. 99
[8] A. Khaled, A. Nemr, A. Sikaily, O. Abdelwahab. Treatment of artificial textile dye effluent containing Direct Yellow 12 by orange peel carbon, Desalination ,2009, 238, 210–232
[9] Md.Z. Alam, E.S. Ameem, S.A. Muyibi, N.A. Kabbashi, The factors affecting the performance of activated carbon prepared from oil palm empty fruit bunchesfor adsorption of phenol, Chem. Eng. J. 155 (2009) 191–198.
[10] R. Gottipati, S. Mishra, Process of adsorption of Cr(VI) on activated carbons prepared from plant precursors by a two-level full factorial design, Chem. Eng.J. 160 (2010) 99–107.
[11] El-Hendawy, A. A.; Samra, S. E.; Girgis, B. S. Adsorption characteristicsof activated carbons obtained from corncobs. Colloids Surf.,A 2001, 180, 209–221.
[12] El-Hendawy, A. A. Surface and adsorptive properties of carbons prepared from biomass. Appl. Surf. Sci. 2005, 252, 287–295.
[13] Tseng, R. L.; Tseng, S. K. Pore structure and adsorption performance of the KOH-activated carbons prepared from corncob. J. ColloidInterface Sci. 2005, 287, 428–437.
[14] K.Y. Foo, B.H. Hameed (2010) "Insights into the modeling of adsorption isotherm systems", Chemical Engineering Journal 2010, 156, 2-10.
[15] S. Langergren, B.K. Svenska Zur theorie der sogenannten adsorption geloester stoffe, Veternskapsakad Handlingar, 1998, 24 (4) p. 1–39.

IOP Conf. Series: Materials Science and Engineering **742** (2020) 012048 doi:10.1088/1757-899X/742/1/012048

The Study on the Method for Maximizing Steviol Glycoside Extract from Stevia Leaves

Judy Retti Witono[1] and Andy Chandra[1]

[1]Department of Chemical Engineering, Parahyangan Catholic University, Jl. Ciumbuleuti no. 94, Bandung, 40141, Indonesia

E-mail : judy@unpar.ac.id

Abstract. One of the 9 types of basic needs of Indonesian is sugar. The main resource as a sugar producer is sugar cane. However, currently with the increasing of population, the area for sugarcane plantation become limited. On the other side, Indonesia, with its wealth of natural resources, has the opportunity to use other raw materials as a source of sweetener. One of the potential sugar sources is Stevia rebaudiana. Its dry leaves contents of steviol glycosides around 5-10%. In addition, this sugar has a therapeutic function as well, since its sweetness level is quite high (200 - 300 times of sugarcane), contains vitamins, minerals and anti-oxidants. The aim of this research is to study the appropriate method that can maximize the extraction of glycoside from stevia leaves. The variables observed during the experiment were temperature generated by microwave and size of leaves. Stevioside separation process was carried out by extraction with using a microwave. Analysis of the glycoside yield was conducted using turbidity meter and colorimeter.

1. Introduction

Steviol glycosides are a group of highly sweet diterpene glycosides isolated from the Stevia rebaudiana leaves. Stevioside constituting about 5–10% in dry leaves [1] and rebaudioside-A are the main component in steviol glycosides which are responsible for its sweet taste and have commercial value [2, 3]. The more detailed components in the Stevia rebaudiana leaves can be seen in table 1 [4, 5].

This is a non-sugar cane sweetener with a sweetness level of 250 - 300 times of cane sugar and is obtained by extracting stevia leaves [6]. Stevia sugar is not intended to replace cane sugar because of its low calorific value, but rather is intended to replace other synthetic sugars as sweeteners which according to various studies are carcinogenic. Because stevia is obtained from plants, its use is safer, non-carcinogenic and non-calorie. Another advantage is that stevia sugar does not cause dental carries, has a low calorific value suitable for diabetics, and does not cause cancer in long-term use.

A commonly extraction method of the steviol glycosides from the stevia leaves consists of extracting dried and powdered leaves with hot water. Another method for the extraction of leaves uses an ethanol-water mixture, acidified water [7], Pressurized Hot Water Extraction (PHWE) [8] followed by an evaporation of the extract. But all of those methods usually are carried out at the high temperature and pressure and also relatively take time. A new method nowadays is developed for plant extraction which can shorter extraction times, lower average temperatures, higher yields, as well as energy and cost saving [9]. This has been tried as well for the extraction of glycoside from stevia leaves as a pre-treatment [10] with the good result.

The goal of this research is to study about the possibility of using microwave not as a pre-treatment but during the extraction.

Table 1. Glycoside found in Stevia rebaudiana leaves

Glycoside	Molecular formula	Sweetening power (with reference to saccharose)	Concentration in leaf (% w/w)	Proportion of glycoside in t (% of total steviol glyco
Stevioside	$C_{38}H_{60}O_{18}$	150–300	4–14	43.1–79.6
Rebaudioside A	$C_{44}H_{70}O_{23}$	250–450	2–4	1.6–9.9
Rebaudioside C	$C_{44}H_{70}O_{22}$	120–500	1–2	0.5–6.0
Steviolbioside	$C_{32}H_{50}O_{13}$	100–125	< 0.4	0.3–3
Rubusoside	$C_{32}H_{50}O_{13}$	100–120	< 0.4	Undefined
Rebaudioside B	$C_{38}H_{60}O_{18}$	300–350	< 0.4	0–0.02
Rebaudioside D	$C_{50}H_{80}O_{28}$	250–450	< 0.4	0–0.4
Rebaudioside E	$C_{44}H_{70}O_{23}$	150–300	< 0.4	5.5–43.2
Rebaudioside F	$C_{43}H_{68}O_{22}$	Undefined	< 0.4	0.04–0.1
Dulcoside A	$C_{38}H_{60}O_{17}$	50–120	0.4–0.7	0.2–0.4

2. Materials and Methods

The raw material used in this study was dried stevia leaves obtained from PT Tigapilar Argo Utama, Solo, Central Java. Stevia leaves need to be given a pre-treatment in order to get maximum results. The initial treatment included washing and separating dried stevia leaves from the dried stems and other impurities.

2.1. Extraction

The moisture content of Stevia leaves was standardized (below 10%) by air drying it in oven at 60^0C. Then the size of leaves was reduced according to the variables determined i.e. -20+30 mesh and -40+60 mesh. The figure of stevia leaves which were used in this experiment can be seen in figure 1.

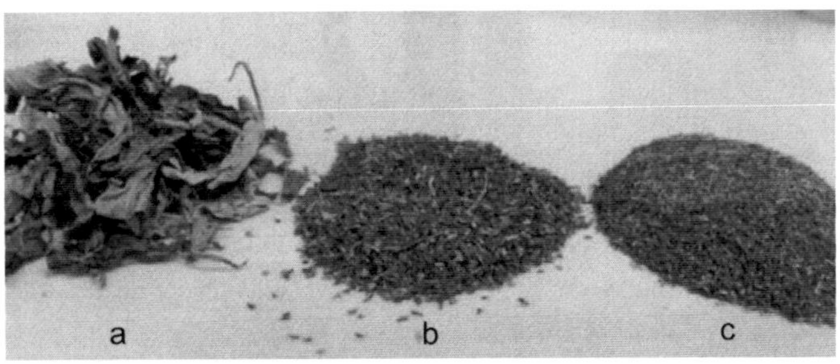

Figure 1. (a) the whole leaves, (b) the -20+30 mesh size leaves and (c) the -40+60 mesh leaves

Instead of using heat to extract the glycoside, it was used a radiation heat generated by microwave oven. The temperature produced by the radiation beam. can be seen from the table 2 below.

Table 2. Relation of time and temperature in the microwave used

Time (min.)	Temperature (^0C)			
	Low	*Medium*	*Med-High*	*High*
0	27	27	27	27
1	28	34	43	48
2	30,5	47	55	59
3	35	53,5	63	72

The experiments were carried out based on the variation of temperature and the size of leaves. The weight ratio of feed and solvent was constant at 1:200. Dried and ground or whole of stevia leaves (1.75 g) was placed in 0.5 L Erlenmeyer flasks and extracted with 350 g of water at different temperature. After cooling, the crude leave extract was filtered and the supernatant was analysed.

2.2. Analysis of the glycoside

The glycoside concentration in supernatant was analysed using turbidity meter (Eutech Instrument TN-100) and colorimeter (Hanna Instrument HI 93727 Colour of Water ISM). The turbidity meter is the instrument to measure the relative clarity of a fluid by measuring the amount of light scattered by glycoside particles suspended in water. The colorimeter is a light-sensitive device used for measuring the transmittance and absorbance of light passing through a supernatant. Prior to use, both instruments must be calibrated. The calibration curve of turbidity meter unit (NTU) vs %-wt. of glycoside and colorimeter unit (PCU) vs %-wt. of glycoside can be seen in figures 2 and 3.

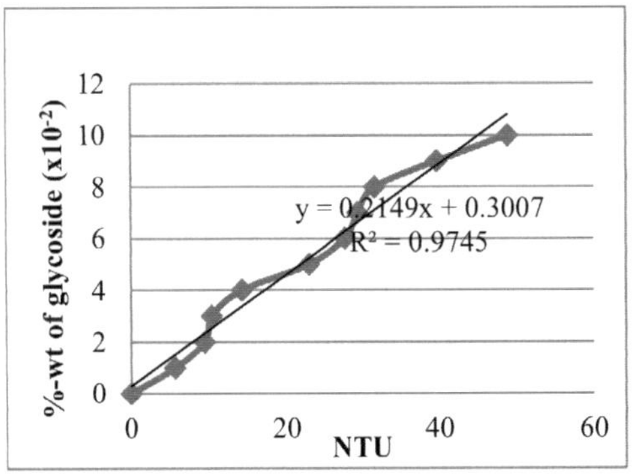

Figure 2. The calibration curve of turbidity meter unit (NTU) vs %-wt. of glycoside

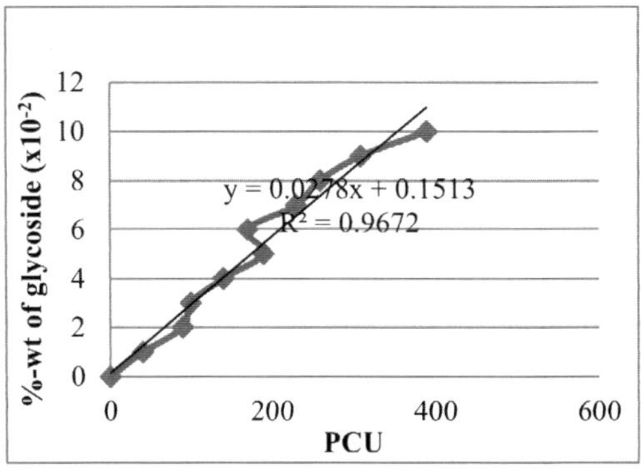

Figure 3. The calibration curve of colorimeter unit (NTU) vs %-wt. of glycoside

3. Results and Discussion

The concentration of glycoside obtained from the stevia leaves extraction was measured by turbidity meter and colorimeter. The results were shown in tables 3 and 4 and figures 4 and 5.

Table 3. Data from turbidity measurement

Temp. (°C)	Size	NTU scale	%-wt. of glycoside ($\times 10^{-2}$)
72	Whole	9,41	2,33
63	Whole	6,12	1,62
55	Whole	2,295	0,87
72	-20+30	10,275	2,51
63	-20+30	6,16	1,63
55	-20+30	3,775	0,92
72	-40+60	12,875	3,07
63	-40+60	8,88	2,21
55	-40+60	5,94	1,58

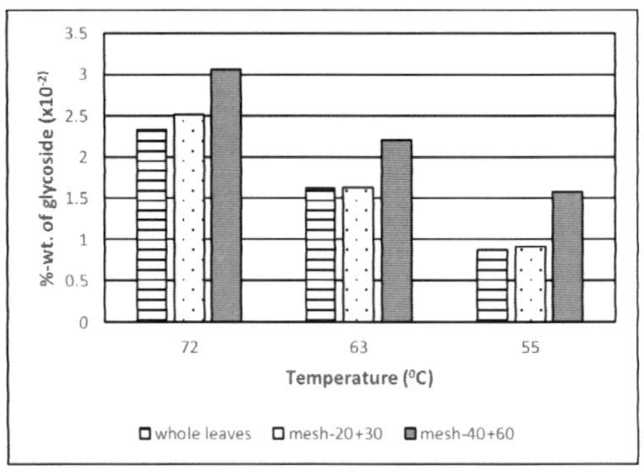

Figure 4. The concentration of glycoside based on turbidity meter measurement

Both of the measurement showed the same tendency. The smaller the leaves size, the more glycosides that can be extracted from the leaves. This is an evident that the grinded leaves construct the higher surface area which give the access to more mass transfer of glycoside from leaves into the solvent. This condition was occurred as well in the glycoside extraction using hot water [8].

From these results can be seen as well that the higher temperature used as a result of different power levels of microwaves, the more glycoside can be extracted. In this experiment, it focused only at the temperature generated from different power levels not the power level itself. And the extraction time was only limited to 3 minutes. However, in relation to the data in table 1, it can be concluded that there is an influence of microwave power level and extraction time to the extraction results.

Tables 3 and 4 show that the glycoside concentration values analyzed using a turbidity meter differ from the glycoside concentration values analyzed using a colorimeter. This is understandable because the turbidity meter measures the concentration based on the light scattered by particles suspended in the solution. And it is known that the two main components in glycoside, namely stevioside and rebaudiosida A, have a water solubility value of 50^{0}C i.e. 6.3 and 6.6 g / L respectively [11]. So that, not all glycosides can be detected in measurements with a turbidity meter. That is why the value of the concentration of glycoside with a turbidity meter is lower than that produced from measurements with a colorimeter.

Table 4. Data from colorimeter measurement

Temp. (°C)	Size	PCU scale	%-wt. of glycoside (×10^{-2})
72	Whole	445	12.52
63	Whole	345	9.74
55	Whole	215	6.13
72	-20+30	495	13.91
63	-20+30	440	12.38
55	-20+30	345	9.74
72	-40+60	550	15.44
63	-40+60	505	14.19
55	-40+60	430	12.11

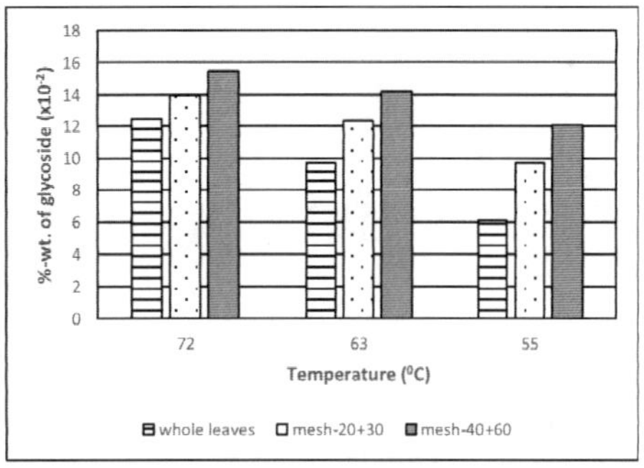

Figure 5. The concentration of glycoside based on colorimeter meter measurement

To prove the existence of steviosida and rebaudiosida-A compounds from extracted samples, an FTIR analysis was also performed. The samples from different runs were compared to the commercial stevia sweetener. From the results of the FTIR analysis in figure 6, it can be seen that for all samples and a commercial sweetener showed a peak at the wavenumber of 3200-3550 cm^{-1} which indicated the presence of OH groups. The infrared spectrum also showed a peak at the wavenumber of 2850-3000 cm^{-1} indicating the CH, CH_2, CH_3 groups and at the wavenumber of 1600-1700 cm^{-1} indicating the C = O group. This can be an evident that the content of stevioside and rebaudioside-A was found in the samples. Whereas the presence of the additional peaks in the samples is predicted due to other substances. Since the samples have not been purified yet.

The advantage of the glycoside extraction process carried out using electromagnetic waves is

1. The extraction temperature can be lower than commercial extraction
2. Extraction time can also be shorter

Therefore, the other active compounds such as antioxidants, chlorophyll etc. which is usually contained therein can be maintained and also break down of glycoside at high temperature can be avoided.

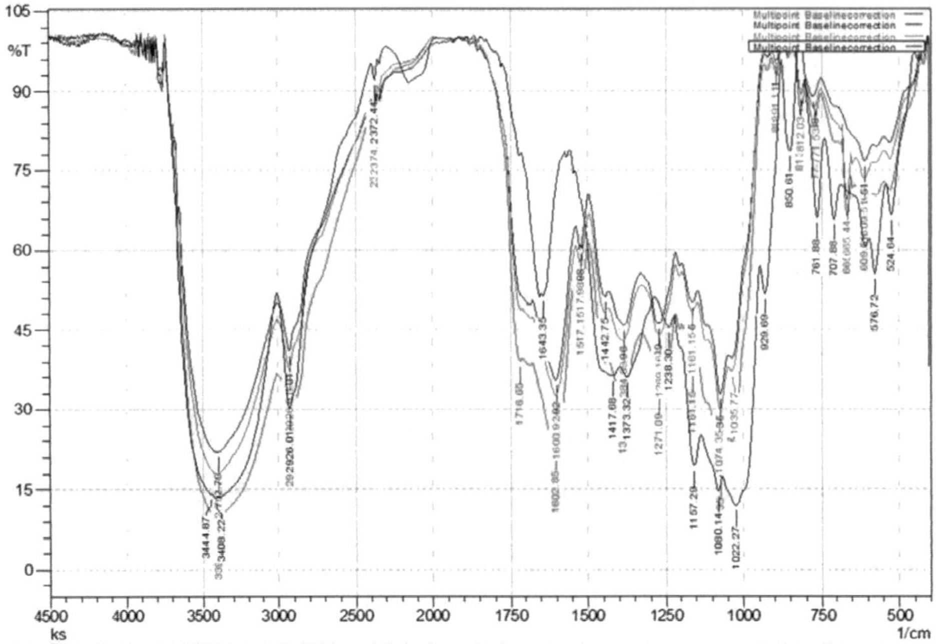

Figure 6. FTIR spectra of samples and commercial stevia sweetener

4. Conclusions

The extraction process of glycoside from stevia leaves using a microwave is a prospect method to maximize the yield. It can save energy and maintain the active compounds.

Acknowledgments

The authors would like to express their thanks to Research Institutions and Community Development - Parahyangan Catholic University (LPPM – UNPAR) which finances this research. Special thanks are addressed to our research group team who made this research can be carried out.

References

1. De S, Mondal S, Banerjee S, and L.P.b.J.W.S. © John Wiley & Sons, Ltd., *Stevioside: Technology, Applications and Health*. 2013
2. Lewis, W.H., *Early uses of Stevia rebaudiana (Asteraceae) leaves as a sweetener in Paraguay.* Economic botany, 1992. **v. 46**(no. 3): p. pp. 336-337-1992 v.46 no.3.
3. Prakash I and V.S.P. Chaturvedula, , 1–28., *Steviol Glycosides: Natural Non-Caloric Sweeteners*, in *Sweeteners, Reference Series in Phytochemistry*, Merillon J M and Ramawat K, Editors. 2016, Springer, Cham.
4. Gonzalez C, Tapia M, Pérez E, Pallet D, and Dornier M, *Main properties of steviol glycosides and their potential in the food industry: a review.* Fruits, 2014. **69**: p. 127-141.
5. Wölwer-Rieck U, *The Leaves of Stevia rebaudiana (Bertoni), Their Constituents and the Analyses Thereof: A Review.* . Journal of Agricultural and Food Chemistry, 2012. **60**(4): p. 886–895.
6. Tijhuis M J, Wapperom D, Wolterink G, van Oosterhout C H M, Temme E H M, van Klaveren J D, Verhagen H, and Fransen H P, *Steviol glycosides in food - exposure scenarios and health effect assessment* in *RIVM Letter Report 350121001/2011*. 2011.

7. Kootstra A M J and Huurman S, *Extraction of steviol glycosides from fresh Stevia using acidified water; comparison to hot water extraction, including purification.* 2017, ACRRES - Wageningen UR.
8. Panja P and Mukhopadhyay M, *Extraction of Natural Sweetener from Stevia Leaves Using Pressurized Hot Water.* Journal of Nutraceuticals and Food Science, 2019. **4**(1:3): p. 1-9.
9. Rosa R, Ferrari E, and Veronesi P, *From Field to Shelf: How Microwave-Assisted Extraction Techniques Foster an Integrated Green Approach. Emerging Microwave Technologies in Industrial, Agricultural, Medical and Food Processing.* 2018.
10. Javad S, Naz S, Ilyas S, Tariq A, and Javed S, *Optimization of the Microwave assisted extraction and its comparison with different conventional extraction methods for isolation of stevioside from Stevia rebaudiana.* Asian Journal of Chemistry, 2014. **26**: p. 8043-8048.
11. Celayaa L S, Kolba E, and Kolba N, *Solubility of Stevioside and Rebaudioside A in water, ethanol and their binary mixtures* International Journal of Food Studies, 2016. **5**: p. 158-166.

IOP Conf. Series: Materials Science and Engineering **742** (2020) 012049 doi:10.1088/1757-899X/742/1/012049

A Systematic Study on the Utilization of Inorganic Salts as Catalyst for the Conversion of Xylose to Furfural

Jenny N M Soetedjo[1,2], Carolus B Rasrendra[3], Hero J Heeres[2]*

[1]Department of Chemical Engineering, Parahyangan Catholic University, Ciumbuleuit 94, Bandung 40141, Indonesia
[2]Department of Chemical Engineering, University of Groningen, Nijenborgh 4, 9747 AG Groningen, the Netherlands
[3]Department of Chemical Engineering, Institut Teknologi Bandung, Ganesha 10, Bandung 40132, Indonesia

E-mail : h.j.heeres@rug.nl

Abstract. The utilization of biomass-waste such as sugar-bagasse,water-hyacynth and palm-oil-fiber as alternative sources for transportation fuels and platform-chemicals is a very active research field. Furfural(FF) is one-of-the-13top platform-chemicals that may be converted to derivatives such as furfuryl-alcohol,furoic-acid and furan with wide applications in the gasoline,diesel and jet-fuel blending-pool. Many studies have been conducted in the mechanism and the kinetics of FF-formation from xylose since the 1940s to maximize FF-yield and to reduce FF-decomposition to undesired-products. Previous studies showed that inorganic-salt gives positive effects on the FF-yield but systematic studies are lacking. Based on it, 60salts were screened in the hydrolysis of 0.1M xylose-aqueous-solution (T=180°C,90minutes,batch). The maximum FF-yield was 53mol% using 5mM-$Fe_2(SO_4)_3$. The highest FF-selectivity was at 84mol% using 5mM-NaCl, though at low xylose-conversion(35 mol%-FF-yield). Some transition metal-chloride i.e.$FeCl_2$,$CuCl_2$,$SnCl_2$ showed interesting FF-yields(48-50%) and selectivity(58-65%) indicating interesting roles of ion Fe^{3+} and Cl^-. Subsequent studies of two salts i.e.$Fe_2(SO_4)_3$ and $FeCl_3$(5mM) in combination with HCl and H_2SO_4(0.1M) were investigated(0.1M-xylose,T=150°C,0-270 min). The result shows that salts increase no FF-yields for H_2SO_4 but increase the reaction-rate. In contrast, $Fe_2(SO_4)_3$ increase no yield nor reaction-rate for HCl-catalyzed-system. In conclusion, the inorganic-salts catalyse xylose-conversion to furfural but best results were obtained using HCl without additional salts.

1. Introduction

Indonesia is the largest palm oil producer in the world with the production up to 41.5 million MT of palm oil in 2018[CITATION Ind19 \l 1033]. Ironically, the increase of CPO production is followed by the increase of waste generated with annual production of waste up to 129-million-m^3 POME (Palm Oil Mill Effluent) and 46-million-ton EFB (Empty Fruit Bunches). Although limited amount of EFB is used as organic fertilizer, most of it is still wasted and becomes one of the major environmental problems in palm oil industry.

EFB, containing 37.3 – 46.5% cellulose and 25.3 – 33.8% hemicelluloses, is actually an excellent source of lignocellulosic biomass which is rich of C6 and C5 sugars that can be converted further into

IOP Conf. Series: Materials Science and Engineering **742** (2020) 012049 doi:10.1088/1757-899X/742/1/012049

high-value-chemicals such as levulinic acid (**LA**) and furfural (**FF**)[CITATION Sud13 \l 1033]. In 2016, NREL (National Renewable Energy Laboratory) of U.S. Department of energy listed **FF** in the top 12 emerging near-term deployment potential biobased platform chemicals[CITATION Bid16 \l 1033]. In fact, the high need of furfural in Indonesia is reflected by its import data of 525,3 tonnes or up to 1,7 million US dollar in 2018[CITATION Bir \l 1033].

Furfural or 2-furaldehyde (**FF**) is an important non-petroleum based chemical building block used for the productions of amongst others furfuryl alcohol, furoic acid and furans that can be extended into wide range of products such as pharmaceuticals, cosmetics, resins, cleaning agent, extracting agent, solvent, vulcanizing agents, flavoring agents, pesticides, antiseptics and disinfectants[CITATION Döb32 \l 1033 \m Cor07 \m Sai82 \m Zei00 \m Bro48 \m HEH12 \m Yan14]. Among its wide application, the current finding of application for furfural as a feedstock to make gasoline, diesel and jet fuel has attracted so much attention[CITATION Yan14 \l 1033].

Furfural for the first time isolated in 1821 [CITATION Döb32 \l 1033] and later produced commercially since the 1920s[CITATION HEH12 \l 1033 \m Bro48]. It is formed by the hydrolysis of hemicellulose into pentose (C5) sugars before further dehydration in **FF**. Studies have been conducted on several acids i.e. sulfuric, hydrochloric, phosphoric, and nitric acids but sulfuric acid is still the one commercially used in the industry[CITATION Sin14 \l 1033]. In fact, a number of process concepts have been implied commercially (Quacker, Agrifurane, Rosenlew, Escher Wyss) but it still has relatively limited **FF** yields (about 50% range) so far [CITATION HEH12 \l 1033]. The main reason is the involvement of decomposition reactions including fragmentation, resinification and condensation reactions from both pentoses and **FF** that form soluble side products such as formic acid, acetic acid together with insoluble solid product known as humin [CITATION Yan14 \l 1033]. The formation of humin decreases **FF** yield and causes operational issues due to the blockage and decline of heat transfer rates as well as the irreversible deposit on the solid catalyst that leads to catalyst deactivation.

Several efforts have been applied to improve **FF** yield by the use of biphasic liquid-liquid systems[19,22,24,25], ionic liquids[26-30] and organic solvents[17,31-35] instead of water and the addition of salts to the reaction mixture[18,21,23]. The addition of inorganic salts have been proven to affect the chemo-selectivity of the reaction. Rasrendra *et al.* has clearly shown that the chemo-selectivity is a clear function of the type of inorganic salt. Using 0.1 M glucose in water at 140°C and 5mM salt, it was shown that Al(III) has driven the yield selection into lactic acid while Zn(II) into 5-HMF (5-Hydroxymethylfurfural)[CITATION CBR10 \l 1057]. In another study, Rasrendra *et al.* showed that $AlCl_3$, Cr and Sn salts has the ability to provoke the isomerisation of d-glucose to d-fructose that increases the chemo-selectivity for the reaction of d-glucose to HMF in DMSO solution[CITATION CBR12 \l 1057].

In fact several studies on salt effect have been conducted for xylose and xylotriose. The effect of chloride salts from several kation groups on xylose and xylotriose was observed by Liu *et al.* and $FeCl_3$ gave an increase in the rate of reaction both on the formation and decomposition of **FF** [CITATION CLi06 \l 1057] and strong effect on hemicellulose removal[CITATION LiL09 \l 1057]. A few years later Marcotullio *et al.* studied the combination of several chloride salts in homogeneous acid catalyzed system using HCl and H_2SO_4. The addition of NaCl, KI and $CrCl_3$ in aqueous systems has been reported in the conversion of xylose to **FF** (see Table 1). The most successful result reported by Marcotullio *et al.* with the **FF** yield of 88 mol% was reported using a mixture of KI and KCl and H_2SO_4 as the catalyst. The addition of 5 wt% of NaCl together with HCl was also shown a promising result with **FF** yield up to 81 mol%. However the studies were still focusing on the chloride salts on HCl system with only one study using the combination of NaCl and H_2SO_4 while the commercial use in industry mostly is still using H_2SO_4. This study is conducted to explore and compare the catalytic effect of other salts on both H_2SO_4 and HCl catalyzed system on the xylose conversion and **FF** yield.

Table 1. The examples of xylose conversion to **FF** by the additional salt in aqueous solution

Entry	C_{Xylose}	T (°C), t	Acid	C_{acid}	Salt	$Y_{FF,max}$ (mol%)	Ref
1.	35 mM	200, 485 s	H_2SO_4	0.05 M	KI–KCl 250-500mM	88	[CITATI

IOP Conf. Series: Materials Science and Engineering **742** (2020) 012049 doi:10.1088/1757-899X/742/1/012049

							ON Mar11 \ l 1057]
2.	35 mM	200, 250 s	HCl	0.05 M	NaCl 5 wt%	81	[CITATION Mar09 \ l 1057]
3.	56 mM	145, 300 min	HCl	0.10 M	$CrCl_3$ 6 mM	39	[CITATION Cho12 \ l 1057]
			HCl	0.10 M	-	29	

2. **Experimental Procedures**

All chemicals were used as received without further purification. Milli-Q water was used for all reactions and analysis procedures. Xylose (99.5%) was from Sigma-Aldrich, FF (99%) was from Fluka-Aldrich. H_2SO_4 (95-97 wt-%), HCl (37 wt-%), formic acid (98% wt-%) and all salts were purchased from Merck GmbH (Darmstadt, Germany).

The reactions were carried out in glass ampoules with an internal diameter of 3 mm, a wall thickness of 1.5 mm, and a length of 15 cm. The ampoules were filled at room temperature with a solution (0.5 cm^3) of 0.1M xylose-aqueous-solution with acid catalyst and salt before subsequently sealed with a torch. A series of ampoules was placed in a rack and subsequently positioned in a constant temperature oven ($\pm$ 0.1°C) which was pre-set at the desired reaction temperature. At different reaction times, an ampoule was taken from the oven and directly cooled in an ice-water bath to quench the reaction. The liquid content was then filtered using a PTFE syringe filter (0.45 mm, VWR, the Netherlands). The particle free aliquot was diluted 7-8 times with water prior to analysis.

The composition of the liquid phase after reaction was determined by HPLC using an Agilent 1200 HPLC equipped with a HP 1200 pump, a Bio-Rad organic acid column (Aminex HPX-87H), and an RID or an UV detector. The mobile phase consists of aqueous sulfuric acid (5 mM) at a flow rate of 0.55 cm^3/min. The column was operated at 60°C. The analysis for a sample was complete in 60 minutes.

The xylose conversion (X_{XYL}), FF yield (Y_{FF}) and FF selectivity (S_{FF}) are mol% based and calculated using equations 1-3:

$$X_{XYL} = \frac{\left(C_{XYL,i} - C_{XYL,t}\right)}{C_{XYL,i}}$$

(1)

$$Y_{FF} = \frac{C_{FF,t}}{C_{XYL,i}}$$

(2)

Here, $C_{XYL,i}$ is the initial xylose concentration (mol/l), $C_{FF,i}$ the initial FF concentration and $C_{XYL,t}$ and $C_{FF,t}$ the xylose and FF concentration at a certain batchtime. For experiments using FF as the feed, the FF conversion (X_{FF}) is defined as:

$$X_{FF} = \frac{C_{FF,i} - C\,(¿¿FF,t)}{C_{FF,i}\,¿} \qquad (3)$$

3. Results and discussion

The studies conducted to observe the effect of anion and cation. The study on the effect of cations on the hydrolysis of xylose is presented in Figure 1 while the study on the effect of anions can be observed in Figure 2.

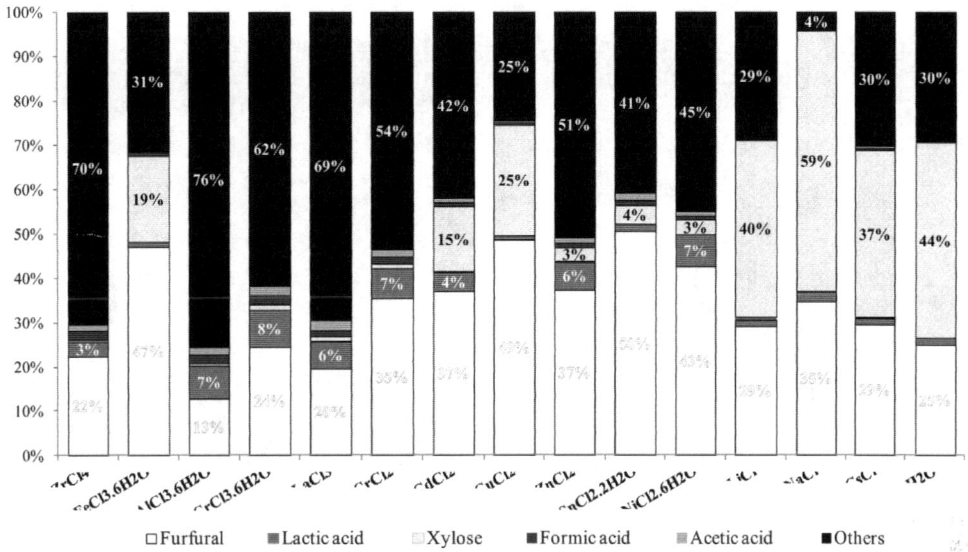

Figure 1. The effect of cations on the hydrolysis of xylose

From Figure 1, it can be seen directly that the cations give different effect on the hydrolysis of xylose. Interesting results were shown on $SnCl_2$, $FeCl_3$ and $CuCl_2$ and which successfully convert xylose almost 50% of **FF**. The last two salts together with NaCl were of the highest interest due to their low selectivity to other decomposition products indicating the important role of Fe3^{+}, Cu^{2+}, Na^{+} and Sn^{2+} in the hydrolysis reactions involved. In fact, NaCl is of the highest interest due to its limited transformation to other products which is only about 5%.

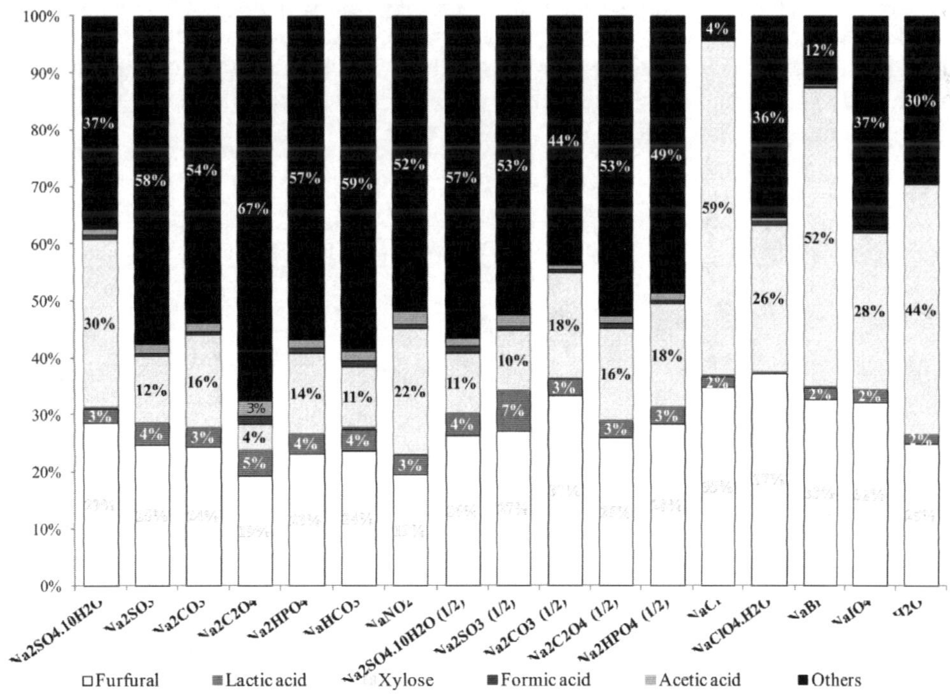

Figure 2. The effect of anions on the hydrolysis of xylose

The effect of anions can be observed in Figure 2. The results show that Cl^- and Br^- can drive the preference of reactions more to **FF** than to decomposition products. Categorizing the results into the same group of cations (see Figure 3), it shows that almost the same yield of **FF** achieved (26-36 mol-%). It confirms the previous results from Marcotullio which indicates that Cl^- in solution seems to be the main feature for the kinetics of xylose reaction and the different cations seem to play only a minor role[CITATION Mar11 \l 1057]. The most interesting result is shown again on NaCl which showed the highest selectivity on **FF**.

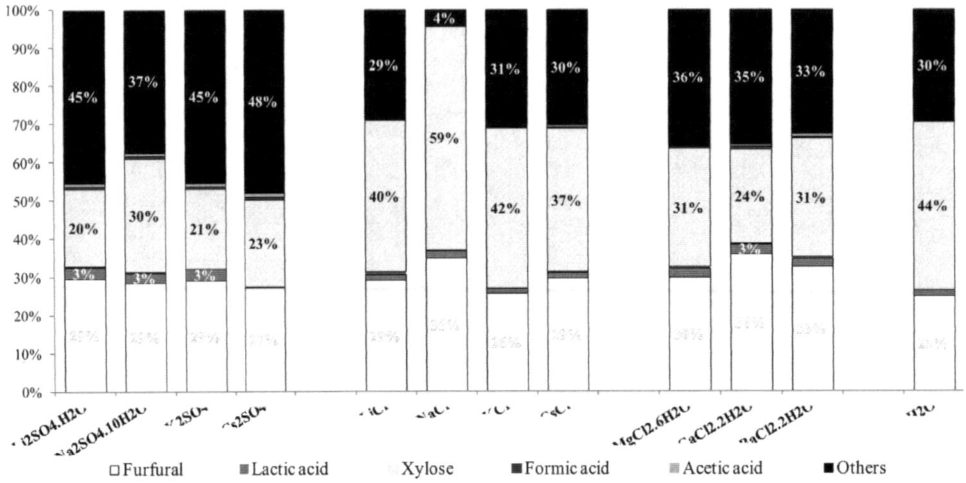

Figure 3. The effect of anions on the hydrolysis of xylose

The highest FF-selectivity was at 84 mol% using 5mM-NaCl, though at low xylose-conversion (35 mol%-FF-yield). The results showed the interesting roles of ion Fe^{3+} and Cl^-. Subsequent studies of two salts $Fe_2(SO_4)_3$ and $FeCl_3$ (5mM) in combination with HCl and H_2SO_4 (0.1M) were investigated on 0.1M-xylose, T=150°C, 0-270 min. The result shows that salts increase the reaction-rate but not the FF-yields for H_2SO_4. In contrast, $Fe_2(SO_4)_3$ increases no yield nor reaction-rate for HCl.

According to the results gained in this works, there should be different roles in the reactions between the Cl^- and H^+ or Fe^{3+}. In fact these results also confirm the acid catalyzed hydrolysis mechanism suggested by Marcotulio as shown in Figure 4.

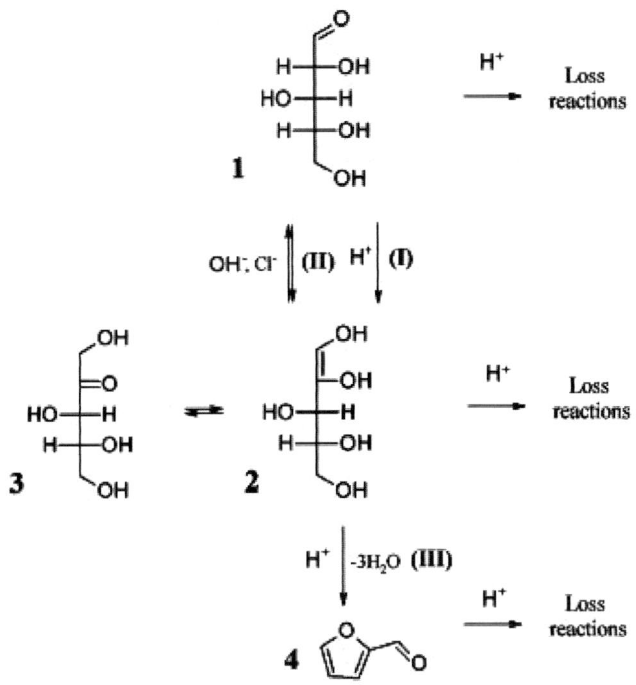

Figure 4. proposed mechanisme of furfural formation from xylose with **1** Aldo form of sugar, **2** 1,2-enediol, **3** keto form of sugar and **4** **FF** [CITATION Mar11 \l 1057]

H_2SO4 as the Bronsted acid donates the H^+ in reaction (I) to form 1,2-enediol or **2** and in reaction (III) to form **FF**. In such mechanism the presence of Cl^- as the Lewis acid of donor of electron favors the formation of the **2**, which can equilibrate with both the aldo **1** and keto **3** form of the sugar, and reacts to form furfural or **4** in presence of an acid. It seems that the speed of reaction in which the Cl^- involved is higher than the one to the loss reactions and resulted in better performance in term of **FF** yield. It may also explain the reason that chloride salt performed better than hydrochloric acid since HCl releases the H^+ which accelerate not only the formation of **2** but also the loss reactions. However in high concentration of chloride salt, there will be high concentration of **2** that may give a driving force to the increase of loss reactions and may give an adversed effect on the yield of **FF**. In fact, the presence of strong acid is needed to favor the selectivity of **FF[CITATION Mar11 \l 1057]**. As for the effect of Fe^{2+} there is no data available to explain its role in the proposed mechanism. However considering the acidic character of Fe^{2+}, it can be assumed to follow the function of H^+. Further involvement of Cl^- and kation such as Fe^{2+} and Na^{2+} in the stabilization of intermediate products was proposed but still need verification and further investigastion[CITATION Mar11 \l 1057].

4. Conclusions

The addition of salts has been proven to affect the mechanism in the hydrolysis of xylose to **FF**. The involvement of Cl^- increase the yield of FF most probably by its catalysis role in the formation of 1,2-enediol while the involvement of Fe^{2+} may be assumed to follow the function of H^+. Another interesting salt is NaCl that can increase the selectivity to FF up to 84mol%. Further investigation and verification is still need to be conducted to understand the role of kation and anion in salts in the mechanism of hydrolysis xylose to **FF** and **FF** decomposition.

Acknowledgement

This research was funded by RISTEKDIKTI and Parahyangan Catholic University and supported by University of Groningen. Authors wishes to acknowledge assistance or encouragement from

IOP Conf. Series: Materials Science and Engineering **742** (2020) 012049 doi:10.1088/1757-899X/742/1/012049

colleagues, special work by technical staff from University of Groningen and Parahyangan Catholic University.

References

[1] IndexMundi 2019 [Online]. Available: www.indexmundi.com. [Accessed 20 December 2019].
[2] Sudiyani Y, Styarini D, Triwahyuni E, Sudiyarmanto, Sembiring K C, Aristiawana Y, Abimanyu H and Han M H 2013 *Energy Procedia* **32** 31-8
[3] Biddy M J, Scarlata C and Kinchin C 2016 *National Renewable Energy Laboratory*, Denver, Colorado, U.S.A.,.
[4] BPS 2018 *Buletin Statistik Perdagangan Luar Negeri Impor or Foreign Trade Statistical Buletin Import* p 35
[5] Döbereiner 1832 *Annalen der Pharmacie* **3** (2) 141-146
[6] Corma A, Iborra S and Velty A 2007 *Chemical Reviews* **107** (6) 2411–2502
[7] Sain B, Chaudhuri A, Borgohain J N, Baruah B P and Ghose J L 1982 *J. Sci. Ind. Res.* **41** (7) 431-438
[8] Zeitsch K J 2000 *The Chemistry and Technology of Furfural and Its Many By-Products* vol. 13, ed 1 (Koln: Elsevier)
[9] Brownlee H J and Miner C S 1948 *Industrial And Engineering Chemistry* **40** (2) 201-204
[10] Hoydonckx H E, Rhijn W M, Vos D E D and Jacobs P A 2012 Furfural and Derivatives *Ullmann's Encyclopedia of Industrial Chemistry* (Weinheim: Wiley VCH Verlag GmbH & Co. KGaA) pp 285-313.
[11] Yan K, Wu G, Lafleur T and Jarvis C 2014 *Renewable Sustainable Energy Rev* **38** 663-676
[12] Singh R, Shukla A, Tiwari S and Srivastava M 2014 *Renewable and Sustainable Energy Rev*, **32** 713-728
[13] Rasrendra C B, Makertihartha I G B N, Adisasmito S and Heeres H J 2010 *Topics in Catalysis* **53** 1241–1247
[14] Rasrendra C B, Soetedjo J N M, Makertihartha I G B N, Adisasmito S and Heeres H J 2012 *Topics in Catalysis* **55** 543–549
[15] Liu C and Wyman C E 2006 *Carbohydr. Res.* **341** 2550–2556
[16] Liu L, Sun J, Cai C, Wang S, Pei H and Zhang J 2009 *Bioresource Technology* **100** 5865–5871
[17] Marcotullio G and de Jong W 2011 *Carbohydrate Research* **346** 1291-1293
[18] Marcotullio G, Cardoso M A T, de Jong W and Verkooijen H M 2009 *International Journal of Chemical Reactor Engineering* **7** A67
[19] Choudhary V, Sandler S I and Vlachos D G 2012 *ACS Catalysis* **2** (9) 2022-2028

AUTHOR INDEX